Studies in the Foundations

Methodology and Philosophy of Science

Volume 3/I

Scientific Research I

The Search for System

Mario Bunge

With 64 Figures

Springer-Verlag Berlin Heidelberg New York 1967

Mario Bunge

McGill University
Montreal/Canada

ISBN 3-540-03994-5 Springer-Verlag Berlin · Heidelberg · New York
ISBN 0-387-03994-5 Springer-Verlag New York · Heidelberg · Berlin

How to Use This Book

This treatise on the strategy and philosophy of science is divided into two volumes of two parts each. Part I, the *Introduction*, offers a preview of the scheme of science (Ch. 1) and some logical and semantical tools (Chs. 2 and 3) that will be used in the sequel. The account of scientific research proper begins with Part II on *Scientific Ideas*. In the beginning was the problem — a subject studied in Ch. 4. Then comes the tentative solution to a problem — that is, a hypothesis (Ch. 5). Next, the hypothesis assumed to reproduce an objective pattern, i.e. the scientific law (Ch. 6). Finally the building and readjustment of systems of hypotheses — i.e., theories — are examined in Chs. 7 (with emphasis on structure and content) and 8 (with emphasis on construction). This closes Vol. I, *The Search for System*. Vol. II opens with Part III, *Applying Scientific Ideas*. Ch. 9 deals with the application of theories to explanation and Ch. 10 with their application to prediction and retrodiction. Ch. 11, on rational action, belongs to the philosophy of technology. Finally Part IV, *Testing Scientific Ideas*, opens with observation (Ch. 12) and goes on to measurement (Ch. 13) and experiment (Ch. 14). The jumping to conclusions from data to hypotheses and conversely (Ch. 15) completes Vol. II, *The Search for Truth*. Look at the structure:

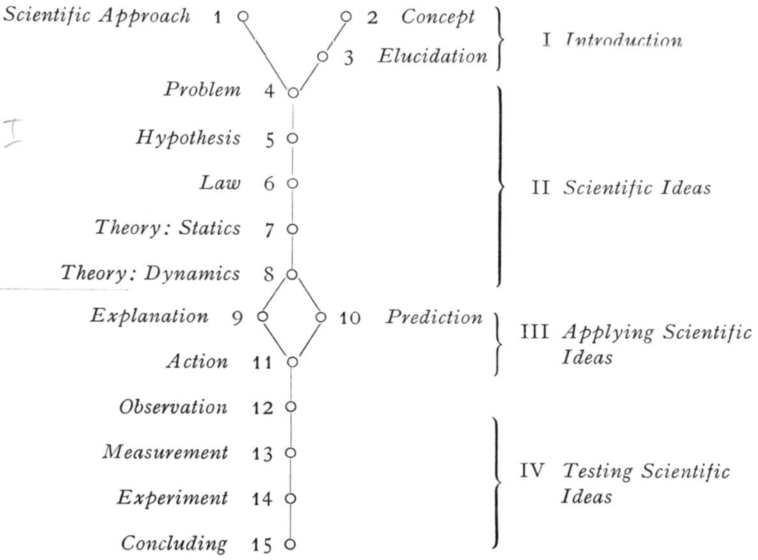

A look at the logical relationships among the chapters as displayed in the previous diagram should help the reader to use the book and understand the philosophy of science it proposes.

The paragraphs between asterisks may be skipped in a first reading. What should not be omitted are some of the problems appended to every section. Their purpose is threefold: to test the reader's understanding of the text, to shake his belief in it, and to invite him to advance the subject. There are more than a thousand such queries. Every problem set has been ordered roughly according to difficulty: the first few are usually exercises whereas some of the last are research problems. In assigning them the instructor should first appraise the background they require.

Quarrels with fellow metascientists have for the most part been avoided in the text and left to the problems. Bibliographical references, too, have been relegated to the problems and to lists at the end of every chapter. As a consequence every section resembles an uninterrupted lecture. With appropriate cuts each volume will cover one semester.

The book has been planned both for independent reading and reference and for use in courses on Scientific Method and Philosophy of Science. Since the text oscillates between an introductory and an advanced level, it may suit a variety of purposes. For one thing, the book might be used as a substitute for formal lectures — which anyhow would seem to have been rendered somewhat obsolete by GUTENBERG. A lively discussion of the text and some of the problems, as well as of the results of exploring the suggested bibliography, should be more interesting and rewarding than a paraphrase.

Warning: any book on our subject is apt to make irritating demands on its reader: he will be asked to sail back and forth between the Scylla of science and the Charibdis of philosophy. The author sympathizes with the traveller but he cannot offer his apologies; instead, he will state the truism that shipwrecking can be avoided either by abstaining from sailing or by training in the skill. May the present book be a deterrent to those who feel seasick at the mere thought of having to learn some science, and a rough guide for those who wish to take a closer look at the beast. Let it be recalled however that no guide is supposed to make the journey for us.

Department of Philosophy
McGill University
Montreal/Canada
January 1967

MARIO BUNGE

Acknowledgments

This text grew from the author's courses and seminars on the methodology and philosophy of science addressed to philosophy and science students at the universities of Buenos Aires, La Plata, Montevideo, Pennsylvania, Texas, Delaware and Freiburg i. Br. from 1957 through to 1966. The successive notes and drafts were first tried on those students and subsequently corrected according to experience. I am indebted to all those students for having acted as an experimental group, and particularly to those who asked embarrassing questions.

I also wish to thank the following persons who made helpful remarks and criticisms although none of them saw the complete manuscript: Professors RICHARD N. ADAMS (anthropology, Texas), HASKELL B. CURRY (logic, Pennsylvania State), CHARLES HARTSHORNE (philosophy, Texas), JOHN P. HARRISON (history, Texas), R. DUNCAN LUCE (psychology, Pennsylvania), MARIO H. OTERO (philosophy, Montevideo), OSVALDO A. REIG (zoology, Buenos Aires), VÍCTOR SANZ (history, Montevideo), GEORGE GAYLORD SIMPSON (zoology, Harvard), and JOSEPH H. WOODGER (biology, London), as well as CARLOS BUNGE (chemistry, Florida), JAMES N. HULLETT (philosophy, Dartmouth), CARLOS IRALDI (psychoanalysis, Buenos Aires), RUBÉN RAMÍREZ MITCHELL (biology, Texas), ROBERT A. SCHWARTZ (philosophy, Rockefeller), GUILLERMO SARMIENTO (ecology, Buenos Aires), and my former assistants JULIO C. COLACILLI DE MURO, HÉCTOR POZZI, MARTHA FOSCHI, and MARTA BUNGE.

To my wife MARTA, who encouraged me to write this book, watched its gestation and criticized it mercilessly, I dedicate it in friendship and love.

Special Symbols

$A \subseteq B$	the set A is included in the set B
$A \cup B$	the union of the sets A and B
$A \cap B$	the common part of the sets A and B
$a \in A$	the individual a is in (or belongs to) the set A
Card (A)	cardinality (numerosity) of the set A
$A \times B$	Cartesian product of the sets A and B
$Cn(A)$	consequence(s) of the set A of assumptions
$=_{df}$	equals by definition
$Df.$	definition
$(\exists x)$	some x (or there is at least one x such that)
e	empirical datum
e^*	translation of e into a semiempirical, semitheoretical language
h	hypothesis
$m(\dot{r})$	measured value of the degree $\dot{r}$
$\overline{m}(\dot{r})$	average (or mean) value of a set of measured values of $\dot{r}$
$P \dashv T$	T presupposes P
p, q	arbitrary (unspecified) propositions (statements)
$P(x)$	x has the property P (or x is a P)
$\{x \mid P(x)\}$	set of the x such that every x is a P
$p \vee q$	p and/or q (inclusive disjunction)
$p \mathbin{\&} q$	p and q (conjunction)
$p \rightarrow q$	if p, then q (conditional or implication)
$p \leftrightarrow q$	p if and only if q (biconditional or equivalence)
Σ_i	sum over i
t	theorem, testable consequence
t^*	translation of t into a semiempirical, semitheoretical language
T	theory
$A \vdash t$	A, therefore t (or A entails t, or t follows logically from A)
$\emptyset$	the empty set
U	the universal set
x	arbitrary (unspecified) individual
(x)	for every x
$\langle x, y \rangle$	ordered pair of the elements x and y

Contents

Part I

Approach and Tools

Part II

Scientific Ideas

Part I

Approach and Tools

It will be convenient to start by taking a panoramic view of the means and ends peculiar to the scientific approach. Such a preview is given in Chapter 1. Like any other brief account of a rich subject this one will have to be presented in a somewhat dogmatic fashion, but it is hoped that the remainder of the book will provide a justification for it. Then we shall need a bridge between science and philosophy: this will be provided by Chapters 2 and 3, which discuss the semantics of scientific concepts and the logic of certain familiar yet tricky conceptual operations, such as classing and defining.

Chapter 1

The Scientific Approach

Science is a style of thinking and acting — indeed, the most recent, universal, and rewarding of styles. As with all human creations, we should distinguish in science the work — research — from its end product — knowledge. In this Chapter we shall take a look both at the overall pattern of scientific research — the scientific method — and at its aim.

1.1. Knowledge: Ordinary and Scientific

Scientific research starts with the realization that the available fund of knowledge is insufficient to handle certain problems. It does not begin from scratch because investigation deals with problems and no question can be asked, let alone answered, outside some body of knowledge: only those who see can see that something is missing.

Part of the background knowledge from which every research starts is ordinary, i.e. nonspecialized knowledge, and part of it is scientific, i.e. it has been obtained by the method of science and can be rechecked, enriched, and eventually superseded by the same method. As research proceeds it corrects or even rejects portions of the fund of ordinary knowledge. Thereby the latter is enriched with the results of science: part of today's commonsense are yesterday's results of scientific research. Science, in short, grows from common knowledge and outgrows it: in fact, scientific research begins at the point where ordinary experience and ordinary thought fail to solve problems or even to pose them.

Science is not just a prolongation or even a mere refinement of ordinary knowledge in the way that the microscope extends the reach of unaided vision. Science constitutes a knowledge of a special kind: it deals primarily, though not exclusively, with unobservable events unsuspected by the uneducated layman, such as the evolution of stars and the duplication of chromosomes; it invents and tries conjectures beyond common knowledge, such as the laws of quantum mechanics or those of conditioned reflexes; and it tests such assumptions with the help of special techniques, such as spectroscopy and the control of gastric juice, which in turn require special theories.

Consequently common sense cannot be an authoritative judge of science, and the attempt to evaluate scientific ideas and procedures in the light of ordinary knowledge alone is preposterous: science elaborates its own canons of validity and, in most subjects, is far ahead of common knowledge, which is more and more becoming fossil science. Imagine a physicist's wife rejecting her husband's theory of elementary particles because it is unintuitive, or a biologist sticking to the hypothesis of the inheritance of all acquired characters because it fits common experience concerning cultural evolution. The moral for philosophers should be clear: Do not try to bring science down to ordinary knowledge but rather learn some science before philosophizing about it.

The radical discontinuity between science and common knowledge in most respects, and particularly as regards method, should not blind us to their continuity in other respects, at least if common knowledge is limited to the opinions held by the so-called sound common sense. In fact, both sound common sense and science attempt to be *rational* and *objective:* they are critical and seek coherence (rationality), and they try to fit the facts (objectivity) rather than indulging in uncontrolled speculation.

But the ideal of rationality, namely the coherent systematization of grounded and testable statements, is achieved by theories — which are the core of science rather than of common knowledge, which is an accumulation of loosely related bits of information. And the ideal of objectivity, namely the building of true impersonal images of reality, can be realized only by transcending the narrow limits of daily life and private experience: by abandoning the anthropocentric viewpoint, by hypothesizing the existence of physical objects beyond our poor chaotic impressions, and by testing such assumptions via intersubjective (transpersonal) experiences planned and interpreted with the help of theories. Common sense can achieve only a limited objectivity because it is much too closely tied to perception and action, and when it does transcend them it is often in the form of myth: science alone invents theories that, while not limited to summarizing our experiences, are tested by the latter.

An aspect of the objectivity shared by sound common sense and science is *naturalism,* i.e. the refusal to countenance nonnatural entities (e.g., disembodied thinking) and nonnatural sources or modes of cognition (e.g.,metaphysical intuition). But common sense, suspicious as it is of the unobservable, has on occasion had a crippling effect on scientific imagination. Science, on the other hand, is not afraid of the unobservables it hypothesizes as long as it can keep them under control: indeed, science has uncommon (yet neither esoteric nor infallible) means for testing such assumptions.

A consequence of critical alertness and of the naturalistic rejection of esoteric modes of cognition is *fallibilism,* i.e. the recognition that our knowledge of the world is provisional and uncertain — which does not exclude scientific progress but rather demands it. Scientific statements, no less than those of common experience, are opinions — only, enlightened (grounded and testable) opinions rather than arbitrary dicta or unchecked gossip. What can be proved beyond reasonable doubt are either theorems of logic and mathematics or trivial (particular and observational) statements of fact, such as "This volume is heavy".

Statements covering immediate experience are not inherently incorrigible but are seldom worth doubting: although they are conjectural, in practice we handle them as if they were certain. Precisely for this reason they are scientifically uninteresting: if common sense can handle them why resort to science? This is the reason why there is no science of typewriting or of car driving. On the other hand, statements covering more than is immediately experienced are doubtful and therefore often worth being checked, rechecked and given a ground. Only, in science doubt is creative rather than paralyzing: it stimulates the search for ideas accounting for the facts in a more and more adequate way. In this way an array of scientific opinions with unequal weight is generated: some are better grounded and tested than others. Accordingly, the skeptic is right when he doubts anything in particular, wrong when he doubts everything alike.

In short, scientific opinions are rational and objective like those of sound common sense — only, much more so. What else, if anything, gives science its superiority over common knowledge? Surely not the substance or subject matter, since one and the same object may be approached either nonscientifically, or even antiscientifically, or in the spirit of science. Thus, e.g., hypnosis may be dealt with ascientifically, as when case histories are described without the help of either theory or experiment. It may alternatively be regarded as a supernormal or even supernatural fact in which neither the sense organs nor the nervous system are involved: i.e., as a result of a direct intermind action. Finally, hypnosis may be approached scientifically, i.e. by framing conjectures about the physiological mechanism underlying hypnotic behavior, and by controlling such assumptions in the laboratory. In principle, then, the object or subject matter does not mark science off from nonscience, even though certain problems, e.g. those of the structure of matter, can hardly be stated outside a scientific context.

If the "substance" (object) cannot be distinctive of all science, then it must be the "form" (procedure) that is: the peculiarity of science must reside in the way it operates to attain a certain end — i.e., in the scientific method and in the aim to which this method is directed. (Caution: 'scientific method' should not be construed as a set of mechanical and infallible instructions enabling the scientist to dispense with imagination: it is not to be interpreted either as a special technique for handling problems of a certain kind. The method of science is just the overall pattern of scientific research.) The *scientific approach*, then, is made up of the *scientific method* and of the *goal* of science.

Let us take a glimpse at the scientific approach — not, however, without first trying our forces on some of the following problems.

Problems

1.1.1. Writers and humanists often complain that science is dehumanized because it eliminates the so-called human elements. Examine this view.

1.1.2. Is science objective to the point of excluding points of view, or does it rather limit the consideration of viewpoints to those which are somehow grounded and testable? For a recent criticism of the "myth" that science is objective, see H. CANTRIL, *The "Why" of Man's Experience* (New York: Macmillan, 1950), Chap. 1. Hint: Make sure to distinguish the psychology of research — concerned with the motives, biases, etc. of the individual investigator — from the methodology of research. See K. R. POPPER, *The Open Society and its Enemies*, 4th ed. (London: Routledge & Kegan Paul, 1962), Ch. 23.

1.1.3. Examine the widespread opinion, shared by philosophers like K. JASPERS, that the conclusions of scientific research are conclusions proper, i.e. final and certain. *Alternate Problem:* Trace the history of the view that genuine science is infallible.

1.1.4. Elucidate the concepts of *opinion, belief, conviction*, and *knowledge*. *Alternate Problem:* Is there any logical relation between *naturalism* (an ontological doctrine) and *testability* (a methodological property of certain statements)? In particular is naturalism necessary, sufficient, necessary and sufficient, or neither for testability? Hints: Distinguish between testability in principle (conceivable testability) and effective testability (the property of a statement of being subjectible to test with the means on hand); and search for counterexamples to the first three theses, i.e., "$T \rightarrow N$", "$N \rightarrow T$", and "$N \leftrightarrow T$".

1.1.5. Traditional philosophy has retained the important distinction drawn by PLATO (*Meno* 97, *Republic* V, 477, 478, *Timaeus* 29, etc.) between opinion or belief (*doxa*) and certain knowledge or science (*episteme*). Opinion, according to Plato, is characteristic of the vulgar as regards every subject matter but is also all we can achieve regarding things transient (physical objects), which *are* not in a complete way since they emerge, change, and pass away: only eternal objects (ideas) can be the subject of perfect knowledge. Discuss this view, distinguishing its relevance, if any, to formal science and to factual science.

1.1.6. Point out the similarities and differences between common knowledge and scientific knowledge. *Alternate Problem:* Given that scientific thinking is unnatural, i.e. comes with difficulty and to only a part of mankind, imagine what would become of scientific research in the aftermath of a nuclear war destroying all scientific centers.

1.1.7. Discuss the view that science is nothing but a systematic continuation of ordinary knowledge. For this view see, e.g., R. CARNAP, "Logical Foundations of the Unity of Science", in *International Encyclopaedia of Unified Science* (Chicago: University of Chicago Press, 1938), I, p. 45, and A. J. AYER, *Language, Truth, and Logic,* 2nd ed. (London: Gollancz, 1953), p. 49.

1.1.8. Philosophers of various shades, from certain medieval schoolmen through the Scottish common-sense realists to the language philosopher G. E. MOORE, have claimed for common sense the right to evaluate scientific theories. Likewise, certain scientists have fought genetics, relativity physics and the quantum theories because they clash with common sense. Discuss this phenomenon. *Alternate Problem:* Freedom of opinion includes the right of everyone to criticize and even ridicule anything. But freedom of research — a companion to freedom of opinion — can be hampered by a hostile public opinion. Can this problem be solved?

1.1.9. L. WITTGENSTEIN and the philosophers of the Vienna Circle have claimed that the criterion of demarcation between science and nonscience (particularly metaphysics) is the *meaningfulness* of the statements that make up science. Accordingly, an analysis of meaning would suffice to decide whether a discipline is scientific or not. Examine this view, see whether it does not upgrade bookbinding and accounting as sciences, and propose your own criterion of demarcation between science and nonscience. *Alternate Problem:* Examine K. R. POPPER'S "The Demarcation between Science and Metaphysics", in P. A. SCHILPP, Ed., *The Philosophy of Rudolf Carnap* (Chicago: Open Court, 1963).

1.1.10. G. W. HEGEL and other philosophers have stated that all sciences except philosophy share the advantage of presupposing either their object or the method guiding the beginning and the ulterior march of investigation. Is it true that the object and the special method of every science are given in advance? Hint: look for counterexamples.

1.2. Scientific Method

A method is a procedure for handling a set of problems. Every kind of problems requires a set of special methods or techniques. The problems of knowledge, by contrast to those of language or action, require the invention or the application of special procedures bearing on the various stages of problem handling, from the very statement of problems all the way down to the control of the proposed solutions. Examples of such *special methods* (or *techniques*) of science are triangulation (for the measurement of large distances) and the recording and analysis of brain waves (for the objectification of brain states).

Every special method of science, then, is relevant to some particular stage in the scientific investigation of problems of a certain kind. The *general method* of science, by contrast, is a procedure applying to the whole cycle of investigation into every problem of knowledge. The best way to learn how the scientific method works is to engage, with an inquisitive attitude, in some scientific research broad enough to ensure that the special methods or techniques do not overshadow the general pattern. (Becoming a specialist in some stage of scientific work, such as measuring, is far from sufficient to acquire a clear grasp of the scientific method: moreover, it may foster the idea that there is a plurality of disconnected methods rather than a single pattern underlying all the techniques.) The next to the best way is to become acquainted with a piece of research — not just with its more or less perishable outcome but with the whole process, starting with the questions that originally prompted the research.

Suppose we ask the question 'Why do different human groups use more or less different languages?'. A simple answer to this question — i.e. an explanation of the empirical generalization that different human groups tend to speak differently — is provided by any myth such as, e.g., the original diversity of ready-made tongues. A scientific investigator of our problem will be suspicious of simple explanations and will start by critically examining the problem itself. In fact, the question presupposes an empirical generalization that might be in need of refinement: what groups are those that speak differently: ethnical groups, social groups, professional groups? Only a preliminary investigation into this prior question can lead us to a more precise formulation of our original problem.

Once such a more precise statement of the problem has been found, a number of guesses will be offered: some regarding the geographical determination of such language differences, others the biological factors, others the social ones, and so on. These various assumptions will then be tested by checking their observable consequences. Thus, e.g., if the kind of occupation is in fact a major determinant of linguistic differences (hypothesis), then occupational groups composed of otherwise similar individuals should speak distinctive jargons (testable consequence).

A number of data will have to be gathered in order to decide which, if any of the above conjectures, is true. And, whenever possible, the data will have to be scientifically certifiable, i.e. obtained and rechecked if necessary by scientific means. For example, random samples of occupational groups will have to be studied in order to minimize the effects of possible bias on the choice of subjects. The merits of the various hypotheses so far proposed will then be evaluated, and in the process some new conjectures may be suggested.

Finally, if the investigation has been careful and imaginative, the solution to the original problem will raise a cluster of new problems. In fact, the most important pieces of research, like the best books, are the most thought-provoking rather than the most thought-blocking ones.

In the above example we may disclose the chief stages of the way of scientific research — i.e. the main steps in the application of the scientific method. We discern, in fact, the following ordered sequence of operations:

1. *Ask well-formulated and likely fruitful questions.*
2. *Devise hypotheses both grounded and testable to answer the questions.*
3. *Derive logical consequences of the assumptions.*
4. *Design techniques to test the assumptions.*
5. *Test the techniques for relevance and reliability.*
6. *Execute the tests and interpret their results.*
7. *Evaluate the truth claims of the assumptions and the fidelity of the techniques.*
8. *Determine the domains in which the assumptions and the techniques hold, and state the new problems raised by the research.*

This cycle is schematically represented in Fig. 1.1.

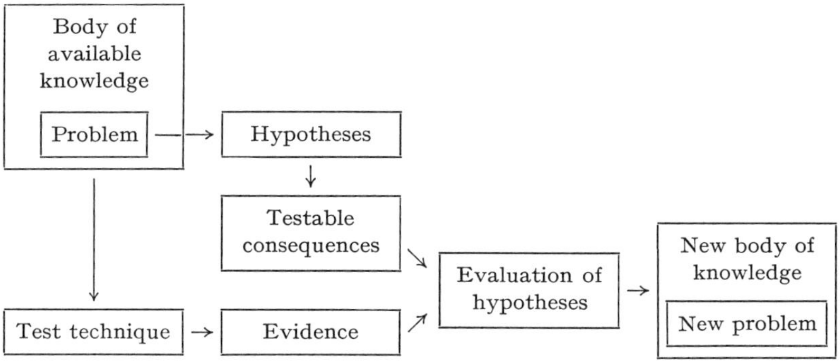

Fig. 1.1. A research cycle. The importance of a scientific investigation is gauged by the changes it induces in our body of knowledge and/or by the new problems it poses

Are there rules for the adequate execution of the above operations? That is, are there effective directions for handling scientific problems? There certainly are some, though nobody has drawn an exhaustive list and everybody should be reluctant to do it after the failure of the philosophers who, from Bacon and Descartes onwards, have professed to know the infallible rules for the direction of inquiry. For the sake of illustration we shall mention and exemplify some quite obvious rules of scientific method; further rules will be found scattered in the rest of the book.

R1. State your problem precisely and, in the beginning, specifically. For example, do not ask just 'What is learning?', but pose a well-circumscribed question such as, e.g., 'Do albino rats learn maze tasks gradually or by little jumps?'.

R2. Try definite and somehow grounded conjectures rather than noncommittal or wild hunches: risk hypothesizing definite relations among clear-cut variables, and such that they do not conflict with most of our scientific heritage. For example, do not rest content with assuming that it is possible to learn on a single trial but assume, e.g., that one-trial learning of orientation in a T-maze has a definite probability.

R3. Subject your assumptions to tough tests rather than to soft ones. For example, in testing the single-trial learning hypothesis do not assign your subjects tasks for which they are already prepared and do not excuse negative results: assign new tasks and accept all the evidence.

R4. Do not pronounce true a satisfactorily confirmed hypothesis: regard it as, at best, partially true. For example, if you have obtained an empirical generalization concerning the probabilities of learning a given task on a single trial, on a second one, and so on, regard your statement as corrigible by further research.

R5. Ask why the answer should be as it is and not otherwise: do not stop on finding data-fitting generalizations but try to explain them in terms of stronger laws. For example, pose the problem of finding the neural mechanism responsible for learning on a first presentation: this will integrate your behavior research with biology.

These and other rules of scientific method are far from being infallible and therefore in no need of improvement: they have grown along scientific research and are — let us hope — perfectible. Moreover, we must not expect the rules of scientific method to replace intelligence by patient drilling. The asking of astute and rewarding questions, the building of strong and deep theories, and the design of delicate and original tests are not rule-directed activities: if they were, as some philosophers have supposed, everyone could conduct scientific research and computers could be turned into investigators instead of being research tools. Scientific methodology can give some hints and it does supply means for spotting mistakes, but is no substitute for original creation and does not spare us all mistakes.

Table manners are more or less conventional and local; consequently it would be difficult to validate or invalidate them in an objective manner, although they are explainable in social and historical terms. What about research manners, i.e. the rules of scientific investigation? They are clearly universal: indeed there is nothing, not even philosophy, as universal as science. But are those rules justifiable? They certainly have a *pragmatic justification:* although they are not infallible, no other rules are known to be better conducive to attaining the goal of science — the building of the truest possible conceptual models of the patterns of things.

But surely that is a poor justification. First, because the application of the scientific method yields, in the best of cases, approximate truths. Second because a rule that is justified by its success but is not integrated with the body of scientific knowledge hangs in the air and cannot meet the challenge of nonscientific procedures — such as divination — for the advancement of knowledge. In other words, we would like a theoretical justification of the scientific method in addition to its pragmatic justification. By the *theoretical justification* of a rule (or norm, or prescription, or direction) we shall understand: (i) the validation of the rule's presuppositions — i.e., showing that what the rule takes for granted is consistent with the known laws; and (ii) showing that the given rule is compatible with the other members of the set of rules — in this case, the scientific method. In short, a rule will be regarded as theorctically justified if, and only if, it is both *grounded* and *systematic* (a member of a consistent system of rules).

In the case of the rules of the scientific method we want them to make up a system of norms based on, or at least consistent with, the laws of logic and the laws of science, not only with the desiderata of research. Thus, the rule "State your problem precisely" clearly presupposes that unique (though perhaps many-membered) answers are being sought: if a plurality of mutually incompatible assumptions were acceptable, the preciseness condition would not be stipulated. In turn, the desideratum of unique solution is required by the logical principle of noncontradiction. At this point the justification of the given rule can stop, because logical principles are presupposed by scientific research, not questioned by it (see Sec. 5.9).

The justification of other rules of the scientific method will be more difficult and may involve tricky philosophical problems — such as whether the scientific analysis of a whole dissolves it without clarification — but anyhow it should be given and it promises to be an interesting work. Unfortunately no theoretical justification of the rules of the scientific method has been attempted: scientific methodology is still in a descriptive, pretheoretical stage. Largely responsible for this neglect seems to be the tacit assumption that anything is good if it works — a strange assumption to make in the case of the scientific method, which is not supposed to yield perfect results. At any rate here is an exciting problem for philosophers that care for living science.

Scientists have not been concerned with either the foundation or the systematization of the rules of scientific procedure: they do not even care to state all the rules they employ. In fact, discussions of scientific methodology seem to be alive only in the beginnings of a science: at least this was the case of astronomy in PTOLEMY's time, of physics in GALILEI's day, of psychology and sociology nowadays. In most cases scientists adopt a trial and error procedure with respect to the rules of research, and those rules that are found effective are silently built in the everyday routine, so that most researchers do not notice them. One does not become method-conscious until the prevailing method is shown to fail.

The scientific method and the goal to which it is applied (objective knowledge of the world) make all the difference between science and non-science. Moreover, both method and goal are philosophically interesting, so that a neglect of either is unjustifiable. Yet it must not be forgotten that a tacit methodology, if sound, is more valuable than a wrong explicit methodology. This must be emphasized at a time when so much space is devoted by psychology and sociology journals to methodological discussions as how to best stop research by prohibiting the use of concepts not applying to directly observable traits. In the face of such dogmatic (theoretically unjustified) and sterile methodological prescriptions it is best to keep in mind what is perhaps the sole golden rule of scientific work: *Audacity in conjecturing, cautiousness in testing.*

To sum up. The scientific method is a mark of science, whether pure or applied: no scientific method, no science. But it is neither infallible nor self-sufficient. The scientific method is fallible: it can be improved both by evaluating the results it leads to and by deliberate analysis. Nor is it self-sufficient: it cannot operate in a knowledge vacuum but requires some knowledge which can in turn be adjusted; and it must be implemented by special methods adapted to the peculiarities of the subject matter. To these techniques we turn next.

Problems

1.2.1. Comment on the following characterization of method (in general) given by the famous Port Royal *Logic* (1662), in *Grammaire générale* [de Port Royal] (Paris: Delalain, 1830), p. 524: "In general we may call method the art of arranging a sequence of thoughts either to discover the truth when we ignore it or to prove it to others when we know it." The art of the discovery of truth was described as analysis or method of resolution; the art of showing the truth to others, synthesis or method of composition. *Alternate Problem:* Why was a new method for the discovery of truth so keenly searched for at the beginning of the modern era? And were the new proposals (such as Bacon's fact collecting and Descartes deduction from clear and distinct a priori principles) successful?

1.2.2. Examine the general characterization of method given by H. MEHLBERG, *The Reach of Science* (Toronto: University of Toronto Press, 1958), p. 67: "A *method* is a statement or a set of statements describing a repeatable sequence of *operations*, such that each individual sequence of operations so described would enable a human individual or group to bring about, either infallibly or in a fair proportion of cases, a repeatable event called the *objective* of the method. [...] If the objective of the method is always an event occurring in some individual object, then the method is said to be *applied to this object*. Thus, in order to drive a nail into a piece of wood, one may hit the head of the nail with a hammer several times in succession. The method consists, then, in a repeatable sequence of hits executed with the hammer in a specified way; the objective of the method is driving a nail into a piece of wood; the object of the method is any system consisting of a nail and a piece of wood". Does this aply to the method of science?

1.2.3. Comment on J. DEWEY's characterization of the scientific method as "a method of changing beliefs by means of tested inquiry as well as of arriving at them". See *A Common Faith*, in D. BRONSTEIN, Y. H. KRIKORIAN and P. WIENER, Eds., *Basic Problems* of *Philosophy* (Englewood Cliffs, N. J.: Prentice-Hall, Inc., 1955), p. 447.

1.2.4. Is *trial and error* a method proper? Be careful to distinguish, within the class of trial and error procedures, hit-or-miss from the methodical examination of possibilities (e.g., hypotheses).

1.2.5. Determine which if any of the following activities and disciplines employ the method of science: speleology (exploration and description of caves), star gazing and describing, descriptive anatomy, bird watching and description, plant and animal collecting and pigeon-holing, personality diagnosis by techniques lacking a pragmatic and/or a theoretical justification, and computer programming and operation.

1.2.6. Analyze and exemplify the various stages in the procedure of a general practicioner faced with a sick patient.

1.2.7. Is sacrifice theoretically justified as a method for producing rain or passing examinations?

1.2.8. Examine the method employed by A. M. AMPÈRE to establish his law of the mutual action of electric currents. See his memoir of June 10, 1822 in *Mémoires sur l'électromagnétisme et l'électrodynamique* (Paris: Gauthier-Villars, s. d.), especially pp. 76—77. *Alternate Problem:* Study the possibility of finding a general methodology (praxeology) applying to all kinds of work, whether intellectual or physical. See T. KOTARBIŃSKI, "De la notion de méthode", *Revue de métaphysique et de morale,* **62,** 187 (1957).

1.2.9. Until recently it was universally taken for granted that the chief rule of scientific method is: "Relevant variables must be altered one at a time". It was assumed that no control of the various intervening factors was effective other than that. As recently as the 1930's it was realized that we never have an exhaustive knowledge of the relevant variables and, even if we did, we could not vary each of them, one at a time, freezing as it were all the others, because there are constant relations (laws) among some of them. Accordingly experiments involving the simultaneous changes of the values of a number of (possibly interacting) variables were planned (factorial design). See R. A. FISHER, *The Design of Experiments,* 6th ed. (London: Oliver and Boyd, 1951). Draw some moral concerning the mutability of the scientific method.

1.2.10. Examine whether, and if so to which extent, the following procedures are used in science. 1. The various methods of deduction. 2. Induction. 3. The hypothetico-deductive method, i.e. the procedure consisting in framing hypotheses and tracing their logical consequences. 4. DESCARTES' methodical doubt (to be distinguish from the skeptic's systematic doubt). 5. HUSSERL's phenomenological method. 6. HEGEL's dialectical method. 7. DILTHEY's empathic comprehension (*Verstehen*).

1.3. Scientific Tactics

The scientific method is the strategy of scientific research: it bears on any whole research cycle and is independent of subject matter (see 1.2). The actual execution of every one of the strategic moves will, on the other hand, depend on the subject matter and on the state of our knowledge regarding that subject matter. Thus, e.g., determining the solubility of a substance in water requires an essentially different technique from the one needed to find out the degree of affinity between two biological species. And the actual solution to the first problem will depend on the state of the theory of solutions, just as the solution to the second one will depend on the state of evolution theory, ecology, serology, and other biological disciplines.

Every branch of science is characterized by an open (expanding) set of problems which it approaches with a set of tactics or techniques. These techniques change much more rapidly than the general method of science. Furthermore, they cannot always be exported to other fields: thus, the historian's craft of testing for the authenticity of a document has no use for the physicist. Yet both, the historian and the physicist, are after truth and search for it according to a single strategy: the scientific method.

In other words, there is no strategic difference among the sciences: the special sciences differ only by the tactics they use to solve their peculiar problems, but they all share the scientific method. This, rather than being an empirical finding, follows from the following *Definition:* A science is a discipline using the scientific method for the purpose of finding general patterns (laws).

Those disciplines which have no occasion to use the scientific method — e.g., because they limit themselves to data gathering — are not sciences, although they may supply science with raw material; this is the case of geography. Nor are sciences those doctrines and practices which, like psychoanalysis, refuse to employ the scientific method (see 1.6).

Scientific techniques may be classed into *conceptual* and *empirical*. Among the former we may mention the tactics for the precise stating of problems and conjectures of a certain kind, and the procedures (algorithms) for deducing consequences from the hypotheses and for checking whether the proposed hypotheses do solve the corresponding problems. (Mathematics provides, of course, the richest set of powerful tactics for stating problems and hypotheses in an accurate way, for deducing consequences from the assumptions, and for checking solutions. But it is of little help in finding problems or in conceiving the nuclei of new hypotheses in the factual sciences. Besides, in the more backward sciences our ideas are not yet clear enough to be susceptible to mathematical translation. Otherwise there is no limitation in principle to the application of mathematical concepts, theories and techniques in factual science: see Sec. 8.2.) As to the empirical techniques, we may recall those for designing experiments, performing measurements, and making instruments for the recording and processing of data. The mastering of most such techniques is a question of training: talent is needed to apply known techniques to problems of a new kind, to criticize the known techniques and, particularly, to invent better ones.

Certain techniques, though not as universal as the general method of science, are applicable in a number of fields. Let us review three such quasi-universal techniques: tree questioning, iteration, and sampling. They all have antecedents in ordinary life and are therefore easily comprehensible.

Tree questioning consists in surveying the set of possibilities (logical or physical, as the case may be) and dividing them step by step into mutually disjoint subsets until the wanted subset (or element) is reached. Suppose the problem consists in finding out which of eight objects has a certain property — e.g., which of the first eight digits has our playmate in mind or which among eight hypotheses is the most likely. If we proceed erratically, i.e. by random trial and error, we shall need a maximum of seven operations (e.g., questions). In using tree questioning,

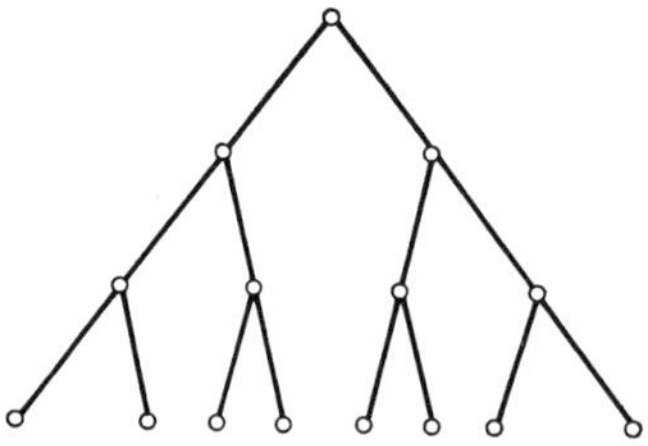

Fig. 1.2. Porphiry's tree questioning for an initial set of 8 objects: orderly subdivision into mutually exclusive alternatives

on the other hand, we proceed in the following way. We divide the field of possibilities (8 objects) into two equal parts and ask whether the wanted object is in the first subset. Since the problem is one of decision (yes or no answer), the reply to this single question will be enough to reduce our initial uncertainty to one-half. We then repeat the procedure until our initial uncertainty has been altogether removed. A total of three questions will solve the problem in our case, as shown in Fig. 1.2. Tree questioning is, then, *methodical trial and error*, to be contrasted with the blind hit-or-miss procedure. In general, for a set of N objects, random questioning requires a maximum of $N-1$ and an average of $N/2$ answers. Tree questioning, on the other hand, requires a maximum of $H = \log_2 N$ bits of information. In our case, $\log_2 8 = \log_2 2^3 = 3$.

Iterative procedures are step by step trials in which a gradual improvement of an approximate solution is obtained: every solution is built upon (is a function of) the preceding one and is better (more accurate) than it. Often the initial stepping stone must somehow be surmised in order for the process to start. When there is no method for finding such a first rough solution (zero order approximation), experience, perseverance and insight will be needed — and a bit of good luck will not be amiss. A familiar example of an iterative procedure is shooting at a fixed target. The information concerning the deviation committed is fed back to the shooter, which enables him to correct the aim in successive steps until the target is hit. In the process the errors, far from accumulating, are used to improve the performance. Iterative procedures, then, are *self-improving:* they can be carried to any desired degree of accuracy, i.e. until the difference between two successive solutions is negligible.

Mathematics has *exact* iterative procedures, i.e. techniques that warrant a *uniform* increase in accuracy; famous examples are NEWTON'S method for computing square roots and PICARD'S method for the approximate solution of differential equations. In every case a sequence of approximate solutions is built on the basis of a fixed relation among two or more members of the sequence, and the sequence has a definite limit. That is, mathematical iterative procedures are convergent.

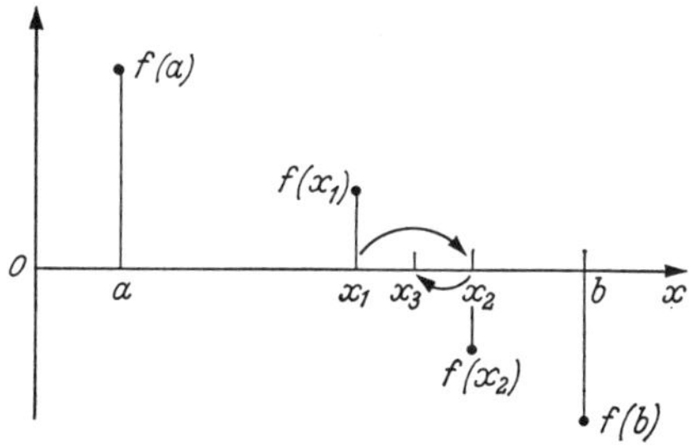

Fig. 1.3. Building increasingly true propositions by the dichotomic method. The exact solution is the limit of the sequence of approximate solutions

Example: Find a solution of the equation $f(x)=0$. Data: $f(\ )$ is continuous and the values it takes at the points a and b have opposite signs (see Fig. 1.3.). Technique: the *dichotomic method.* First guess: the given function has a zero midway between a and b, i.e. $x_1=(a+b)/2$. Test: compute $f(x_1)$. There are two possibilities: either $f(x_1)$ is zero, in which case the problem is solved, or it is different from zero. In the latter case there are again two possibilities: either $f(x_1)$ has the sign of $f(a)$ or that of $f(b)$. Suppose the former is the case; then the zero of the function will lie between x_1 and b. Try the simplest guess: $x_2=(x_1+b)/2$. If $f(x_2)=0$, the problem is solved. Otherwise $f(x_2)$ has either the sign of $f(b)$ or that of $f(x_1)$. Suppose the former is the case. Try $x_3=(x_1+x_2)/2$, compute $f(x_3)$ and proceed as before. In this way a sequence is built each term of which equals the average of the last two. Either one of the members of the sequence solves the problem or the sequence approaches the exact solution, i.e. the solution is the limit of the sequence. In the former case an exact solution is obtained, in the latter approximate solutions to any desired degree of accuracy are obtained. Notice that the concept of *partial truth* is involved in iterative procedures. We shall meet this concept again, in Secs. 10.4 and 15.2.

Our third example of a quasi-universal special method of science is *random sampling,* i.e. the extraction of a small subset from an original set or population (which may be infinite) in such a way that the selection does not depend on the properties of the individuals, but is blind to them and consequently unbiased. Random sampling is what we are supposed to do when we try a free sample of some merchandise or when we control the quality of a manufactured good without testing all the items. Random sampling is used also when a hypothesis is subjected to the test of experience: we check the hypothesis against a small number of data relevant to it and chosen without bias from a potential infinity of data.

Tree questioning, iterative procedures and random sampling are as many specializations of the *method of successive approximations,* which is characteristic, though not exclusively, of science. In pure logic there is no place for this method because exact solutions (or exact proofs of the absence of solutions) are sought. But in factual science and in large areas of numerical mathematics *approximate solutions* are all we can get, whence the method of successive approximations is indispensable.

The great interest of the method of successive approximations for the theory of knowledge (epistemology) is that it constitutes a clear reminder of the following points. First, scientific research proceeds *gradually,* and indeed in such a way that even the right insights that may be hit upon by chance from time to time are the outcome of previous research and are eventually subjected to correction. Second, scientific research yields, at least in relation with the world of facts, *partial truths* rather than complete and accordingly final truths. Third, the scientific method, in contrast with the haphazard groping of common sense and of uncontrolled speculation, is *self-correcting:* it can recognize mistakes and can attempt to obtain higher-order approximations, i.e. truer answers.

Other tactics of science are less universal: they must be discussed with reference to specific scientific problems and theories. Thus, the X-rays technique for the identification of chemical compounds requires the application of wave optics to the diffraction of waves by crystal lattices: such a theory alone allows us to interpret the observed rings in the X-rays diagrams, which would otherwise be meaningless signs, since the rings bear no resemblance to the atomic configurations about which they convey information.

In general, the special methods of science are somehow *grounded* in scientific theories, which are in turn tested by such techniques. This is true even of such an elementary technique as weighing with a two pan balance: it presupposes statics and, particular, the law of the lever. Scientific techniques and instruments are never consecrated by success alone: they are designed and justified with the help of theories. The possibility of theoretically justifying whatever special method is employed in science renders it neatly different from the various pseudosciences, which employ groundless procedures, such as divination by looking at a lamb's liver or at ink blots, or by listening to the recounting of dreams.

The design and justification of the special techniques of science belongs to the special sciences. Although every scientific technique raises philosophical problems regarding inference, such problems must be discussed in the context of the respective disciplines. Unfortunately these questions are as a rule either neglected or treated without philosophical competence, as a result of which the very nature of scientific techniques and of the results they yield are widely misunderstood. For example, if the question of the theoretical validation of the empirical techniques of science were in a more mature state everyone would realize that empirical information is not weighed in a theoretical vacuum: every piece of evidence must be judged in the light of the theory employed in the design and implementation of the technique used to gather that information. Just as no factual theory stands by itself, so no datum constitutes an evidence for or against a theory unless it is gathered and interpreted with the help of some scientific theory. In particular, no alleged information obtained through extrascientific means (e.g., mediumnism) can count as evidence against scientific theories or in favor of nonscientific theories. There is no test of science independent of science. This does not entail that the results of science are above criticism, but that only the internal criticism of science is legitimate. A moral for philosophical critics of science: First bow, then strike.

Let us now explore some consequences of the thesis that science is methodologically one despite the plurality of its objects and associated techniques.

Problems

1.3.1. Point out the differences between the techniques and the general methodology of a given scientific discipline. See P. Lazarsfeld and M. Rosenberg, Eds., *The Language of Social Research* (Glencoe, Ill.: The Free Press, 1955), pp. 9—10.

1.3.2. Comment on and illustrate the fourteen search principles proposed by E. Bright Wilson, *An Introduction to Scientific Research* (New York: McGraw-Hill, 1952), pp. 140ff.

1.3.3. Does biology need special methods of its own in addition to those of physics and chemistry? If so, why?

1.3.4. Examine the steps in a typical pharmacological sequence, as described by C. D. Leake, "The Scientific Status of Pharmacology", *Science*, **134**, 2069 (1961).

1.3.5. Geology has always employed physical *concepts* ("deformation", "pressure", "transport", "heat", "melting", "solidification", etc.). But the use of physical *theories* (mechanics, hydrodynamics, thermodynamics, etc.) came only late, and the application of physical *methods* was not tried until our century. In particular, experimental geology (the simulation of geological processes in the laboratory) is a new-born. Use this example, and if possible others as well, to illustrate and expand the thesis that a discipline acquires only gradually a scientific status, and usually does it by adopting some of the ideas and special methods of a mature contiguous science.

1.3.6. In what does the comparative method consist, in which sciences is it used, and why? *Alternate Problem:* Examine the iterative methods and discuss their relevance to the theory of knowledge. See, e.g., E. WHITTAKER and G. ROBINSON, *The Calculus of Observations*, 4th ed. (London and Glasgow: Blackie & Son, 1944), Secs. 42—45.

1.3.7. F. BACON thought he had invented routine procedures for conducting scientific research: *Novum Organum*, (1620), repr. in *The Philosophical Works of F. B.*, ed. by J. M. ROBERTSON (London: Routledge, 1905), Aphorism LXI, p. 270: "the course I propose for the discovery of sciences is such as leaves but little to the acuteness and strength of wits, but places all wits and understandings nearly on a level". What did BACON have in mind: the scientific method or a set of data gathering and comparison techniques?

1.3.8. Examine the claim that psychology cannot employ the objective methods of science because the subject (the investigator) and the object (the subject of research) are one and the same (or because the object of investigation is a part of the cognitive subject).

1.3.9. The layman's study of a subject's personality involves placing himself in the subject's shoes, the better to understand his behavior. This procedure has been called the *method of sympathetic understanding* (empathy, *Verstehen*) and was advocated by W. DILTHEY and R. G. COLLINGWOOD as the proper method of psychology and history. Examine this claim. See T. ABEL, "The Operation Called 'Verstehen' ", *American Journal of Sociology*, **54**, 211 (1948), W. H. WALSH, *An Introduction to Philosophy of History* (London: Hutchinson, 1958), esp. p. 58, and M. BUNGE, *Intuition and Science* (Englewood Cliffs, N. J.: Prentice-Hall, Inc., 1962), pp. 10—12.

1.3.10. In the course of the history of philosophy the following precepts concerning the philosopher's use of science have been advanced.

(i) Philosophy can find no use for either the methods or the results of science; (ii) philosophy can use some results of science but none of its methods; (iii) philosophy can use the general method of science rather than its results; (iv) philosophy can use both the method and the results of science. Expound your own opinion and argue in favor of it. *Alternate Problem:* From a logical point of view all possible scientific questions and their answers are "there" as just as many actual infinities. Examine the view that the method and the techniques of science constitute an operator converting questions into answers. (Symbolically, $M: Q \to A$.)

1.4. Branches of Science

By contrasting the general method of science to the special methods of the particular sciences we have learned this. First, the scientific method is a way of handling intellectual problems — not things, instruments, or men; consequently it can be employed in all fields on knowledge. Second, the nature of the subject matter dictates the possible special methods of the corresponding subject or field of research: object (problem system) and technique go hand in hand. The diversity of the sciences is apparent as long as their objects and techniques are focused on; they vanish as soon as the underlying general method is disclosed.

The first and most remarkable difference among the various sciences is the one between the *formal sciences* and the *factual sciences*, i.e. between those dealing with ideas and those dealing with facts. Logic and mathematics are formal sciences: they refer to nothing in reality and therefore cannot use our transactions with reality (i.e., experience) in order to validate their formulas. Physics and psychology are among the factual sciences: they refer to facts supposed to happen in the world and accordingly have to resort to experience in order to test their formulas.

Thus, the formula "x is blue" or "$B(x)$" for short, is true of certain things: i.e., it becomes a definite true proposition if the variable x is assigned as a value the name of something actually blue, like the Aegean sea; and it is false of most other things, i.e. it becomes a false proposition for most other values of the object variable x. On the other hand, "x is blue and x is not blue", or "$B(x) \& -B(x)$" for short, is false for every value of x, i.e. in all circumstances. Consequently its negate, "It is not so that x is blue and x is not blue" is true and its truth is independent of fact; in particular, it does not depend on experience (the realm of fact in which man takes part). In short, "$B(x)$" is the skeleton or form of a simple factual idea (if we stick to the interpretation of the predicate 'B' as designating the property of blueness). On the other hand "$-[B(x) \& -B(x)]$" (read 'it is not so that x is B and x is not B')

is the structure of a formal idea, in this case a logical truth: its truth value does not depend on the particular values that x may take on; moreover, it is independent of the interpretation we may wish to attach to the sign 'B'.

Logic is concerned, among other things, with the structure of both factual and formal ideas; but whereas in the former case logic is insufficient to find out truth values, in the latter case logic and/or mathematics are all we need to validate or invalidate any such pure ideas. In short, formal science is *self-sufficient* as regards both content and method of proof, whereas factual science depends on fact for content or meaning, and on experiential fact for validation. This explains why complete formal truth is attainable whereas factual truth is so elusive.

The form of ideas may be said to constitute the proper subject of formal science. An equivalent way of characterizing formal science is to say that it is concerned with *analytic formulas*, i.e. formulas that can be validated by rational analysis alone. Consider, for example, the statement that, if A and B are sets, then if A is (properly) included in B, then B is not included in A. The truth of this statement does not depend on the kind of set and is not established by studying sets of real objects: the formula belongs to the theory of abstract (nondescript) sets: it is purely formal and, consequently, universal — i.e. applicable whenever there is any talk of sets, kinds, or species, whether they be of numbers or of plants. There are a number of kinds of analytic formulas. For our immediate concern the most important are those which are true (or false) by virtue of their logical form, and those which are true (or false) on the strength of the meanings of the symbols occurring in them. The former set — syntactical analyticity — is exemplified by "If x, y and z are numbers, then: if $x=y$ then $x+z=y+z$". The second — semantical analyticity — is instanced by "Synthetic formulas are all those and only those formulas which are not analytic". Formal science contains only analytic formulas, whereas factual science contains, in addition, *synthetic formulas*, i.e. formulas that cannot be validated by reason alone.

The neat dichotomy between formal and factual science should not blind us to the fact that conceptual knowledge of all kinds (as distinct from habits, skills, and other kinds of nonconceptual knowledge) consists of ideas: logic is a set of ideas and so is theoretical physics. All ideas, however concrete their reference may be, have some definite form or other. Thus, the form of "x is blue" is the same as that of "x is prime", i.e., a subject-predicate schema: "$P(x)$". Likewise, "x is lovelier than y" and "x is larger than y" involve a binary predicate: both are schemas of the form "$L(x, y)$" or, more specifically, "$x>y$". Moreover any given formula, no matter what its content may be, can be transformed into a logically equivalent formula: thus, the simple statement p may be converted without either gain or loss into $--p$ (double negation), $p \,\&\, t$, and $p \vee -t$, where 't' stands for any tautology (logical identity). On the other hand contents, if any, are rigid. In particular, no synthetic formula follows from analytic formulas and no analytic formulas follow from synthetic ones: the mathematician cannot infer anything about the world on the strength of his mathematical knowledge alone and, likewise, the physicist can establish no mathematical theorem on the basis of his factual knowledge.

Since every formula has some logical form or other — and is on occasion nothing but a logical form — we must expect to find fibers of formal science everywhere in the body of science, even if such fibers are not apparent. What rigidity the body of knowledge may have is due to the logical and mathematical structures embedded in it rather than to the facts it deals with or the evidence by means of which its truth claims are weighed. For, after all, the scientific knowledge of facts is always partial, indirect, uncertain and corrigible, whereas forms are our own make and we are free to freeze them. In short, what hardness is to be found in science lies in its formal structure: data and hypotheses are soft, i.e. corrigible.

It does not follow that objective *facts* are soft, i.e. changeable at will: for better or for worse most cannot. It does follow that factual science presupposes and contains certain formal theories which it does not question and cannot subject to doubt because facts are irrelevant to pure ideas. (Nor does it follow that formal theories are incorrigible: they are perfected without pause in their own formal contexts — only, not as a result of attempts made to better match facts and accordingly not with the same special methods of factual science.) In short logically, though not psychologically, factual science *presupposes* formal science. (We shall dwell on this in Secs. 5.9 and 15.6.)

Within formal science various orderings can be tried; our subject being factual science we shall not touch that question. As to factual science we shall adopt the ordering displayed in the diagram below. The diagram seems methodologically sound, in the sense that it suggests what disciplines any given science presupposes. Yet alternative orderings are possible and boundaries between contiguous disciplines are hazy and of little practical avail. Moreover, it would be foolish to place much emphasis on the problem of classifying the sciences, once a favorite pastime of philosophers and now a subject for science administrators and librarians. A more interesting subject awaits us: the aim of research.

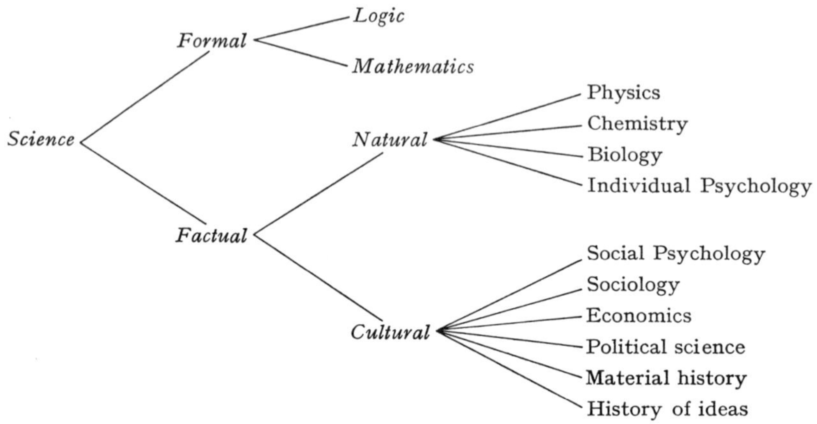

Fig. 1.4. A classification of the sciences. The reader is invited to rearrange the items and fill in some of the gaps

Problems

1.4.1. Give two examples each of factual ideas and formal ideas. Show, furthermore, that thay are indeed factual and formal respectively.

1.4.2. Many statements in factual science can be rigorously proved by deduction from premises (e.g., axioms of a physical theory). Does it follow that they have no factual content?

1.4.3. If a formula is a priori, i.e. independent of experience, then it is analytic, i.e. its validation is a purely logical affair. Does the converse hold, i.e. is every analytic formula a priori? Or is it possible to have analytic a posteriori formulas, i.e. formulas that can be derived by purely logical means on the strength of previous assumptions yet cannot be validated, as to truth value, apart from experience? See M. BUNGE, *The Myth of Simplicity* (Englewood Cliffs, N. J.: Prentice-Hall Inc., 1963), Ch. 2.

1.4.4. Many mathematical theories have been built largely in response to needs of daily life or of factual science, both pure and applied. Does this prove that mathematics deals with facts? And does it prove that mathematics is tested by being applied?

1.4.5. ARCHIMEDES and other mathematicians have employed mechanical devices for proving mathematical theorems. Does this show that mathematics can be pursued as a factual science, or that the alluded proofs were not mathematical proofs at all but rather heuristic procedures?

1.4.6. If logic and mathematics are not concerned with reality, why are they applicable? Hint: examine whether formal science is applied to reality or rather to our ideas concerning reality.

1.4.7. Certain formulas, such as "If p, then: if q then p" (i.e. $p \rightarrow (q \rightarrow p)$) and "For every x, either x is P or x is not P" (i.e. $(x)[P(x) \vee - P(x)]$) are *universally* true: the former holds for all values assigned to the propositional variables p and q, and the latter for all values of the individual variable x and the predicate variable P. From this it has been concluded that logic holds for the most general traits of all objects, whence it would be a kind of general ontology or even "*Une physique de l'objet quelconque*" (F. GONSETH). Hint: begin by establishing whether logic is really concerned with objects of any kind or rather with ideas of any kind.

1.4.8. The label *empirical science* is more frequent than *factual science* in English speaking countries. Why? And are they mutually incompatible designations, or do they point to different aspects of science: to its object (the world of facts) and to the way it validates its truth claims (experience)?

1.4.9. Analyze the relations between any two contiguous branches of science — e.g., climatology and geophysics, geology and physics, zoology and palaeontology, anthropology and archaeology, history and sociology, economics and sociology.

1.4.10. Improve on the classification of the sciences offered in the text. Make sure you use a definite classing criterion. Hint: do not try to include all sciences, because some new science is surely being born while you are wrestling with the problem. *Alternate Problem:* Granting that the classification of the sciences is a somewhat stale problem, does it follow that the boundaries among the sciences are altogether artificial and arbitrary, or do they correspond to objective differences in subject matter and special method? And does the classification of the sciences have any relevance to *ontology*, the discipline dealing with basic categories such as object, space, time, and change?

1.5. Goal and Scope of Science

Methods are means devised to attain certain ends. To what ends are the scientific method and the various techniques of science employed? Primarily, to increase our knowledge (intrinsic or cognitive goal); derivatively, to increase our welfare and power (extrinsic or utilitarian goals). If a purely cognitive aim is pursued *pure science* is obtained. *Applied science* (technology) employs the same general method of pure science and several of its special methods, only applied to ends that are ultimately practical. If these utilitarian goals do not match with public interest applied science can degenerate into impure science — a matter for the sociology of science to investigate.

As regards goals we have, then, the following partition:

Science
- *Pure* (e.g., psychology)
- *Applied* (e.g., pedagogy).

The main branches of contemporary applied science are:

Applied Science (Technology)
- *Physical Technologies* (e.g., electrical engineering)
- *Biological Technologies* (e.g., medicine)
- *Social Technologies* (e.g., operations research)
- *Thought Technologies* (e.g., automatic computation)

It is sometimes held that there is no such division of the sciences into pure and applied because all science is ultimately directed to satisfying needs of some sort or other — but this opinion overlooks the goals of either and is at a loss to explain the differences in outlook and motivation between the investigator who is after a new law of nature and the investigator who is after a new thing: the former wants to understand things better, the latter wants to improve our mastery over them. At other times the difference is granted but it is claimed that applied science is the source of pure science rather than the other way around. This view too is mistaken: there must be knowledge before it can be applied — unless it happens to be a knowhow or skill rather than conceptual knowledge, in which case it is practical from the start. (See Sec. 11.1. for details.)

What is true is that action — industry, government, education, etc. — often *poses problems* that can be solved by pure science alone. And if such problems are worked out in the free and disinterested spirit of pure science, the solutions to such problems may eventually be applied to practical ends. In short, practice is a source — alongside with sheer intellectual curiosity — of scientific problems. But giving birth is not rearing. A whole cycle must be performed before anything comes out from practice: Practice→Scientific Problem→Scientific Research→Rational Action. This was most the frequent pattern until about mid-nineteenth century, when physics gave birth to electrical engineering: from then on technology proper — rather than prescientific craftmanship — became firmly established. Since then, intellectual curiosity has been the source of most, and certainly of all important, scientific problems; technology has often followed in the wake of pure research, with a decreasing time lag between the two. If the external goals of science are exaggerated curiosity and freedom of research — freedom to doubt received ideas and try new ones, even if they do not look socially useful — are stamped on. The immediate result is the languishing of pure science, which ultimately leads to technological stagnation. The most practical policy is to abstain from making practical demands on pure research.

The primary target of scientific research is, then, the *advancement of knowledge*. This is the case even with applied research, such as the investigation of the effect of drugs on pathological conditions; only, in such cases not just knowledge but useful knowledge is sought. Now, there is research for the sake of knowledge, but there is no such thing as knowledge in itself: knowledge is always *of* something — e.g., of the ageing of stars, or of men. The central goal of research in pure factual science is, by definition, to improve our knowledge of the world of facts; that of applied scientific research, to improve the control of man over facts.

Does that mean that scientific research aims at *mapping facts*, at writing as it were a huge cosmography containing the description of every event in nature and culture? Clearly not. First, because an exhaustive description of even our little finger would be practically impossible given the number of its constituents and the variety of events that occur within it during one second; and even if it were possible it would be uninteresting. Second, because no description of a real system can be reasonably complete unless it employs the laws of the system, since laws are what constitute the essence of whatever exists: a sheer description of appearances will miss the essential traits of the system. But once the laws are known little interest remains for a detailed description. Third, because we are interested not only in actuals but also in possibles — the seeds of the future — and again laws alone can give us a knowledge of possibilities. Fourth, because no description can assist us either in explaining what happens or in predicting what may happen: scientific explanation and prediction are based on law statements which in turn interlace in theories. The understanding of the world, in short, is achieved with the help of theories, not of catalogues. Consequently, the exhaustive mapping of every single bit of reality — or even of every item of human experience — is not sour grapes: it is no grapes at all.

What factual science seeks is to map the *patterns* (laws) of the various domains of fact. The conceptual reconstruction of an objective pattern is a scientific law (such as the law of inertia); a system of such law statements is a scientific theory (such as NEWTON's theory of motion).

Rather than a cosmography, then, factual science is a cosmology: a conceptual reconstruction of the objective patterns of events, both actual and possible, whereby their understanding and forecast — hence their technological control — is made possible.

When scientific techniques are applied to data gathering without finding general patterns, embryonic science — *protoscience* — is produced. And when the goal of mature science is pursued but its method and techniques are not employed, nonscientific *speculation* is at stake, whether in the form of philosophy of nature or of traditional metaphysics (the ontology which is neither inspired nor controlled by science). Nonscientific speculation thrives on the backwardness of science proper: thus, philosophical psychology and philosophical anthropology are strong wherever the corresponding scientific disciplines are still in a protoscientific stage — which is not surprising, as they are both easier and more interesting than the aimless collection of isolated items of information. In short, there is no science proper unless the scientific method is applied to the attainment of the goal of science: the building of theoretical images of reality, and essentially of its web of laws. Scientific research is, in short, the search for pattern.

(The terms 'world' and 'reality' are avoided by some philosophers on the ground that they denote metapyhsical concepts: they claim that all there is to know is our own experience, and consequently the sole legitimate goal of science is to account for the sum total of human experience. This view — radical empiricism — does not account for the very existence of most sciences, namely those dealing with empirically inaccessible objects such as the atoms inside our skulls. Science attempts to explain facts of any kind, including those comparatively few in which man gets involved — experiential facts. As a matter of fact experience is not the sole or even the main object of research, hence not the sole *referent* of scientific theories: experience, if scientific, is an indispensable *test* of those theories but does not provide the content or meaning of all of them. Moreover, in order to explain human experience — the object of the sciences of man — we need some knowledge of the natural world of which we are a part: and this world, mostly unseen and untouched, is gradually being mapped by testable theories that go beyond what can be experienced.)

Science, then, tends to build *conceptual mappings of the patterns of facts* — i.e., factual theories. But mythology too offers models of the world, both to understand and to better bear it. Why should we prefer scientific theories to mythological speculations? One is tempted to reply: because scientific theories are true reconstructions of reality. But a glimpse at the endless convulsions of science, in which most theories are caught in some error or other and just a few of them are pronounced unbelievably true, yet not quite, should convince us that scientific research does not attain complete truth. What right do we have, then, to believe that science is better off than mythology — especially if science, too, invents concepts, such as "field", "neutrino", and "natural selection", which can be associated with no sensory experience?

Must we conclude that mythology and science supply just different but equally legitimate mappings of reality? Clearly not: science does not claim to be true, hence final, incorrigible, and certain as mythology does. What science claims is (i) to be *truer* than any nonscientific model of the world, (ii) to be able to *test* such a truth claim, (iii) to be able to *discover its own shortcomings*, and (iv) to be able to *correct its own shortcomings*, i.e. to build more and more adequate partial mappings of the patterns of the world. No extrascientific speculation is as modest and yields as much.

What enables science to achieve its goal — the building of progressively truer partial reconstrutions of reality — is its method. By contrast the nonscientific speculations about reality (i) do not usually ask proper questions, but rather problems with false or untestable presuppositions, such as "How and when was the universe created?"; (ii) they do not propose hypotheses and procedures both grounded and checkable but offer groundless and usually untestable theses as well as uncontrollable (inscrutable) means for finding out their truth (e.g., revelation); (iii) they do not design objective tests of their theses and of their alleged sources of knowledge but resort to some authority; (iv) they accordingly have no occasion to contrast conjectures and procedures with fresh empirical results: they remain content with finding illustrations of their conceptions for persuasion purposes rather than for the sake of test — as shown by the eagerness with which they explain every negative evidence away; (v) they give rise to no new problems — their whole point being to put an end to inquiry by providing a ready made set of answers to every possible or permitted question.

Science, on the other hand, yields problematic but improvable conceptual reconstructions of reality. Actually it does not provide a single model of reality as a whole but a *set of partial models* — as many as theories dealing with different aspects of reality, such a variety depending not only on the richness of reality but also on the assortment and depth of our own conceptual outfit. Research does not start with such synthetic views of chunks of reality but arrives at them through *rational and empirical analysis*.

The first step in analysis, whether scientific or not, is the discrimination of components at some level — e.g., the distinction of organs, or of functions, in an organism. In a second stage the relations among these components are found — and this already provides a first picture of the whole, i.e., the synoptic conceptual picture that had been sought. Once such a model of the system (set of interrelated items) is on hand it can be used as a tool for a deeper analysis, the outcome of which is hoped to be a more adequate synthesis. Proceeding the other way around, i.e. starting with grandiose synthetic views instead of working in a piecemeal, analytic way, is characteristically nonscientific.

Scientific research, then, does not end up with a single final and complete truth: it does not even seek a single world-embracing formula. The outcome of research is a set of more or less true, as well as partially interconnected statements (formulas) concerning different aspects of reality. In this sense, science is pluralistic. In another sense it is monistic: it attacks all fields of knowledge with a single method and a single goal. The unity of science does not reside in a single all-embracing theory, nor even in a unified all-purpose language, but in its singleness of approach.

The process of reconstructing the world with ideas, and of testing every partial reconstruction, is endless — notwithstanding the groundless hopes that the definitive theory is just around the corner. Research does not cease to discover holes in its mapping of the world. Hence, science cannot have an ultimate goal, such as building a complete and flawless cosmology. The goal of science is rather the *ceaseless perfecting* of its chief products (theories) and means (techniques), as well as the subjection of more and more territories to its sway.

Are there *limits* to this expansion of the object of science? That is, are there problems of knowledge that cannot be attacked with the method and the goal of science? The unavoidable temporary limitations determined by our ignorance are not in question; nor are extrinsic limitations meant, such as those imposed by ideological, political, or economic power. What is being asked is whether there are any objects of knowledge inherently recalcitrant to the scientific approach. An optimist would think that, since the history of science shows an increasing domain of facts covered by science, we must believe that this expansion will never stop — unless we manage to blow ourselves up. Yet no past experience and no historical trend, however suggestive, is demonstrative: problems might come up, for all we know, which could prove impermeable to the scientific approach.

The preceding conclusion need not lead us to pessimism with regard to the scope of the scientific approach: between pessimism and optimism there is place for realism. A realistic appraisal would seem to be as follows. Firstly, we can hope that every problem of knowledge will eventually be shown to be either partially solvable or unsolvable by the means (special methods) and with the data available to science at any given moment. Secondly, no method more powerful than the method of science has been found, and every successful effort in such a search has resulted in perfecting the scientific method; in particular, the attempts to grasp reality directly, without working (e.g., by perception, empathy, or pure speculation), have all failed and, moreover, we can explain why they were bound to fail: namely, because most facts are beyond experience and must therefore be hypothesized. Thirdly, the scientific method and the special techniques by which it is implemented are not final: they have evolved from more rudimentary ancestors and they will have to be improved if we want to obtain better results. Fourthly, since what is peculiar to science is not a definite object (or set of problems) but rather an approach (method and goal), anything is turned into a scientific subject, i.e. into an object of scientific inquiry, if treated with the method of science in pursuit of the goal of science — even if such a treatment is unsuccessful. In short, we cannot and do not wish to warrant the success of the scientific approach to problems of knowledge of any kind: science is not a panacea: we are just making the more modest claim that the scientific approach is the best available.

There is one object though — one might be tempted to rejoin — that factual science does not study: namely, science itself. Yet clearly the study of science can be approached scientifically and is so done on occasion: we have, in fact, a number of immature sciences of science. If science is viewed as a peculiar activity of individuals and teams, the psychology of science can emerge; this discipline will study, among other things, the cognitive drive, the psychological processes of hypotheses generation, mental rigidity among scientists, etc. Viewing science in its social context can lead to the sociology of science, i.e. the study of the social factors that prompt and those which inhibit research, the role of science in the planning and control of human action, and so on. And if science is studied as an aspect of cultural evolution, the history of science emerges, i.e. the study of the origins and development of a line of research, of the changes of scientific outlook, and so forth. The above are *external* approaches to science, in the sense that they do not analyze and criticize either the method or the outcome of research but take them for granted. Moreover, the psychology, the sociology and the history of science are factual (empirical) sciences of science: they handle large masses of empirical data.

The *internal* approach to science has, since its inception, been a philosophical subject. It is philosophers — and occasionally scientists on holidays — who have studied the general pattern of scientific research, the logic of scientific discourse, and the philosophical implications of method and outcome. This internal study of science bears on scientific knowledge apart from its psychological origin, cultural setting and historical evolution, whereas the external approach is concerned with the activities of the men involved in the production, consumption, waste, and corruption of science: the external sciences of science are as many branches of the sciences of culture. The internal study of science, on the other hand, steps above its object, in the semantical sense that it is a discourse on a discourse. Just as a statement about a statement is called a metastatement, so the internal study of science may be called *metascience*, itself part of the theory of knowledge (*epistemology*).

Metascience may be divided into three parts: the *logic* (syntax and semantics) of science, concerned with problems such as the structure of scientific theories and the empirical import, if any, of scientific concepts; the *methodology* of science, dealing with the general method of science and the techniques by which it is implemented, such as, e.g., randomization; and the *philosophy* of science, studying the logical, epistemological, and ontological commitments and upshots, if any, of scientific research. These problem domains have some roots in the past but they have only recently been approached scientifically. Moreover, these fields are unequally advanced: whereas the formal logic of science, particularly the syntax of theories, is an exact science, the methodology and the philosophy of science are still limited, in the main, to describing and analyzing science: only occasionally they produce theories proper, such as the theory of the probability of hypotheses, but then such theories usually apply to oversimplified models of science rather than to real science. Metascience, in short, is chiefly a *protoscience* rather than a fully grown science: it adopts a scientific approach but has so far produced few scientific results.

Anyway, in addition to science *tout court* we have the science of science:

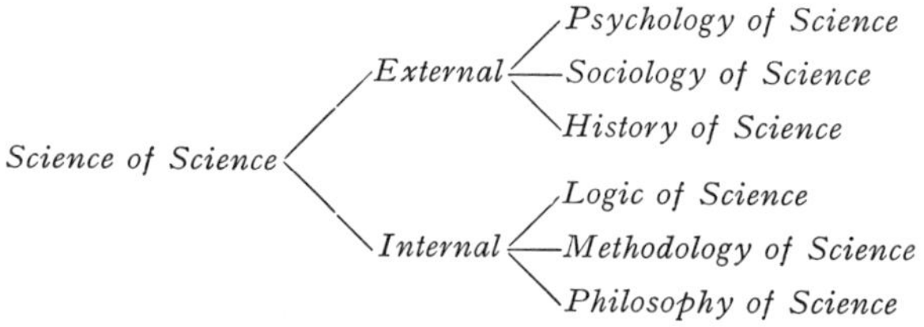

In conclusion, however limited the outcome of the scientific approach may be, no *inherent* limitations to it have been shown to exist and, moreover, it is only from within science that such limitations can be correctly appraised: all of nature and all of culture, including science itself, can be made to fall under the domain of science. There certainly are subjects that have so far not been approached scientifically — for instance, love — either because no one has as yet realized their existence or because they have not attracted the curiosity of investigators or, finally, because of external circumstances such as prejudice — e.g., the tenet that certain human experiences cannot be approached scientifically but must remain private. Such tenets have in their favor not only the weight of tradition but also a mistaken conception of science: most often, its identification with physics. These prejudices are among the last bastions of obscurantism; they are rapidly crumbling down: we are having scientific studies of aesthetic experience and even of the subtle manipulations of men's minds by means of stale ideologies, such as those which discourage the scientific study of man.

The successes of the scientific approach, as well as its independence with regard to subject matter, account for the expansive power of science, which now occupies territories previously occupied by the humanities — e.g., anthropology and psychology — and is continually exploring new territories. The same factors account also for the increasing importance of science in modern culture. Since the Renaissance the centre of culture has steadily shifted from religion, art, and the classical humanities to science — formal and factual, pure and applied. It is not only that the intellectual results of science and its applications to good and bad ends are being acknowledged by even the least cultured painter: an even more important and welcome change consists in the spread of a scientific *attitude* towards problems of knowledge and towards problems whose correct solution requires some knowledge. This is not to say that science is gradually absorbing all of human experience: that we shall end by loving and hating scientifically just as we can cure and kill scientifically. Except scientific research itself, human experiences are not scientific, not even if they take advantage of scientific knowledge: what can be and should be scientific is the study of any such experience.

Important changes in outlook and behavior, both individual and collective, can be expected from a widespread diffusion of the scientific attitude — not however from a popularization of just some results of scientific research. The universal adoption of a scientific attitude might render us wiser: it would make us more cautious in receiving information, in keeping beliefs and in making forecasts; it would render us more stringent in testing our opinions and more tolerant to other people's opinions; it would make us more eager to freely inquire into new possibilities and readier to get rid of consecrated myths; it would enhance our trust in experience guided by reason and our confidence in reason checked by experience; it would stimulate us to plan and control action better, to select aims and to search for norms of conduct consistent with such ends and with available knowledge rather than with habit and authority; it would foster the love of truth, the willingness to acknowledge error, the thrust to perfection and the understanding of the inevitable imperfection; it would give us an ever young world-view founded on tested theories instead of a die hard untested tradition; and it would encourage us to hold a realistic view of human life: a well-poised rather than either an optimistic or a pessimistic view. These may seem remote or even improbable effects, and at any rate scientists could never, by themselves, bring them about: a scientific attitude demands a scientific training, which is desirable and possible only in a scientifically programmed society. But this much may be granted: that the growth of the relative importance of science in the body of culture has already born some such fruits in a limited scale, and that the program is worth being tried, especially in view of the limited success of alternative programs.

To conclude: the scientific approach has no known inherent limitations; it is in a process of rapid expansion and is yielding increasingly true partial pictures of the world outside and inside man — not to speak of the tools for its control. (Should anyone hold that the scientific approach does have inherent limitations we would ask him to sustain his claim — by conducting a scientific investigation of the problem.) By virtue of its spiritual power and its material fruits science has come to occupy the center of modern culture — which is not to say the culture of our days. In fact, it would be foolish to forget that, alongside higher culture, folk culture still lingers. And pseudoscience occupies in contemporary urban folk culture a position similar to the one held by science in higher culture. It should prove both instructive and amusing to take a look at what is often smuggled for science although it lacks both the method and the goal of science. To this subject — folk science — we now turn.

Problems

1.5.1. Draw a distinction between the goals of science and those of individual scientists — which may include the obtainment of fame, power and riches. Explain why individuals with purely egoistic goals can make distinguished contributions to pure (disinterested) science.

1.5.2. Can we conclude, from the objectivity of scientific research, that it is impersonal? If not, i.e. if research does involve the whole person even when performed by teams, does it follow that it cannot attain objective truth — that the objectivity of science is therefore mythical, as has been held? See Problem 1.1.2 and M. Polanyi, *Personal Knowledge* (Chicago: University of Chicago Press, 1959).

1.5.3. Management, advertisement, warfare, and other actitivities can be conducted either empirically (in the traditional way) or scientifically, i.e. with the assistance of experts with scientific knowledge and with a scientific attitude. Are scientific management, advertisement, and warfare sciences? If so, why? If not, what do they lack?

1.5.4. Is scientific knowledge a means or an end? Begin by completing this question: a means and an end are terms of a triadic relation involving also a subject. *Alternate Problem:* Means and ends come in pairs. Change the goal and you may have to change the means. Apply this to the proposal of investigating theological problems with the scientific method.

1.5.5. Expand and illustrate the thesis that science is self-corrective i.e. that it is criticized and improved from within. *Alternate Problem:* Does the self-corrective character of science render philosophical criticism invalid and/or ineffective?

1.5.6. Describe and exemplify scientific analysis of the two kinds: *factual* (e.g., chemical) and *conceptual* or theoretical analysis (e.g., the analysis of forces into imaginary components along coordinate axes).

1.5.7. In what senses is science analytic: logically, methodologically, or ontologically? (*Logical analyticity:* the property of a statement of being determinable as true or false with the sole help of an analysis of its logical structure or of the meanings of its terms. *Methodological analyticity:* the property of a procedure of decomposing an object either in fact or in thought, instead of leaving it as a block; such an analysis can be into parts, properties, and relations. *Ontological or metaphysical analyticity:* the doctrine that the world is an aggregate or a system of smaller units.)

1.5.8. Expand the thesis that a scientific conceptual synthesis is not independent of analysis, but rather an outcome of analysis. *Alternate Problem:* Is history a science or a protoscience?

1.5.9. Examine the following theses concerning the unity of science. (i) The unity of science resides in its object: reality. (ii) The unity of science resides in its objective, namely recounting the history of what there is. (iii) The unity of science consists in having or in striving for a single language — be it a sense-data language (sensationalism), an observation language (empiricism), or the language of mathematics (Pythagoreanism). (iv) The unity of science consists in the ultimate reduction of all factual science to physics (physicalism). (v) The unity of science resides in its singleness of approach (method and goals). *Alternate Problem:* Discuss the following conflicting proposals concerning the goal of factual science. (i) The aim of science is the complete adaptation of our thought to our experiences (physicist E. MACH). (ii) The aim of science is the creation of a world-view completely independent of the investigator (physicist M. PLANCK). *Alternate Problem:* Comment on the "Declaration of Interdependence in Science", *Science*, **111**, 500 (1950), in which the unity of method and goal of all the sciences are stated.

1.5.10. Discuss any of the following preliminary work on the extension of the scientific approach to subjects traditionally covered by the humanities. (i) N. Rashevsky, *Mathematical Biophysics*, 3rd ed. (New York: Dover, 1960), Vol. II, Ch. XVI on aesthetics, and *Mathematical Biology of Social Behavior*, 2nd ed. (Chicago: University of Chicago Press, 1959), Appendix IV on history. (ii) M. Bunge, "Ethics as a Science", *Philosophy and Phenomenological Research*, **22**, 139 (1961) and "An Analysis of Value", *Mathematicae Notae*, **18**, 95 (1962). *Alternate Problem:* Try to explain why philosophical anthropology, philosophical psychology, political philosophy and the philosophy of education are still taught alongside the corresponding sciences (pure or applied).

1.6. Pseudoscience

Ordinary knowledge can develop in either of the following three directions. (i) *Technical knowledge:* the specialized but nonscientific knowledge that characterizes the arts and crafts. (ii) *Protoscience* or embryonic science, as exemplified by careful but aimless observation and experiment. (iii) *Pseudoscience:* a body of beliefs and practices whose practicioners wish, naively or maliciously, pass for science although it is alien to the approach, the techniques, and the fund of knowledge of science. Still influential pseudosciences are dowsing, psychical research and psychoanalysis.

Science is not unrelated to technical knowledge, protoscience, and pseudoscience. In the first place, science makes use of the artisanal skills, which are in turn often enriched by scientific knowledge: In the second place, science makes use of some of the raw data gathered by protoscience — though most of it are useless because insignificant. In the third place, occasionally a science has grown from a pseudoscience, and sometimes a scientific theory has been stiffened into dogma to a point where it has ceased to correct itself and has become a pseudoscience. In short, many are the flow lines between science and its neighbors:

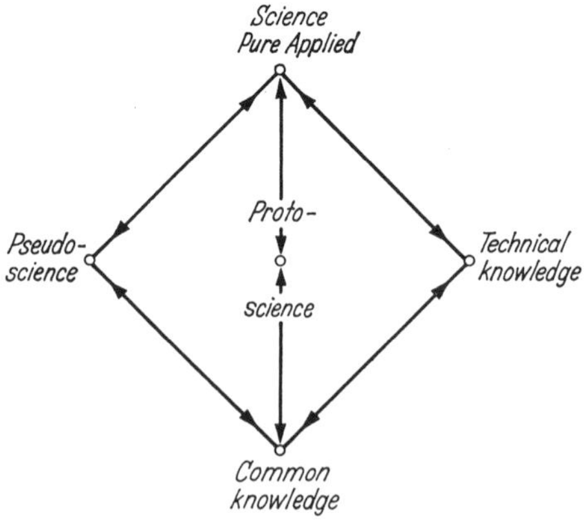

What is wrong with pseudoscience? Certainly not just that it is basically false, since anyway all our factual theories are at best partially true. What is wrong with pseudoscience is, first, that it refuses to *ground* its doctrines and could not do it because pseudoscience makes a total break with our scientific heritage — which is not the case of scientific revolutions, all of which are partial since every new idea has to be gauged by means of others that are not questioned in the given context. Second, pseudoscience refuses to *test* its doctrines by experiment proper; moreover, it is largely untestable because it tends to interpret all data in such a way that its theses are confirmed no matter what happens: the pseudoscientist, like the fisherman, exaggerates his catch and neglects his failures or excuses them. Third, pseudoscience lacks a *self-correction mechanism:* it cannot learn from either fresh empirical information (which it swallows without digesting it), new scientific discoveries (which it despises), or criticism (which it rejects indignantly). It can make no progress because it manages to interpret every failure as confirmation and every criticism as an attack. Differences of opinion among its sectarians, when such deviations arise at all, lead to endless sect splitting rather than to progress. Fourth, the primary aim of pseudoscience is not to set up, test, and correct systems of hypotheses (theories) mapping reality, but to influence things and men: it has, like magic and like technology, a primarily *practical aim* rather than a cognitive one but, unlike magic, it presents itself as science and, unlike technology, it does not enjoy the backing of science.

Our first example of a pseudoscience will be dowsing or, more generally, *rhabdomancy*. The thesis of rhabdomancy is that certain particularly sensitive subjects can unconsciously and directly feel underground inhomogeneities, such as ore mines and oil basins. The technique of rhabdomancy is to use a hazel-twig or a pendulum as an indicator of such a feeling. Summarily, the pattern would be this: Geological Accident → Unconscious Reception → Involuntary Bodily Motions → Pendulum Oscillations → Perception of Oscillations. Some rhabdomants claim that the first link in this chain may also be a tumor or a car ailment.

What can be wrong with rhabdomancy? First, both the thesis and the technique of rhabdomancy are not *grounded* on the bulk of scientific knowledge, according to which a direct action of physical bodies upon mental states is impossible: both a physical agent and it action on a biological mechanism are needed, for the simple reason that mental functions are proper of highly developed nervous systems, which are in turn physical systems. On the other hand, the standard geological prospecting techniques (e.g., with the help of artificially produced seismic waves) are based on well-known physical laws: the mechanism of their operation is known and this is why they are regarded as reliable. Second, the thesis of rhabdomancy is *untestable* or nearly so for either or both of the following reasons: (a) it involves neither a definite mechanism nor a definite law, so that it is difficult to ascertain what can be argued about and what experiments could conceivably refute the thesis; (b) if the dowser makes a correct guess of, say, an underground water vein, the thesis is pronounced confirmed; should he fail to correctly signal water he might defend his belief either by saying that there is water, only deeper than the drilling can reach, or by humbly acknowledging that he made a mistake: he took for indicators what were just signs of fatigue, or of excitement. No geologist can ever achieve such a hundred per cent degree of confirmation.

Notice that experience is irrelevant to the refutation of rhabdomancy. First, because this belief is empirically untestable. Second, because a dowser with a descriptive knowledge of the terrain can beat a geologist equipped with scientific instruments and laws but with no equivalent knowledge of the locality. Consequently rhabdomancy must either not be discussed or disposed of by metacientific argument: by showing that its thesis and its technique fail to be grounded and testable — two requisites of scientific ideas and procedures.

Our second example will be *parapsychology* or psychical research, the modern names for spiritualism, mediumnism, cartomancy, and other archaic beliefs and practices. This doctrine holds the existence of certain phenomena, such as telepathy (thought transmission), clairvoyance (seeing at a distance), precognition (future seeing), and telekinesis (mental causation of physical phenomena). These alleged facts are ascribed by parapsychology to extrasensory perception (*ESP*) and other supranormal abilities which it does not profess to explain. Parapsychology is somewhat elusive not only because it deals with nonphysical entities (such as ghosts) and events (such as telepathy) but also because it makes no detailed, hence no definitely testable assertions about either action mechanisms or regularities — but this makes it all the more suspicious to the critical metascientist. Let us spell out this complaint.

Firstly, parapsychologists do not state and treat their theses *as hypotheses*, i.e. as corrigible assumptions concerning unperceived events: by calling the alleged anomalies cases of extrasensory perception the parapsychologist commits himself a priori to a definite assumption which he will henceforth try to illustrate rather than substantiate. Secondly, the theses of psychical research are *loosely stated* and have *little content:* they are just assertions about the existence of certain rare events with no intimation about the possible mechanism for the production, propagation and reception of psychical messages. Of course, no physical mechanism is or could be accepted by parapsychology, since it would automatically place the whole field within the reach of physics and psychology: explanations in terms of subliminal cues or in terms of special new waves miss the whole point of parapsychology. The sole "interpretation" of the alleged anomalies that a parapsychologist can admit is that they are nonphysical and nonnormal: as soon as he attempts to be more specific he risks being refuted.

Thirdly, the vague theses of parapsychology are *nonnaturalistic* and *ungrounded*. More than this: they collide head-on with scientific knowledge. The latter, in fact, suggests the following generalizations: (i) no event lacks a physical basis, (ii) the mind is not a substance, however subtle, that can abandon the body, propagate over space and act on matter: 'mind' is a name for a complex system of functions or states of the nervous system; (iii) no effect preexists its cause and, in particular, no message can be received before it has been sent — as required by precognition. The inconsistency of *ESP* with science voids it of empirical support, because empirical information alone is no evidence at all: for a datum to become an evidence in favor or against a scientific hypothesis it must be interpreted in the light of some set of theories. Since parapsychology lacks a theory althogether it must accept the interpretations proposed by normal science; but *ESP* impugns the competence of the latter to handle the alleged anomalies it deals with, hence it cannot accept any data, not even those it collects itself. In short, *ESP* can marshal no evidence in its own favor.

Fourthly, the observations and experiments conducted by parapsychologists have been shown to be *methodologically invalid* a number of times: (i) many of them have been exposed as frauds; (ii) they are not repeatable, at least in the presence of unbelievers, and there is considerable disagreement among psychical researchers concerning the "facts"; (iii) parapsychologists tend to reject contrary evidence: this they do, e.g., by selecting lucky runs and stopping as soon as randomness reappears; (iv) they often misapply statistics, for example when they apply it to nonrandom samples (selected subsequences of trials) as if they were strictly random, much in the same way as vitalists refute materialism by showing how small the probability is that an organism will emerge spontaneously from the "random" encounter of a myriad of atoms.

Fifthly, even though the theses of parapsychology are separately testable — though barely so — parapsychologists tend to combine them in such a way as to make *the whole set insensitive to test,* hence immune to criticism on the strength of experience: if a set of guessings is consistently about chance the subject will be said to be tired, or to resist belief, or even to have lost his paranormal ability — which is unrelated to other abilities, so that it can be manifested by above-chance performances alone, never by a personality analysis, let alone by a neurophysiological investigation; if the subject does not read the right card or message but the next one in a sequence, he is said to exhibit the phenomenon of forward displacement, which is in turn interpreted as a clear case of precognition; if he fails to move the die or to blow the trumpet at a distance, again a momentary inhibition or even a final loss of ability can be resorted to. In this way the gang of parapsychological theses is rendered unassailable and, by the same token, the testing techniques of science become pointless in their regard: conspiracy defeats checking.

Sixthly, parapsychology is guilty of not having produced, in 5,000 years of existence, *a single empirical regularity*, let alone a law statement embodied in a theory. It has produced neither certified facts nor laws: it is not even an untested but promising young theory: it is not a theory at all, as the few theses of the doctrine are elusive and they gather for purposes of mutual defence against criticism rather than for the logical derivation of testable consequences. In other words, psychical research has not attained the goal of science — and has never aimed at it.

Our last example of a pseudoscience will be *psychoanalysis* — not to be confused with either psychology or psychiatry (the technology associated with psychology). Psychoanalysis claims to be both a theory and a therapeutical technique. As a theory it would be acceptable if shown to be true enough; as a technique, if shown to be effective enough. But in order to sustain either the claim to truth or the claim to efficiency a body of ideas and practices must subject itself to the canons of grown up pure and applied science — at least if it wishes to pass for science. Psychoanalysis fails to pass the science tests.

Firstly, the theses of psychoanalysis are *alien to, and often inconsistent with psychology, biology, and anthropology*. For instance, the whole doctrine is alien to learning theory, the most advanced chapter of psychology. The hypothesis of an unconscious racial memory has no foot in genetics; the assertion that aggressiveness is instinctive and universal contradicts ethology and anthropology; and the hypothesis that every man harbors an Oedipus complex is contradicted by anthropological findings. If these were secondary points of the doctrine it would not be grave: they are important ones and, what is more, psychoanalysis cannot resort to science in order to get spare parts for its worn out doctrine, because it presents itself as an independent rival science.

Secondly, some psychoanalytic hypotheses are *untestable:* for example, those of infantile sexuality, of the existence of disembodied entities within the self (the id, the ego, the superego), and of sleep as representing a return to the maternal womb.

Thirdly, those theses of psychoanalysis that are testable have been illustrated but never *tested* by psychoanalysts with the help of the standard testing techniques; in particular, statistics plays no role whatsoever in psychoanalysis. And when they have been tested by psychologists they have failed. Examples: (i) the conjecture that every dream is a wish fulfilment has been tested by asking subjects with objectively known urges, such as thirst, to report their dreams: there is a very low correlation between urges and dreams. (ii) According to the catharsis hypothesis exposure to films showing brutal behavior should result in a discharge of aggressiveness: experiment has shown the opposite result (R. H. WALTERS et al., 1962). (iii) No significant correlation between early feeding habits and toilet training, on the one hand, and personality traits on the other, has been found in follow-up studies (W. H. SEWALL, 1952 and M. A. STRAUS, 1957). (iv) When control groups have been formed to gauge the influence of psychoanalytic therapy on neuroticism no favorable influence was found, the percentage of cures being sometimes below the percentage of spontaneous remissions (H. H. W. MILES et al., 1951; H. J. EYSENCK, 1952; E. E. LEVITT, 1957); on the other hand, the reconditioning technique is successful in most cases, on top of which it is backed up by learning theory (J. WOLPE, 1958).

Fourthly, although some psychoanalytic conjectures are individually testable and have occasionally been tested, *as a gang they are untestable*. For example, if the analysis of a dream's content does not show it to be the imaginary fulfilment of a wish, the psychoanalyst will argue that this only proves the subject has strongly repressed his wish, which so remains beyond control; similarly, if a man fails to exhibit the Oedipus complex he is said to have repressed it, perhaps for fear of castration. In this way the various members of the gang protect themselves mutually and the doctrine as a whole remains unassailable by experience.

Fifthly, psychoanalysis, in addition to absorbing every evidence that would normally (in science) be regarded as unfavorable, *resists criticism*. Moreover, it disposes of criticism by the *ad hominem* argument that the critic is exhibiting the resistance phenomenon thereby confirming the psychoanalytic hypothesis concerning resistance. Now, if neither argument nor experience can conceivable shake a doctrine, then it is a dogma and not a science. Scientific theories, far from being perfect, are either hopeless and accordingly forgotten, or perfectible and accordingly corrected.

The above completes our schematic account of mancies that wish to be taken for sciences. More detailed metascientific analyses of pseudoscience are desirable for various reasons. First, in order to help the younger sciences — particularly psychology, anthropology and sociology — to get rid of pseudoscientific beliefs. Second, to help people acquire a critical attitude in the place of gullibility. Third, because pseudoscience is a good testing ground of metascience and, in particular, of criteria by which science is marked off from nonscience: metascientific doctrines should be gauged, among other things, by the amount of nonsense they allow for.

To contemporary science, on the other hand, pseudoscience offers very little. A few of its untested conjectures might be worth trying — when they are testable at all; some of them might have an element of truth after all, and even by establishing that they are false some knowledge would be gained. But the most important problem pseudoscience poses to science is this. What are the psychological and social mechanisms whereby archaic superstitions, such as the belief in premonitions and the belief that dreams tell the hidden truth, have managed to survive down to the atomic age? Why do not superstitions and their more bulky developments, the pseudosciences, just wither away as their logic is shown to be faulty, their methodology either too naive or too malicious, and their theses inconsistent with the best available data and theories of science?

Problems

1.6.1. Pseudoscientists often advertise their lore by pointing out that scientist *A* or philosopher *B* believes it. What kind of an argument is this? And is this a test of the pseudoscience under patronage or rather of the given thinker's scientific attitude?

1.6.2. Why do ghosts never show up in Piccadilly Circus or in Times Square? Why are mediums, sensitives and visionaries more and more scarce? Why do astrologists never look back to their past prophecies to compute the percentage of hits? Why are their genuine guesses reasonable to begin with, i.e. such that any well-informed normal person could make them? Why do not psychoanalysts employ the statistical techniques of control of qualitative hypotheses: just because they do not master these techniques? Why do not healers advertise the frequency rather than the total number of alleged cures? Why do not parapsychologists and psychoanalysts make definite predictions?

1.6.3. Report on any of the following works on psychoanalysis. H. J. EYSENCK, "Psychoanalysis: Myth or Science?", *Inquiry*, **1**, 1 (1961). H. J. EYSENCK, Ed., *Handbook of Abnormal Psychology* (London: Pitman Medical Publishing Co., 1960), Chap. 18. E. NAGEL, "Methodological Issues in Psychoanalytic Theory", in S. HOOK, Ed., *Psychoanalysis, Scientific Method and Philosophy* (New York University Press, 1959). W. H. SEWALL, "Infant Training and the Personality of the Child", *American Journal of Sociology*, LVIII, 150 (1952). J. WOLPE, *Psycho-*

therapy by Reciprocal Inhibition (Stanford: Stanford University Press, 1958), passim. L. BERKOWITZ, *Aggression* (New York: McGraw-Hill, 1962). *Alternate Problem:* Report on the papers alluded to in the text; locate them by perusing the *Psychological Abstracts*.

1.6.4. Report on any of the following articles on parapsychology. W. FELLER, "Statistical Aspects of ESP", *Journal of Parapsychology*, **4**, 271 (1940). R. ROBINSON, "Is Psychical Research Relevant to Philosophy?", *Proceedings of the Aristotelian Society*, Suppl. Vol. XXIV, 189 (1950). J. L. KENNEDY, "An Evaluation of ESP", *Proceedings of the American Philosophical Society*, **96**, 513 (1952). G. SPENCER BROWN, "Statistical Significance in Psychical Research", *Nature*, **171**, 154 (1953). G. R. PRICE, "Science and the Supernatural", *Science*, **122**, 359 (1955) and the ensuing discussion in the same journal, **123**, 9 (1956). C. E. M. HANSEL, "A Critical Analysis of the Pearce-Pratt Experiment", *Journal of Parapsychology*, **25**, 87 (1961) and "A Critical Analysis of the Pratt-Woodruff Experiment", *ibid.*, **25**, 99 (1961). See the defence of parapsychology by M. SCRIVEN, "The Frontiers of Psychology", in R. G. COLODNY, Ed., *Frontiers of Science and Philosophy* (Pittsburgh: University of Pittsburgh Press, 1962).

1.6.5. Could parapsychology and psychoanalysis improve by a more definite statement of their hypotheses, a better logical organization, and additional empirical data, as is often claimed by their less fanatical supporters?

1.6.6. Astronomy was born from astrology, chemistry from alchemy and medicine from shamanism. Can we conclude that every pseudoscience gives birth to or at least shades into a science and, consequently, that new sciences will eventually evolve from parapsychology and psychoanalysis?

1.6.7. Comment on either of the following statements. (i) S. FREUD, *Introductory Lectures on Psychoanalysis*, 2nd ed. (London: Allen & Unwin, 1929), p. 16: psychoanalysis "must dissociate itself from every foreign preconception, whether anatomical, chemical, or physiological, and must work throughout with conceptions of a purely psychological order". (ii) R. H. THOULESS, quoted by S. G. SOAL and F. BATEMAN, *Modern Experiments in Telepathy* (London: Faber and Faber; N. Haven, Conn.: Yale University Press, 1954), p. 357: "I suggest that the discovery of the *psi* phenomena has brought us to a [...] point at which we must question basic theories because they lead us to expectations contradicted by experimental results [...] we must be ready to question all our old conceptions and to distrust all our habits of thought". *Alternate Problem:* In what are the pseudosciences different from the normal heresies withing science?

1.6.8. Perform a metascientific analysis of any of the following doctrines: phrenology, graphology, homoeopathy, osteopathy, *Rassenkunde.* Find out whether they share the method and the goal of science. *Alternate Problem:* Make a study of miraculous healing (by faith, confession, logotherapy, patent medicine, etc.) and its special logic. Show, in particular, whether the latter involves (i) the fallacy of the *post hoc ergo propter hoc* (after that, whence because of that); (ii) the disregard of alternative hypotheses (e.g., suggestion); (iii) the neglect of unfavorable cases or their conversion into favorable ones by the addition of *ad hoc* hypotheses (e.g., bewitchment or insufficient faith).

1.6.9. What should be examined in order to determine whether a given doctrine is scientific or not: its use of a special jargon, its employment of empirical procedures (e.g., observation), its apparent practical success, the quantity and quality of its followers, or the methods it employs, its continuity with the bulk of science, and its goal?

1.6.10. Homoeopathy claims to cure with certain highly diluted natural products. If the concentration of any homoeopathic medicine is computed, a figure of the order of one molecule per cubic centimeter is obtained. Is this enough to dispose of homoeopathy or would it be necessary to conduct experiments? In any case, what kind of an argument would that be? *Alternate Problem:* Study the psychology of gullibility. For the protection of belief in the face of disconfirmation, see L. Festinger, H. W. Riecken and S. Schachter, *When Prophecy Fails* (Minneapolis: University of Minnesota Press 1956). For an amusing story of medical quackery in the U.S.A., see S. Holbrook, *The Golden Age of Quackery* (New York: Macmillan, 1959).

Bibliography

Ackoff, R. L.: Scientific method: Optimizing applied research decisions, Ch. 1. New York and London: John Wiley & Sons 1962.

Bunge, M.: Metascientific queries, Chs. 1 to 3. Springfield (Ill.): Ch. C. Thomas 1959.

Cohen, M. R.: Reason and nature, 2nd ed., Chs. 3 and 4. Glencoe (Ill.): The Free Press 1953.

—, and E. Nagel: An introduction to logic and scientific method, Chs. X and XX. New York: Harcourt, Brace & Co. 1934.

CONANT, J. B.: On understanding science. New Haven (Conn.): Yale University Press 1947.

GARDNER, M.: Fads and fallacies. New York: Dover 1957.

MEHLBERG, H.: The reach of sience, Part II. Toronto: Toronto University Press 1958.

NASH, L. K.: The nature of the natural sciences, Boston: Little, Brown & Co. 1963.

WEST CHURCHMAN, C., R. L. ACKOFF, and E. L. ARNOFF: Introduction to operations research, Ch. 1. New York and London: John Wiley & Sons 1957.

WILSON, Jr., E. BRIGHT: An introduction to scientific research, Ch. 3. New York: McGraw-Hill Book Co. 1952.

Chapter 2

Concept

Unlike inborn patterns of behavior and unlike know-hows, scientific knowledge is entirely *conceptual:* it consists of systems of concepts interrelated in definite ways. (Example of concept: "greater than". Example of a conceptual system: the proposition "Lawrencium has a greater atomic weight than Nobelium". Example of a higher order conceptual system: the theory of statics.) Scientific research, on the contrary, does involve skills that are only partly conceptualized: the laboratory and field know-hows and even skills employed in the handling of concepts.

The concept is the unit of thought; accordingly the theory of concepts should be the philosophical equivalent of the atomic theory. Concepts, like material atoms, are not given in experience but must be sought by analysis. Analysis of what? Clearly, of the linguistic expressions of knowledge, since conceptual knowledge comes wrapped in signs: words, symbols, diagrams, etc. In order to get access to the ideas of science we have to pierce through the languages of science. This perforation is performed with the assistance of philosophical analysis, a tool for disclosing the structure and elucidating the meaning of conceptual systems.

This chapter and the next are devoted to a philosophical analysis of scientific concepts — that is, to the logic and the epistemology of concepts. The logic of concepts has two parts: the *syntax* of concepts, which studies their structure, and the *semantics* of concepts, which studies their connotation and denotation, if any. We shall see that the syntax and the semantics of concepts are intertwined, if only because the domain in which a concept legitimately applies is determined by its connotation. Finally, the epistemology of concepts is concerned with studying their function in the process of knowledge, and is often hardly distinguishable from the semantics of concepts. In this study we shall make moderate use of the elements of modern formal logic and semantics, but no acquaintance with these disciplines will be expected from the reader.

2.1. Scientific Languages

Unlike mystics and occultists, scientists objectify their thoughts by means of signs that can be perceived and understood by anyone who cares to. In this way they facilitate their own work and they offer it to public control and use. In other words, the conversion of personal knowledge into scientific knowledge is accompanied by the representation of the former by sets of conventional material marks (signs) belonging to one or more languages. Our access to scientific knowledge is, therefore, through sets of artificial signs designed to convey ideas — rather than feelings, as is the case of the artistic languages. It will therefore be convenient to review some notions of *semiotics*, the science of signs.

Some languages are more or less spontaneous historical creations: these are the *natural languages*, such as English. A natural language serves primarily the purposes of elaborating, recording and communicating common knowledge. No piece of science can dispense with ordinary language but none can do without a language of its own. Every science builds an *artificial language* of its own that includes signs borrowed from ordinary language but is characterized by signs and sign combinations introduced along with the peculiar ideas of that science. Both natural and artificial languages are not only communication tools but also tools for thinking.

In short, a first partition of languages is as follows:

Languages
- *Natural* (e.g., English)
- *Artificial*
  - *Conceptual* (e.g., arithmetical notation)
  - *Nonconceptual* (e.g., musical notation)

(Some philosophers think that a science is just an artificial language, whence book titles such as *The Language of Physics* and *The Language of Sociology* for works devoted to the corresponding metascience. But that is taking cans for canned food. Science works out sign systems and works with them but only in so far as they materialize our ideas concerning non-linguistic objects, such as chemical binding. Furthermore, science is communicable but it does not serve the purpose of communication. Science, in short, has a language but is not a language: it is a body of ideas and procedures expressed in a number of languages. Accordingly, a philosophical examination of science limited to analyzing its languages will miss what is peculiar to science: the quest for objective truth.)

When speaking or writing about a body of signs (a language) we step outside the object of our inquiry: we employ a higher-level language from the height of which, as it were, we look down to the former. The language we talk about is called the *object language*; what we say about it we say in a *metalanguage*. Thus, when we say that a certain proposition (statement) p is true, we express a *metastatement* concerning the statement p. If p is expressed by a sentence belonging to a certain object language, then the corresponding metastatement will be expressed by a sentence of a metalanguage. This level difference may be marked by enclosing the object symbol between quotation marks, namely thus:

The sentence 's' expresses the proposition p [2.1]

The proposition "p" is true [2.2]

The sentence '[2.2]' belongs to a metalanguage of s. [2.3]

Notice that in mentioning linguistic objects, such as sentences, we enclose them in simple quotes, whereas conceptual objects, such as the propositions expressed by sentences, are enclosed in double quotes when mentioned. Alternative conventions are, of course possible: what matters is to adopt a consistent and simple one and to stick to it — which is easier said than done.

In principle there is no limit to the number of language levels, as suggested by [2.3] which belongs to a metametalanguage of s; in turn, the object language to which s belongs might be a metalanguage. The distinction of language levels avoids confusion and paradox. (Recall the semantical paradox of the liar who says 'I am lying'.) The distinction occurs and should be shown everywhere in science, where we find *metatheorems* (theorems about theorems), *metalaws* (laws concerning laws), and *metarules* (rules of rules) — not to speak of mere comments about theorems, laws, and rules, which remarks belong to a higher level language than the one to which its objects belong.

Like every other human creation, language can be studied both in itself (internal study) and as a social object (external study). The latter approach is the one adopted by the psychologist, the anthropologist, the sociologist and the historian interested in language as a cultural phenomenon. Such an approach, valuable as it is, has no particular relevance to our goal: we are not so much concerned with the *uses* of signs in real social life (the object of *pragmatics*, the union of the above mentioned empirical sciences) as with the *structure* of signs (the object of *syntax*) and with their *relations to ideas and things* (the object of *semantics*). The reason for selecting syntax and semantics as the proper tools for investigating the scientific languages is this. We are ultimately interested in the ideas and procedures of science rather than in the historically conditioned ways in which such ideas and procedures are expressed by the various scientific communities: we must accordingly focus on whatever remains invariant under cultural changes such as the shift from one natural language to another, and even under changes in artificial languages such as those of notation.

Take, for instance, the sentence

The greater the production volume the greater the cost of the total product. [2.4]

Alternative equivalents of this sentence are its translations into other natural languages: in this way a set of linguistic items (sentences) is produced, every one of which expresses one and the same idea — in this case a certain proposition belonging to economics. The idea under consideration is most nakedly, exactly and universally expressed in mathematical language, for example thus:

$$z = m \cdot y + n, \qquad [2.5]$$

where 'z' designates in this case the cost of the total product, 'm' the production cost per unit, 'y' the production volume and 'n' the fixed overhead costs. We know, of course, that [2.5] is too simple to be true: [2.5] is only a first (linear) approximation to a more complex relationship between y and z — but this is not our business.

An unlimited number of alternative readings (interpretations) of the same signs 'y', 'z', 'm' and 'n' can be given: in fact, the same formula [2.5] may be used to express physical, biological, psychological, etc. relations: a variety of specific contents may be poured into the form [2.5]. Every interpretation of a mathematical form is determined by a set of *meaning assumptions* and designation rules such as "'y' designates [stands for] the production volume". If no fixed interpretation of the symbols is adopted — i.e. if no designation rules are laid down — a void schema remains that asserts nothing about the world and is therefore of the concern of mathematics. If only '$=$' in [2.5] retains its usual meaning (identity), whereas the remaining symbols are uninterpreted, [2.5] is an *abstract* formula; if the letters are interpreted as numerical variables and '$+$' as arithmetic addition, [2.5] is attached an arithmetical interpretation: it is no longer abstract but it is still a *formal* formula, i.e. one with a purely mathematical meaning.

The internal study of any linguistic expression, such as [2.4] or [2.5], bears both on its form and on its content, the form being determined by the peculiar way in which the signs are combined and the content by what they say — if they say anything at all, which they may not, as is the case when no particular meaning is attached to the component signs. The permissible sign combinations are ruled by formation rules (or grammatical rules in an ample sense of 'grammar'); their meanings by designation rules.

Formula [2.5], though more accurate than the corresponding verbal sentence [2.4], is still an incomplete analysis of the latter: production costs, production volumes, and the like are not things but properties of things of a kind (merchandises). Accordingly we should replace the expression 'production volume' by 'production volume of x', where 'x' designates an arbitrary item of a definite kind, such as a set of typewriters. Formula [2.5] could therefore be completed to read

$$z(x) = m(x) \cdot y(x) + n(x). \qquad [2.6]$$

Here, '*x*' designates the individual or *object variable*, a variable ranging over the set of individual objects (e.g., typewriters) concerned. A number of scientific advances have consisted in realizing that something was a property rather than a thing, or a relation rather than an inherent property. Recall the debunking of heat as a thing, caloric, and its conversion into a property; the discovery that length is a relational rather than an intrinsic or absolute property of bodies, and that the mind is a system of functions rather than a substance. In all these cases the deeper logical analysis was an aspect of a deeper scientific analysis: in general, the logical analysis of a formula requires substantive knowledge and this is why there are no final logical analyses of formulas with a factual content.

As it stands, even with the addition of designation rules, [2.6] does not designate a proposition proper, i.e. a definite statement that can be true or false to some extent. In fact, as long as the values of the object variable x and the numerical variables y, z, m and n are not fixed, formula [2.6] is a schema with blanks: it expresses a *propositional function*, only more complex than "x is blue" (see 1.4). It might well be that [2.6] were meaningless for actions, or that it held for typewriters but not for alternative values of x, for small values of the production volume y but not for large ones, and so on. In any case [2.6] is an indefinite formula, a schema or matrix out of which a number of sentences can be generated, every one of which will designate an exactly or approximately true (or false) proposition. Schemas expressing propositional functions are called *sentential functions* or open sentences.

A sentential function such as $P(x)$ may be such either because the value of the object variable x is left indefinite and/or because the *predicate variable* P has not been fixed — e.g., through a designation rule such as: The predicate '$P(\ )$' designates the concept "prime number". If 'P' designates a predicate constant rather than a predicate variable, i.e. if the value of P is fixed, then x is the sole variable left in the sentential function $P(x)$. If x takes on a particular value, say the constant c designating a certain individual, such as the number 3, then we get the sentence: $P(c)$ — e.g., '3 is a prime number'. In short, a sentential function becomes a sentence, and correspondingly the denoted propositional function becomes a proposition, if all its object variables and all its predicate variables take specific values — i.e., upon *specification.*

There is another way in which a propositional function may become a true or a false idea, i.e. a proposition: namely, by *generalization*. This is achieved by prefixing to it a *quantifier*. Thus, the propositional function "x is blue" can be turned into the true proposition "some x are blue" or into the false proposition "all x are blue". The prefix 'for some x'

(or 'for at least one x') is an *existential quantifier;* 'for every x' (or 'for all x') is a *universal quantifier.* We shall make use of the quantifiers indicated in the following list.

Table 2.1. *List of quantifiers*

Name	Symbol	Read
Indefinite existential quantifier	$(\exists x)$	There is at least one x such that
Definite existential quantifiers	$(\exists x)_n$	There are exactly n x such that
Bounded existential quantifier	$(\exists x)_U$	There are x in U such that
Unbounded universal quantifier	(x)	Every x is such that
Bounded universal quantifier	$(x)_{x \in U}$	Every x in U is such that

For $n=1$ the definite existential quantifier becomes $(\exists x)_1$, read 'There is exactly one x such that'; mathematicians often designate this singularizer by '$\exists!$'. The subscript occurring in the symbol for the bounded universal quantifier, namely '$x \in U$', is read 'x is in U' of 'x belongs to U', where 'U' designates the *universe of discourse,* i.e. the domain of individuals referred to by the whole statement.

A number of propositions can then be formed with a single individual variable x and a single predicate variable $P(\ )$, either by assigning definite values to both or by prefixing quantifiers. The following table exhibits the kinds of propositions that result; they are ordered as to extension or generality.

Table 2.2. *Kinds of elementary propositions*

Propositions	*Individual*	*Singular*		$P(c)$ with P and c constant
		Existential		$(\exists x)_1\ P(x)$ or $\exists!\ P(x)$
	Indefinite			$(\exists x)\ P(x)$
	General	*Particular*		$(\exists x)_n\ P(x)$ with $n>1$
		Universal	*Bounded*	$(x)_{x \in U}\ P(x)$
			Unbounded	$(x)\ P(x)$

Propositions are the most important but not the sole objects of interest found in scientific knowledge or, for that matter, in any body of ideas. We also find *proposals* (e.g., "Let us assume p"), *problems* (e.g., "Is c a P?"), *rules* (e.g., "Do A to get B") and other ideas which are neither true nor false. (Incidentally, the logic of proposals, problems, rules, promises, threats, and other such conceptual objects lacking a truth value is practically nonexistent.) What we do not find in the body of scientific knowledge are *advices* ("It is advisable for x to do y"), *requests* ("Please, x, do y") and *commands* ("Do x"). These kinds of objects we do meet in the course of research, as of any other action, but not in the outcome of research.

In short, the kinds of (well-formed and self-contained) expressions we find in the scientific languages are these:

Table 2.3. *List of well-formed and self-contained expressions*

Sign (linguistic object)	Designatum (conceptual object)	Example
Sentential function	Propositional function	x is prime
Sentence	Proposition	c is prime
Proposal sentence	Proposal	Let c be prime
Question	Problem	Is c prime?
Rule sentence	Rule	Do A to get B

The above are all *well-formed formulas:* they obey more or less explicit rules of composition or formation. Such rules prohibit the engendering of linguistic monsters such as 'x is' (for being incomplete) and 'x is heavier' ("heavier" being a two-place or dyadic relation and not an intrinsic property). In addition to the *rules of formation,* or rules for the constitution of well-formed formulas (*wffs*), syntax studies the *transformation rules* of the various languages, i.e. the prescriptions governing the admissible transformation of formulas into one another. For example, every equivalence — such as "If x and y are numbers, then $x=-y$ if and only if $x+y=0$" — may be handled as a rule permitting the substitution of one member of the equivalence (e.g., "$x=-y$") by the other member (e.g., "$x+y=0$"). (But equivalences are not rules.)

In the scientific languages the formation and transformation rules are seldom explicitly stated: they must be dug out when a piece of discourse is either analyzed or formalized. In the case of formalization (in the sense of laying all the cards on the table) an exhaustive list of both formation and transformation rules is given. (The formalization of theories will be studied in Sec. 8.3.) Such a listing tops the syntactical analysis of the given fragment of discourse. Elementary logic is a universal syntax studying the formation and transformation rules of all conceptual languages; a deeper syntactical analysis requires substantive knowledge.

Every complex well-formed formula contains at least one term standing for a concept. Thus, "x is a prime number" involves the specific concept of prime number and the generic concept of having a property, or of belonging to a set of elements all of which are characterized by a property. An analysis of well-formed expressions will acordingly involve their decomposition into terms and inter-term relations (syntactical analysis) as well as a disclosure of the relations between them and the concepts they designate (semantical analysis). In short, the internal analysis of a language is both syntactical and semantical and in either case it focuses on concepts and their linguistic representations. This chapter and the following are precisely concerned with some traits of such a dual analysis in so far as it can help us to understand science. But before going into more detail we should touch on certain general problems concerning the languages of science.

From the point of view of the sociology of science, a scientific language is just a professional jargon. From a metascientific viewpoint a scientific language is not on the same footing as the London burglar's argot, which can be learned by reading EDGAR WALLACE's novels — and this because the languages of science are built to express extraordinary rather than ordinary knowledge. Every science uses, in addition to ordinary language, expressions and transformations of these which make sense only in the context of some theory. Thus, the expression 'The alpha-particles ejected from the ion chamber were accelerated in the cyclotron up to 100 MeV', though a simple descriptive sentence, makes no sense outside the context of atomic physics: in order to understand the phrase it is necessary to grasp certain theories rather than just taking hold of the *Oxford Dictionary*. A complete internal (syntactical and semantical) analysis of a piece of scientific discourse, even of a simple one like the above phrase, requires an adequate understanding of a background body of specialized knowledge.

In science, in contrast with art, the main determinant in the idea-sign partnership is the idea. It is true that symbols act as it were on thought because they are ready-made artifacts and in this capacity they channelize thought and occasionally even drag ideas to an unwanted extent. Yet ultimately the power of signs lies in their capacity to stand for the ideas we want them to stand for. No scientific revolution has been effected by purely linguistic changes, but an influential idea can bring about important changes in the language of a science, which may eventually be propagated to contiguous sciences. Witness the language changes prompted in biology by DARWIN's evolution theory and, more recently, by the geneticist's adoption of certain key concepts of information theory, such as "information" and "code", as well as by the social scientists hearing about decision theory — by virtue of which every one of us has been turned into a decision-maker. The diffusion of originally specific scientific terms and expressions takes place to the extent to which they are felt to provide new insights — i.e. to the extent to which the linguistic diffusion accompanies a knowledge exportation. In short scientific languages are created, modified and diffused along with scientific theories and procedures; consequently their study cannot be undertaken independently from the latter.

As scientific knowledge deviates from common knowledge (see Sec. 1.1), so its languages deviate from ordinary language. Just as common knowledge is impotent to evaluate scientific knowledge, so ordinary language is much too coarse a tool for investigating the delicate special languages of science, such as the language of quantum mechanics. Rather than being a tool of analysis ordinary language is therefore an object to be analyzed — an analysandum — and even cleansed with the help of finer languages, in the first place elementary logic. Such an analysis is on its way since the turn of the century and it has shown, on the one hand, the psychological subtleties (e.g., the rhetorical overtones) of ordinary language, and on the other hand the limitations and inaccuracies of the natural languages and of the logic (ancient and medieval) tied with them. Thus, e.g., usage has consecrated the phrase 'He can do all sorts of things', although one can only perform individual acts of definite kinds.

The limitations and imperfections of ordinary language have called forth, of course, the invention of the artificial languages of science. Likewise the imperfections of classical (Aristotelian) logic — associated with the European tongues — has prompted the creation of modern logic (symbolic logic and semantics), which most philosophers of science use nowadays as tools for their analyses of science. To further traits of these tools we turn next.

Problems

2.1.1. Characterize and illustrate the concept of artificial language. Be careful to distinguish an artificial language of science, such as that of analytical mechanics, from the *ad hoc* toy languages sometimes invented by philosophers either to better investigate the traits of real languages or to elude difficult problems. Favorite among such toy languages is the one consisting of a finite set of constants (designating individuals) and a final set of unary predicates (designating intrinsic properties).

2.1.2. Choose a fragment of scientific literature and show what expressions in it belong to ordinary language and what to artificial languages.

2.1.3. Take a scientific text and fish out of it a couple of metalinguistic expressions, a couple of proposal sentences and a couple of rule sentences. *Alternate Problem:* Discuss the popular view that every piece of scientific language is a set of declarative sentences.

2.1.4. List the peculiarities of the scientific languages in contrast with the characteristics of ordinary language.

2.1.5. Comment on CONDILLAC'S dictum "A science is just a well-built language", and on H. POINCARÉ's belief that the function of the scientist is just the creation of a clear and concise language to express concrete facts. Take into account, (i) that a number of mutually incompatible theories are expressible in one and the same language, and (ii) that a formula expressible in a language, e.g. "2=1", may not belong to the corresponding theory.

2.1.6. Take an experimental report and determine whether it is couched in a sense-data language or whether it involves terms not designating human experience. Determine next: (i) whether a purely sensationalist language might suffice for scientific purposes, (ii) whether, if such a language existed, it could be public instead of private, and (iii) whether such a language could express universal objective statements, i.e. propositions going beyond what any individual subject could experience.

2.1.7. The terms 'red', 'rough' and 'painful' take part in the description of certain experiential facts: they may accordingly called *phenomenal terms*. On the other hand, 'thing', 'length' and 'living' take part in the description and explanation of objective facts, whether they can be experienced or not: they may be called *physical object terms*. 'I see red' is a phenomenalist sentence, whereas 'There is a red flag over there' is a physical object sentence. Some philosophers, notably the early positivists, have claimed (i) that there is a *phenomenalist language* — not just a phenomenalist vocabulary, which clearly does exist, and (ii) that such a phenomenalist language is or ought to be the *basic language* of science, in the sense that all physical object terms, in particular the theoretical ones (such as 'temperature') are or ought to be constructed out of phenomenalist terms by purely logical means. Discuss these claims. Try to build a scientific term, such as 'distance' or 'mass', out of phenomenalist terms. And determine what advantages such a reconstruction would have for a subjectivist (subject-centered) philosophy, such as empiricism, if the vocabulary but not the syntax of science were thus reducible to the subject's experience.

2.1.8. Look at the following fragment taken from J. Z. YOUNG, *The Life of Mammals* (Oxford: Clarendon Press, 1957), p. 572: "The essence of a hormone, in the exact sense, is that it constitutes a specific chemical signal. We have seen that the concept of signalling implies a set of instructions to act in a certain manner. The signals ensure correct action because they are sent out in a controlled pattern or code, transmitted to a distance and 'decoded' by receivers by the selection of some of the possible actions of the latter. The glands of internal secretion are able to act in this way because the hereditary instructions of the genes ensure that release of their products is so controlled that it occurs under appropriate conditions. The signals are decoded by certain tissues that are sensitive to them and are hence known as *target organs*". Search in the scientific literature for further examples of the renewal of a scientific jargon under the influence of new theories. *Alternate Problem:* It has long been noticed that certain terms are backed up, as it were, by scientific theories, as a consequence of which many a controversy over "the right" name of a thing involves a dispute over rival theories. This was certainly the case with the "dephlogisticated air" discovered by PRIESTLEY but only correctly recognized by LAVOISIER and consequently rechristened by him as 'oxygen'. Discuss and illustrate the theoretical commitment of scientific terminology. See W. WHEWELL, *Novum Organum Renovatum*, 3rd ed. (London: Parker, 1858), pp. 264ff. and 294ff., and N. R. HANSON, *Patterns of Discovery* (Cambridge: University Press, 1958), pp. 54ff.

2.1.9. Discuss the place of syntactical and semantical analysis in science and/or philosophy. See C. W. MORRIS, *Foundations of the Theory of Signs*, Vol. I, No. 2 of the *International Encyclopedia of Unified Science* (Chicago: University of Chicago Press, 1939), especially Sec. vii; R. CARNAP, *The Logical Syntax of Language* (London: Routledge and Kegan Paul, 1937), especially Part v; M. BUNGE, *The Myth of Simplicity* (Englewood Cliffs, N. J.: Prentice-Hall, 1963), Ch. 1.

2.1.10. Sketch a programme for investigating the logic of proposals and show its relevance to the study of scientific hypothesis. *Alternate Problem:* Expand on the relation between equivalence and rule hinted at in the text.

2.2. Term and Concept

Consider the sentence 'DARWIN was a scientist': it is one of the many possible linguistic expressions of the corresponding proposition (see Sec. 2.1). Each word in this sentence is a *term*, i.e. a linguistic unit. But not every term independently designates a concept and not every concept refers by itself to a trait of reality. In fact, the concepts involved in the proposition in question are "DARWIN", "class membership" (designated by the ambiguous 'was a') and "scientist of the past". 'DARWIN', linguistically a proper name and logically an individual constant, designates the concept "DARWIN", which in turn represents the once living individual DARWIN. 'Was a' is a linguistic form of 'to be a', which in this case designates membership in a certain class, namely, the class of scientists of the past. And 'scientist of the past' designates the concept "scientist that lived before our time"; it is a phrase (a complex sign or expression) that designates a single concept.

An exact rendering of the given sentence is: 'DARWIN belongs to the class of scientists of the past'. This sentence can be symbolized '$c \in P$', where '$\in$' designates the individual DARWIN, '$\in$' designates the relation of class membership, and 'P' the class of scientists of the past. An equivalent symbolization is '$P(c)$', read 'c is a P', or 'c satisfies P', or 'c is a value of the argument of the predicate P'.

Certain philosophers would hold that 'scientist of the past' is a fiction without a real reference or counterpart, all classes being just names for arbitrarily grouped individuals. This is certainly the case of some class terms, e.g. 'nice people', but not of all. Certain sets, or rather certain collections homogeneous in some respects (e.g., biological populations) are regarded as real even if, as in the case of the biologists of the past, they are extinct. The physicist who speaks of solid bodies, the biologist who investigates the habits of porpoises, and the sociologist who deals with social strata, do not take a nominalist position but believe in the existence of natural kinds, hence of objective groupings based on objective similarities, common ancestry, or other traits. Moreover, scientists do not restrict the term 'reality' to cover only what exists at the present moment relative to the subject's frame of reference. We have learned to call all real whatever exists somewhere in the four-dimensional spacetime continuum.

Three levels or realms are involved in the above discussion: the linguistic, the conceptual and the physical levels; they are schematically represented in Table 2.4. Notice that, in keeping with the convention adopted in Sec. 2.1, we distinguish linguistic from conceptual entities, when mentioning them, by enclosing linguistic entities, such as the numeral '3', in single quotes, whereas conceptual entities (e.g., the number "3"), are enclosed in double quotes.

Table 2.4. *Linguistic, conceptual, and physical levels*

Linguistic Level (Terms & phrases)	'Darwin was a scientist' (Sentence)	'c' = 'Darwin' (Term)	'$\in$' = 'belongs to' (Term)	'P' = 'scientist of the past' (Predicate)
Conceptual Level (Concepts & propositions)	"Darwin was a scientist" (Proposition)	"c" = "Darwin" (Concept)	"$\in$" = "class membership" (Concept)	"P" = "Scientist of the past" (Concept)
Physical Level (Facts, things, properties, etc.)	The fact that Darwin was a scientist	The man Darwin		The collection of scientists of the past

Two kinds of relation among the three levels concerned are of interest to us: designation and reference. The relation of *designation* holds between some members of the linguistic level and their correlates, if any, on the conceptual level. Thus, we say that '*c*' designates "DARWIN" and call the concept "DARWIN" the *designatum* of the term '*c*'. The relation of designation is an asymmetrical sign-idea relation; in general it is not a one-to-one relation. The relation of *reference* obtains between some members of either the linguistic or the conceptual level and their correlates, if any, in the physical level. For example, we say that DARWIN, the man, is the *referent* of both '*c*' and "*c*"; or, equivalently, that '*c*' and "*c*" represent DARWIN on their respective levels. The relation of reference is an asymmetrical relation between some linguistic and conceptual entities and their physical referents, if any. The union of the two relations, that of designation and that of reference, may be called *denotation*. Table 2.5 summarizes the above.

Table 2.5. *Designation and reference*

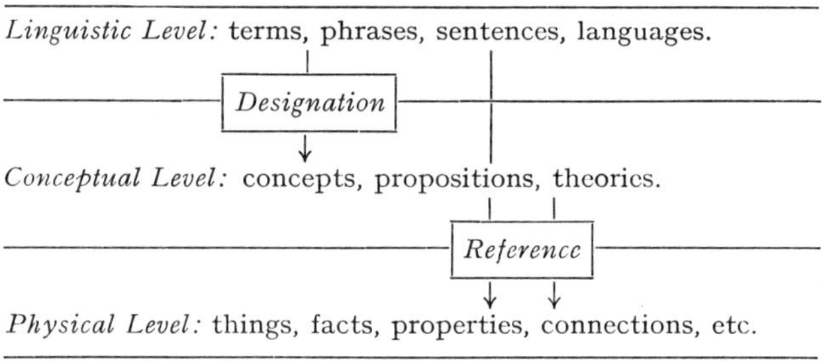

Not all terms designate nor all terms and concepts refer. Thus 'it', '!' and '(' designate no concepts: they are parts of meaningful expressions but have no meaning by themselves: they are called *syncategorematic terms*. That not all terms and concepts have an external reference is far from obvious: according to certain varieties of empiricism and materialism all ideas have nonconceptual (empirical or material) referents. Our discussion of analytic formulas in Sec. 1.4 showed, on the contrary, that science handles signs and concepts with no external reference — in short, formal symbols. Thus, we shall say that the concept of (pure) number and all the concepts subsumed under it — the subclasses of reals, integers, primes, and so on — have no reference. It is only denominate numbers, such as "10 meters" and "10 atoms", that have referents — namely, properties of material systems; but by the same token they are not studied in mathematics, a formal science. We shall also say that the numeral (number name) '7' designates the concept "7", which in turn refers to nothing outside itself. Since in mathematics only such formal concepts occur, one dispenses with double quotes when mentioning them: thus, one says that 7 is prime, instead of writing ' "7" is prime', because there is no danger of confusing the concept 7 with anything real. It is only with concrete concepts that confusion may arise: think of confusing atoms with either the concept of atom or the word 'atom'.

In general: the concepts of formal science, or *formal concepts*, are referenceless. All those which are not formal may be called *nonformal*. Many of them will be concrete or factual terms: namely, those which are intended to have a real referent, even if it should turn out that they have no real counterparts. Thus, "centaur" may be regarded as a concrete concept because it was originally intended to denote a concrete entity. Nonformal concepts are often called *descriptive terms*, which suggests that whatever is not formal is descriptive. Since there are also interpretive, normative, and other kinds of concept, we shall not adopt that name. In addition, the substitution of 'descriptive *term*' for 'non-formal *concept*' involves a commitment to a nominalist philosophy in which there are no concepts and propositions but only terms and sentences. We do not accept this philosophy because we need the distinction between ideas and their linguistic expressions (e.g., between scientific theories and their corresponding languages), and because we must face the fact that every deep scientific change involves not only the introduction of new symbols but also the *reinterpretation* of old symbols, a process in which certain existing signs are made to designate new concepts — as exemplified by 'mass' upon the Newtonian revolution and by 'species' upon the Darwinian revolution.

How are we to find out which terms shall count as designating concepts and which as purely rhetorical? A straight rule for finding the terms which designate concepts is this: "First get rid of the inessential ingredients of the expression concerned. Then lay bare its logical structure. Finally pick out the logically complete symbols". Example: 'Indeed, none of NEWTON's predecessors built rational mechanics'. First step: eliminate 'indeed'. Second step: 'For every x, if x is a predecessor of c, then it is not so that x built d'. Third step: '$(x)[P(x, c) \rightarrow - B(x, d)]$'. The formal (logical) concepts involved in this formula are: "universality", designated by '(x)', "conditional", designated by '$\rightarrow$', and "negation", designated by '$-$'. Being logical concepts they are all universal. On the other hand, the nonformal (or extralogical, or subject matter, or specific) concepts are: the individual concepts "NEWTON" and "rational mechanics", and the two-place functions (binary relations) "predecessor" and "build". (The concept of time might have been dug out as well had we cared to perform a faithful translation of the verb tense.) The term 'of' and the apostrophe ' ' ' in the given sentence are seen to be logically as supernumerary as 'indeed'; moreover, they do not occur at all in certain languages.

The formal/nonformal dichotomy of concepts cuts across a number of alternative groupings. We shall find the one shown in Table 2.6 particularly useful.

Table 2.6. *Kinds of concepts*

1	*Individual Concepts* ("*c*", "*x*")		
2	*Class Concepts* ("copper", "living")		
3	*Relation Concepts*	*Noncomparative*	*Relations Proper* ("∈", "between")
			Operators ("&", "+")
		Comparative ("≦", "better adapted than")	
4	*Quantitative Concepts* ("population", "length")		

Individual concepts apply to individuals, whether definite (specific) or indefinite (generic); "NEWTON" is a definite or determinate individual concept, whereas "*x*" is an indefinite or indeterminate individual concept (i.e. one denoting an arbitrary individual). What is regarded as an individual will depend on the level of analysis: an individual on a given level may be a system or just an aggregate of lower level individuals. Thus, e.g., for some purposes rational mechanics will be regarded as an individual among other individual theories, and at other times it will be treated as a set of formulas, hence as a class concept.

Class concepts apply to sets of individuals — as in the case of "copper", which refers to the set of all possible copper samples — or to sets of sets — as in the case of "living", which embraces all biological species. The structure of class concepts is the one-place or monadic predicate $P(\ )$.

Relation concepts apply to relations among objects (individuals or sets) of some kind; thus, "twice" applies to numerical variables and to their particular values. Operators are a special kind of relations, namely those which yield another member of a set: thus, whereas '$A \subset B$' symbolizes the relation of inclusion of A in B, '$A \cap B$' stands for the operation yielding the common part of A and B. Comparative relations are those which, like $\leqq$, permit us to order individuals or sets.

Every relation concept has the structure of a many-place or multiple-variable predicate: $P(\ ,\ )$ if binary, $P(\ ,\ ,\ )$ if ternary, and so on. The number of blanks (places or variables) of a predicate is called its *degree*. Binary or two-place relations (logically: second degree predicates) are the most important. They can be characterized by the possession or the lack of the following formal properties. (i) *Connectedness* in a given set: if x and y are any members of the set, then either $P(x, y)$ holds or $P(y, x)$ does; for example, $\leqq$ is connected in the set of real numbers, whereas $\subseteq$ is not connected in the set of all sets. (ii) *Uniqueness:* every element x is paired to a single other element y; if the converse holds as well, the relation is one-to-one or *biunique.* "Twice" is unique both ways; "square of" is unique, not so its converse, namely "square root of", which maps every positive number x into a couple of numbers: $\sqrt{x}$ and $-\sqrt{x}$. (iii) *Symmetry:* if $P(x, y)$ then $P(y, x)$; brotherhood is a symmetric relation, in contrast to parenthood. (iv) *Reflexivity:* $P(x, x)$, as in the case of $\leq$. And (v) *Transitivity:* if $P(x, y)$ and $P(y, z)$, then $P(x, z)$.

A relation which pairs the element of two sets in a unique way is called a *function.* The two sets may be identical. A function need not establish a 1:1 pairing: it may be a many-one correspondence; but it cannot be a one-many correspondence. The set of x's (the set over which x ranges) is called the *domain* of the relation (or function in particular);

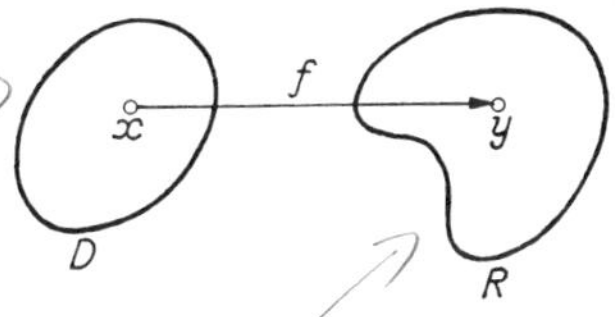

Fig. 2.1. A function f maps its domain D into its codomain or range R, whence it may be writen 'f: $D \rightarrow R$', or '$y = f(x)$'

the set over which y ranges is called the codomain or *range* of the function. A function mirrors or *maps its domain into its range* in such a way that each element of its domain has a single image in its range (uniqueness condition); but the function may be such that several individuals of its domain are mirrored by one element in its range. Functions are often written in the form of equations, namely thus: $y = f(x)$, read: 'the value of f at x equals y'. An intuitive representation of a function is shown in Fig. 2.1. First caution: the members of the domain and the codomain of a function need not be mathematical objects. Second caution: in particular, the codomain may coincide with the domain, as when a set of numbers is mapped into itself by a *numerical function* such as "$y = a x$".

Functions are the structure of quantitative concepts or *magnitudes,* also called quantities. For example, temperature is a function T that maps the set of bodies (generic representative: σ) into the set of real numbers (generic representative: t). The independent variable σ is the *object variable* whereas t is the *numerical variable* or *value of the function* T at the point σ. Being a function, the two variables occurring in temperature can be separated in the usual way: "$T(\sigma) = t$", read 'the temperature of σ equals t'. The numerical variable t occurring in the temperature function is equal to the number of temperature units on a given scale, e.g. KELVIN's. When the scale and the unit system are fixed by the context we need not write them explicitly; otherwise it should be indicated by a special symbol, say s. In short,

$$T(\sigma, s) = t. \qquad [2.7]$$

Similarly with length, duration, force, concentration, and all other magnitudes: their structure is that of a function of one or more variables — at least one of them an object variable (see. Sec. 2.3). Caution: distinguish the function f from its numerical values y, i.e. the mapping from the image it yields.

The logical analysis of quantitative concepts (magnitudes) as functions or mappings (1:1 or many-one) enables us to *distinguish a property from its numerical values.* And this enables us, in turn, to handle the concept without fixing numbers and conversely, if necessary, to deal with numbers without caring for the properties they measure. Thus, e.g., we may speak of weight without specifying the value of its numerical variable; or, conversely, we may deal directly with the latter, as when we make weight computations. The same analysis enables us, furthermore, to realize that the numerical formulas of factual science, in particular the quantitative law statements, do not contain the *whole* concepts involved but only their *numerical* variables: they contain the numerical values of the functions, not the functions. Thus, when we symbolize the formula "Weight = Mass × Acceleration of gravity" we do not write the three magnitudes in full but only their numerical variables w, m and g respectively. That is, we do not put '$W = M \cdot G$' but '$w = m \cdot g$'. In other words, what are multiplied are the *numbers* m and g, not the entire concepts "M" and "G": it would be senseless to subject non-mathematical entities to mathematical operations and this is why we extract, of the whole concepts, their mathematical components.

The preceding discussion on the logical structure of concepts is summarized in Table 2.7.

Table 2.7. *Logical structure of concepts*

Kind of Concept	*Logical Structure*
Individual Ex.: Mars, John Doe	Individual constants ($a, b, \ldots$) and variables ($x, y, \ldots$)
Class Ex.: Solid, Man	Monadic predicates P (constant or variable) or the corresponding classes: $C_1 = \{x \mid P(x)\}$
Relation Ex.: Part of, Between	Polyadic predicates (higher-degree predicates): $F(x, y)$, $F(y, y, z), \ldots$ or the corresponding classes: $C_2 = \{\langle x, y\rangle \mid F(x, y)\}$, $C_3 = \{\langle x, y, z\rangle \mid F(x, y, z)\}, \ldots$
Quantitative Ex.: Numerosity, Mass	Numerical functors: $F(x) = y$, $F(x, y) = z, \ldots$

*These various kinds of concept differ in *logical strength:* out of quantitative concepts relation concepts may be obtained, the latter generate class concepts and these individual concepts, whereas the converse process is impossible without further ado. Thus, from a table of distances to a given point we may construct an ordered sequence in respect of closeness from the point concerned, i.e. we may derive a sequence of relation statements such as "1 is the closest to 0", "2 is the next closest to 0", and so on. From this relational concept of closeness we may derive a class (or absolute) concept of closeness by conventionally choosing a point on this side of which all points will be classed as close. Finally, by intersecting classes we may obtain individuals.

*Individual and class concepts are then, in a sense, poorer or weaker than either relation or quantitative concepts. But this does not entail that we can dispense with individuals and classes, if only because, in order to construct relations, individual or class concepts are needed. Thus, e.g., relations hold always among individuals, or among classes, or among individuals and classes. And quantitative concepts involve individuals variables or constants, which are in turn predicated. Consider "The age of the planets of our solar system is about 5,000 million years", which can be recast as "For every x, if x is a planet of our solar system, then the age of x in years is about 5,000 million". This statement involves a generic individual variable (x), two class concepts ("planet" and "our solar system"), a quantitative concept ("the age of x in years equals y") and an individual constant occurring in the latter ("5,000 million"). In short ,the stronger concepts do not always enable us to dispense with the weaker ones: they may be built out of the latter.

*Nor it is possible to proceed the other way around, i.e. to eliminate the stronger in favor of the weaker concepts. What can be done is to establish definite relations among them. In particular, it is possible to define relations in set-theoretical terms. Thus, a binary relation R may be regarded as the set of *ordered pairs* $\langle x, y\rangle$ of its relata. This set is included in the set of all ordered pairs x, y such that x is in X and y is in Y, i.e., $X \times Y$, called the *cartesian product* of X and Y. That is $R \subset X \times Y$. For example, the relation "square of" in the set of positive integers may be regarded as the infinite set of couples: $^2 = \{\langle 1,1\rangle, \langle 2,4\rangle, \langle 3,9\rangle, \ldots\}$. With this representation of relations as classes of couples, triples, etc. — not as classes of individuals — the theory of relations is reduced to the theory of sets — as far as form goes.

*The representation of relations as classes does not entail their elimination but a shift of focus which brings to the fore the formal traits of relations — which is exactly what is desired in logic and mathematics, and what is to be avoided in factual science. Consider, for example, a toy universe constituted by just three objects, named a, b, and c, among which only two relations hold: "heavier than" and "hotter than", and such that the heavier happen to be the hotter. Although the two relations are different, their set-theoretical representation is the same because both effect the same pairings of individuals. Thus, if a is both heavier and hotter than b, which is in turn heavier and hotter than c, then the two relations will be mirrored by the single set $H=\{\langle a, b\rangle, \langle b, c\rangle\}$ — as was to be expected, since the form of the two relations is the same. The "reduction" of relations and other concepts to class concepts *(extensionalization)* is good strategy in formal science because it accompanies the display of form and increasing abstraction. To preach extensionalism in factual science would be improper for exactly the same reason: because in factual science content is as important as form. But we are already stepping on the territory covered by the next section.

Problems

2.2.1. Identify the concepts in the propositions expressed by the following sentences. (i) 'The morning star is a planet'. (ii) 'All planets revolve around at least one star'. (iii) 'There are many planetary systems'. *Alternate Problem:* Spot the incorrect sentences in the following set: '2 abbreviates 1+1', '2 abbreviates '1+1''. "2' abbreviates 1+1', "2' abbreviates '1+1".

2.2.2. Mention some class concepts, relation concepts and quantitative concepts belonging to the same semantical family, as in the cases of (i) "fluid", "more fluid than" and "fluidity", and (ii) "heavy", "heavier than" and "weight".

long longer than longitude
late later than time

2.2.3. Symbolize "The position of x is $y°$ West and $z°$ North" and distinguish the object variable and the numerical variables.

2.2.4. Construct a predicate of degree 3, i.e. a three-place functor, subsuming the following propositional functions: "The zygote x develops in the environment y", "y is the environment of the organism z", and "The zygote x develops into the organism z". See J. H. Woodger, *Biology and Language* (Cambridge: University Press, 1952), pp. 126—127.

2.2.5. Dig up the logical structure of the force concept in Newtonian mechanics. Take into account that forces are exerted between pairs of bodies and depend on the reference frame and, in general, on time. Also, remember that a force is, mathematically, a vector in a three-dimensional space and that such an object is in turn an ordered triple, i.e. $F=\langle F_1, F_2, F_3\rangle$, where the F_i's are the components of the force.

2.2.6. (i) Define the class concept of fiancée in terms of the relation concept of engagement. (ii) Form the class concept of rotation out of the quantitative concept of rotation ("rotation of x around y through $z°$"). *Alternate Problem:* Disclose the logical structure of the concepts of absolute and conditional probability and build out of them the class concept "probable".

2.2.7. According to traditional usage, concepts are the designata of nouns, adjectives, and terms that can be converted into such alone. Is this grammatical rule adequate for recognizing concepts?

2.2.8. Analyze the sentence 'John Doe is somewhere' and decide whether 'John Doe' and 'somewhere' designate concepts. For a discussion of the ambiguous name 'John Doe', see P. SUPPES, *Introduction to Logic* (Princeton, N. J.: Van Nostrand, 1957), p. 81.

2.2.9. Discuss FREGE's assertion that the occurrence of the singular definite article is always a sure symptom of reference to an object, whereas the indefinite article accompanies always a concept word. See G. FREGE, *The Foundations of Arithmetic* (1884; N. York: Harper, 1960), Sec. 51. *Alternate Problem:* Examine thingism ("There are only individual things") as a possible rationale of extensionalism (anti-intensionalism).

2.2.10. The logic of concepts is sometimes called *logic of terms*. What is the rationale of this designation? And is it justified if the distinction between linguistic and logical categories is kept? Furthermore, the very term 'concept' has fallen in disuse. Is this a sign of a decline of conceptual thinking, as it has been suggested? *Alternate Problem:* Let S be a sign and R an operator such that $RS=$ "S"; i.e., the effect of R on S is the concept referred to by S. Show that, if S designates a concept with no real counterpart, then $R=I$, where I is the identity operator. What happens if S has a real referent?

2.3. Extension and Intension

Every concept has both an intension or connotation and at least one extension or denotation or domain of applicability. The intension of "life" is the set of properties characterizing living beings, i.e. metabolism, self-regulation, adaptation, goal-striving, etc.; and the extension of "life" is the class of living beings present, past and future. The intension of "soul" is made up of immateriality, separability from the body, body-directing, etc.; the extension of this concept is the empty set: no actual object satisfies it. No concept proper lacks an intension, but a genuine concept may have an empty extension. The intension and the extension of concepts are determined, in science, by theoretical and empirical research.

The *intension* $I(C)$ of a concept C is, then, the set of properties and relations P_i subsumed under the concept or which the concept, so to speak, synthesizes. In short,

$$I(C)=\{P_1, P_2, \ldots, P_n, \ldots\}, \qquad [2.8]$$

where the P_i are assumed to be possessed by the objects falling under the extension of C. If $I(C)$ happens to be a unit set, the concept is identical with its intension. The P_i may be formal or nonformal. Thus, in "even number" evenness is a property assigned to the concept of even number: there is no even number beyond the concept "even number". On the other hand, the property of being a unit of living matter is assigned to the referent of "cell", not to the cell concept.

The intension of concepts behaves inversely to their extension: the more properties are assembled the less individuals will share them. In other words, the intension of general concepts is included in or is at most identical with the intension of the corresponding specific concepts. Symbolically,

$$\text{If} \quad (x)(Sx \rightarrow Gx), \quad \text{then} \quad I(G) \subseteq I(S), \qquad [2.9]$$

where 'S' designates a species, 'G' a genus, and '$\subseteq$' the relation of class inclusion. The intensional difference between the genus and the species (the scholastic *differentia specifica*) is, then, the difference between the above sets, i.e. the complement if $I(G)$ in $I(S)$:

$$\textit{Differentia}\,(S, G) = I(S) - I(G). \qquad [2.10]$$

For instance, if $G=$"triangle", and $S=$"right-angled triangle", then one of the members of *Differentia* (S, G) is "one angle is a right angle".

A sufficient condition for the unequivocal determination of the intension of a concept is that a complete description or analysis of either the concept or its referent is at hand. In fact, suppose a given C can be exhaustively described or analyzed as having the properties $P_1, P_2, \ldots, P_n$, so that we may write the equivalence.

> For every x, x is a C if and only if x is a P_1, and x is a P_2, ... and x is a P_n,

or, more briefly,

$$(x)[Cx \leftrightarrow P_1 x \,\&\, P_2 x \,\&\, \ldots \,\&\, P_n x]. \qquad [2.11]$$

Then, clearly, the intension of C is the set of all P_i. The existence of complete descriptions is the sole guarantee for the exhaustive determination of the intension of concepts, but it is not necessary for a concept to have a fairly definite intension.

Although a complete description of actual existents, and even of certain constructs (notably infinite sets) is beyond human power, an accurate determination of the set of *characteristic* or peculiar (not necessarily essential) properties is often possible, at least to ensure an unambiguous application of a concept. Suppose that, out of all the properties that make up the intension of C, only a small number, m, are peculiar to C. That is, suppose $P_1, P_2, \ldots, P_m$ are both necessary and sufficient for the unequivocal distinction between C and any other concept. Let us call *earmarks* of C these peculiar properties essential to our handling C, though perhaps inessential for the referent of C. We shall say that the set of earmarks of C makes up the *core intension* of C:

$$I_{\text{core}}(C) = \{P_1, P_2, \ldots, P_m\}. \qquad [2.12]$$

Clearly, the core intension is included in the total intension: $I_{\text{core}}(C) \subseteq I(C)$, since m is smaller than $n\,(\leqq \infty)$, the indeterminate number of properties actually possessed by the referent of C. Thus, e.g., most zoologists agree that the concept of mammal has only three or four earmarks, all of them of an osteological character and related to the lower jaw-middle-ear complex. This leads them to regard these earmarks as defining the class concept "Mammalia", although none of them is obviously related to, let alone identical with, the possession of mammary glands.

The core intension of a concept, made up of its earmarks, although insufficient for a complete characterization, supplies what we shall call a *working definition* of the concept, namely thus:

$$(x)\,[C\,x =_{wdf} P_1 x \;\&\; P x_2 \;\&\; \ldots \;\&\; P_m x]. \qquad [2.13]$$

Recall that the m defining properties may be just a few of those making up the total (perhaps unknown) intension, and that they may be far from essential: a working definition of a concept does not provide the essence of the concept's reference but is just a tool for classing. Yet, of course, the less superficial or derivative the properties chosen as earmarks the deeper knowledge will the working definition embody and the more natural will be the grouping of individuals made with its help.

Having the core intension of a concept, whether embodied in a working definition or specified in a less clear way, is both necessary and sufficient for determining the domain of application or extension of the concept. The *extension* of a concept is the set of all objects, real or unreal, to which the concept can apply. The extension of a concept may be an infinite set, as in the case of "natural number", a finite but unlimited set, as in the case of "organism", or a finite and limited set, as with "digit". It may even consist of a single member (e.g., the extension of "3" is 3) or of none (e.g., "clairvoyant"). We designate the empty class by '∅' and the universal class by 'U'.

The extension $E(C)$ of a class concept may accordingly be defined as

$$E(C) =_{df} \{x \mid C(x)\}, \qquad [2.14]$$

which can be read: 'the extension of C equals, by definition, the set of objects satisfying the condition $C(x)$, or having the property C'. In other words, the extension or denotation of C is the set of objects with the properties that characterize C — or at least with the earmarks of C. In the case of concepts with real referents it may pay to analyze their extension further into the set $A(C)$ of actuals falling under the concept and the set $P(C)$ of possibles satisfying the function $C(x)$: $E(C) = A(C) \cup P(C)$. For example, the total extension of "man" is made up of the present human population (including our enemies) and the set of all past and future men. The actual extension of a class concept is often called a collection, aggregate or population, whereas the total extension is usually named a class, but the terminology in this field is unsettled.

The foregoing considerations will allow us to introduce the notion of concept generality. A concept C' will be said to be *more general* than a concept C — or C will be said to be *subsumed under* C' — if and only if the intension of C' is included in the intension of C or the extension of C is included in the extension of C'. *Symbolically:

$$C' > C \leftrightarrow [I(C') < I(C)] \vee [E(C) \leq E(C')]. \qquad [2.15]$$

In addition,

$$[I(C') < I(C)] \rightarrow [E(C) \leq E(C')], \qquad [2.16]$$

and

$$[I(C') = I(C)] \rightarrow [E(C) = E(C')]. \qquad [2.17]$$

The equality sign in [2.16] holds for concept pairs such as "rational animal" and "social rational animal", which are believed to have the same extension although the former is more general than the latter. The converse of [2.17] is not true: for example, "ghost" and "wizard" have the same extension — the empty set — but manifestly different intensions. This suggests that the meaning of a concept is not determined by its extension alone.

*The foregoing can be generalized to relation concepts and quantitative concepts, i.e. to concepts the structure of which is a many-place predicate. Whenever a concept involves more than a single variable we may have to specify its various *partial extensions*, which will be the ranges of its various variables. Thus, for a binary relation concept $C(x, y)$, we have

$$E_1[C(x,y)] = \{x \mid (\exists y)\, C(x,y)\} \text{ and } E_2[C(x,y)] = \{y \mid (\exists x)\, C(x,y)\}. \qquad [2.18]$$

$E_1(C)$ is usually called the *domain* of C and $E_2(C)$ the *converse domain* of C or its counterdomain. The union of the two is called the *field* of the relation: $F(C) = E_1(C) \cup E_2(C)$. And the *total extension* of the binary predicate $C(x, y)$ may be defined as the set of ordered couples of its arguments, i.e.

$$E(C) =_{df} \{\langle x, y\rangle \mid C(x, y)\}. \quad [2.19]$$

This set of ordered couples is the *Cartesian* (or *cross*) *product* of the partial extensions of $C(x, y)$:

$$E[C(x, y)] = E_1[C(x, y)] \times E_2[C(x, y)]. \quad [2.20]$$

This is the set of all possible pairings of the values of the x-variable to the y-variable. The generalization of these notions to higher degree predicates is straightforward; but, of course, there is no point in introducing special names for $E_3[C(x, y, z)]$ and the other partial extensions.

*In the case of quantitative concepts or magnitudes it is useful to distinguish the *object extensions* (the various ranges of the object variables) from the *numerical extensions* (the various ranges of the numerical variables). Consider the concept of melting point in a given temperature scale, the structure of which is "$M(x, s) = y$". The object extension of "melting point" is the set of all solids; its numerical extension reckoned in the Kelvin scale-cum-unit system is, roughly, the temperature interval [1° K, 5,000° K]. To say, then, that 'melting point' denotes a number or even a subset of the set of real numbers; is a half-truth; the other half is that it denotes a set of bodies and a set of conventions for assigning numbers.

*We are now in a position to take a closer look at the *logic of magnitudes*. We saw in Sec. 2.2 that the temperature of a body σ, reckoned in a scale-cum-units system s, can be analyzed thus:

$$T(\sigma, s) = t, \quad [2.7]$$

where t is a real number. In other words, T is a function from the body-scale complex to reals. Adopting an extensionalist point of view we can as well say that we first pair the object variable σ to a scale variable s and then pair this ordered couple to a number t, thus forming the pair $\langle\langle\sigma, s\rangle, t\rangle$. We next abstract from any particular couple and build the set of all pairs. This set is the extension of the general concept of temperature. Briefly,

$$E(T) = \{\langle\langle\sigma, s\rangle, t\rangle \mid \sigma\in\Sigma \ \& \ s\in S \ \& \ t\in R\}. \quad [2.21]$$

*For example, imagine a toy universe consisting of just two bodies called a and b, each of which may have only one of three temperature values on the Kelvin scale: 0, 10 and 100 absolute degrees. We have then the following couples:

$$\langle\langle a, °K\rangle, 0\rangle, \langle\langle a, °K\rangle, 10\rangle, \langle\langle a, °K\rangle, 100\rangle,$$
$$\langle\langle b, °K\rangle, 0\rangle, \langle\langle b, °K\rangle, 10\rangle, \langle\langle b, °K\rangle, 100\rangle.$$

Every ordered pair represents a possible state of our toy universe (relative to the given scale) and the set of all couples is, according to [2.20], the Cartesian product $\Sigma \times S \times R_0$ of the set Σ of bodies, the set S of temperature scales, and a subset R_0 of real numbers. That is, in our toy universe $E(T)=\{a, b\} \times \{s_1, s_2, \ldots, s_n, \ldots\} \times \{0, 10, 100\}$. In general, the extension of the function T from $\Sigma \times S$ to R_0 is

$$E(T)=\Sigma \times S \times R_0.$$

In building the Cartesian product we abstract from particular bodies and particular degrees reckoned in special scales and units, and focus on their pairings, i.e. on the function or mapping itself, which is not a set but a regular correspondence.

*In general, any magnitude M involving an object variable and a numerical variable — such as length or population density — can be analyzed as a function M from a physical set $P \times S$ to a numerical set $R_0 \subseteq R$. A suggestive notation for this correspondence is '$M: P \times S \to R_0$'. The extension of this concept equals the product of the domain $P \times S$ by the range R_0 of the magnitude:

$$E(M)=P \times S \times R_0, \tag{2.22}$$

where 'P' designates the object-domain of M, i.e. the set of physical objects referred to by M, 'S' stands for the set of all existing and conceivable scale-cum-unit systems, and 'R_0' names a subset of real numbers. Often P itself will be a set of n-tuples. Thus the duration or lapse between two point events e and e' may be regarded as a relation between the ordered pair $\langle e, e' \rangle$, a time scale s, and the number t of time units in that scale: $D(e, e', s)=t$. Hence the extension of the duration concept is, on this level of analysis, the Cartesian product of the set $P=E \times E$ of pairs of events $\langle e, e' \rangle$, the set S of all conceivable time scales, and the set R^+ of nonnegative reals: $E(D)=E \times E \times S \times R^+$.

*The fineness of such an *extensional analysis* (not definition) of quantitative concepts will depend on the context. Thus in classical mechanics every length and mass value depends only on the thing concerned and on the chosen scale, whereas in relativistic mechanics it depends also on the velocity of the body relative to a given frame of reference. That is, instead of the phrase 'the length (mass) of σ' we now find expressions like 'the length (mass) of σ in motion, relative to the reference frame K, with velocity v'. In short, whereas the mass functor in classical mechanics is: $M(\sigma, s)=m$, in relativity we have: $M(\sigma, v, k, s)=m$. The new function is a fifth degree relation involving two numerical and three nonnumerical variables. Calling Σ the set of bodies, V the set of velocity values — i.e. the $[0, c]$ interval — K the set of reference frames, S the set of scale-cum-unit systems, and R^+ the set of nonnegative reals, the extension of "relativistic mass" turns out to be: $E(M)=\Sigma\times V\times K\times S\times R^{+}$. In this case the domain is $\Sigma\times V\times K\times S$ and the range R^{+}.

*The above extensional analysis illumines the structure of magnitudes but it does not constitute a *definition* and it says nothing about their peculiar nonlogical properties. The peculiarities of each magnitude are determined by the law statements in which they occur. Hence it is not true that a magnitude or, in general, a relation can be given either intensionally (by enumerating its properties) or extensionally (by showing its partial extensions), much less in a purely extensional way. Both procedures are complementary to one another: a purely intensional treatment would deprive us of the assistance of logic and mathematics and would leave us ignorant of the empirical import, if any, of the concept in question, whereas a purely extensional treatment would leave us in the dark as to the difference between, say, mass and length, which in physics have the same extension.

*In general, the determination of the *meaning* of a symbol is made *jointly* by the intension and by the extension of the concept it designates. Therefore meaning questions such as "What do you mean by 'x'?" are ambiguous and each deserves a couple of answers, one regarding the intension and other the extension of the concept involved. For example, the question 'What *are* organisms?', which may be rephrased as "What is *meant* by 'organism'?", must be answered by listing the earmarks of organisms (self-regulation, metabolism, reproduction, etc.) and by listing typical species of organism.

*Let us sharpen the concept of meaning by summarizing the above ideas in a definition. We shall stipulate that the *meaning* of a sign standing for a given concept is the pair intension-extension of that concept. That is,

$$\text{If 's' designates } C\text{, then } \mathit{Meaning}\,(s)=\langle I(C), E(C)\rangle. \qquad [2.23]$$

In science one very often handles the core intension $I_c(C)$ of a concept, determined by its earmarks according to formula [2.13]; and usually a subset $E_c(C)$ of the domain of application of a concept is known, the remainder of the extension $E(C)$ being left for future research. This suggests introducing the concept of *core meaning* as a subconcept of [2.23]:

$$\text{If 's' designates } C\text{, then } \mathit{Core\ Meaning}\,(s)=\langle I_c(C), E_c(C)\rangle. \qquad [2.24]$$

Since two pairs are identical if and only if the corresponding members are identical, two signs, s and s', will have the same meaning — will be *synonymous* — if and only if they designate concepts with the same intension and the same extension:

$$\text{'}s\text{'}\ \mathit{Des}\,C\ \&\ \text{'}s'\text{'}\ \mathit{Des}\,C' \rightarrow \{s\,\mathit{Syn}\,s' =_{df} [I(C)=I(C')\ \&\ E(C)=E(C')]\}. \qquad [2.25]$$

*If two terms designate concepts with the same extension but different intensions or conversely, they have different meanings. For instance, "equiangular triangle" has the same extension as "equilateral triangle" in Euclidean geometry, but this does not entail the synonymity of the corresponding terms. In fact, the equal extension of the concepts concerned follows from a theorem to the effect that all equiangular triangles are equilateral and conversely. This is an equivalence of the form: "$(x)\ (Ax \leftrightarrow Lx)$". The predicates '$A$' and '$L$' do not stand for the same property and consequently cannot be exchanged without a shift of meaning. What does remain invariant upon an exchange of A and L in any proposition containing either is the *truth value* of the proposition. If logically equivalent propositions were also semantically equivalent there would be no point in establishing them.

*The above synthesizes two doctrines of meaning: the *extensionalist* doctrine, according to which the meaning of a concept is identical with its extension, and the *intensionalist* doctrine, according to which (a) extensions are reducible to intensions and (b) the meaning of a concept is identical with its intension. Extension and intension being two complementary aspects, extensionalism and intensionalism are onesided viewpoints. Extensionalism is legitimate in formal science, where it has aided in the enterprise of basing all of mathematics on set theory, which focuses on extensions. Sets of individuals, of subsets, of couples, and so on can be set up arbitrarily in formal science, i.e. without regard to the nature of the objects involved. Only the possession of common properties, aside from the property of belonging to the given set, is what gives a set whatever unity it may have and entitles it to be called a *class*. The mathematician and the logician may pay no attention to the unity that distinguishes a natural class or any other objective grouping from an arbitrary set. But the semanticist and the factual scientist cannot afford to ignore intensions: not only do they rarely handle extensive or denotative characterizations of classes, but they are content when they can characterize a class by some of its properties, i.e. intensionally, and when they find a sample of it. For them extensions and intensions come in pairs*.

Let us now apply what we have learned about the logic of concepts to their methodology, by examining the most elementary conceptual operation in science: classification.

Problems

2.3.1. Determine the intension and the extension of any scientific concept. *Alternate Problem:* Restate [2.10] with the help of the set-theoretical theorem: $A - B = A \cap \overline{B}$; and compute the Boolean sum of $I(S)$ and $I(G)$ recalling that $A + B =_{df} (A - B) \cup (B - A)$ and $I(G) - I(S) = \emptyset$.

2.3.2. Discuss the following passage from G. FREGE in *The Foundations of Arithmetic* (1884; New York: Harper 1960), Sec. 53: "By properties which are asserted of a concept I naturally do not mean the characteristics which make up the concept. These latter are properties of the things that fall under the concept, not of the concept. Thus 'rectangular' is not a property of the concept 'rectangular triangle'; but the proposition that there exists no rectangular equilateral rectilinear triangle does state a property of the concept 'rectangular equilateral rectilinear triangle'; it assigns to it the number nought". *Alternate Problem:* Distinguish the extensional from the intensional interpretation of 'species" and find out whether the biologist dispenses with either.

2.3.3. How would you determine what an Oxford student "is", i.e. the connotation and the denotation of the concept "Oxford student"? *Alternate Problem:* Does the biologist determine ("define", in the jargon) a species by enumerating its members, without caring for intensions — as the nominalist philosopher would have it? Hint: Recall how species membership is determined.

2.3.4. Determine the connotation and the denotations of "distance" in a one-dimensional space such as the straight line. *Alternate Problem:* Compare the concepts of circle and ellipse. Which is more general? Which is richer?

2.3.5. Determine the connotation and the extension(s) of "mass". *Alternate Problem:* In the formula

$$d(P, P') = [(x - x')^2 + (y - y')^2]^{\frac{1}{2}}$$

for the distance between any two points P and P' on the Euclidean plane, what constitutes the connotation and what are the extensions?

2.3.6. (i) Examine the concepts of core intension and working definition in the light of the Aristotelian doctrine of concepts. (ii) Link the concept of working definition with that of definite class. *Alternate Problem:* In Aristotle's system "eternal", "necessary" and "perfect" are coextensive, which they are not in alternative systems. Work out the contextual character of extension and intension: relativize the formulas of this section to a system.

2.3.7. Perform an extensional analysis of a concept with the structure of an n-place predicate $P(x_1, x_2, \ldots, x_n)$. *Alternate Problem:* In contemporary science the extension of "vital force" is empty. Does this condemn the concept of vital force as meaningless or rather as jobless?

2.3.8. EINSTEIN's formula "$E = mc^2$" is often regarded as implying that "mass" and "energy" have the same extension and nearly the same intensions just because the two concepts are invariably related by the aforementioned law statement. Is this interpretation correct? Hint: Begin by disclosing the object variable(s) and find a class of systems with energy but with no mass. *Alternate Problem:* Is it possible to determine the connotation of scientific concepts aside from the systems in which they occur? And is it possible to determine the extension of scientific concepts having an external referent without employing scientific data?

2.3.9. Give an extensional account of intension by interpreting the predicates occurring in [2.8] as classes. Find the limits of this interpretation of intension as a class of classes. *Alternate Problem:* Find the scope of the principle of extensionality, according to which "If two predicates have the same extension then they are identical"; or, in terms of classes, "Two classes are identical just in case they have the same members". This axiom is essential to set theory.

2.3.10. Examine the current status of the problem of synonymity. Do not fail to distinguish the pragmatic from the semantic sense of 'meaning'. See L. LINSKY, Ed., *Semantics and the Philosophy of Language* (Urbana, Ill.: University of Illinois Press 1952), especially C. I. LEWIS, "The Modes of Meaning", N. GOODMAN, "On Likeness of Meaning", and B. MATES, "Synonymity"; R. CARNAP, *Meaning and Necessity*, 2nd ed. (Chicago: University of Chicago Press, 1956); W. V. Quine, *Word and Object* (New York and London: Technology Press of the M.I.T. and John Wiley, 1960). See whether the formulas [2.23] to [2.25] in the text solve any of the difficulties pointed out to by the aforementioned authors. *Alternate Problem:* Examine the possibility of defining the meaning of the term 'meaning' as the set of all ordered intension-extension pairs of concepts.

2.4. Partition, Ordering and Systematics

Thus far we have dealt with concepts from a logical point of view; we shall now focus on their methodological functions. Regarded from a methodological viewpoint concepts are tools for distinguishing items and grouping them: they enable us to perform conceptual and empirical analyses and syntheses. In particular, individual concepts help us to discriminate among individuals, and class concepts enable us to establish classifications. Certain relation concepts make comparison and ordering possible; and quantitative concepts are the core of measurement.

Determinate individual concepts have a high resolving or discriminating power but no power of synthesis or systematization. Generic individual concepts (individual variables) have no resolving power since they denote nondescript individuals of a kind, but on the other hand they are the root of generalization or synthesis: the substitution of variables for specific constants marks the start of generalization. Class concepts allow for both synthesis and discrimination among sets. Relation concepts enable us to make even finer distinction and also to establish links among concepts. Finally, quantitative concepts lead to the subtlest of all and the most exact of discriminations; and, when combined with one another in law statements, quantitative concepts yield the tightest systematization of ideas. No wonder that conceptual work in factual science starts with variables and classes and ends up with magnitudes.

Classification is the simplest way of simultaneously *discriminating* among the elements of a set and grouping them into subsets — i.e., of analyzing and synthesizing. What it classes is called the *universe* or domain of discourse — or U for short. The universe of discourse may be any set whatever: a class of individuals or a class of sets; it can be discrete (countable) or continuous (nondenumerable), composed of things (e.g., a population of organisms), of facts (e.g., births), of properties (e.g., wavelengths), or of ideas (e.g., numbers). The type of classification will depend on the goal (cognitive or practical) and on the relations among the members and subsets of the given set (U). But certain principles of classification are of a logical nature, i.e. independent of aim and subject matter; consequently they must be dealt with first.

One of the principles of correct classification is that the characters or properties chosen for performing the grouping should be stuck to throughout the work: a shift from, e.g. skeletal to physiological characters in the classification of vertebrates will produce not only different classes but also different systems of classes, i.e. alternative classifications. Another rule of correct classification is that the subsets of the same hierarchical rank (e.g., biological species) should be exhaustive and pairwise disjoint, i.e. should jointly cover the whole field and should have no members in common: this rule requires modification in the case of evolutionary taxonomy (see Sec. 3.1). A third rule is not a logical but a methodological one, namely, the various classifications of one and the same universe of discourse should be coincident (as regards the extensions) if they are to be natural rather than artificial groupings. The three rules are often violated either because of logical negligence or because of real difficulties such as those posed by borderline cases.

The most elementary form of classification is *partition*, i.e. pigeonholing: it consists in distributing the elements of the universe of discourse among a number of pairwise disjoint classes or niches standing in no systematic relationship among one another. The simplest partition is, of course, the *dichotomy*: it is so simple that it almost always occurs as a first stage in analysis. A single concept will induce a dichotomy. Thus, e.g., the concept "edible" will by itself induce the dichotomic partition of all living beings (universe of discourse U) into edible (A) and unedible (not-A, or $\bar{A}$, the complement of A in U):

$$A=\{x \mid A(x)\},\quad \bar{A}=\{x \mid -A(x)\},\quad U=A\cup\bar{A},\quad A\cap\bar{A}=\emptyset$$

where '$\emptyset$' designates the empty set. Such was probably the first classification of organisms: namely, by their most primitive use. Then slightly more sophisticated — both finer and less anthropocentric — classifications evolved, such as the partition by habits, namely into aquatic, amphibians, terrestrials, and aerials; this was a tetrachotomy:

Aquatic	Amphibians	Terrestrial	Aerial

$$A_1\cup A_2\cup A_3\cup A_4=U,\quad A_i\cap A_j=\emptyset \text{ for } i\neq j.$$

A beginning of order can be seen in this primitive ecological classing; still, the groups stand side by side with no logical relationships among the corresponding concepts, and the order is extrinsic (geographical) rather than intrinsic (biological).

Next to partition, in order of complexity, comes the *ordering* of the given universe by means of some antisymmetric and transitive relation holding among any two members of the set. Take, for instance, the ordering of chemical elements; this ordering is performed in respect of the atomic number characterizing each element. The problem of ordering the 103 known elements can be thus summarized:

Universe of discourse = set of chemical elements.
Respect or property chosen = atomic number.
Ordering relation: the arithmetical functor "$\leqq$".

The solution to this problem is the strict simple or *linear ordering*

$$H-He-Li-Be-B-C-\ldots-Cf-Es-Fm-Md-No-Lw.$$

(This simple-looking sequence of symbols is the result of an incredibly complex experimental and theoretical work, since atomic numbers are anything but superficial phenomenal properties.)

The above ordering of chemical elements is correct but not illuminating enough: it says nothing about the natural groupings of elements. The desideratum of grouping in pure science — as opposed to applied science — is not the simplest and most discriminating (or analytic) ordering or partition but an arrangement combining a high *resolving power* with the highest possible connectedness or *systemicity*, tending to represent the objective connectedness (if any) of the members of the set. Such a desideratum is not achieved, in the case of chemical elements (or, for that matter, of biological species), by either pigeonholing or chain-like (linear) ordering. A solution is afforded, in the case of chemistry, by a two-dimensional array in which linear order with respect to atomic number is combined with partition with respect to valence and other chemical properties related to valence. This, of course, is MENDELEEFF's *periodic system*, a systematic classification of atomic species the members of which are classes assembled in wider classes (groups or periods) of similar chemical properties, so that each is a *natural* kind. Thus, e.g., all the "inert" gases are placed in a single column (group), and groups of elements with similar properties recur with a period of eight.

The deepest and consequently the most fruitful groupings in science are neither partitions nor orderings but what we shall call *systematic classifications*, in which one or more *relations* link classes together, these relation concepts denoting objective relations. A systematic classification is not a mere boxing or place-and-name assignment (cataloguing) like NOAH's mythical grouping of animals: it is the outcome of an operation whereby concepts — and their referents if any — are related to one another in such a way that some sort of connectedness or system is made apparent. And the best systematic classification is, in turn, the one which effects the most *natural* — least arbitrary, least subjective — of groupings.

The best known systematic classification type is that of biology, in which every major group (taxonomic category or rank) is divided into subordinated groups, every one of which is further subdivided, and so on until a set is reached which is not made up of subsets but of concrete populations or of individuals. The Linnean hierarchy in zoology is essentially this:

Kingdom (e.g., Animalia)	T_7
Phylum (e.g., Chordata)	T_6
Class (e.g., Mammalia)	T_5
Order (e.g., Primates)	T_4
Family (e.g., Hominidae)	T_3
Genus (e.g., *Homo*)	T_2
Species (e.g., *H. Sapiens*)	T_1

More recent taxonomic systems involve even finer subdivisions.

The relation holding among the various ranks or levels is the class inclusion relation $\leq$, which connotes proper inclusion ($<$) or identity($=$). The relation among most ranks is one of proper inclusion. But in some cases a genus or some other taxon is regarded as made up of a single species (or the corresponding subordinate rank); in these cases the identity relation holds: Species = Genus. (But this is a controversial point.) In short, the major taxonomic groups are held together, in the Linnean classification, by the relation $\leq$ in the following way:

$$T_1 \leq T_2 \leq T_3 \leq T_4 \leq T_5 \leq T_6 \leq T_7 .$$

The ordering relation $\leq$ has the following properties reviewed in Sec. 2.2: (i) It is *asymmetrical*, i.e. if $A \leq B$, then $-(B \leq A)$. (ii) It is *reflexive:* $A \leq A$ — whereas, on the other hand, the proper inclusion relation $<$ is irreflexive, i.e. $-(A < A)$. (iii) It is *transitive:* if $A \leq B$ and $B \leq C$, then $A \leq C$. Relations such as $\leq$, which are asymmetrical, reflexive, and transitive, generate a *partial ordering* of the given universe of discourse. A *complete* or simple ordering resulting in a chain is generated by asymmetrical irreflexive and transitive relations, such as the one of proper inclusion.

*Now, most ranks are composite: thus, a given phylum may include subphyla and usually includes various classes; a given class is in turn divided into several orders, and so on. Such a distinction amounts to subdividing a taxonomic level T_i into "horizontal" or brethren pigeonholes or taxa $T_{i1}, T_{i2}, \ldots, T_{in}$. Thus if a category of the order genus, T_2, consists of three species, we will designate these by T_{11}, T_{12}, and T_{13}: these will be the sets subordinated to the set T_1. In general, for the i-th rank, $T_{i+1} = \bigcup_{j=1}^{n} T_{ij}$ (logical sum of sets over j from 1 to n) subject to the condition that no brethren groups shall have common members, i.e. $\bigcap_{j=1}^{n} T_{ij} = \emptyset$ (logical product of sets over j from 1 to n). The first formula says that the partition is exhaustive, the second that the parts are pairwise disjoint or mutually exclusive. (Borderline cases will be dealt with in Sec. 3.1.)

Consider, for example, a universe of discourse U of the order rank, subdivided into two families, F_1 and F_2, one with three, the other with two genera (G_i), each of which includes in turn from two to three species (S_i). A Linnean classification of the given original set consists in the following set of eight statements:

$$F_1, F_2 < U,$$

$$G_1, G_2, G_3 < F_1; \quad G_4, G_5 < F_2,$$

$$S_1, S_2 < G_1; \quad S_3, S_4 < G_2; \quad S_5, S_6, S_7 < G_3; \quad S_8, S_9 < G_4; \quad S_{10}, S_{11}, S_{12} < G_5 .$$

Useful ways of graphically displaying a *hierarchy* like this are the *Euler-Venn* diagrams familiar from set theory and the *trees* familiar from the theory of relations. (See Figs. 2.2 and 2.3.) The two are just alternative visual representations of one and the same mesh of logical relations; that is, they are logically equivalent. Moreover, they visualize purely *logical* relations among sets; in particular, an Euler-Venn diagram does not represent a set of part-whole relations among things, and a tree does not represent in this case a genealogical tree or lineage.

Note the chief formal and semantical traits of a *hierarchy*, i.e. a taxonomic system. (i) Every hierarchy has a single *beginner*, which in biology is the highest rank taxon taken into consideration (an order in the above example). (ii) The relation among the members of different

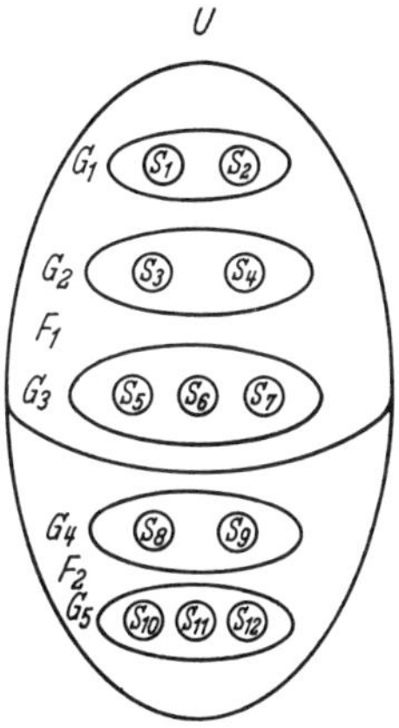

Fig. 2.2. Euler-Venn diagram of example in text

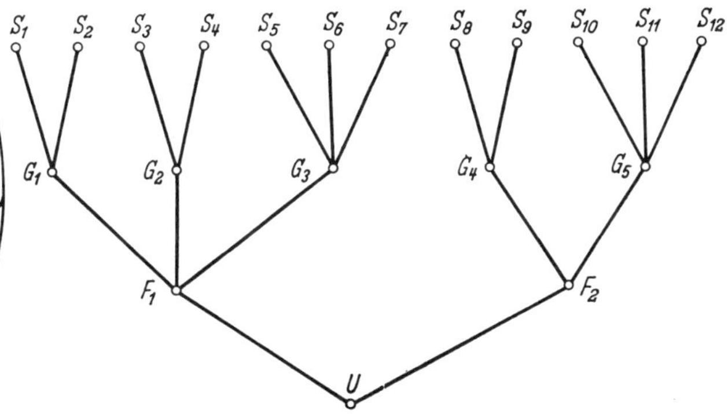

Fig. 2.3. Hasse diagram of example in text. Read from bottom to top

ranks is usually *one-many*. (iii) The groups, however divergent from one another, which are removed from the beginner in a given number of steps, belong all to the same *rank* or level in the hierarchy. (iv) All the taxonomic categories are *classes*. (v) Except the lowest rank sets (species), taxonomic categories are *classes of classes:* only species are sets of things or, rather, sets or populations of individual organisms. (vi) Only species concepts, such as, "*Homo sapiens*", have immediate real referents, namely populations. All higher taxonomic concepts refer to lower rank sets, i.e. to further concepts, not to extraconceptual entities (organisms): they are, in short, *concepts of concepts.* Not pure concepts, though, but concepts which, in a mediate way, point to concrete objects, namely individual organisms.

*The old philosophical controversy over the reality or unreality of classes ought to be renewed in the light of the hierarchy concept. It is possible to take a conceptualist stand as regards classes of the species level and, at the same time, a seminominalist position with regard to all higher rank classes — a seminominalist rather than a fully nominalist position because a natural classification will reflect real, e.g. biological, connections among such higher order ranks, as we shall see in the next section.) *

A systematic classification, then, consists in the organization of a bunch of concepts, e.g. biological taxa, into a hierarchy. A hierarchy goes far beyond a catalogue since it is based on the subordination or subsumption of concepts: it establishes a system — not a system of propositions (a theory) but a *sytem of concepts*. The symbols on each level of a hierarchy do stand for propositions; thus, e.g., '$F_1, F_2 < U$' abreviates the conjunction "F_1 is included in U and F_2 is included in U". But the symbols on the next level of the hierarchy — e.g., '$G_1, G_2, G_3 < F$' — stand for propositions that are not logically derivable (deducible) from the former: for all we know there could be four or five genera instead of just three in any given case. In a hierarchy *concepts*, not propositions (e.g., law statements) are made to hang together in a logically definite way, namely via the relation of class inclusion.

A taxonomic system, then, is not a theory (a system of propositions) but a *system of concepts* and an associated set of hypotheses. The latter are used in setting up the classification but they do not occur in the latter; for example, "The more numerous the common external morphological characters of two species the closer they are." A theory may or may not *back up* a given taxonomic system: i.e., a systematic classification may be constructed with the help of and justified by a biological theory — as is the case with evolutionary or phylogenetic systematics — or not. This has to do with the epistemological and methodological status of a taxonomic system, not with its logical status, and is a point that will be pursued further in the next Section.

Like partitions, hierarchies involve groupings. Unlike partitions, they superimpose a partial ordering upon the units (sets) resulting from the partitions, in such a way that the units are made to hang together in a precise way. No wonder that biologists — in contrast with collectors — tend to prefer to speak of *systematics* rather than of *taxonomy*, which is noncommittal. Now, different taxonomic systems bring about more or less tight systematizations: if species (chemical, biological, social, etc.) are related not only by means of logical (set-theoretical) relations but, in addition, through concrete relations such as "heavier than", "descendent from" or "dependent from", then tighter and deeper systems will result, as will be seen in the sequel.

Problems

2.4.1. Report on the classification of Primates given by G. G. Simpson, "The Principles of Classification and Classification of Mammals", *Bulletin of the American Museum of Natural History*, **85**, 1 (1945).

Alternate Problem: Propose a classification of numbers, or of any other set.

2.4.2. State symbolically the properties of the relations that engender (i) partial orderings and (ii) strict partial orderings. See P. SUPPES, *Introduction to Logic* (Princeton, N.J.: Van Nostrand, 1957), Sec. 10.5. *Alternate Problem:* Whatever is edible is so for some species or set of such, and even so perhaps in part rather than wholly. Does this invalidate the dichotomy of the universe of things into edible or unedible?

2.4.3. What is the ground for the widespread use of dichotomies in ordinary as well as in scientific knowledge? Is it rooted in the nature of things (e.g., in that reality consists of sets of pairwise contradictory things and properties) or is it due to the possibility of generating the dual of any concept by means of logical negation (formation of the complement $\bar{A}$ of a set A)? *Alternate Problem:* The whole of elementary logic can be written with the sole use of negation and disjunction. Does this have anything to do with dichotomies?

2.4.4. Since the establishment of a classification requires previously available concepts (which may be refined in the course of working out a classification), is it correct to regard classification as a procedure of concept *formation*, or is classification rather an *occasion* for the introduction and elucidation (refinement) of concepts? *Alternate Problem:* Is concept formation a problem for logicians or rather for psychologists?

2.4.5. Is it possible to order any set? *Alternate Problem:* Can the nominalist admit that a dog is *in re* a mammal? If not how is he to interpret the phrase 'Dogs are mammals'?

2.4.6. Examine the concepts of relative chronology and absolute chronology as employed, e.g., in archeology and prehistory. *Alternate Problem:* Suppose N biological properties are known and it is desired to class organisms according to their properties. Suppose further that these N properties are (i) qualitative (presence/absence), (ii) on the same footing (equally weighty), and (iii) seemingly mutually independent. How many different species are logically possible? Hint: nevermind the number of organisms.

2.4.7. Comment on the opinion that the periodic system of the elements is just "an easily surveyed assemblage of facts". *Alternate Problem:* Define in set-theoretical terms the relation of belonging to the same species (e.g., "contemporary"). Hint: Use the (equivalence) relation

C/G induced by a partition $G=\{S_1, S_2, \ldots\}$ in a heterogeneous collection $C=\bigcup C_i$, with $\bigcap S_i=\emptyset$.

2.4.8. Does classification require a knowledge of the essential qualities of the members of the universe of discourse? *Alternate Problem:* Are species discovered or invented? See E. MAYR, "Concepts of Classification and Nomenclature in Higher Organisms and Microorganisms", *Annals of the New York Academy of Sciences*, **56**, 391 (1952). Hint: Distinguish species from "species".

2.4.9. Given that no two individuals of the same biological species are identical in all respects, how are the diagnostic characters (earmarks) of the species to be determined? (i) By a priori imagining an ideal type (archetype) of which actual specimens are but more or less imperfect copies, as the Platonists and the *Naturphilosophen* claimed? (ii) By random choice? (iii) By finding the modes (most frequent characteristics or values) of the greatest population at hand? (iv) By finding the more constant specific differences?

2.4.10. Why is classification so much more important in biology than in physics? Because biologists have unduly stressed the systematic approach at the expense of the search for laws and the building of theories? Caution: classification is of little importance in even less advanced disciplines, like history.

2.5. From Pretheoretical to Theoretical Systematics

Classing and ordering may be superficial or deep, anthropocentric or objective, scientifically barren or fruitful. A practically valuable classing or ordering may be superficial and anthropocentric, as is the case of the partition of herbs into medicinal and nonmedicinal; or it may be shallow and ethnocentric, like the partition of men into white and colored. Focusing on a single character will not result in a deep and objective grouping unless the chosen character happens to be essential, i.e. a source property or relation determining a whole group of other characters, as is partly the case of atomic number in the systematics of chemical species, or with viviparity in the case of zoological species.

In other words, a cluster of *interrelated* properties will yield a more objective or natural arrangement than one or two earmarks, and essential or source properties will lead to deeper insights than phenomenal (observable) earmarks. For, after all, the desideratum of any classification of sets of concrete objects is, in pure science, an objective or *natural* systematics, i.e. the disclosure and display of natural kinds and their objective relationships. It is only in technology — e.g., in agronomy — that artificial classifications are valuable. Now, properties are represented by concepts, and the interrelation of a set of properties is represented by logical relations among the corresponding concepts; these relations among concepts are stated by propositions that are in turn organized in bodies of theories. Therefore, every natural systematics is backed up by some theory, which can in turn correct and enrich the former.

Biological systematics based on exosomatic (external, directly observable) characters was the first *objective* classification of organisms: it was objective or natural in the sense that it did not hinge on our wants, habits or whims, but was based on real similarities in a number of respects. Yet this primitive systematization — called *alpha systematics* — is not entirely reliable and sheds little light on the biological interrelations of species. Thus, the same torpedo form is shared by fishes and cetaceans; bats, birds and some insects are winged — and so on. Criteria of external similarity are often objective but not *deep* enough: they frequently, but not always, indicate the belonging to one and the same natural kind. The older systematics has therefore been called artificial in contrast with the new (phylogenetic) systematics. The name, however, is unfair: an artificial or nominalistic classing or ordering is the one employing artificially chosen earmarks or even artificial earmarks, such as the alphabetic ordering of a set of persons, or the classing of chemical compounds by color (unless the physical rather than the physiological concept of color is used). Apha systematics can be as natural as phylogenetic systematics since both hinge on objective traits; only, the former is far less deep than the latter, and it is less deep because it is *pretheoretical*.

A hypothesis underlying alpha sytematics is that the more numerous the common external morphological characters of two species the closer or more alike these are. But this hypothesis is true to a first approximation only. In fact, different groups have often practically the same morphological characters and, conversely, within the same species large morphological variations can be found; hence, morphology alone is not a reliable criterion of biological relatedness. This is why systematics is nowadays the joint work of morphologists, ecologists, geneticists, physiologists, biochemists, and so on rather than the concern of collectors. In short, proper classing requires ideas as much as it requires observations.

What is strange is not that a hypothesis presupposed by the older systematics has turned out to be defective, but that it should hold in such a high percentage of cases. Evolution theory explains this as follows: form similarity is a consequence of either common ancestry or convergent processes of adaptation to a common environment. The new biology, then, not only improves on the older one but also explains the latter's successes and failures — which it can do because it embraces theories, in particular the theory of evolution through natural selection. (Scientific explanation, as we shall see in Ch. 9, is performed in the bosom of theories.) According to evolution theory external morphological characters are not too reliable diagnostic characters or *indicators* suggesting but not warranting deeper seated properties: they are manifestations, hence symptoms from a methodological viewpoint, of largely imperceptible but inferable phylogenetic, physiological, biochemical, ecological and other affinities. (For indicators or objectifying indices, see Sec. 12.3.) Degree of kinship is, in evolution theory, the chief determinant of the systematic closeness of biological groups: similarity is a consequence, hence an indicator, of kindred.

From a methodological point of view the situation is this. The first operation of both the classical and the evolutionist systematist is the *identification* of individuals — i.e. establishing relations of the form "$x \in S$". But whereas the older systematist focused on external earmarks, the phylogenetic systematist takes into account, in addition, further characters among which he establishes a hierarchical order. (The identification of individuals amounts to *grouping* them into species or subspecies on the basis of the symmetric relation of similarity in certain respects. Only, the similarity as determined by the contemporary sytematist is deeper than the mere external similitude to which the classical systematist paid attention.) The second operation is to *order* the various groups thus formed in a genealogical tree, i.e. to order them by means of the nonlogical relation which DARWIN called propinquity of descent. This is an antisymmetric, nonreflexive and transitive relation. It is not a merely logical relation like the one of inclusion, which was enough to build Linnaean hierarchies (see Sec. 2.4), but a biological relation of time priority and parentage. It establishes a *strict partial ordering* of biological groups along the time axis (see Fig. 2.4).

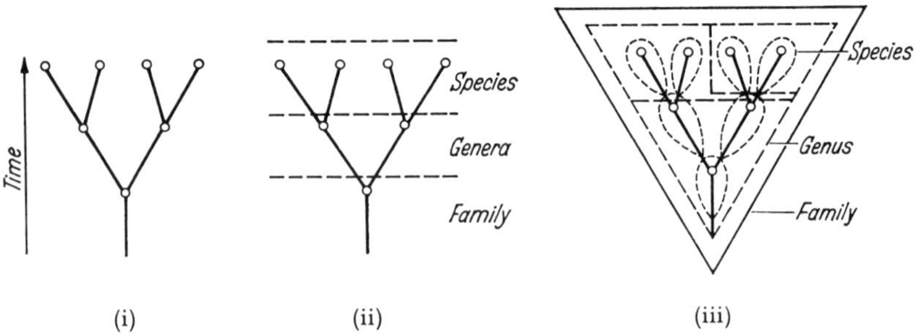

Fig. 2.4. Relations between phylogeny and taxonomy. In (i) a hypothetical phylogenetic tree (dendrogram) is represented. In (ii) an incorrect partition is performed. In (iii) a correct classification is given. (After G. G. SIMPSON)

This new way of ordering biological groups, based on the relation of descent, has several effects. First, genealogical trees representing sections of the hypothesized history of organisms are introduced alongside taxonomic hierarchies; that is, a natural process and not just a logical subordination — which might be arbitrary — is made the kernel of systematics. Second, each branching in the tree of descent corresponds to a biological diversification that has reached the point of speciation (new species formation); biological theory explains the emergence of diversity with the help of gene mutation and the consequent selection of genetic combinations by interaction with the environment (adaptation). Third, a closer connectedness or systematization of the data and hypotheses is obtained, since in addition to the symmetrical relation of similarity (holding among the members of a group and among related groups) and the asymmetrical, reflexive and transitive relation of inclusion $\subseteq$ holding among the various taxa, the antisymmetrical, irreflexive and transitive relation of descent now holds among some species. The new systematics is, therefore, much more tight or systematic than the older. But there is no reason to think that it will be the last: biochemistry is promising to provide an even tighter systematics, by supplementing the above mentioned relations with relations of affinity among the proteins characterizing the various species. ('Supplementing' rather than replacing, since organisms cross a number of levels of organization.)

Chemical systematics has had a similar evolution from a pretheoretical to a theoretical stage. The various chemical elements can be arranged in a number of ways: according to color, taste, specific weight, boiling point, atomic weight (determined by the number of constituents in the atomic nucleus), atomic number (equal to the number of electrons), and so on. Sensible properties are discarded because they lead to anthropocentric and superficial — though sometimes useful — groupings. And among nonsensible properties macroscopic characters, such as specific weight or boiling point, are less appropriate than microscopic characters because they derive from the latter. Source properties, such as the atomic number and the number of electrons in the outer atomic shells, are chosen in the deepest of all natural chemical systematics — the periodic system.

It is important to realize that the periodic system presupposes the *atomic theory* much as phylogenetic systematics presupposes evolution theory. A mere glance at a contemporary periodic table will substantiate our claim: each pigeonhole in it is basically characterized by a set of numbers representing the numerical values of quantitative properties introduced by the atomic theory and otherwise unthinkable or meaningless. Thus, in the pigeonhole occupied by Carbon we find the following numbers: 6, the atomic number or number of electrons in the Carbon atom; 12.011, the atomic weight of a random sample of Carbon; 2, the number of known isotopes; and 2—4, the number of electrons in the first and second shells respectively. These are all deep-seated, nonphenomenal properties.

Just as in the case of biological systematics, the progress in chemical systematics was away from phenomenalism and from nominalistic (artificial) groupings. And, just as in the case of biology the outline of a genealogical tree suggests looking for missing links, so in the case of the chemical system the prediction of the properties of previously unknown elements and isotopes was possible. Moreover, a prediction of the possible transmutations of elements can be made by just looking at the numbers characterizing neighboring elements. The periodic system is, then, far more than a summary of facts occurring in the chemical laboratory: it is part of the atomic theory and it summarizes our theoretical knowledge concerning the composition and structure of nuclei and atoms.

Partitioning alone, ordering alone, and measuring alone may each be highly analytical, i.e. discriminating — depending on the fineness of the operations. But a high resolving power is not the sole desideratum of arrangement: systematization, i.e. synthesis, is equally desirable. And systematization is brought about by a *combination of partition, ordering and, if possible, measurement in the light of theory.* Partition, order and measurement are only parts of the process of analysis and synthesis, which begins with discrimination or distinction, proceeds with description, and culminates with theory.

Theory, an end by itself, is also *the* means for advancing to deeper and deeper systematics. The reason for this is that theory alone can introduce in a nonarbitrary way nonobservable (diaphenomenal) properties, through concepts that far from referring to superficially observable earmarks denote essential or source properties that are usually hidden to the senses and can only be hypothesized. The deeper systematics being those involving some theory, the deepest of all will be those based on the deepest theory — that is, on the theory whose key concepts are the farthest from immediately observable phenomena. (For the concept of theory depth see Sec. 8.5.) And any such systematics, far from being external to theory, will summarize, illustrate and help theory, which will in turn explain it. Should the theory be found defective — a discovery that is bound to occur sooner or later — the systematics accompanying it may have to be mended or abandoned. Systematics, in short, is an aspect of scientific systematization: it will be protoscientific if backed by no theory, scientific proper if some testable theory underlies it. Systematics, then, is not a special science, let alone a superscience: to the extent to which it is theoretical rather than pretheoretical, systematics embodies not only net results of empirical investigation but also net results of theoretical research. By the same token it helps to guide both.

Problems

2.5.1. Discuss the original classing of vitamins in terms of objectiveness and depth. Recall that vitamins were originally grouped according to their occurrence in food rather than according to their chemical structure. Thus, the group *B* vitamins (thiamine, riboflavine, nicotinic acid, etc.) have nothing in common except their joint occurrence in certain foods. *Alternate Problem:* Which come first: the operations of identifying an individual as a member of a group and the ordering of groups, or the formation of the very concepts of such groups? Hint: Adopt an evolutionary viewpoint.

2.5.2. Decide which of the following partitions of biology, the one by kind of organism or the one by problems and methods, gives the deepest insight into modern biological research:

Partition according to objects	*Partition according to problems*
1. Botany Subdivisions: bryology, mycology, dendrology, forestry, etc., etc. 2. Zoology Subdivisions: entomology, ichthyology, herpetology, ornithology, etc., etc.	1. Morphology (form & structure) Anatomy Histology Cytology 2. Physiology (function) Physiology proper Embryology Genetics 3. Ecology (habitat) 4. Ethology (behavior) 5. Phylogeny (evolution)

Alternate Problem: How do we decide in what respects should organisms be compared?

2.5.3. From the datum that two given species are similar, what can be inferred (i) in the case of morphological ("artificial") systematics and (ii) in the case of phylogenetical ("natural") systematics? *Alternate Problem:* Why is virus classification in a quandary? See N. W. PIRIE, "Principles of Classification Illustrated by the Problem of Virus Classification", *Perspectives in Biology and Medicine*, V, 446 (1962).

2.5.4. Consider the genealogical tree of Fig. 2.5. '*A*', '*B*', '*C*' and '*D*' represent contemporary species; '*E*', '*F*' and '*G*' the species from which the former originated. Is *B* more closely related to *A* than to either *C* or *D*, or conversely? Notice (i) that *B* is less removed from *A* than from both C and D; (ii) the emergence of B occurred shortly after that of A; (iii) A and B did not originate in a common stock. (Adapted from W. HENNIG.)

2.5.5. Which of the diagrams in Fig. 2.6 is the correct interpretation of the phylogenetic tree of the previous problem? (Same source.) *Alternate Problem:* Does the dendogram provide a complete representation of phylogeny? See D. L. HULL, "Consistency and Monophyly", *Systematic Zoology*, **13**, 1 (1964).

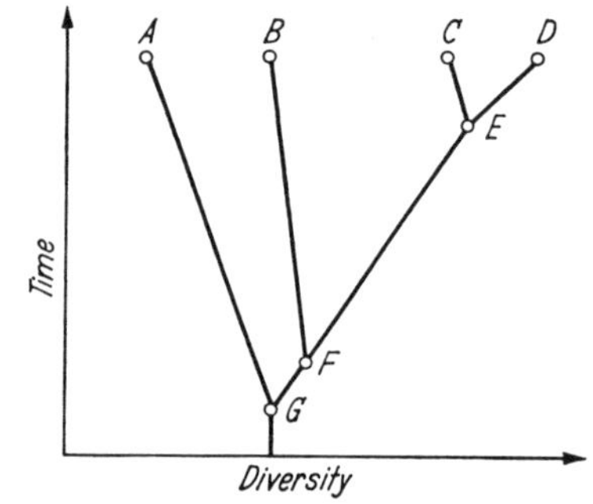

Fig. 2.5. Genealogical tree of Problem 2.5.4

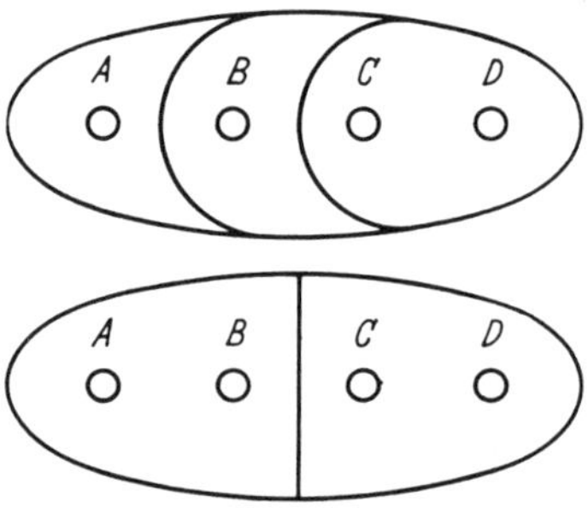

Fig. 2.6. Euler-Venn diagrams corresponding to Fig. 2.5

2.5.6. Is biological systematics (i) an end in itself and, indeed, the central goal of biological research, or (ii) just a way of storing and displaying (partitioning or ordering) the factual raw material to be processed by physiologists, geneticists, biochemists, and so on, or (iii) as much a part of biology as the periodic table is a part of physics and chemistry, and serving both as a record of findings and as a tool of research? Accordingly, is systematics a special discipline or a part of a scientific discipline? *Alternate Problem:* Analyze the following statements: "One can see a species", "The species one sees depend upon his opportunities for observation", "A species is a simple thing", and "A species is a formal unit". (Found in a recent scientific paper.)

2.5.7. Discuss the proposal (J. HUXLEY, 1958) that each taxon should be construed as a group characterized both by a common ancestor (a *clade*) and as a unit with a number of characters in common as a result of evolution (a *grade*). *Alternate Problem:* Discuss in detail the approaches of the pure scientist (e.g., the biologist) and the applied scientist (e.g., the agronomist) toward taxonomy.

2.5.8. Examine any of the current classifications of the so-called elementary particles. Does it involve observational concepts or theoretical concepts? And is it a table of data or a system? *Alternate Problem:* Study the classification of illnesses according to their causes and according to their symptoms (e.g., fever), or discuss classifications of personality based on anatomophysiological characters (such as HIPPOCRATES' four humors).

2.5.9. Are partitions, orderings, and systematic classifications perfectible? Discuss the cases of ideal and material objects separately and exclude the obvious case of dichotomy. *Alternate Problem:* Speculate on the impact that the discovery of "noble" gases compounds (1962) may have on the periodic table, and draw a moral concerning the evolution of systematics.

2.5.10. Examine the following theses sketched in the text: (i) that *phenomenalism* ("All empirical predicates are phenomenal or reducible to such") and *nominalism* ("Every set results from an arbitrary grouping and is therefore a mere name") must go hand in hand, and (ii) that the progress of science has been away from both phenomenalism and nominalism, toward the disclosure of deeper and deeper transobservational properties (denoted by theoretical concepts) and objective natural kinds. *Alternate Problem:* Perform a detailed critical analysis of the current discussions on the concept of species. See E. MAYR, Ed., *The Species Problem* (Washington: American Association for the Advancement of Science, 1957) and P. C. SYLVESTER-BRADLEY, Ed., *The Species Concept in Paleontology* (London: Systematics Association, 1956), and peruse recent volumes of biology journals.

2.6. Systematics of Concepts

In 2.2. we classed concepts according to their logical form, namely into individual, class, relation and quantitative concepts. We also classed them, according to their semantical status, into formal or pure and nonformal (including concrete) concepts. Different classifications are obtained if alternative classing criteria are chosen. In this section we shall focus first on the functions, then on the systematic import of concepts.

If the *functions* of concepts in science are chosen as the *fundamentum divisionis*, the perfectible classification of Table 2.8 is obtained.

Formal concepts provide the elements for the formal skeleton (e.g., a calculus) of factual systems, such as classifications and theories. Their study belongs to formal science but they can be introduced in any field of factual science: they are not tied to a specific subject matter but are related to our approach and proficiency. Thus, the applicability of set theory to zoological systematics does not depend on animals but on the zoologist's realization that set theory is the formal calculus of systematics and, consequently, the more explicitly he uses it the more accurate and shorter his papers will be. The spread of formal concepts leads not only to stronger systematization and higher accuracy within each field of science but pushes the conceptual integration of science ahead.

Table 2.8. *Functions of Concepts*

Concept	*Formal*	*Basic:*	those providing the rational cement, like "and" "all", "set", "square root", "distance", "group".
		Metalogical:	those occurring in analyses and theories of formal theories; e.g., "well-formed formula", "provable", "axiom", "theorem", "theory".
	Nonformal	*Descriptive:*	those enabling us to describe experiential matters of fact, like "body", "red", "near", "event", "between", "liquid".
		Interpretive:	those occurring in the interpretation of descriptions, like "species", "atom", "birth rate", "motivation", "inhibition".
		Prescriptive:	those occurring essentially in norms, rules, and conventions — e.g., those designated by verbs in the imperative form.

Nonformal concepts are, of course, those which enable us to account for the world and to plan our investigation of it. They supply, so to say, the flesh by enabling us to refer to facts (descriptive and interpretive concepts) or to our decisions and acts (prescriptive concepts). (Such a reference is fairly direct in the case of descriptive and prescriptive concepts but indirect in the case of interpretive concepts, such as "electric field".) In "This ape is sad" only descriptive concepts occur. "This ape is sad because he is homesick" contains one interpretive concept. And "Homesick apes ought to be entertained with *TV*" contains a prescriptive concepts. Descriptive concepts occur, of course, in descriptive contexts; interpretive concepts are dominant in theoretical contexts; and prescriptive concepts are frequent in methodological discourse. But concepts of the three kinds may occur jointly in one and the same proposition, such as "In order to detect isotopic mass differences a mass spectrometer must be used".

A second partition of concepts we shall need is the one regarding their *systematic import*, i.e. the role they play in systematization. We may divide concepts, in this regard, into *extrasystematic* — like "soluble" and "conducting" — and *systematic* — like "solubility" and "conductivity". The most ordinary among the ordinary or extrasystematic concepts are the *ostensive* ones, i.e. those referring to directly observable individuals, properties, or relations, such as "body", "sticky", and "beneath". Science cannot dispense with extrasystematic concepts: if there were no links between systematic and extrasystematic concepts, scientific theories would be untestable und unintelligible. Only, such extrasystematic concepts are not typical of science: they are presupposed and elucidated by science, but science is built with specific, technical concepts that cannot be defined in terms of extrasystematic concepts.

Now, scientific systematization can consist of either systematics or theory. Correspondingly systematic concepts may be taxonomic and/or theoretical. *Taxonomic concepts*, such as those of species, rank, and hierarchy, can all be elucidated with the help of descriptive concepts and elementary logic. This is not the case of *theoretical concepts*, i.e. of concepts *introduced or elucidated by a theory*, such as "electric field" (introduced by electromagnetic theory), "natural selection" (introduced by Darwinian evolution theory) or "subjective utility" (introduced by utility theory). The specification of the meaning of theoretical terms (signs designating theoretical concepts) requires, in addition to logic, the special theory concerned and scientific experience: logic will show the structure of the concept, theory its connotation, and scientific experience (observation, measurement, and/or experiment) will supply the extension. Moreover, sometimes it will not be easy to decide whether a given concept is extratheoretical or theoretical: it may have become a member of common knowledge after having been with us for some time. But no matter what the historical vicissitudes of the concept, we shall adopt the following *Definition*: A concept will be called theoretical at a given time if and only if, at that time, it belongs to some theory.

Since theoretical concepts are the core of science and pose the most interesting epistemological problems, we shall do well to analyze them by resorting again to classification, a simple technique of conceptual analysis. The classification shown in Table 2.9 will be used in the sequel.

Table 2.9. *Systematics of Theoretical Factual Concepts*

Theoretical Factual Concepts
- *Generic*
  - *Ontological:* the proper subject of ontology. Ex.: "matter", "system", "event", "process", "evolution", "mind", "law", "randomness".
  - *Metascientific:* the proper subject of metascience. Ex.: "hypothesis", "law statement", "test", "confirmation", "systematics".
- *Specific*
  - *Observational:* referring to observable entities or properties. Ex.: "body", "position", "juxtaposition", "stimulus", "response", "number of reinforcements".
  - *Nonobservational:* not denoting experiential entities or properties. Ex.: "center of mass", "learning", "valence", "political power".

Generic theoretical concepts are those that permeate the whole of science: they are not peculiar to a given factual science. *Ontological* concepts, such as "event" and "real", are involved in or stand behind any factual theory even though the theory may not contain them in explicit form. Such concepts often pass for ordinary (presystematic) or for excessively metaphysical, which is a pity, because then every sort of ontological vagary is permitted. The neglect of ontology protects half-baked metaphysics: we have not the choice of making metaphysical commitments or of avoiding them, but of adopting a good or a bad metaphysics (see 5.9). *Metascientific* concepts, designated by terms such as 'evidence' and 'problem', occur in the remarks accompanying a statement, in the discussion of its validation and domain of validity, and on a multitude of other occasions. As in the case of ontological concepts, the best way of having muddled metascientific ideas — of which no scientist escapes anyway — is to abstain from analyzing and codifying them.

Specific theoretical concepts, by contrast to the generic ones, are the property of special theories or of groups of such. *Observational* concepts are those denoting directly observable objects, such as "body" and "stimulus". Theories referring to systems that are at least in part accessible to direct observation — such as, e.g., classical mechanics — may, but need not, contain observational concepts; these are often taken over from ordinary knowledge: theory elucidates them. But theories dealing with unobservable objects — such as the theory of the electromagnetic field in the vacuum or a theory of concept formation — may contain no observational concepts. Observational concepts will always occur, however, in the statements purporting to test a theory, whether it involves observational concepts or not.

Nonobservational concepts are characteristic of the initial assumptions (axioms) of a theory; some such theoretical concepts go down to the theorem level. Thus, e.g., in Newtonian mechanics "point particle", "mass" and "force" — the three undefined concepts typical of the theory — are non-observational. They are presented without previous notice in the axioms (i.e., they are primitive or undefined notions) and they refer to no observable trait of experience. It is rather the other way around: a large section of experience can be described and, what is more interesting, interpreted with the help of such nonobservational concepts. The basic concepts of mechanics are then at once descriptive and interpretive; only, they do not accurately describe actual facts but rather more or less idealized reconstructions of facts, both actual and possible. Some of the nonobservational concepts of Newtonian mechanics fail to occur in the lowest level theorems, where only the theoretical construct "point mass" occurs conspicuously (being the object to which statements in point mass mechanics refer) alongside the concepts of distance and duration.

Nonobservational concepts have in turn been divided into intervening variables and hypothetical constructs. *Intervening variables* are concepts mediating or intervening between observational concepts. Such are "center of mass" in mechanics, "entropy" in thermodynamics, and "learning", "drive" and "habit strength" in behavioristic psychology. Intervening variables are assigned no property, whether empirically accessible or not, and might therefore be dispensed with: they are just computational auxiliaries linking the observable properties of a system (see Fig. 2.7). *Hypothetical constructs*, on the other hand, refer to non-

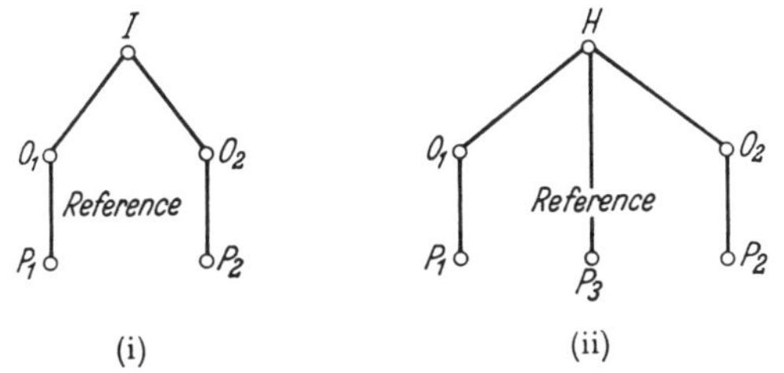

Fig. 2.7. Intervening variables and hypothetical constructs. (i) The intervening variable I mediates between the observational concepts O_1 and O_2 that denote the properties P_1 and P_2 respectively, but I refers to no property. (ii) The hypothetical construct H mediates between the observational concepts O_1 and O_2 and, in addition, has an intended objective referent, namely property P_3

observable but inferable entities or properties. For example, "atomic weight" in atomic theory and "production cost" in economy are non-observable yet not ghostlike, because they occur in statements that can be tested with the help of other statements closer to experience; besides, their referents are supposed to be real.

The difference between hypothetical constructs and intervening variables is neither logical nor epistemological: they can have as complex a structure as desired and both must be invented, not just drawn from observation. The difference between them is *semantical*, i.e. it resides in the referent assigned to either. Whereas intervening variables are assigned no referent other than the concrete system as a whole (e.g., "center of mass" refers to a body), hypothetical constructs are assigned precise, real though unobservable referents (things, properties or relations). Thus, e.g., in standard electromagnetic theory potentials are intervening variables whereas field strengths are hypothetical constructs; that is, potentials are not assigned an independent physical meaning and in principle they are dispensable, whereas the concept of field intensity is the core of that theory.

The difference between intervening variables and hypothetical is not absolute but *relative* to the theory in which they occur as well as to the philosophical position adopted. One and the same concept may be handled as a hypothetical construct in one approach and as an intervening variable in an alternative approach. Thus, e.g., "habit", "drive" and "mental state" are at best intervening variables for behaviorist psychology (e.g., C. HULL's), whereas in alternative approaches (e.g., E. C. TOLMAN's) they can be hypothetical constructs referring to real properties or systems of properties of the nervous system. In general the externalist or phenomenological approach will assign some kind of reality to the referents of observational concepts alone, and it will regard all nonobservational concepts as suspicious auxiliary devices mediating or intervening among observational concepts. On the other hand the internal or representational approach, the one seeking to disclose the mechanisms of systems rather than just describing their external properties, will recognize both kinds of concept and, instead of accepting all intervening variables as nothing but useful symbols, will try to explain some of them in terms of hypothetical constructs. An example of such a deeper analysis is the neurophysiological foundation of behavior psychology.

A more detailed discussion of nonobservational concepts and, in particular, of the intervening variable-hypothetical construct feud, must be deferred to Chs. 3, 5 and 8. We must now attack the problem of concept refinement.

Problems

2.6.1. Distinguish, in the following text, the formal, descriptive, interpretive and prescriptive concepts. "When I say that I (the *ego*) was at Grand Central Station yesterday at noon, it is true that from the nature of things a certain object which I speak of as *my* body, i.e. the body of the ego, was there at the time stated. But this is merely an implication, i.e. it is not a simple identification of the ego with the body, otherwise the phrase *my body* would be pointless. Once I become accustomed to think of the ego as something of the nature of a coördinate reference frame, the matter is perfectly clear. The thing that counts, in the depiction of the world in *me*, is the position of my *reference frame* relative to the external world. To say that *I* was at the station yesterday at noon is to say that this reference frame was thus situated. The body attached to this reference frame is in that sense *my* body. And clearly, the ego so defined is something immaterial." A. J. LOTKA, *Elements of Mathematical Biology* (N. York: Dover, 1956), p. 374.

2.6.2. Illustrate the concept of systematic concept in its two categories: taxonomic and theoretical.

2.6.3. Exemplify generic and specific theoretical concepts. *Alternate Problem:* Are observational concepts just packages of sense-data?

2.6.4. Point out the metascientific and the ontological concepts involved in the text of Problem 2.6.1.

2.6.5. Decide which of the following concepts are observational and which are nonobservational: "input", "output", "mean free path", "electric current", "reaction velocity", "catalyzer", "hormone", "vitamin", "gene", "muscle", "afferent", "metabolism", "conditioned reflex", "inhibition", "intensity of optical stimulus", "habit", "drive", "motive".

2.6.6. Draw a list of observational concepts, intervening variables, and hypothetical constructs occurring in a scientific theory of your choice.

2.6.7. Are the so-called "observables" of atomic physics, such as electron position, momentum, and energy observational concepts in the strict sense of 'observational' employed in the text? *Alternate Problem:* Are the previously mentioned concepts intervening variables or hypothetical constructs?

2.6.8. Show how statements concerning a given intervening variable might be translated into statements concerning a hypothetical construct. Study, if desired, the case of the pair "potential difference" — "electric field intensity".

2.6.9. Propose a criterion, other than a definition, for distinguishing intervening variables from hypothetical constructs. Be careful to specify that the distinction between the two is contextual and not absolute.

2.6.10. Analyze the presystematic and the systematic concept of motion. Show that, whereas the former is generic, the latter is specific; point out the qualitative and the quantitative concepts of motion in the theories of motion (the various mechanical theories); and find out what is gained and what is lost in the successive refinements of the concept of motion. *Alternate Problem:* Establish the relations between observational and descriptive concepts, and between theoretical and interpretive. Cautions: "number of reinforcements" is both observational and interpretive, and "intention" is interpretive but, so far, not theoretical.

Bibliography

BUNGE, M.: Foundations of Physics, Ch. 1. Berlin-Heidelberg-New York: Springer-Verlag 1967.

CARNAP, R.: The logical syntax of language. London: Routledge and Kegan Paul 1937.

CARNAP, R.: Introduction to semantics. Cambridge (Mass.): Harvard University Press 1942.
— Introduction to symbolic logic and its applications. New York: Dover 1958.
— R.: Meaning and necessity, 2nd ed. Chicago: University of Chicago Press 1956.
FREGE, G.: Translations from the philosophical writings of, ed. by P. GEACH and M. BLACK. Oxford: Blackwell 1960.
GREGG, J. R.: The language of taxonomy. New York: Columbia University Press 1954.
HALMOS, P. R.: Naive set theory. Princeton: Van Nostrand 1960.
HEMPEL, C. G.: Fundamentals of concept formation in empirical science, Vol. II, No. 7 of the international encyclopedia of unified science. Chicago: University of Chicago Press 1952.
— Introduction to problems of taxonomy. In: J. ZUBIN (Ed.), Field studies in the mental disorders. New York: Grune & Stratton 1961.
HENMUELLER, F., and K. MENGER: What is length?. Philosophy of Sci. **28**, 172 (1961).
HENNIG, W.: Grundzüge einer Theorie der phylogenetischen Systematik. Berlin: Deutscher Zentralverlag 1950.
— Systematik und Phylogenese. In: Bericht über die Hundertjahrfeier der deutschen Entomologischen Gesellschaft Berlin. Berlin: Akademie Verlag 1956.
JEVONS, W. STANLEY: The principles of science (1874), Ch. XXX. New York: Dover 1958.
LEWIS, C. I.: The modes of meaning. Philosophy and phenomenological research **4**, 236 (1943).
McCORQUODALE, K., and P. E. MEEHL: Hypothetical constructs and intervening variables. Psychol. Rev. **55**, 95 (1948).
MARTIN, R. M.: Truth and denotation. Chicago: University of Chicago Press 1958.
MARX, M. H.(Ed.): Psychological theory: Contemporary readings, Ch. III. New York: Macmillan 1951.
MENGER, K.: On variables in mathematics and in natural science. Brit. J. Philos. Sci. **5**, 134 (1954).
MORRIS, C.: Foundations of the theory of signs, Vol. I, No. 2 of the International Encyclopedia of Unified Science. Chicago: University of Chicago Press 1938.
NAGEL, E.: Logic without metaphysics, Chs. 4 and 5. Glencoe (Ill.): The Free Press 1956.
QUINE, W. V.: World and object. New York and London: The Technology Press of the M. I. T. and J. Wiley 1960.
SIMPSON, G. G.: Principles of animal taxonomy, Chs. 1 and 2. New York: Columbia University Press 1961.
SUPPES, P.: Introduction to logic, Chs. 9 and 10. Princeton (N.J.): Van Nostrand 1957.
TÖRNEBOHM, H.: A logical analysis of the theory of relativity, Secs. 28, 29 and 31. Stockholm: Almqvist & Viksell 1952.
WOODGER, J. H.: Biology and language. Cambridge: University Press 1952.

Chapter 3

Elucidation

Three diseases plague and may forever plague our conceptual outfit: shortage of rich concepts, abundance of poor ones, and vagueness of all except the strictly formal ones. Philosophers can do little to enrich the stock of scientific concepts and to eliminate the unfit among them: the growth and selection of the concept population is part of the evolution of science. On the other hand, philosophical analysis can be effective in the critical examination of scientific concepts. This criticism can be destructive, as when the use of nonobservational concepts is condemned in the name of prescientific philosophies; or constructive, as when conceptual vagueness is spotted and the attempt is made to decrease it, i.e. to make concepts more definite. That is, although philosophers do not usually conceive scientific concepts they may help in rearing them. This assistance is the more valuable because certain anachronistic ideas prevail in scientific quarters concerning the ways scientific terms are given a meaning. One such untenable tenet still popular among scientists is that all scientific concepts should be defined to start with (Aristotelian prejudice) and, moreover, by reference to operations, if possible of an empirical character (operationalist prejudice). We shall see that meanings are specified and refined in a number of ways and that operational definitions are logically impossible, but that there are certain sign-object correspondences (referitions) that discharge the function often assigned to operational "definitions".

3.1. Vagueness and Borderline Cases

'Field', 'pile' and 'freedom' are *ambiguous* terms because each of them designates various concepts. Ambiguity is a pervasive characteristic of signs; even mathematical signs can be ambiguous when out of context. Ambiguity is ambivalent: on the one hand it enables us to economize signs and thereby keep down the size of our several technical vocabularies; but on the other hand ambiguity shelters confusion. Fortunately, it can always be removed, in part or entirely, with the adjunction of further signs. Thus, 'electric field', 'atomic pile' and 'political freedom' are less ambiguous than the original nouns, and letters can be less ambiguous than the cables preceding them. On the other hand vagueness or blurredness has no positive aspect and is a conceptual rather than a linguistic disease, hence it is rather more difficult to cure. "Small", "bald" and "hot" are vague concepts because their very intension is blurred and correspondingly their extension is indefinite as long as no stipulations (conventional criteria) are laid down for determining their domain of applicability.

Since every concept has an intension and an extension (see Sec. 2.3), vagueness can be intensional or extensional. *Intensional vagueness* consists in a partial indefiniteness of the intension — whether or not it is intentional. "Organism" and "machine" suffer from intensional vagueness: the properties they connote are not exhaustively determined — this being a major source of confusion in contemporary discussions on the scope of cybernetics. In order to shrink the aura of vagueness surrounding these two concepts we need more advances in the theory of organisms and in the theory of machines as well as more extensive philosophical analyses: a linguistic ukase will hardly satisfy all contending parties. *Extensional vagueness* consists in a partial indetermination of the extension of a concept. As we shall see, this defect can be cured more easily than intensional vagueness — which explains the current preference for the extensional approach. Yet, since the task of science is never finished, we may confidently expect that some vagueness, whether intensional or extensional, large or small, will always blur the more interesting concepts — among them the very concept of vagueness. In fact, it is likely that no natural kind (species, class) is known as a definite or clear-cut class, and it is not always clear whether such a vagueness or indeterminacy is a feature of human limitation or is an objective trait.

*The concept of *intensional vagueness* can be analyzed in terms of the set $I(C)$ of properties, both known and unknown, connoted by C, and the subset $K(C) \subseteqq I(C)$ of the known properties included in the intension of C. We define the intensional vagueness $V_i(C)$ of a concept C as the difference between $I(C)$ and $K(C)$, i.e. as the set of the properties belonging to I but not to K:

$$V_i(C) =_{df} I(C) - K(C). \qquad [3.1]$$

Since only $K(C)$ is well-determined, the concept of intensional vagueness is itself vague. And when we employ the core intension $I_c(C)$ (see Sec. 2.3) instead of the full known intension $K(C)$, we allow for even a larger vagueness zone, as $I_c(C) \subseteqq K(C)$. A desideratum of scientific research is, of course, to shrink the difference between $I(C)$ and $K(C)$, i.e. to reduce the aura of $K(C)$ (see Fig. 3.1).

*On the other hand, the concept of *extensional vagueness* can be made quite definite. A reliable test for the extensional definiteness of a concept is its performance in dichotomy: if a concept C allows us to unambiguously and exhaustively perform a partition of any set into the subset C of all the members with the property $C(x)$ and the complementary subset $\overline{C}$ of all those elements which do not satisfy the function $C(x)$, then the extension or denotation of C is entirely definite. If, on the other hand, there are a number of borderline cases that might as well be

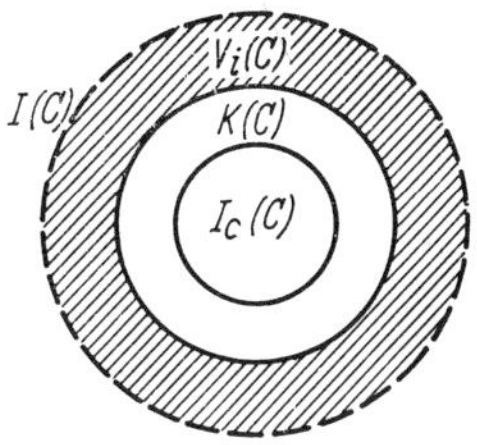

Fig. 3.1. Intensional vagueness: the ill-limited shaded zone. The ovals picture sets of properties, not of individuals

grouped with C or with $\overline{C}$ — as is the case of men in the process of becoming bald — then the extension of C is to some extent indefinite and the proportion of borderline cases will measure this extensional vagueness.

*The above proposal can be made more precise. For definite, i.e. nonvague concepts, the overlap of the extension $E(C)$ and of its complement $E(\overline{C})$ is empty, i.e. $E(C)\cap E(\overline{C})=\emptyset$; consequently, the size or measure of this common part is zero: $M[E(C)\cap E(\overline{C})]=0$. Such are the clear-cut concepts handled by set theory. But for vague concepts the measure $M[E(C)\cap E(\overline{C})]$ of the overlap of the extensions of C and not-C will be a non-negative number n, and the ratio of n to the measure N of the universe $U=E(C)\cup E(\overline{C})$ will measure the vagueness of C. That is, we may define the extensional vagueness $V_e(C)$ of a concept C as

$$V_e(C)=_{df}\frac{n}{N}\,. \qquad [3.2]$$

Correspondingly, the concept of extensional definiteness can be defined as

$$D_e(C)=1-V_e(C)\,. \qquad [3.3]$$

Since $\overline{X}\cap\overline{\overline{X}}=\overline{X}\cap X=X\cap\overline{X}$, we infer that $V_e(\overline{C})=V_e(C)$ and correspondingly for the extensional definiteness. That is, extensional vagueness (definiteness) is shared to the same extent by each member of every pair of opposite concepts.*

If the extension of a concept is a finite set, its extensional vagueness will be simply the percentage of borderline cases, and actual counts or random samplings will supply estimates of such a fraction n/N of unclear cases to the total number of cases. For example, if students are graded from 0 to 10 ,the dividing point between good and bad students being 5, then the vagueness zone will be roughly the interval (4,6). Since the measure of this set is 2, the vagueness of either "good student" or "bad student" relative to the given grading procedure is $\frac{2}{10} = \frac{1}{5}$. In other cases the vagueness fringe itself may be ill-limited, and correspondingly the fraction of borderline cases somehow indeterminate. In such cases the upper bound or outermost contour of the vagueness fringe may be taken, so that an upper bound of the degree of vagueness is determined.

*The *total vagueness* of a concept may now be introduced as follows:

$$V(C) = \langle V_i(C), V_e(C) \rangle. \qquad [3.4]$$

This, in turn, will allow us to define the concept of haziness of meaning. If a concept C is blurred so will be the term 't' which designates it, even if t is an unambiguous sign, i.e. even if 't' designates just C. Since a concept can be intensionally and/or extensionally vague, the corresponding term will be indeterminate to the same extent. But an indeterminateness in the intension or in the extension of a concept will — according to our definition [2.23] in Sec. 2.3 of 'meaning' — be reflected in a haziness of the corresponding term. It will therefore be in keeping with our previous formulas to define *meaning indefiniteness* through

If 's' designates C, then: *Meaning Indefiniteness* (s)

$$= V(C) = \langle V_i(C), V_e(C) \rangle. \qquad [3.5]$$

Let us now examine the methodological import of concept vagueness.*

What is the relation between intensional and extensional vagueness? Intensional vagueness is a necessary but not a sufficient condition for extensional vagueness: if a concept is extensionally vague then it is also intensionally vague but the converse need not be true. In fact, intensional vagueness is consistent with a fairly definite extension. Thus "catalyst", "cancer" and even "talent" are intensionally still quite vague but their extension can be made quite definite with the help of practical criteria (decision rules) enabling us to decide whether a given object falls under the extension of each of these concepts. Such practical criteria consist in regarding a few earmarks (see Sec. 2.3) as *sufficient* for the inclusion of something in a class — as when the ability to reason is chosen as a test for humanness. Although such earmarks allow for an unequivocal identification or classing, they may be far from important and consequently, if taken alone, they may supply superficial characterizations. For example, the mere presence of a dentary-squamosal joint is usually employed as a *practical* criterion for identifying a fossil as a mammal even though it is not well known why that osteological characteristic should be related to viviparity and other essential properties of mammals. Similarly acidity tests and intelligence tests, by disclosing the presence of certain sufficient conditions, help us determine the extension but not the intension of the corresponding concepts.

Extensional vagueness can also be reduced by performing finer partitions. Thus, the introduction of the transition category of mediocre student between those of bad and good student will reduce the extensional vagueness of the latter concepts — not however the total vagueness of the concept of student rank. In fact, if we grade a set of students from 0 to 10 we may assign bad students the interval [0, 4), mediocre students the interval [4, 6], and good students the interval (6, 10]. But now instead of a single middle vagueness interval, say (4, 6), we will have two narrower vagueness fringes: there will be a penumbra of, say, $\frac{1}{10}$ between "bad" and "mediocre" and a similar zone between the transition species and "good student" (see Fig. 3.2). An even finer partition

Fig. 3.2. Finer partitioning redistributes vagueness

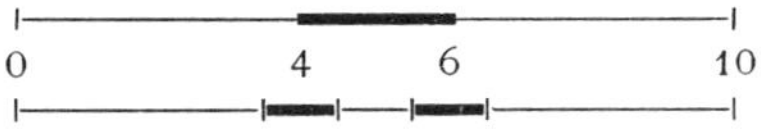

will further decrease the vagueness of the specific categories — say A, B, C, D, F — without, however, changing the vagueness of the whole, in this case the generic concept of student standing. In general: finer partitions decrease the extensional vagueness of the specific categories subsumed under the principal category without necessarily decreasing the vagueness of the latter. And mind that we cannot merely replace the principal category by the union of all the finer categories subsumed under it: we need both the original concept (the universe of discourse) and the components into which it is analyzed.

Sometimes conceptual vagueness mirrors an objective haziness or indeterminateness, not in the sense that facts are hazy but in the sense that there are often *transition forms* among natural kinds. Such transition forms prevent a sharp demarcation, originate conceptual vagueness, and may even ruin classifications. Two conspicuous instances which have not ceased posing difficulties are the transition elements in chemistry and the intermediate biological species.

If chemical elements are arranged according to the number of electrons in the outermost shell, as they usually are, then difficulties arise when that shell begins to form before an inner shell is complete — which is not surprising since inner shells had not been taken into account in the classification. Thus, e.g., iron, cobalt and nickel have each two electrons in their outer shell; hence, according to an older view they should be placed in a single column (group VIII) in the periodic table, but this would ruin the ordering with respect to the atomic number. In this case the familiar trick of performing a finer partition succeeds: group VIII is subdivided into three subgroups.

Biological transition species pose a harder problem. First, because the gradation is less discontinuous than that of the chemical elements. Second, because the vagueness of the concepts involved is so much greater than in the case of chemistry, where some essential or source properties — such as atomic number and the distribution of electrons among shells — are fairly well known. Take, for instance, the problem of classing therapsids (*Cynodontia* and *Bauriamorpha*). This is an extinct transition group placed between reptiles and mammals. If some purely skeletal characters are taken as definitory (as earmarks), *Therapsida* will be classed as *Reptilia*. But if some (plausibly inferred) physiological characters are chosen as earmarks, *Therapsida* can be grouped with *Mammalia*. Both moves have been proposed and no agreement has been reached. Moreover, no agreement will be reached on this point, nor on any other taxonomic problem, unless the *criteria* of classification are previously stated and agreed on. In fact alternative classifications, all of them logically correct, are always possible on alternative criteria. A discussion of the possible consequences — e.g., the complication — that the grouping of a given transition form with this or that class would have, without a previous agreement on the *fundamentum divisionis*, is a waste of time.

*Let us pose in general the problem of *classing transition forms*. Let C_1 and C_2 be two class concepts, e.g. those of reptile and mammal, and let C_{12} be a third concept, referring to a group which, like *Therapsida*, is somehow placed between C_1 and C_2. We stipulate that C_{12} is a *transition group* if and only if its intersection with the common parts of C_1 and C_2 is nonempty (i.e., $C_{12} \cap (C_1 \cap C_2) \neq \emptyset$). Suppose further that each of the three concepts is neatly defined by its earmarks (see Sec. 2.3), namely thus:

$$C_1 x =_{wdf} R_1 x \;\&\; R_2 x \;\&\; \ldots \;\&\; R_r x$$

$$C_{12} x =_{wdf} T_1 x \;\&\; T_2 x \;\&\; \ldots \;\&\; T_t x$$

$$C_2 x =_{wdf} M_1 x \;\&\; M_2 x \;\&\; \ldots \;\&\; M_m x .$$

(The numbers of earmarks, r, t and m, need not be the same: closely related species may show a fairly complete homology between different characters but higher rank categories, such as classes, will in general be nonhomologous and, consequently, the number of their earmarks will differ. Besides, the number of earmarks depends not only on the object of study but also on our knowledge; thus, e.g., we are certainly better acquainted with fruit flies than with men and with either than with therapsids.) The data of our problem are of this kind: the transition form C_{12} shares such and such characters with C_1 and such and such other characters with C_2. In short, we are given the intersections of the intensions of the three concepts involved. The problem is: Where shall we place C_{12}? *

*An answer to this problem requires a detailed examination of the relations among the intensions of the three concepts concerned. Since we have defined "intension" as a set (of properties) we may facilitate the discussion by drawing Euler-Venn diagrams in which ovals symbolize clusters of properties, not sets of individuals. The cases which satisfy our definition of "transition group" are illustrated in Fig. 3.3; they are:

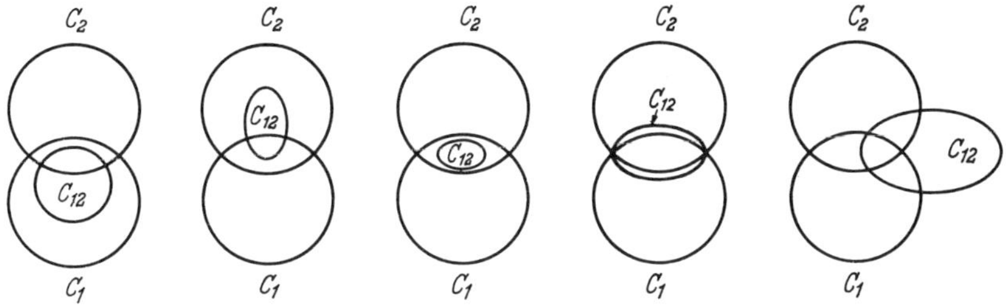

(i) (ii) (iii) (iv) (v)

Fig. 3.3. The possible relations (in intension) among the transition group C_{12}, its nearest ancestor, and its nearest descendant

(i) The core intension of C_{12} is a proper part of the core intension of C_1: C_1 has all the earmarks of C_{12} and, in addition, some of its own (i.e., $r > t$).

(ii) The core intension of C_{12} is a proper part of the core intension of C_2: C_2 has, in addition to the earmarks of C_{12}, some of its own ($m > t$).

(iii) The core intension of C_{12} is included in the intersection of the core intensions of C_1 and C_2 (i.e., $C_{12} \subseteqq C_1 \cap C_2$, $t < r$, $t < m$).

(iv) The core intension of C_{12} includes the intersection of the core intensions of C_1 and C_2 (i.e., $C_1 \cap C_2 \subseteqq C_{12}$).

(v) The core intension of C_{12} overlaps partially with the intersection of the core intensions of C_1 and C_2 but none of the above conditions is fulfilled.

*So far it has not been found out which of the above logical possibilities is realized in nature; consequently no decision concerning the relative places of *Reptilia*, *Therapsida* and *Mammalia* should be forced. It is only after the above-mentioned class relationships have been established, at least provisionally, that a grounded decision can be taken. Such a step will eventually be taken by applying one of the following decision rules:

(i) If the possibility (i) above is realized, count therapsids as reptiles.

(ii) If the possibility (ii) above is realized, group therapsids with mammals.

(iii) If any of the remaining possibilities (iii) to (v) is realized, keep *Therapsida* as a separate group.

*Even if a clear-cut decision could be made at this time it might not be final, because the criterion of classification may not be altogether adequate; moreover, there may be even more than one criterion or *fundamentum divisionis* involved, as is so often the case with different ranks in a hierarchy. Thus, e.g., it would be desirable to complete morphological considerations with alternative, independent considerations, especially since we know that morphology alone is insufficient, hence misleading. Among such supplementary considerations the likelihood of the several evolutionary lines accompanying the alternative groupings should be taken into account. (Thus, if therapsids are classed with reptiles, it must be concluded that mammals have evolved from them in a parallel way along from four to nine different and independent lines — polyphyletic ancestry. If, on the other hand, therapsids are included among mammals, then the monophyletic origin of the latter follows. Now, the hypothesis that the hazy border between reptiles and mammals was crossed independently by four to nine different lineages of mammal-like reptiles seems unlikely: since the probability of such a joint occurrence equals the product of the single transition probabilities — every one of which is smaller than unity — it is extremely low. This theoretical remark, which is far from conclusive, speaks in favor of either transferring *Therapsida* to *Mammalia*, or keeping the former as a class by itself. Whether this argument carries any weight or not, it is philosophically irrelevant: it was intended to illustrate the kind of considerations that might help to solve the problems posed by transition forms. As we say in Sec. .2.2, taxonomy is not an independent science and it cannot advance in isolation from theory.)

Be that as it may we conclude that, before proposing any solution to a problem concerning the placement of a transition group in biological systematics, the following tasks should be performed: (i) determining the core intensions of the major concepts concerned; (ii) finding their logical relations; and (iii) discussing the above in the light of general principles of biological theory (concerning, e.g., the feasibility of the various conceivable lines of descent accompanying the alternative relations). This alone should suffice to show that the work of the taxonomist can become logically and theoretically as sophisticated and involved as desired.

To conclude. The meaning of signs can be sharpened, and the vagueness of the corresponding concepts can be reduced, if not entirely, substantially and in a number of ways. Extensional vagueness can be shrunk by introducing finer partitions or by adopting practical criteria which need not go to the heart of the matter, and by executing the corresponding empirical operations. And intensional vagueness can be reduced by logical analysis and by theoretical research — particularly by multiplying the constant relations (law statements) in which the given concept occurs and by displaying the structure of the theory in which it is embodied. Consequently the meaning of a sign — which according to our view is composed of the intension and the extension of the corresponding concept — cannot be sharpened by empirical operations alone. The meaning of nonformal concepts is specified in a gradual way by joint theoretical and empirical research. Let us take a closer look at the several elucidation procedures.

Problems

3.1.1. In order to ascertain whether a telephone is in working order no specialized knowledge is required: some know-how is enough. Some amount of telephone engineering is needed, on the other hand, in order to determine the connotation of "telephone in working order". Describe this situation in the terms elucidated in the text. *Alternate Problem:* Discuss the relevance of ambiguity and vagueness to problem statement. Hint: Begin by analyzing questions such as 'What is x?' and 'How much is some?'.

3.1.2. The traditional logic of concepts holds that the intension and the extension of concepts are inverse to each other in the sense that the widest concepts (e.g. "object") are the ones connoting the fewer properties. Determine (i) whether "centaur" exemplifies the doctrine and (ii) whether the existence of concepts with a fixed denotation and a variable intension (exemplify!) fit the doctrine. For a criticism of the principle and a peculiar solution to the problem see C. I. LEWIS, "The Modes of Meaning", *Philosophy and Phenomenological Research*, **4**, 236 (1943).

3.1.3. Would you hold that any person must be either bald or nonbald? If not, would you conclude that the law of the excluded middle is not universally true, or would you rather infer that the law applies only when the concepts involved are precise or clear-cut? See B. RUSSELL, "Vagueness", *The Australasian Journal of Psychology and Philosophy*, **1**, 84 (1923), and S. KÖRNER, "Deductive Unification and Idealisation", *British Journal for the Philosophy of Science*, XIV, 274 (1964).

3.1.4. Discuss the following proposals for measuring the vagueness of a proposition in terms of the vagueness of the concepts occurring in it. (i) The vagueness of a proposition equals the vagueness of the vaguest of all the concepts occurring in it. (ii) The vagueness of a proposition equals the sum of the vagueness of its constituent concepts divided by the number of concepts. Consider, to begin with, the following simple proposition forms: $p_1 = c \in A$ and $p_2 = A \subset B$, where A and B are classes and c is an individual. Suppose further that formal concepts are entirely definite, i.e. $V(\in)=V(\subset)=\langle 0, 0\rangle$. If it is further assumed that c is nonvague, both proposals yield the same total vagueness for p_1, namely $V(p_1)=V(A)$. On the other hand, the vagueness measures differ for p_2. In fact, calling $V(A)=\langle a_1, a_2\rangle$, $V(B)=\langle b_1, b_2\rangle$, and assuming that $a_1 \geqq b_1$ and $a_2 \geqq b_2$, the first proposal yields $V(p_2)=V(A)=\langle a_1, a_2\rangle$ whereas the second yields $V(p_2)=\langle \frac{1}{2}(a_1+b_1), \frac{1}{2}(a_2+b_2)\rangle$.

3.1.5. Distinguish the semantical concept of vagueness from the psychological concept of obscurity and relate the two concepts. *Alternate Problem:* Sometimes a scientific writer will deliberately employ an evasive (vague) phraseology, not for intellectual dishonesty but, on the contrary, because the use of more definite expressions would require more precise knowledge, which happens to be missing. Thus, for instance, one speaks of the similarity or affinity of organisms, or of the de Broglie wave *associated* with an electron — not however *identical* with it or *guiding* it, because the identity hypothesis and the pilot wave hypothesis have led into trouble, whereas the concept of association is non-committal as long as the nature and mechanism of the association is left indefinite. Offer further examples of deliberate honest vagueness and draw some consequence concerning the relations of language to knowledge. Finally, contrast this kind of vagueness with the obscurity of certain philosophies.

3.1.6. Plant and animal species often consist of several subspecies differing among one another genetically, morphologically and ecologically. In other words, many species are complex or *polytypic*. Now, a standard criterion for grouping individuals in a given subspecies is this: if at least 75 per cent of the specimens of the population on hand can unequivocally be distinguished from those of an adjacent population, then they constitute a subspecies. Determine, on the basis of this decision rule, the maximum degree of extensional vagueness of the concept "subspecies".

3.1.7. Does the vagueness of concepts related to transition forms originate solely in our ignorance or does it also somehow correspond to reality? And would it be possible to eliminate such a vagueness by introducing quantitative concepts of, say, mammalness and reptility? *Alternate Problem*: Does the existence of borderline cases render set theory inapplicable in taxonomy?

3.1.8. Report on the discussion between L. VAN VALEN, *Evolution*, **14**, 304 (1960), C. A. REED, id., 314 and C. G. SIMPSON, id., 388, concerning the origin of mammals and the classing of species intermediate between reptiles and mammals. *Alternate Problem:* Would it be possible to set up a calculus of not-clear cut sets, i.e. such that $A \cap \bar{A} \neq \emptyset$ in it? And would it be helpful in taxonomy?

3.1.9. Should biological systematics be *monophyletic* (single origin of each category), *polyphyletic* (multiple origin of each category) or eclectic? And what kind of consideration should dominate in this discussion: logical, empirical, or theoretical? See M. BECKNER, *The Biological Way of Thought* (New York: Columbia University Press, 1959), pp. 73 ff.

3.1.10. Should the approach to taxonomy be predominantly empirical or speculative? In particular, should the ordering of diagnostic characters according to importance be based on the extent of constancy among the members of a group, or should it fit an ideal type (archetype, *Bauplan*) imagined a priori, independently of the search for constancy? Or is this a wrong alternative and is there the possibility of combining the empirical search for constancies with theoretical considerations explaining them? *Alternate Problem*: It is not always possible to determine whether any given statement of a class is true or not. When a class of statements is such that no clear-cut partition between the true and the false is possible, it is said to be an *indefinite class*. Study this problem of the vagueness of the concept of statement set.

3.2. Sharpening

Concepts are begotten and reared in a number of ways: by constructing classes (e.g., "mammal"), by grouping the latter into wider classes (e.g., "vertebrate"), through the discovery of relations (e.g., "descent"), by invention (e.g., "evolution") and so on. None of these ways is methodical: i.e. there are no known standardized procedures (techniques) of *concept formation*. At most there are contraceptive prescriptions, such as "Do not go beyond observation". Concepts are formed spontaneously alongside the growth of knowledge, whether common or specialized: after all concepts are pills of knowledge. In this way more or less vague concepts are born and develop.

Once a tolerably vague concept has been conceived it may be desirable and possible to *elucidate* it, i.e. to sharpen its meaning. For such a sharpening of signs and concepts philosophy and science have built definite procedures, i.e. techniques. The techniques for conceptual elucidation may be classed into three groups: (i) *interpretation* by reference to what the sign or the concept stands for; (ii) *analysis* — e.g., definition; (iii) *synthesis*: the building of an ordered set of statements (theory) in which the concept concerned occurs either as a building block (undefined concept) or as a defined idea. The elucidation procedure consisting in embedding a concept in a theory will be examined in Ch. 7; in the present chapter interpretation and analysis will be studied.

Just as living beings show traces of evolutionary processes, so any system of scientific concepts exhibits the stages of its own evolution from primitive to advanced forms. In scientific and technological contexts we may, in fact, distinguish three levels of concept as regards refinement: (i) *concepts borrowed from common knowledge* (e.g., "muscular force"); (ii) *refinements of common knowledge concepts* (e.g., "force" in physics); (iii) *newly introduced concepts* (e.g., "entropy"). No piece of scientific discourse dwells entirely on the lowest conceptual level: if it did it would not go beyond common knowledge (see Sec. 1.1). And no science dwells entirely on the upper level: if it did it would be unable to establish contact with experience, hence to explain it and to profit from it. Every branch of science, whether pure or applied, contains statements establishing links among concepts belonging to the various levels of concept sophistication; such links do not wipe out their peculiarities: for example, they are not definitions enabling us to reduce all scientific concepts to ordinary concepts. (For the relations between high-level concepts and observational concepts, see Secs. 8.4, 15.3 and 15.6).

That scientific progress is often marked by the invention of radically new ideas has been denied on various grounds, of which two will be recalled here. It is said, in the first place, that thought can at most reflect reality (primitive materialism) or experience (primitive empiricism), and that consequently there is no invention or creation of ideas. Yet it is only factual theories proper, i.e. systems containing theoretical concepts (see Sec. 2.6), that can account for reality and its proper part, experience. A less well-known objection to the possibility of conceptual novelty has been advanced by some of the inventors of quantum mechanics, which is characterized by a newness hardly attained by previous theories. They have held, paradoxically enough, that a factual theory cannot introduce radically new concepts because the results of observation and experiment are largely described in ordinary language terms. This objection stems from the widespread confusion between what a theory says and how the theory is put to the test. This confusion between meaning and testability is at the root of the operationalist philosophy of science (see Sec. 3.6). Moreover, it is not altogether true that the description of scientific observations and experiments can always be done with ordinary concepts alone. A geneticist who manipulates nucleic acids will not be able to avoid theoretical terms in describing the results of his experiments, much less in explaining them. Moreover the very design and interpretation of his experiments will require technical concepts, because both the planning of experiments and the reading of instruments involve hypotheses and systems of hypotheses. The purity of science is not attained by remaining on the common sense level but by grounding and testing the conjectures.

A concentration on the results of empirical operations with neglect of the ideas behind them will lead to a distorted picture of scientific knowledge and to a vulgaristic epistemology, according to which science is just refined common sense (see Sec. 1.1). Moreover, it may lead to scientific stagnation. An example of the latter is the current reluctance of many physicists to invent new concepts to cope with the stubborn unsolved problems posed by the unusual behavior of elementary particles and atomic nuclei. It may well be that, just as the transition from macrophysics to microphysics required the invention of new ideas, the much-needed further transition to the newly discovered levels of organization will require the introduction of radically new ideas. After all, matter at these levels has properties which the available theories do not enable us to understand. New levels, new concepts; and conversely: once new ideas are on hand further new levels may be hoped to be discovered. And sometimes the ideas that guided research to discover new fields are found to hold in it at least to a first approximation.

Indeed, a common way (but not a method) of conceptual progress is the *exportation* of ideas (in particular concepts) from their original context. For example, the concept of stress has profitably been exported from physics to psychology and sociology. But, unless carefully performed, such an exportation of technical ideas may result in shear confusion. This was the case when the term 'field' was borrowed from physics by some biologists ("morphogenetic [organ-forming] field") and psychologists (K. LEWIN and his school) with the implication that they were "like" physical fields of force but with no indication concerning the nature, structure, and measurement of the field. To be fruitful, the exported term must cover at least the original concept and must suggest either fruitful new problems or must be assimilated by a scientific theory in the new field: it must not be used metaphorically or to give the appearance of a scientific approach or to cover conceptual indigence. In isolation, signs are neither good nor bad.

Some concepts are not exportable: they designate things, properties, or facts peculiar to a given field. Thus, an exportation of the concept of life may lead to animism; and the concept of machine, if expanded excessively, leads to a mechanistic ontology which misses the richness of levels. Other concepts are inextensible in the sense that they are *rigid:* they are class concepts, such as "cat" and "rest", which cannot give rise to relation concepts, let alone quantitative concepts (see 2.2). The concept of rest is parochial for that reason: we must say of a body either that it is at rest or not at rest relative to a reference frame; we cannot say that it is more, or less, at rest than something else, let alone that it is in slow or in quick rest: to say so would be self-contradictory. On the other hand its dual, the concept of motion, is *flexible:* it can be qualified (e.g., "rapid motion") and it can be quantified (e.g., "moving at 1,000 Km/hr"). A switch from the concept of rest to the concept of motion is all we need to escape the paradoxes of the pre-Socratics regarding change: we do not generate the contradiction "the arrow is at a given place and is not at that same place" once we replace the static concept of being by the dynamic concept of motion: in fact, we now can say that the arrow moves through a given place. Moreover, the concept of motion permits us to eliminate the concept of rest through an explicit definition, namely thus: "x is at rest relative to $y =_{df} x$ does not move relative to y". (Or, if the quantitative concept of motion is preferred: "$R(x, y) =_{df} [V(x, y) = 0]$".) Of course, this shift from the rigid "rest" to the flexible "motion" is not an isolated and purely linguistic operation: it has accompanied a deep change in physical theory and even in world outlook — namely, the replacement of Aristotelian physics, for which rest was more basic than motion, by the dynamic Galilean physics. Which, incidentally, goes to show that in factual science the selection of basic concepts is not arbitrary and may involve a transformation of the categorial framework. In science concept elucidation is contextual and gradual, and it is not the work of lexicographers, logicians, or even experimentalists, but rather of theoreticians.

In the attempt to elucidate a concept a *deformation* of it may occur, as a result of which some characteristic notes of its original intension are lost — for better or for worse. Thus, e.g., the concept of mind has been changed by modern psychology to something bearing little resemblance to the original concept, which made mind an ingredient of the immaterial and immortal substance called soul: we now tend to regard mind as a system of bodily functions (see Sec. 1.6). This concept deformation has been part of the progress of science. Other concept deformations may not be progressive and may not be prompted by the needs of science but rather by philosophical or even sociological tenets. Examples of non-illuminating concept deformations were the attempts to reduce mass to an acceleration ratio thereby doing away with matter (see Sec. 3.3), and the proposal to regard biological progress as mere increased adaptation — in keeping with conformist ideals.

Whether progressive or not, the concept deformation that occurs so often in science can have a philosophical motivation. And it is most radical when it involves a level *reduction*, i.e. when it shows that the laws characterizing one level are *the same* as the laws peculiar to a putatively different level. This was, e.g., the case of the concepts of optics when this discipline was shown to be a chapter of electromagnetic theory. If no such identity is established but on the other hand the *deduction* of a set of laws from another set referring to a lower level of organization is performed — as when chemical laws are deduced from physical laws in conjunction with specifically chemical hypotheses concerning the composition and structure of molecules — then no genuine concept *reduction* is effected: something at least as valuable is attained, namely a concept elucidation and the corresponding linking of two different levels of reality. The distinction between reduction and pseudoreduction is never clearly made yet it is important not only for philosophy but also for science, because reductionism is double-edged: on the one hand it stimulates the exportation of knowledge, concerning the basic levels, to the higher levels; but on the other hand it inhibits the search for new, emergent properties and laws rooted to those characterizing the lower levels but not identical with them. In any case, concept deformation is performed in theoretical contexts and sometimes it involves metaphysical (ontological) considerations and problems.

Some concepts are particularly helpful for performing concept extension and elucidation. Conspicuous among these sharpening tools are the mathematical concepts of set and probability. As soon as properties are mentioned in some context — and they cannot fail to be mentioned — the corresponding sets of individuals having such properties can be thought of. For example, the class $\mathscr{P}$ generated by the property symbolized by the unary predicate 'P' is: $\mathscr{P}=\{x\,|\,P(x)\}$. And once sets have been formed the discourse can be both expanded and made more definite with the help of set theory. Thus, e.g., the slippery concept of life — which is so often treated as if it denoted a peculiar substance — can be tamed with the help of set-theoretical concepts in either of the following ways. One possible move is *direct extensionalization*, i.e. the definition of the life concept as the set of living things: "Life$=_{df}\{x\,|\,x$ is alive$\}$". This is no great gain: it only prevents us from falling into the mistake of regarding life as an entity separate from the function of living. A more enlightening elucidation of the concept of life is by *indirect extensionalization*, i.e. by analyzing "living" as the conjunction of certain earmarks $P_1, P_2, \ldots, P_n$ and then proceeding to build the corresponding classes $\mathscr{P}_1, \mathscr{P}_2, \ldots, \mathscr{P}_n$ and finally taking their common part: $\mathscr{L}=\mathscr{P}_1\cap\ldots\cap\mathscr{P}_n$. This second procedure requires, of course, some basic biological knowledge going beyond the mere description of plant and animal life: it is a definition that encapsulates a body of fundamental knowledge.

Given any two properties, P and Q, we may then set up the corresponding sets and inquire, for example, whether $\mathscr{P}$ is included in $\mathscr{Q}$ or conversely. We may next introduce the new sets formed by the operations of union (i.e. $\mathscr{P}\cup\mathscr{Q}$) and of intersection ($\mathscr{P}\cap\mathscr{Q}$). We can go beyond this qualitative stage and measure these sets. In particular we may find the measure of the intersection $\mathscr{P} \cap \mathscr{Q}$ relative to the measure of $\mathscr{P}$, i.e. we may evaluate the ratio $M(\mathscr{P} \cap \mathscr{Q})/M(\mathscr{P})$. If this number satisfies certain requirements (the axioms of the probability theory), it can be interpreted as the probability that a member of the set $\mathscr{P}$ is also a member of $\mathscr{Q}$. As soon as this is performed a further calculus, the theory of probability, can be applied to expand the field and refine the concepts involved. New concepts can then be formally introduced and mutually related. A couple of examples will show how powerful this elucidation technique is.

The tricky concept of *possibility* can be elucidated in terms of the probability concept, namely so: "x is possible $=_{df}$ y is a nonempty set such that every member of y has a certain probability and x belongs to y". Even shorter: anything is possible which is probable. If the probability in question — a positive number less than unity — is allowed to take on the zero value, the refined concept of possibility is seen to subsume the concept of impossibility much as rest had been made a particular case of motion: just as "rest" has been identified with "zero velocity motion", so "impossibility" is now identified with "zero probability". Concept refinement through quantification may then lead to the dissolution of age-long oppositions of concepts: many such alleged strifes of opposites are seen to result from a coarse analysis which perceived only the extremes and ignored the rich gamut in between.

Similarly the intuitive or presystematic concept of *fallibility* may be both refined and at the same time extended to an open class of objects: tools, machines, institutions, theories, friends, etc. In fact, the fallibility of x can be defined as the probability of x's failure in a given respect during a given time interval. The numerical value of the fallibility of something can then be measured in simple cases by the actual long run percentage of failures (divided by the total time lapse or number of time units). And the *reliability* of an object can then be defined as the complement of its fallibility, i.e. as the probability that the object will operate without a failure in the given respect over a given period: $R(x) = 1 - P(x)$. In such a tearless way whole masses of important concepts can be subject to logic, computation, and measurement.

The progress of science has depended largely on the ability to exploit powerful concepts devoid of empirical content, particularly those that are found ready-made in logic and mathematics. Biologists, psychologists, sociologists, management experts, strategists and even philosophers are nowadays making rapid progress to the extent to which they discover that key concepts such as those of set, order, function, and probability can be used as tools for concept refinement and simultaneous theory construction. (Until recently scientists in the behavioral sciences have been using the probability concept only in connection with the *test* of statistical hypotheses such as those of the form "The probability of a member of a class $\mathcal{P}$ being also a member of a class $\mathcal{Q}$ equals the fraction p". They are now thinking in probability concepts, i.e. they are trying statements concerning behavior in which the concept of probability occurs essentially. In other words, probability is being used in the construction of theories of behavior, not only in their empirical test.) Something similar is bound to happen to those engaged in classifying objects of various kinds: they will learn that set theory, lattice theory and other qualitative (nonnumerical) mathematical theories provide the proper formal skeleton of systematics (e.g., biological taxonomy), and with this realization much work which is still protoscientific will become scientific. The view that mathematics is just a tool for computing numerical values is still influential but is rapidly waning: it is being realized that, apart from its intrinsic value, mathematics is valuable for factual science because it is an instrument for refining concepts, stating problems, formulating conjectures, deriving logical consequences and consequently for building theories. In short, mathematics is a tool for improving thought and its expression (see Sec. 8.2). Dictionaries and language philosophers cannot hope to compete with mathematics in the task of concept sharpening.

We have been taking for granted that well-reared concepts are preferable to unruly ones, but the opposite — namely the maximum vagueness of concepts — is desired by obscurantists. To an illuminist, concept refinement is both an end in itself and a means for further conceptual advancement. It is also commended for stopping quarrels and favoring agreement, but this is a naive view: the more vague the concepts involved in a context the easier it is *both* to start controversy and to come to terms, whereas the more definite the concepts are made the more shades of opinion will become manifest. (Just think of the many ways in which OCKHAM'S rule, "Entities are not to be multiplied without necessity", has been read.) Concepts are not refined in order to decrease the amount of dissent in the world but to increase the fruitfulness of research and argument.

Concept refinement is one aspect of the march of knowledge. The evolution of knowledge, in fact, does not consist in piling up new information — as some cartoons of science have it — but in a creative and selective process that pivots around the invention and improvement of ideas. In this process concept refinement plays an indispensable part: it does not replace the original invention of an obscure embryonic idea but it helps to develop it. This must be insisted on because the role of concept refinement has been exaggerated by some philosophers and neglected by many scientists. The former bias is understandable: after all, philosophers are chiefly concerned with ideas and are supposed to be professional conceptual analysts. And the exaggeration in the opposite sense is not surprising either in view of the widespread belief that science, in contrast to philosophy, is not concerned with symbols and their designata but with "solid facts" alone. Did not modern science and modern philosophy begin largely as a revolt against the hollow verbosity of a decadent scholasticism? The motto of the Royal Society of London still is *Nullius in verba* — "There is nothing in words"; this was the password of the Baconian philosophy, which purported to be "a philosophy of works, not of words". But this, of course, is a delusion: science is a system of ideas expressed by signs, and many problems in scientific research are disputes over the meanings of signs, whether verbal or not.

The recognition that a given dispute hinges around the specification of the meaning of a symbol may reorient the argument. Thus, the discussions concerning the question whether viruses are alive, whether bees have a culture, or whether automatic computers think, require a clarification of the meaning of the term 'life', 'culture', and 'thought' respectively. Very often, redefinitions must be proposed which attach more or less new meanings to old terms. If an altogether new concept is actually introduced a fresh term may have to be coined to name it. But neologisms must be introduced parsimoniously in order to prevent the unnecessary swelling of vocabularies. The multiplication of symbols may have no other goal that concealing vacuity — as is so often the case with protoscience and pseudoscience. Yet an excess of parsimony may lead to trouble. Thus, from the 1650's to the 1850's the term 'force' covered at least four different though related concepts: those of muscular force, mechanical force, energy, and power. Long and involved arguments, partly based on the ambiguity of the term, might have been saved if three new terms had been introduced. The final clarification did involve such a multiplication of words but was not the result of either philological search or logical analysis: it involved the discovery of the conservation of energy, which brought the concept of energy to the fore. Because, by the way, the more important concepts in factual science are those occurring in law statements; therefore the mere establishment and interrelationship of law statements is an effective procedure for sharpening the meaning of scientific symbols. This was not understood until recently.

In fact, during two millenia the Aristotelian doctrine has prevailed according to which (i) definition is *the* way meanings are attached to signs (words and nonverbal symbols) and (ii) the perfect form of definition is by genus and difference — as exemplified by "Two is the smallest even number", the class of even numbers being the genus and 'the smallest' designating the specific difference between 2 and the other members of its genus. Nowadays we grant the existence of a number of ways in which signs and the corresponding ideas can be made definite. We may specify the meaning of a sign in a more or less complete way and in a number of fashions: by exemplification, by informal and partial description of the sign's designatum (the so-called real definition), by classification, by definition, by building a theory, and prehaps in still other ways. We have realized that definition is but one such procedure and certainly a much less important one than theory building. Moreover, we have learned that the attempt to define every concept leads to circularity — as exemplified by dictionary definitions — and that the way to avoid such a vice in a given context is to start by admitting a set of undefined (primitive) concepts, which may be clarified by remarks and examples but mainly by the role they fulfil in the system, and which help define all the other concepts in the system. Yet the traditional prestige of definition is such that every procedure by which the content and function of symbols is rendered *definite* to some extent is too often called a definition.

Let us close the preceding general remarks on concept sharpening by grouping elucidation procedures from a semantical point of view. The meaning of a sign may be specified either in terms of further signs or by reference to nonlinguistic objects. In the former case we may speak of a *sign-sign* relation, in the latter of a *sign-fact* relation. A definition, such as "Insulator $=_{df}$ non-conductor", is clearly a sign-sign relation. On the other hand, "'*H*' is the symbol for hydrogen", far from dwelling entirely on the linguistic level, establishes a relation between a linguistic object and a physical object. Such meaning-giving relations between signs and nonlinguistic objects shall be named *referitions*. We shall distinguish two kinds of referition: sign-experient referitions — in which reference is made to experience — and sign-physical object referitions, which point to objective signs or events. Thus, "Let '$V(x, y)$' denote the subjective value (utility) assigned by a subject x to an object y" is a sign-experient referition. On the other hand, "Let 'E_n' denote the n-th energy level of the atom" is a sign-physical object referition. As we shall see, the very existence of referitions of the latter kind is denied by certain philosophies.

Let us first study definitions, then referitions.

Problems

3.2.1. Examine the account of concept formation given by SEXTUS EMPIRICUS in *Against the Professors*, Bk. III, in *Works*, transl. R. G. BURY (London: Heinemann, 1949), vol. IV. Note particularly his criticism of the empiricist conception of geometrical concepts involved in pseudodefinitions like "The surface is the limit of a body" and "The line is the limit of a surface". *Alternate Problem:* Discuss the Aristotelian doctrine of concept formation, adopted later on by empiricists, according to whom every concept is formed by abstraction from some experiential data.

3.2.2. Report on any of the following writings on elucidation: W. WHEWELL, *Novum Organum Renovatum*, 3rd ed. (London: Parker, 1858), pp. 50ff.; A. KAPLAN, "Definition and Specification of Meaning", *Journal of Philosophy*, XLIII, 281 (1946); R. CARNAP, "On Explication", Ch. I of *Logical Foundations of Probability* (Chicago: The University of Chicago Press, 1950); M. BUNGE, *Intuition and Science* (Englewood Cliffs, N.J.: Prentice-Hall, 1962), pp. 107ff. and *The Myth of Simplicity* (Englewood Cliffs, N. J.: Prentice-Hall, 1963), Ch. 1, "Metascientific Elucidation".

3.2.3. Trace the progressive elucidation of some scientific concept and point out its deformations if any.

3.2.4. Discuss the elimination of the problem of the vagueness of the pairs "tall" and "short" effected by replacing them by the single quantitative concept of height, the numerical variable of which ranges over a continuum of values between zero and infinity. Next discuss quantification in general as a procedure of concept refinement.

3.2.5. The vagueness of "body" in the context of ordinary knowledge is much reduced in physics, where "body" can be defined as anything that satisfies approximately the laws of classical mechanics. Give further examples of this kind and discuss theorification as a procedure of concept refinement.

3.2.6. Why do we seize on, and name, triangles, squares, spheres and other simple figures rather than any of the more complex and irregular forms given in experience? Could we not have started geometry with the latter?

3.2.7. Study the function of concepts that make wide generalizations possible, such as those of zero, empty set, unit set, improper point, and improper line.

3.2.8. Comment on the following text by W. HEISENBERG, cofounder of quantum mechanics, in *Physics and Philosophy* (London: Allen & Unwin, 1959), p. 55:" The concepts of classical physics are just a refinement of the concepts of daily life and are an essential part of the language which forms the basis of all natural science".

3.2.9. The concept of mass is elucidated in the contexts of the various theories of mechanics. Now, any precise measurement of the mass of a body will yield an inexact number, such as, e.g., $(1 \pm .001)g$. In other words, the numerical value of the mass of a body is known empirically, if at all, with some error. Is this an indication that the theoretical concept of mass is vague? Or that an indeterminateness occurs in the application of the theoretical construct "mass"? Or does it rather show that the numerical values of the exact concept of mass are known only to within an error?

3.2.10. Biological evolution can be characterized by two properties: *direction* and *rate*. While the concept of rate of evolution has been satisfactorily elucidated, the concept of direction of evolution is still vague. (The rate of evolution of a given family can be defined as the number of genera originated per million years.) Study this situation and attempt to elucidate the concept of direction of evolution. See G. G. Simpson, *The Meaning of Evolution* (1949; N. Haven: Yale University Press, 1960), Ch. XI. *Alternate Problem:* In the case of the preceding problem an elucidandum ("evolution") was elucidated in terms of two concepts one of which ("direction of evolution") is not clearer than the one it purports to sharpen. This situation is not new in science; thus the heat concept was elucidated in terms of the concepts of energy and temperature long before the latter was given a satisfactory elucidation (in the context of statistical mechanics). Draw some lessons.

3.3. Definition

Let us, first of all, make it clear that we shall not deal here with any of the popular connotations of 'definition', such as description, identification, classification, or measurement: we shall be concerned with a special technical operation bearing on signs: definition is, in fact, a *sign-sign correspondence* (see Sec. 3.2). In this strict sense a definition is a purely conceptual operation whereby (i) a new term is *formally introduced* into some sign system (such as the language of a theory), and (ii) the meaning of the newly introduced term is *specified* to some extent — i.e. to the extent to which the meaning of the defining terms is sharp.

Note, in the first place, the relativity of both the introduction and the meaning specification to a sign system: dictionary definitions are usually framed in the context of common knowledge whereas scientific terms are usually defined, if at all, in the context of scientific systems. Outside their proper context definitions may become pointless. In the second place, the introduction of a new term by way of a definition is *formal* in the sense that the new term may have emerged naturally and only after a long illegal existence may have been officially recognized and trimmed. Thirdly, definitions can make meanings definite on condition that the defining signs have some meaning — which does not seem to be the case of, say, "Ex-sistent *Dasein* is the letting-be of what-is" (M. HEIDEGGER).

The newly introduced term is called the *definiendum* — that which is defined — and the expression that defines it the *definiens*. For example, in "Phylogenesis $=_{df}$ Evolution of the species", the definiendum is "phylogenesis" and the definiens the right-hand member of the definition. The definiendum shall be *new in the system* concerned but it may be an old acquaintance from alternative contexts where it can occur with the same meaning or with a similar one. The definiens terms must, of course, preexist the definition, either by virtue of previous definitions or for having been adopted as ultimate definers in the context.

To start defining at all, a set of undefined or *primitive* concepts is needed. For example, in PEANO's system of arithmetic the concepts of natural number, successor of a natural number, and zero, are taken as primitives in addition to generic (logical) concepts such as those of identity, negation, conjunction, and universality, which are used in tying the specific primitives up to form statements and to transform the latter. In statics "reference system", "position", "body" and "force" can be taken as mutually independent primitives in addition to certain generic (logical and mathematical) concepts such as those of vector and vector addition. And in kinematics "time" occurs in the place of "force" in the preceding set of undefined concepts. In every factual context we must, in short, accept two batches of basic concepts: (i) a set of *generic* (logical and/or mathematical) primitives and (ii) a set of *specific* (subject-matter) primitives. Usually the former are silently taken over from formal science, and the latter are attached a meaning (in the system) by way of referitions that are mostly hinted at rather than explicitly stated.

To say of a concept that it is undefined or primitive in a given context is to characterize it negatively, hence incompletely. What is most important about primitives is not so much that they are definers and can therefore infuse meanings into other signs, as that they are the *building stones* of theories. In fact, every basic assumption or axiom of a theory is just a statement made up of generic and specific primitives alone. An important (metatheoretical) function of definition is, accordingly, that it points to a partition of the set of concepts of a system into the *basic* and the *derivative* ones. (Which shows, incidentally, that the proper place for a formal discussion of definition is the chapters on theory rather than this semantical introduction.) Accordingly one does not usually define the most important or basic concepts but rather the less important ones: those which, in principle, might be replaced by their corresponding definiens. Anyhow for a concept, as for an organism, the important thing is to be born: official recognition, either as a primitive or as derivative, is of secondary importance.

As regards logical form two kinds of definition proper or *nominal definition* must be distinguished: explicit and implicit. In explicit definition the definiendum and the definiens are neatly separated by a special sign; in implicit definition the two terms are distinguishable but inseparable: they are, so to say, integrated.

The standard forms of *explicit definition* are the following. (i) "$A =_{df} B$", i.e. A equals B by definition. Example: "Phenomenal $=_{df}$ Appearing to someone's senses". (ii) "'A'='B'", i.e. 'A' names the same object as 'B' does; or also: 'A' and 'B' are synonymous. Example: "'2'='1+1'". (iii) "$A=(\imath x)\,B(x)$", i.e. A is the sole x which satisfies the condition $B(\;)$, or has the property B. Example: "Temperature is the magnitude measured by thermometers". (iv) "$(x)\,[A(x)\leftrightarrow B(x)]$", i.e. A is true of every x if and only if B holds for every x. Example: "For every x, y and z in the set of numbers, $x/y=z \leftrightarrow x=y\cdot z$".

*The first two kinds of definition are *metalinguistic* or inter-level definitions: the sign '$=_{df}$' raises the definiendum to a metalanguage of the language in which the definiens resides, and the inverted commas in (ii) indicate that it is not the concepts A and B themselves but their respective names 'A' and 'B' that are at stake. Consequently, whereas in all other cases a definition may be interpreted either as bearing on signs or as bearing on the designated concepts, definitions of the class (ii) apply to signs only. Finally, the definitions of the last two kinds, (iii) and (iv), are *intralinguistic* or same-level definitions in the sense that both the definiendum and the definiens belong to be same language level: in fact, both the definite description symbol '$\imath$' and the equivalence symbol '$\leftrightarrow$' can relate only expressions belonging to the same language level — e.g. the object language, in which no reference to further linguistic objects is made. It is doubtful whether equivalences should be regarded as definitions proper: firstly, because they may not establish intensional identity and secondly because they can often be proved (they are not always conventions).*

In an *implicit definition* the definiendum does not occur alone on one side of an equivalence relation but is part of a more complex sign. Ex. 1: "Two bodies have the same electric potential (or the same temperature) if and only if no electricity (correspondingly: no heat) flows through a third conducting body connecting them". This equivalence defines the concept of equal electric potential (or of equal temperature). Ex. 2: The relation "square root" can be defined implicitly and intensionally in terms of the converse relation, "square", as: "$(y=\sqrt{x}) \leftrightarrow (y^2=x)$"; but it can also be defined, in an extensional way, explicitly: "$\sqrt{\ } =_{df} \{\langle x, y\rangle \mid y^2=x\}$", where '$\langle x, y\rangle$' stands for the ordered couple formed by x and y. The implicit definitions consisting in equivalences of some sort state conditions that are both *necessary and sufficient* for the introduction of a symbol. They shall be called *simple* implicit definitions. There are several other kinds of implicit definition, which allow some play between definiendum and definiens.

Conditional implicit definitions have a conditional form as, e.g., in "If C, then: A if and only if B)", where the clause C states a condition for the equivalence — a sufficient condition, of course. Thus, the implicit definition of arithmetical division, "$x/y=z \leftrightarrow x=y \cdot z$", is valid for numbers but not for every other mathematical concept. If the condition that x, y, and z be numbers — in short, "$x, y, z \in N$" —, is explicitly stated in order to indicate the scope of the definition, then we obtain a conditional implicit definition of arithmetical division: "$x, y, z \in N \rightarrow (x/y=z \leftrightarrow x=y \cdot z)$". It could be argued that every *complete* definition, whether implicit or explicit, is ultimately conditional, and that the antecedent can legitimately be dispensed with only when the context points to it anyhow.

Conditional definitions are sometimes rejected because any statement of the form "$C \rightarrow (A \leftrightarrow B)$" is automatically true, regardless of the satisfaction of the biconditional $A \leftrightarrow B$, if the antecedent C is false. But such vacuous, hence useless truths, can be avoided with the additional stipulation that C is being presupposed, i.e. taken for granted. That is, instead of '$C \rightarrow (A \leftrightarrow B)$' we may write: '$C \,\&\, [C \rightarrow (A \leftrightarrow B)]$'. With this proviso we can request that the condition of any definition be made explicit whenever necessary.

Recursive definitions are very common in logic and mathematics but still rare in factual science. A recursive definition introduces a term by relating it to one or more contiguous terms of a denumerable set or sequence. Thus, star brightness is defined by recursion thus: "$B_n = 2.5\, B_{n+1}$", where 'n' refers to the star magnitude. Recursive definitions are limited to members of denumerable sets, whence they could not be used for defining magnitudes. In addition they are incomplete because they require the independent specification of one of the members, e.g. the first; thus, in the case of star brightness the recursive definition leaves B_1 undetermined. This is not really a shortcoming if only the concept of relative brightness is wanted.

Inductive definitions, which contain recursive definitions as a subclass, are more definite than the latter because they fix the value of the first member of the sequence. An example of an inductive definition is that of the sum of natural numbers in PEANO'S system of arithmetic by means of the two expressions: "$n+O=n$" and "$n+Sk=S(n+k)$", where 'S' designates the function "successor of".

The above kinds of implicit definition give satisfactory characterizations of the definiendum. A highly ambiguous specification of meaning is the one provided by a set of postulates (axioms), i.e. the so-called contextual or *postulational definition*, which is alleged to apply to the primitives of a theory. The building of an axiom system certainly introduces the basic symbols and stipulates the basic relations among them, but it can hardly be regarded as a *definition* proper although it effects the best possible *elucidation*. In fact, any set of axioms can be interpreted in a number of ways and in every interpretation or model the primitives acquire a peculiar meaning — on condition that the appropriate interpretation rules are *added* to the axiom system explicitly or implicitly. The postulates *limit* the interpretation possibilities but they do not characterize the primitives of a theory in an unambiguous way: they do not make their meanings definite. The specification of the meaning of the primitives of a scientific theory is made via referitions (see Secs. 5 and 6).

What a set of axioms does characterize in an unambiguous way — provided the necessary referitions or correspondence rules are added — is the theory's *object* or *key concept*. Thus, the concept of set can be characterized by laying down the axioms of set theory (e.g., "$A \cap B = B \cap A$") and saying that anything satisfying these axioms is a set. Similarly the concept of electromagnetic field might be defined as stating that it is the referent of certain law statements (e.g., MAXWELL'S equations), the meaning of which is partially specified by a set of referitions (e.g., "'j' designates the current density"); in short, the concept of electromagnetic field (not its referent) is defined as that which satisfies a certain theory. Definitions of this kind are called *axiomatic definitions* and they are often used in mathematics. But if they are to be used in factual science a word of caution must be appended, namely, that the definition of a concept will not count as a description of the concept's referent. A second warning is that any given factual theory can provide only a temporary and contextual (hence partial) definition of its key concept; a change of theory may involve a modification of the key concept. This situation has no exact parallel in formal science, where the object of inquiry coincides with its key concept.

We may now deal with some important misunderstandings concerning the nature and function of definition in science. The first is the belief that no inquiry should begin unless its object is defined. For instance, an investigation on the nesting habits of mocking-birds should start by defining "mocking-bird", "habit" and "nest" because — so the contention runs — otherwise we would not know what we are talking about. This requirement is, of course, absurd if only because (i) we cannot define the most important terms, namely those which function as the building blocks (primitives), and (ii) we often start with vague concepts which are gradually elucidated through the inquiry itself — which would not be the case if the language of a science were supplied to it from the outside. What the propounders of the rule probably had in mind was that the object of inquiry must be *identified* to begin with. Clearly, if someone who has never seen a mocking-bird before is asked to investigate its nesting habits, he will not be able to gather much reliable information. But identification need not be based on definition: identification can be made with the help of descriptions and empirical tests.

A second widespread mistake is that *any equation* can serve the purpose of defining one of the members of the formula. That this is erroneous can be gathered from the following counterexample: "$3=\sqrt{9}$" would not be accepted as a definition of "3", nor would "$3=\log_2 8$", although both equations are true. If we want a definition of "3" we must begin by choosing some arithmetic system in which "3" occurs and find out which are the specific (extralogical) primitives of this theory. Once the primitives have been identified the defined symbols are introduced in an orderly way — which can be altered in an alternative theory. For example, in PEANO's theory we shall introduce "3" by means of the chain of definitions: "$1=_{df}S0$", "$2=_{df}S1$", $3=_{df}S2$"; but it would be possible, only inconvenient, to change the definition and even to choose "3" as a primitive. Exactly the same must be done in any axiomatized factual theory: it is only after listing the primitive concepts and the primitive formulas (i.e. the axioms) of the theory that we can set the task of determining the status of a given concept in the theory. To proceed arbitrarily, in isolation from a well-determined context, is absurd because the status of being primitive or defined is conferred by the theory as a whole, not by some arbitrarily chosen formula of it. In short, equations do not warrant definitions — unless they happen to be definitory equations; equations warrant only substitutions and computations. This point is important enough to deserve a more detailed discussion.

Magnitudes, such as density, offer peculiar difficulties. "Density" (the concept, not the property referred to by it) is usually defined as the ratio of mass to volume. More precisely the *average density*, designated by '$\bar{d}$', can be introduced in the following way:

$$\bar{d}=_{df}\frac{m}{v}. \qquad [3.6]$$

But clearly we cannot divide concepts — unless they happen to be numbers. What we do subject to arithmetic operations is the *numerical variables* involved in magnitudes rather than the whole magnitudes. Thus, we divide values of functions, not the correspondences themselves. That is, the symbols occurring in [3.6] cannot designate the full concepts of average density, mass, and volume, but their respective numerical variables. But then [3.6] does not supply a *complete definition* of the concept of average density: all it gives is a definition of the numerical variable of the average density in terms of the numerical variables of the mass and the volume of an arbitrary piece of matter. In other words, "average density" is a function of an object variable x which can take on as values names of specimens (or of classes) of substances; the numerical values of this function are the (nonnegative) real numbers $\bar{d}$. In short, $\bar{D}(x) = \bar{d}$. When we compute density values with the help of formula [3.6] we do not use the whole concepts behind the formula but just their numerical presentatives. Hence [3.6] is only a *partial definition* of "density". Moreover, there is no complete definition of the density concept and, what is worse (or rather better), none is needed. In many theories the density concept is adopted as a primitive. The meaning of 'density' is specified in an indirect way by the whole bunch of law statements in which its numerical variable occurs, and any one of them can be used to *compute* densities in terms of other magnitudes — but only [3.6] will give a partial definition of the density concept.

For theoretical purposes [3.6] is insufficient: we need the far stronger and more basic concept of *point density*, or density at a point of a material system. [3.6] cannot comply with this request because the volume of a point is zero and the division by zero does not exist. The introduction of the more general concept of point density permits then to obtain the concept of average density as a derived concept. *The numerical variable of the point density, if defined at all, is defined as follows: $d =_{df} dm/dv$. If $m = kv$, i.e. if the system is homogeneous, the point density coincides numerically with the average density. Otherwise the average density must be introduced in terms of the point density through the formula: $\bar{d} = (1/v) \int d(P) \cdot dv$, where '$dv$' stands for a volume element at P.*

The definability of the weaker concept "average density" in terms of "point density" entails that the former is dispensable for *theoretical* purposes; but it is indispensable for purposes of comparison of the calculations with measurement. In fact, the experimenter does not handle mass points but bodies with a nonvanishing volume: he measure average densities rather than point densities. (Similarly with velocity, acceleration, pressure, current intensity, and most other magnitudes.) The two concepts are needed and they are related both conceptually and through the empirical operation of measurement. The latter supplies, in fact, the following bridge: measuring the average density provides an *estimate* or approximate value (for a sufficiently small volume) of the exact but empirically inaccessible point density. The relation between the numerical variables of the two density concepts is

$$d \cong m(\bar{d}), \qquad [3.7]$$

where '$m(\bar{d})$' designates the empirically measured value of $\bar{d}$ and '$\cong$' symbolizes approximate equality. This situation is common in science, where two concepts of a given magnitude are needed: a *refined* and a *coarse* one, the former for theory, the latter for experience.

From the preceding discussion the following conclusions can be drawn. Firstly, there are no *full* definitions of magnitudes: only their numerical arguments can be defined by equations — as long as they are not arguments of primitive concepts, in which case the equations that may relate them serve the purpose of computation, not of definition. This fact should not be mourned except by those who identify "undefined" with "indefinite", "indeterminate" or "vague". Secondly, we never compute magnitudes, such as lengths of forces: all we do is to compute their numerical values or numerics. Thus, when we compute m from f and a by means of "$m=f/a$", what we do is to subject the numerical variables or quantities f and a to calculation. This is one of the reasons NEWTON's second law of motion, "$f=ma$", cannot be used to define the mass concept in terms of the concepts of force and acceleration: just as every other law statement involving magnitudes, NEWTON's law is a relation among the numerical variables of certain magnitudes; the other reason is that "mass" and "force" are primitives in Newtonian mechanics. Our third conclusion is that, strictly speaking, we never measure full magnitudes but only their numerical components. Again, this should not be a tragedy except for those who mistake computation and measurement for definition.

The third mistake we must face regards the proper way of introducing *dispositional terms* such as 'visible' or 'soluble', i.e. terms denoting potential rather than actual properties and relations. The received doctrine is that the way of introducing such terms is by way of a bilateral "reduction" statement of the form characterizing conditional definitions, i.e. "$C \rightarrow (A \leftrightarrow B)$" or, in case A, B and C are all one-place predicates bearing on the single object variable, x,

$$(x)\,[C(x) \rightarrow (A(x) \leftrightarrow B(x))]. \qquad [3.8]$$

The operationalist interpretation of this formula is as follows: "For every x, if x is subjected to the test condition C (e.g., put in water) then x is assigned the attribute A (e.g., solubility) just in case x exhibits the behavior B (e.g., dissolves)." The rationale of this proposal concerning the introduction of dispositionals is the empiricist prescription that predicates are all to be introduced by reference to empirical procedures. Statements like [3.8] would effect the "reduction" of the newly introduced concept A to the observational predicates B and C.

The above is an artificial problem typical of philosophies of science unconcerned with real science. It is artificial because scientists are not usually after the chimera of reducing theoretical terms to pretheoretical terms, and because they bypass the task of directly refining qualitative terms such as 'soluble'. What is done instead is to build quantitative theoretical concepts such as "degree of solubility" (or "acidity", "conductivity", "permeability", and so on). The corresponding qualitative concepts are subsequently defined, if at all, in terms of the stronger quantitative concepts. In the case of solubility the following chain of definitions can in fact be established:

(1) *Solubility* (of substance x in solvent y at temperature t and pressure p) $=_{df}$ *concentration* (of x in a saturated solution of y at t and p).

(2) x is *soluble* (in y_0 at t_0 and p_0) $\leftrightarrow$ [*Solubility* (of x in y_0 at t_0 and p_0) $> s_0$], where the subscript '$_0$' indicates a particular value, or rather an interval, of the numerical variable affected by it; in particular 's_0' is a conventional value of the degree of solubility. The magnitude "concentration" occurring in the first definition is in turn elucidated in a partial way through an explicit definition.

Something similar is done with all dispositionals that are not adopted as primitives. For example, in the theory of magnetism the dispositional "magnetic" is introduced in terms of the absolute permeability, the numerical variable of which is in turn defined by: $\mu =_{df} B/H$", where B and H are primitives of MAXWELL's theory. Once the numerical variable of the permeability has been so defined, the dispositional concept "magnetic" is defined in the following way: "x is magnetic $=_{df}$ the permeability of x is much larger than unity". Here again the definiens terms are quantitative and one of them is dispositional itself.

The following lessons can be derived from the preceding story. First, instead of painfully trying to refine a qualitative concept, scientists may prefer to introduce a stronger, quantitative concept right away. Second, rather than attempting to elucidate concepts with an objective referent with the help of anthropocentric concepts (such as "test condition C" and "observed behavior B"), as demanded by some philosophers, scientist try to frame elucidations with the help of concepts referring to physical objects. Third, the sharpening process is not always from the qualitative to the relational to the quantitative, but can be inverted. Fourth, some dispositional qualities or potentialities, such as solubility, can be reduced to actual qualities such as concentration. Fifth, elucidation is not done in a theoretical vacuum: in science definitions presuppose or involve laws and are built in the bosom of systems.

Consequently it should not be surprising that definitions suffer the fate of theories and, in general, of knowledge. The history of scientific ideas should be instructive in this respect, particularly for those who believe that definition is wholly arbitrary. Take, for instance, the concept of acid. BOYLE had identified (not defined) acids by a set of reactions: he had *tests*, hence *criteria* of acidity but, lacking a theory concerning the constitution and function of acids, he could offer no theoretical definition of "acid". (But at least he did not mistake his acidity tests for "operational definitions" of acidity.) LAVOISIER and others, one century after BOYLE, tried to catch the essence of acidity by finding some elementary constituent of all acids, but failed. The substantialist approach was eventually replaced by the functional one, and "acid" was defined as that which breaks down in aqueous solution to give positively charged hydrogen ions. Finally this definition was refined by introducing a quantitative concept of acidity, resting on three law statements. *First law: "All acid formulas are of the form: HA". Second law: "All acids break down in the form: $HA = A^- + H^+$". Third law: "At equilibrium, the ratio of concentrations, namely $[H^+] \cdot [A^-]/[HA]$, is a constant". The value of this constant, called K_{HA}, is adopted as the degree of acidity: strong acids are characterized by a large value of that constant. On the basis of the preceding definition, universal and quantitative acidity tests can be worked out: action becomes guided and explained by theory.*

Defining scientific concepts, then, is not always a purely linguistic business but often embodies theory and empirical information, so that changes in either may force changes of definitions. Consequently definitions must not only be *formally correct* but also *materially adequate*, both in the epistemological sense that the knowledge they presuppose or involve must be substantially correct, and in the pragmatic sense that the definitions conform, at least roughly, to actual expert (not ordinary) usage. This being so, the nature and function of definition in science cannot be assessed by logic alone: it involves epistemological and historical issues as well.

The function of definition in science will be discussed in the next section after certain somewhat delicate logical and semantical problems are cleared up.

Problems

3.3.1. Point out the definiendum and the kind of definition in the following cases.

1. $p \to q =_{df} -(p \,\&\, -q)$.
2. $(x = \log_b y) \leftrightarrow (y = b^x)$.
3. $x \neq 0 \to x^0 =_{df} 1$.
4. $(0! = 1) \,\&\, [(n+1)! = (n+1) \cdot n!]$.
5. $B = \text{curl}\, A$, where B is given.
6. Something is a group if and only if it satisfies the following axioms [here a list of postulates of group theory follows].
7. A genus is the union of its species.
8. The term 'hot' is synonymous with the expression 'having a high temperature compared to the temperature of the human body'.
9. The fittest organisms are the best adapted to their surroundings.
10. "Knowledge is true opinion" (PLATO).

3.3.2. Analyze the following definitions

1. An instant is the frontier between two time intervals or moments.
2. $0 =_{df} (\imath y)\,[(x)\,(x + y = x)]$.
3. "Sensation is the feeling of first things" (W. JAMES).
4. "The meaning of a term is its use" (L. WITTGENSTEIN).
5. "Intelligence is what intelligence tests measure" (several operationalist psychologists).

3.3.3. Analyze the expressions, often found in the scientific literature, 'The function f defined by $f(x) = y$', and 'By means of a position measurement we define the particle's position'. *Alternate Problem:* Consider an infinite series the individual terms a_n of which are unknown, only the ratio of any two successive terms being given. Suppose further that this ratio is constant, i.e. independent of n: $a_{n+1}/a_n = c$. Point out what kind of definition this equality supplies and show that it suffices to determine the sum of the series (up to the initial term a_1 which is left indeterminate).

3.3.4. Painters and aestheticians have argued whether white and black are colors. Similarly, mathematicians in the past argued whether 0, $\sqrt{2}$ and $\sqrt{-1}$ are numbers. Propose a way out. *Alternate Problem:* Clever experiments have been designed to find out whether the learning of conceptual subjects involves both understanding and recall. Were the experiments necessary?

3.3.5. Report on the first modern treatment of definition, PASCAL'S small treatise "De l'esprit géometrique" (1657?), in *Oeuvres complètes* (Paris: Ed. du Seuil, 1963). The need for primitives is here shown and "real definitions" are criticized for not being free (conventional) but controvertible: "definitions are made only to designate the things one names, not to show their nature" (p. 350). *Alternate Problem:* Report on the views on definition expounded by W. WHEWELL in *Novum Organum Renovatum*, 3rd ed. (London: Parker, 1958), pp. 30—40, where he maintains that many controversies in the history of science have been battles over definitions, but "these controversies have never been questions of insulated and *arbitrary* Definitions, as men seem so often tempted to suppose them to have been" (p. 36). "When a definition is propounded to us as a useful step in knowledge, we are always entitled to ask what Principle it serves to enunciate" (p. 37).

3.3.6. Examine the following opinions concerning NEWTON's second law of motion, "$f = ma$". (i) The law may be regarded as a definition of "force" in terms of "mass" and "acceleration". (ii) The law can be regarded as a definition of "mass". (iii) The law can be regarded as the source of an "operational" definition of relative mass in terms of relative acceleration, in the case of a pair of bodies interacting with the same force, since in this case by NEWTON's third law $f_1 = -f_2$, whence (by the second law) $m_1 a_1 = -m_2 a_2$, from which it follows that $m_1/m_2 = -a_2/a_1$. For discussions on this problem, see H. MARGENAU, *The Nature of Physical Reality* (New York: McGraw-Hill Book Co., 1950), Ch. 12, E. HUTTEN, *The Language of Modern Physics* (London: Allen &Unwin, 1956), Ch. iii, Sec. 1, P. SUPPES, *Introduction to Logic* (Princeton: Van Nostrand, 1957), and M. BUNGE, "MACH's Critique of Newtonian Mechanics", American Journal of Physics, **34**, 585 (1966).

3.3.7. Find a satisfactory definition of "biological species" which could be used as a criterion for the unambiguous identification of biological individuals. *Alternate Problem:* Elucidate the concept of belonging to the same species S. Show that it is an equivalence relation on S (reflexive, symmetric and transitive). And show that species are equivalence classes.

3.3.8. To a first approximation the melting point of a substance may be defined as the temperature at which it melts: the quantitative concept of melting point is thereby made to depend on the theoretical concept of temperature and on the observational concept of melting. Does that definition presuppose any law? And what about the fact that some substances have no single melting point but melt within temperature intervals as large as five degrees? Should we change the definition of "melting point" or could we replace this somewhat vague concept by a stronger one? *Alternate Problem:* Frame a definition of "happiness" and check whether it is materially adequate in some culture. Try to make it quantitative and see whether it can be universal (cross-cultural).

3.3.9. Examine the so-called *denotative definitions*, consisting in naming some or all of the members of the extension of a concept, as in the case of "Frege and Russell are logicians". Granting that exemplification and even complete enumeration when possible are procedures for the specification of meaning, can they be regarded as definitions? And what conditions must be met by denotative characterizations in order to be exhaustive? *Alternate Problem:* K. MENGER has proposed to define "length in inches", or briefly "l_{in}", as the class of all pairs $\langle x, l_{in}(x)\rangle$. See, e.g., F. HENMUELLER and K. MENGER, "What is Length?", *Philosophy of Science*, **28**, 172 (1961). Taking into consideration that the domain B of the function "l_{in}" so defined is the set of all bodies, and the range R^+ of the function is the set of all nonnegative reals, we may reword the proposed definition as: "$l_{in} = B \times R^+$". Does this analysis qualify as a definition, i.e. does the formula unambiguously specify the meaning of "l_{in}" or does it rather determine its total extension? In general: do denotative (extensive) characterizations of magnitudes, however exhaustive, qualify as definitions? Hints: Recall the definition of "definition" and see whether there is exactly one property common to all the members of the set $B \times R^+$.

3.3.10. Study the relation between definitions and criteria, first in formal science (e.g., convergence criteria), then in factual science (e.g., stability criteria). *Alternate Problem:* Could every dispositional quality be elucidated in terms of actual qualities, thus inverting the Aristotelian primacy of potentiality over actuality?

3.4. *Problems of Definition

Let us now tackle some problems concerning definition. In the first place let us ask what properties identify a good definition. Some of these properties were mentioned in the preceding sections but shall be looked at now from a different angle.

A first formal requirement is that definitions should be *consistent* both internally (self-consistent) and with the corpus in which they occur. That a contradiction may easily creep into a definition is clear: in an equivalence like "$A \leftrightarrow (B \,\&\, C)$", C might be inconsistent with B, whence A would be false no matter what truth value B were assigned. For example, an archaic philosopher might wish to define 'is moving' as 'is at rest and is not at rest'. It is almost as clear as this that a definition might inadvertently clash with a part of the body of knowledge in which it occurs. Thus, in the above equivalence 'B' might stand for "c is dead" and 'C' for "c is thinking". The conjunction of these propositions is allowed by logic but prohibited by factual science: thinking is *factually inconsistent* with being dead. In short, then, a proper definition is internally and externally, logically and factually consistent.

In the case of operation symbols, formal consistency is in part ensured by the *uniqueness* of the operation, which must in turn be warranted by some previous formula (axiom or theorem). Suppose we introduce the operation symbol '$\#$' in the set of numbers by means of the implicit definition: "$(x\#y=z)\leftrightarrow(x^2=y)\ \&\ (z<y)$". By choosing $x=2$, $y=4$ and $z=0$, a triple satisfying the second hand member, we obtain: $2\#4=0$ in the first hand member. But the triple $x=2$, $y=4$, $z=3$ satisfies the equivalence as well and yields $2\#4=3$. Replacing this in the former result we obtain: $0=3$, an instance of the contradiction "$0\neq 0$". That is, $\#$ is not a unique operation.

A second syntactical requirement is that explicit definitions shall set up identities or formal equivalences, so that the definiendum can in all cases be *exchanged* with the definiens with preservation of the truth value. (Propositional equivalences, i.e. statements of the form "$p\leftrightarrow q$", may be interpreted or used as definitions even though, by themselves, they are not linguistic conventions: they usually occur as theorems in some theory.) Thus, in the arithmetic of integers 'The successor of one' can be replaced by '2' in every formula: the two signs designate the same concept. On the other hand, 'Two is a successor of one' cannot be so used: being an indefinite description, not a definite one, it does not establish a full equivalence.

A stronger version of the exchangeability condition is usually proposed: it is claimed that in every definition the definiendum and the definiens ought to be freely exchangeable not only without either change or gain of truth value (*salva veritate*) but also without shift of *meaning*. When these two conditions are satisfied — as they are by "$2=_{df}1+1$" — the definition effects just an abbreviation of discourse and entails the *eliminability* of the defined symbol. But this pragmatic function is not what makes definitions interesting: definitions are most valuable when they introduce *new concepts*, i.e. when they involve changes in meaning. Thus, consider the definition of the quotient of integers in terms of their product by means of "$x/y=z\leftrightarrow x=y\cdot z$". All those triples of integers which satisfy the left-hand member satisfy also the right-hand member; but in the definiendum a new concept, with properties of its own, occurs which is absent from the definiens: namely, the concept of fraction (rational number). Consequently we do not impose on definition the double condition of extensional and intensional invariance but only the former: we require exchangeability in the extensional sense, not in the intensional sense. We shall return to this point.

The semantical partner of the requirement of extensional equivalence or exchangeability is the condition of *non-creativity*, which commands to introduce no new hypotheses by way of definition. (We do not adopt the strong version of the non-creativity condition, according to which no new concept should be introduced by way of definition.) In other words, a definition shall not increase the content of a system; the whole content of a system should be in its premises (axioms and auxiliary assumptions, such as data), and definitions should only facilitate the working out of the logical consequences of such premises. In brief: if a "definition" is necessary to prove a statement, then it is not a definition. Let us clear up two misunderstandings in this respect.

In the first place, the definitions of a theory are treated, for purposes of deduction, as premises on an equal footing with the theory's hypotheses: in fact, a number of branches in a deductive tree are apt to be definitions: while in some cases definitions enable us to *reword* premises, in others

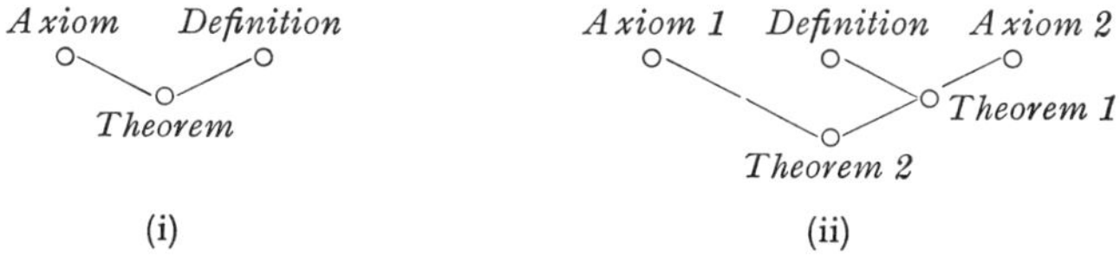

Fig. 3.4. A common situation: theorems deduced from prior assumptions and definitions. (i) Reformulation. (ii) Bridging

they enable us to build *bridges* between some of them (see Fig. 3.4). But in every other respect definitions have a status of their own, which is lower than that of the initial assumptions (axioms): it is epistemologically lower because they supply no new knowledge. In short, definitions proper are *non-creative*.

In the second place, the axioms of a theory contribute to elucidate the meaning of the theory's basic (primitive) concepts, but this does not entail that the axioms of a branch of science, such as mechanics, are just *disguised definitions* — as conventionalism holds. If they were they could not be contradicted by experience and there would be no point in trying to improve them: in fact, the history of science would resemble the Shakesperean idiot's tale. Moreover, *any* formula may be used for the partial specification of the meaning of some of the terms occurring in it, but the point is that not every specification of meaning amounts to a definition. Thus, the singular statement "This is a book on metascience" might be used to specify the meaning of 'metascience' but it is not a definition, as it establishes no equivalence between signs (in fact, it is a referition). In short, postulates and definitions are not exchangeable but an extralogical postulate will do more than supply information about the world: it will also cast some light on the meaning of the terms occurring in the sentence expressing it.

W. V. QUINE and N. GOODMAN have proposed an ingenious device by which synthetic propositions (e.g., the axioms of a factual theory) can apparently be eliminated in favor of definitions. Suppose we have a theory of heat flow containing the extralogical primitive "F" having the structure of a binary relation. The meaning of 'F' may be specified in the following way: If any two objects, x and y, stand in the relation F — that is, if $F(x, y)$ holds — then heat flows from x to y, and if $-F(x, y)$ then heat does not flow from x to y. Let us now postulate that there are pairs of bodies such that no heat flows between them, i.e.

$$(\exists x)(\exists y)[-F(x, y) \,\&\, -F(y, x)] \quad \textit{Existence postulate.} \qquad [3.9]$$

This axiom warrants that a certain nonempty set exists, namely the set of pairs of bodies among which no heat flows. This assumption backs up or justifies the introduction of the concept of thermal equilibrium, designated by the two-place predicate 'E':

$$E(x, y) =_{df} -F(x, y) \,\&\, -F(y, x) \quad \textit{Definition.} \qquad [3.10]$$

We next *feign* to forget about the physical existence postulate [3.9] and start the following reasoning. Instead of regarding "F" as a primitive we start with "E". Next we postulate that E is symmetrical, i.e. we assert the following law statement:

$$(x)(y)[E(x, y) \to E(y, x)] \quad \textit{Law statement.} \qquad [3.11]$$

Finally we remark that this postulate is redundant. In fact, the symmetry of E follows from the definition [3.10] as is easily seen by exchanging x and y in it and remembering the commutativity of the conjunction. At first sight we conclude, then, that the law [3.11] can be replaced by the definition [3.10] and that the corresponding empirical test procedures can be replaced by pencil and paper operations. With some patience the whole of factual science could be so reformulated as a set of definitions.

If the above procedure were flawless the principle of non-creativity would be violated and, what is more, factual science might be rebuilt so as never to require empirical tests: the checking of logical correctness would take their place. Since this consequence sounds unlikely something must be wrong with the above procedure. What is it? Simply that [3.11], cut off from the preceding considerations, is not a physical law but an empty formula: only an existence postulate warranting that the concept "E" has a real referent (a nonempty extension) can transform [3.11] into an extralogical postulate. Such an existence postulate could be: "$(\exists x)(\exists y)E(x, y)$", or, even shorter, "$E \neq \emptyset$" which, by virtue of the definition [3.10], is equivalent to the postulate [3.11]. (That an existence postulate must be adjoined to every universal formula in order to get a universal law proper will be argued in 6.4.) We conclude, firstly, that *isolated* generalizations, not backed up by (explicit or tacit) existential statements, can be replaced by definitions, but that such statements do not qualify as extralogical postulates (e.g., laws) proper. Secondly, that no term should *arbitrarily* be introduced in science either as a primitive or as a defined symbol: it must either fulfil a useful syntactical function or have a possible denotation, i.e. a possible real referent. (This stricture does not hold, of course, for language games.)

A second semantical requirement is that the connotation (intension) of the defined term should match with its denotation (extension) — or, stated negatively, definitions should not consecrate unnecessarily Pickwickian senses. For example, if a definition of "mind" is set up, it shall not be assignable to oysters or to computers, because neither of them are entitled to have mental functions. This is not a matter of usage or convention but of theory: we refuse to tax oysters and computers with a mind not because of any linguistic habit, i.e. not on pragmatic grounds, but because oysters and computers lack the organ of mental functions, namely, a brain. This borders upon the next requirement.

A third semantical requirement refers to definitions that enlarge the meaning of a term already in use: the generalized concept should reduce to the narrower one in the latter's proper field. For example, if "temperature" and "entropy" are defined for nonequilibrium states, these enlarged concepts shall have to coincide with those of equilibrium temperature and equilibrium entropy. The requirement under consideration is a sort of correspondence principle for concepts, which regulates their generalization. (For the concept of concept generality, see 2.3.) This will do for the semantical requirements on definition.

A pragmatic requirement is that definitions should be *fruitful*, either because they save time (practical fertility) or because they establish relations among concepts and contribute thereby to systemicity (theoretical fertility). For example suppose that, instead of the (implicit) definitions of the basic trigonometric functions, namely "$\sin\vartheta =_{df} y/r$" and "$\cos\vartheta =_{df} x/r$", we were to propose either of the following pairs:

$$\begin{aligned} &\sin\vartheta =_{df} 2^{-\frac{1}{2}}(x \pm y), \qquad &&\sin\vartheta =_{df} 2^{-\frac{1}{2}}(x \pm iy) \\ &\cos\vartheta =_{df} 2^{-\frac{1}{2}}(x \mp y) &&\cos\vartheta =_{df} 2^{-\frac{1}{2}}(x \mp iy) \end{aligned} \qquad [3.12]$$

with $i =_{df} \sqrt{-1}$. The four pairs of definitions satisfy the formal conditions of consistency, exchangeability, and non-creativity. Moreover, they are consistent with the Pythagorean theorem, "$x^2 + y^2 = r^2$", the supreme principle in plane trigonometry. Finally they are fertile in a trivial sense: they give rise to a number of theorems analogous to those of the usual plane trigonometry. Yet the four new theories using any of the alternative definitions and the Pythagorean theorem, while formally correct and isomorphic to the usual theory, will miss the whole point of the latter, which is to be a simple tool for relating and computing angles in terms of the sides of right-angled triangles.

Again, suppose physicists had chosen to baptize the expression '$m^5 v^{\frac{8}{7}}$' with a special name. They would not have been justified because that expression plays no role in fundamental theory although it may well occur in some of its applications. What are given special names are the concepts of linear momentum "mv" and kinetic energy "$\frac{1}{2}mv^2$": these are fruitful concepts and so are the corresponding definitions. They are fruitful because they occur in law statements of fundamental theories and they denote physical properties. If a symbol performs no syntactical function and designates no property, science has no use for it. In short, definitions are not framed capriciously but in the face of definite theoretical or practical needs.

Let us finally lay down a condition with an ontological and a methodological import, namely: Whenever possible *define the higher by the lower or peer*. In particular: Do not define physical concepts in terms of physiological or psychological concepts. Consider the following definition of "gas": "A body is a gas [or is gaseous] if and only if it affects neither our sense of touch nor our muscular sense but can be sensed if inhaled in the absence of air". This definition is formally correct and didactically useful but it would not be accepted in physics because it is anthropocentric: it reduces a low level property to a set of high level properties (sensory or secondary qualities). Ever since the beginning of the modern period a tacit condition on scientific definitions has been that the ontological status of the definiens should be lower than or equal to that of the definiendum. A naturalistic world view underlies this requirement — which goes to show, once more, that in science definitions are not always purely linguistic conventions and that science is not philosophically neutral.

The above injunction does not refer to the epistemological status of the terms related by a definition. The definiens need not be more immediate, epistemologically less abstract than the definiendum. It is only when a definition is framed with a predominantly didactic aim, i.e. when used mainly as an *explanation of meaning*, that the definiens is to be closer to experience or common knowledge than the definiendum. (Exclusive attention to this psychological function of definition is a source of misunderstandings.) But when "water" is defined as that substance which has the composition formula "H_2O", a low level (epistemologically speaking) concept is defined in high level terms. Definitions allow us sometimes to climb the ladder of abstraction, others to descend it. Which shows that, although the definiendum is logically equivalent to the definiens, there need be no epistemological equivalence between the two. Accordingly the exchangeability of the definiendum and the definiens (syntactical requirement) does not entail the *dispensability* of the defined symbol: both terms may be needed for different purposes.

We need 'water' to deal with experience, we need 'substance with the composition formula "H_2O"' to understand experience; and we need the equivalence established by the definition in order to link theory with experience. Moreover, the statement "Water is the substance with the composition formula 'H_2O'" may be regarded either as a nominal definition *or* as a "real definition", i.e. characterization: in the first case "water" (the concept) will be at stake, the substance water in the second case. In fact, the proposition may be taken to establish an equivalence between two concepts or it may be regarded as an assumption that might be false — and that would certainly have been regarded as false in the beginnings of modern chemistry, when the formula of water was taken to be "HO". In general, what a statement is depends not only on its structure and referent but also on the function it performs, and this depends in turn on the aims of its user. Only believers in the magical power of signs can hold that signs have a value inherent in them.

Again, consider the concept of linear momentum, which in Newtonian mechanics is introduced by means of the explicit definition "$p =_{df} mv$". Syntactically 'p' can always be exchanged for 'mv' (as long as we restrict ourselves to Cartesian coördinates). But 'p' cannot be *eliminated* because it has a meaning of its own: it occurs by itself in the law of conservation of linear momentum and in some cases the numerical value of p can be measured directly, i.e. independently from mass and velocity measurements — for example, in the case of a charged particle in a magnetic field. We need the three concepts, "m", "v", and "p", each in its own right, because they refer to distinct properties of physical systems. That this is so is further confirmed by the fact that, in more general theories, a new, more general concept of linear momentum is introduced independently from v.

Definition is misunderstood when the *exchangeability* of concepts is interpreted as the *eliminability* of at least half of them. In fact, the whole point of definition is the introduction of *new symbols* (which occasionally designate new concepts) in such a way that *no new formulas* are produced (in keeping with the principle of non-creativity). Thus, we can always exchange the formula "$x-y=z$" with the equivalent statement "$x=y+z$" and, in fact, we use this equivalence to define (implicitly) subtraction in terms of addition. But the formal equivalence of the statements does not preclude the introduction of the *new concept* of negative number — and this is why we take the trouble to set up the definition, namely in order to formally introduce that new concept. Negative numbers are not eliminable (dispensable) just because they are definable in terms of positive numbers: they substantially enrich the class of numbers, first by themselves, then by leading (jointly with the square root operation) to imaginary numbers. Only the subtraction *operation* is theoretically (not pragmatically) dispensable. Once a new concept has been introduced, whether as a primitive or by means of a definition, it may acquire a "life" of its own and may render a whole new branch of research possible.

It is time to assess the place of definition in science. Definitions are neither all-important as was believed formerly nor trifles, as the doctrine of eliminability holds. The functions of definition in science seem to be chiefly the following.

(i) *Formation (introduction) of new signs.* This may be done mainly with the end of abbreviating or simplifying expressions, as when '&' is used as an abbreviation for 'and'. The notational economy that is often achieved by this means enables us better to grasp complex concepts as units: just think of the psychological advantage of dealing with '$P_n(x)$' instead of with '$a_n x^n + a_{n-1} x^{n-1} + \cdots + a_0$'.

(ii) *Formal introduction of new concepts* on the basis of old concepts. Think of the infinity of concepts generated out of the sole basic (primitive) concepts "zero" and "successor" (or, alternatively, "zero" and "+1").

(iii) *Specification of meaning:* defining is a way of determining the meaning of terms which may have been in presystematic usage before. It is not the sole or even the best procedure, if compared with the building of a whole theory embodying the concept, but it is an effective procedure.

(iv) *Interrelation of concepts:* by linking concepts, definition contributes to organization or systemicity.

(v) *Identification of objects:* together with descriptions (in particular "real definitions") definitions supply criteria for the recognition of objects. Thus, to ascertain whether a given number is a multiple of 7 we divide it by 7 and see whether an integer results, thereby using the definition of 'multiple of 7'.

(vi) *Logical hygiene:* ambiguity and vagueness can be reduced if the terms are defined. But definitions cannot, of course, eliminate whatever ambiguity and vagueness there is in the primitive symbols.

(vii) *Precise symbolization* of certain concepts, and consequently exact analysis of them. Definitions can be used for the symbolization of slippery concepts, such as that of formal existence. Thus, we may say that *x exists formally in the system S* (language or body of knowledge) if and only if x is a primitive (symbol or formula) of S, or x satisfies a definition $D(x, S)$ in S:

$$x \text{ exists in } S =_{df} x \text{ is a primitive of } S \vee D(x, S). \qquad [3.13]$$

In this formula we have made use of the idea that *definability is relative* to a system or context rather than inherent in the definiendum. The primitives of a system are undefinable in it but they may eventually be defined in alternative systems. This triteness has been disputed by philosophers such as G. E. MOORE who claim that there are absolutely undefinable terms, namely those which designate aspects of immediate experience, such as 'yellow' and 'good'. The rationale of this claim is the dream of constructing the world from experiential primitives. That this is just a nightmare is shown elsewhere (Secs. 2.6, 3.7, 5.4, 8.1, 8.5, etc.); that the thesis of absolute undefinability is wrong, and precisely in relation with yellowness and goodness, is shown by the fact that these concepts can be defined in contexts other than ordinary knowledge. For instance, 'yellow' can be defined in physiological optics as the predicate designating the sensation produced in the human eye by electromagnetic waves in a certain wavelength interval; and 'good' may be defined, in the context of value theory, as positively valuable. What is true is (i) that these terms are primitive in ordinary language, and (ii) that a definition of 'yellow' will not help the color blind having color experience, just as a definition of 'good' might not help the value blind in case he existed. Definitions do not replace experience and convey no new knowledge (save psychologically) — nor do they claim to, even though they embody some experience and some awareness of experience. To proclaim the absolute undefinability of certain terms, i.e. the impossibility to define them in any context, is sheer dogmatism. Definability is relative to context and undefinability must be *proved* before it is proclaimed; moreover, it can be proved by PADOA's technique for proving the independence of concepts (see 7.6).

This ends our account of definition; our next task is to study interpretation and interpretation procedures, i.e. referitions.

Problems

3.4.1. Complete, with the help of logic textbooks, the list of conditions for proper definition laid down in the text and justify them. *Alternate Problem*: Propose an example of a biconditional whose two members fail to mean the same thing, and see whether it qualifies as a definition.

3.4.2. Examine the rule "Begin every discourse by defining your key terms", or the weaker rule "Leave no basic term undefined". *Alternate Problem*: Show that the definition of angle between two vectors must be preceded by the proof that the inner product of the vectors is less than or equal to the product of their lengths (Schwarz inequality). Draw some moral.

3.4.3. Examine the so-called real definitions, i.e. characterizations of objects by enumeration of a number of their properties. Are they definitions? And do they supply the essence of the object referred to by them?

3.4.4. Analyze the "definitions" by necessary conditions and those by sufficient conditions, with particular reference to "living" (by contrast with "inert") and "man" (by contrast with "automaton"). Do they satisfy the requirement of exchangeability? Apply the result of your inquiry to an analysis of the following statement by biologist G. WALD in "Innovation in Biology", *Scientific American*, **199**, No. 3 (1958), p. 113: "Biology long ago became convinced that it is not useful to define life. The trouble with any such definition is that one can always construct a model that satisfies the definition, yet clearly is not alive". *Alternate Problem:* Find out what are the conditions for definitions not satisfied by working definitions (see 2.3).

3.4.5. Do definitions bear on signs or on concepts? *Alternate Problem:* Are nominal definitions pragmatically arbitrary (e.g., unmotivated)?

3.4.6. Examine the doctrine that every definition is a purely linguistic convention which neither presupposes knowledge nor involves it, its only functions being contributing to the neatness of language and the economy of discourse. Trace the origin of this doctrine back to T. HOBBES' *De Corpore* (1655) and find out whether it was superior to the Aristotelian doctrine which it opposed.

3.4.7. Synonymity is often regarded as reducible to definability. That is, the equivalence "x is synonymous with y if and only if x is definable by y or y is definable by x" is often asserted. Would this be consistent with the view that "synonymity" is a semantical concept whereas "definability" is a syntactical one? *Alternate Problem:* Are equivalences definitions or do they underlie some definitions?

3.4.8. Examine the conventionalist doctrine that the axioms of physics are disguised definitions, hence conventional and empirically untestable. See H. POINCARÉ, *Science and Hypothesis* (London: Walter Scott, 1905), Ch. VI, E. LE ROY, "Un positivisme nouveau", *Revue de métaphysique et de morale*, **9**, 143 (1901), and P. DUHEM, *The Aim and Structure of Physical Theory* (1914; New York: Atheneum, 1962), pp. 208ff. *Alternate Problem:* Examine W. V. QUINE and N. GOODMAN'S "Elimination of Extra-Logical Postulates", *Journal of Symbolic Logic*, **5**, 104 (1940), and W. V. QUINE'S "Implicit Definition Sustained", *Journal of Philosophy*, LXI, 71 (1964).

3.4.9. For a battery-fed metallic circuit at ordinary temperatures Ohm's law "$e = Ri$" holds to a first approximation. In this formula R is a constant characteristic of every kind of conductor and is called the material's electric resistance. The usual definition of "R" is by means of Ohm's law, namely thus: "$R =_{df} e/i$". Recall that, in this formula, 'R' occurs as nothing but an abbreviation of 'e/i'. Is this correct? Why do we interpret the sign 'R' as denoting the physical property of resistance? Is the interpretation supplied by the definition? And does the definition enable us to dispense with the resistance concept?

3.4.10. Report on recent developments in the theory of definition. See in particular A. Tarski, "Some Methodological Investigations on the Definability of Concepts" (1934), in *Logic, Semantics, Metamathematics* (Oxford: Clarendon Press, 1956), and K. L. de Bouvère, *A Method in Proof of Undefinability* (Amsterdam: North-Holland, 1959). *Alternate Problem:* Can every implicit definition be converted into an explicit one? See A. Robinson, *Introduction to Model Theory and to the Metamathematics of Algebra* (Amsterdam: North-Holland, 1963), pp. 117—118.

3.5. Interpretation

We interpret a fact when we explain it, and we interpret an artificial sign (a symbol) when we find or stipulate what it means in a certain context. And an artificial sign means, if anything, what it stands for, i.e., its designatum. The *designatum* of a symbol is, in turn, a conceptual or a physical object or, more generally, a set of objects. Accordingly, meaningful symbols are those which designate ideas or facts, whereas meaningless symbols designate none. The designation relation may be one-one or one-many: in the latter case it is ambiguous. Any of the designata of an ambiguous symbol may be called a *sense* of it. Thus, one of the senses of 'fossil' is "remains of a living being", another is "colleague unwilling to retire". In particular, a unit sign or *term* is meaningful if it designates a nonempty set; the designation is unambiguous if the set is a unit set. And a *sentence* will be meaningful if it stands for a set of propositions; the sentence will be ambiguous unless it stands for a single proposition and meaningless if it stands for no proposition.

Meaning is contextual, i.e. relative rather than intrinsic and absolute. For example, the Spanish word 'tonto' is meaningless in English: it stands for no idea in this context. But 'tonto' means in Spanish the same as 'silly' does in English. Likewise, whether a sentence is meaningful or not depends on the context in which it occurs. Thus, 'The moon is sad' is meaningless in astronomy, which does not contain the sadness concept. But in poetry, by virtue of the conventions regulating metaphors, the above sentence is an ambiguous designation of a set of propositions, one of which is "It makes me sad to look at the moon now".

In other words: the same concept C is designated by 'tonto' in Spanish and by 'silly' in English, i.e. C is the meaning of s in S and also the meaning of e in E: $C = Mean(s, S) = Mean(e, E)$. On the other hand, $Mean(s, E) = Mean(e, S) = \emptyset$. In the case of the poetical sentence s, designating the set p of propositions, we have: $p = Mean(s, P)$, but $Mean(s, A) = \emptyset$. In general, for a sign s in a sign system or language L, we shall have a set i of ideas, which may be an empty set, corresponding to s: $i = Mean(s, L)$. Change the context L and the idea behind s, if any, may change as well. Not even the symbols of logic have an absolute meaning: they may have different meanings in different logical theories.

The preceding sentences have dealt with one of the concepts of meaning covered by the term 'meaning': namely, with one of the terms of the relation of designation: we have now, in fact, identified the meaning of a sign with the idea it stands for. There are, however, alternative concepts of meaning, i.e. other senses of 'meaning'. In Sec. 2.3 we defined the meaning of a term as the pair formed by the intension and the extension of the concept designated by the term. A third concept of meaning relevant to our discussion is tied up with the nature of the extension. We have seen (Secs. 1.4 and 2.1) that ideas may be pure or not pure: they may be self-contained or they may point to nonideal objects. The idea of number is of the first kind, whereas the idea of atom is of the second kind. In a third sense, then, a sign will be meaningful if and only if it designates an idea which, in turn, has a nonideal referent; referent-less signs, on the other hand, are meaningless in this sense. All three senses of 'meaning' are relevant to metascience.

It will therefore be convenient to introduce a fourth, more *general concept of meaning* subsuming the former three, one bearing on symbols, the ideas they stand for, and the latter's referents if any. This concept of meaning will be introduced effectively (not formally) by means of the following chart.

Class 1 is that of *meaningless* symbols, i.e. signs without designata and, a fortiori, with no referent. Class 2 is that of *formally meaningful* symbols, i.e. signs that stand for logical or mathematical ideas. Thus, e.g., '·' is formally meaningful in arithmetics, where it stands for or means the concept of arithmetic product; in this case the analysis of meaning stops here because the concept "·" refers to no empirical operation and to no event, at least in the context of arithmetic. Class 3 is that of *empirically meaningful* symbols, i.e. those designating ideas that, in turn, have an empirical reference; in particular, empirically meaningful concepts denote experiences or traits of experience. Thus, 'mirage' is empirically meaningful: it designates a concept the denotation of which is a set of phenomena, but it has no objective reference since mirages occur in human subjects and not in the physical world. Notice that our table does not include the class of meaningful signs which designate no idea, as is the case of 'Ouch!'. In science ,which is a body of ideas, they have even less use than meaningless signs which, being noncommittal, may conventionally be ascribed any meaning. Class 4 is

Table 3.1. *Kinds of meaning*

	(1) Meaningless	(2) Formally meaningful	(3) Empirically meaningful	(4) Objectively meaningful	(5) Fully meaningful	Relation
Sign	○	○	○	○	○	Designation
Idea (Concept, proposition, theory)		↓ ○	↓ ○	↓ ○	↓ ○	Empirical reference
Experience			↓ ○		↓ ○	Objective reference
Objective fact				↓ ○	↓ ○	
Example	'Existence'	'&'	'Pain'	'Valence'	'Table'	

that of *objectively meaningful* but empirically meaningless signs, i.e. those designating ideas about facts or things beyond experience yet reputed real. Strictly speaking they should be called *allegedly objectively meaningful* signs: after all there is no warrant that they are all truly objective. Many scientific ideas fall in this class. For example, 'free particle' is factually meaningful, in the sense that theories of free particles purport to refer to such objects, a claim substantiated by the fact that there are values of x for which the function "x is a free particle" is approximately true. But 'free particle' names no class of experiential items because it is impossible to experience anything with entirely free particles: as soon as we establish an information-gathering intercourse with them they cease to be free; we experience at most with approximately free particles. Finally, class 5 is that of symbols that are both *empirically and objectively meaningful*, as is the case of names of classes of perceptible objects. Notice that we have not included in 4 and 5 those signs purporting to designate physical objects about which we have not the slightest idea.

A full analysis of meaning will disclose which of the above five classes the symbol concerned belongs to. It will also take into account that meanings are contextual, to the extent that the membership of a given sign may change as a result of research. Thus, an initially meaningless sign may eventually be assigned a meaning of some sort or other: this has happened, e.g., with inscriptions originally believed to be natural and then deciphered as language signs. Conversely a sign may be deprived of meaning; for example, the arithmetical signs '·' and '+' lose, when adopted in abstract theories, their original meaning and even lose any specific meaning. Again, developments in observation techniques may attach an empirical meaning to a symbol initially assigned an objective meaning only: think of the molecules now rendered indirectly visible through the electron microscope.

For better of for worse the analysis of meaning is not philosophically neutral. An ordinary language philosopher will probably find no use for formal meaning and for objective meaning: ordinary language is a thing-language closely tied to daily experience and has few words for pure ideas and for theoretical concepts. And a radical empiricist will find no use for objectively meaningful but empirically meaningless symbols (class 4). Yet they are characteristic of science. Take, for example, the sentence 'The solar system was formed about five billion years ago': strictly speaking it is empirically meaningless since it corresponds to no set of experiences. One may, of course, invent a fictitious observer imagined to have witnessed the great event; or one may try to save the sentence by saying that, if an observer had been there at the time, he would have witnessed the event — a counterfactual statement. But nonexistent, hence unobservable, observers can impart no observational content to a sentence. The invention of superhuman (or perhaps supernatural) and nonempirical entities is a suicidal move for an empiricist to make. Consistent empiricists must therefore request the elimination of all empirically meaningless symbols. But such a mutilation would kill science.

Yet the requirement of empirical meaningfulness, murderous as it is in its extreme version, has a sound root, namely the wish to avoid altogether meaningless and accordingly untestable expressions. The predicates 'meaningful' and 'testable' are in fact related through this thesis: *If a formula is empirically testable then it is empirically meaningful.* That is, testability is sufficient for meaning, though not necessary. If the converse proposition is asserted at the same time, i.e. if testability is declared to be necessary for empirical meaningfulness, one ends up with the equivalence of the two, i.e. with the thesis that a sentence is empirically meaningful if and only if it is testable. And if one adds that empirical meaning is the only meaning there is, one ends up with the *verifiability doctrine of meaning*, according to which a sentence is meaningful (meaning "empirically meaningful") if and only if it is verifiable—or, more generally, testable. But the equation of testability and meaning is disastrous for the following reasons.

In the first place the doctrine decrees the synonymity of 'meaning', a semantical term, with 'testability', a methodological term. Yet interpretation in empirical terms is prior to actual empirical tests: in order to design and execute the test of a hypothesis we must discover what it commits us to on the level of experience — i.e. what empirical consequences it entails. Suppose a theoretician hypothesizes the existence of a new class of particles which he christens *epsilons*. An experimentalist setting out to test this hypothesis will use it, in conjunction with bits of accepted theories, in the design of some experiments or observations and he will determine beforehand what kind of empirical results he should expect in case epsilons in fact exist, and what if they do not. In short, he must possess at least part of the intension of the concept "epsilon" before he plans to find out whether its extension is nonempty; but for a sign to be meaningful it is necessary and sufficient for it to designate a concept with a nonempty intension. A second argument against the verifiability doctrine of meaning is that tests bear on whole statements (and on sets of statements at that), not on concepts, so that the doctrine provides no means for specifying the meanings of terms and their designata. In fact, we can test the conjecture "That liquid is an acid", but not either "liquid" or "acid". Since only statements can be made to correspond to operations, the demand (made by operationalism) that every concept be correlated with a set of operations, and if possible measurements, is unfulfillable.

*The confusion between testability, empirical meaning, and objective meaning is responsible for much of the present controversy over the foundations of quantum mechanics. Take, for instance, the phrase 'The electron is at the place x at the instant t'. According to the verifiability doctrine of meaning, still popular among physicists, the phrase is meaningless as long as no position measurement is effectively made at t such that it either locates the electron at x or fails to do so. In other words, the meaning of the phrase would be attached by the operation designed to test it. This is a clear case of confusion between meaning and test. The phrase has an objective meaning all the time; moreover, although it is not empirically meaningful it does entail observational consequences and is empirically meaningful in an indirect way — as long as it is not meant to refer to an isolated particle; this is why can we find out under what conditions it is testable. It would be otherwise if the concept of hidden position or position in itself (unperturbed by experiment) were at stake: a sentence to the effect that an electron had such a hidden position would be objectively meaningful but empirically meaningless and accordingly untestable. In our classification of meanings, 'observable position' belongs to class 5 whereas 'hidden position' belongs to class 4.

The attempt to fill every scientific formula with an empirical content leads to populating the universe past and future with an infinite staff of unobserved observers, while it would be more sensible to recognize right away that theoretical constructs have no empirical meaning and this is why they go beyond ordinary knowledge. That attempt also leads to puzzles. Thus, for instance, we remarked a while ago that 'free particle' is empirically meaningless though objectively meaningful. Consequently, a theory of free particles cannot be assigned an empirical interpretation; in particular, the theory's statements concerning the position and the momentum of a free particle cannot be interpreted as observation statements, i.e. as statements concerning results of observing (or, rather, measuring) the position and the momentum of a free particle. Yet most physicists, starting from the assumption that a particle is free and, particularly, free from interactions with measuring devices, claim that certain consequences of such a theory of free particles — such as HEISENBERG'S uncertainty relations for this case — must be interpreted as referring to the particle's interaction with a macroscopic observing device — or perhaps even with the observer. Once this semantical miracle has been performed a bonus miracle occurs: the theory, which to start with made no assumption concerning the measuring apparatus, let alone about the observer (since it was supposed to be a physical, not a psychological theory), predicts the amount of the disturbance no matter what particular apparatus is smuggled in the "interpretation" of the formulas. (See Sec. 7.4 for a more detailed examination of this puzzle.)

The current debate over the foundations of quantum mechanics is largely a debate over the interpretation of its symbols and a controversy between those who emphasize the need for *testability* and those who emphasize the factual (objective) content or external *reference* of the theory. The former are eager to secure the test of the theory and are apt to lose sight of its objective referent to the point of refusing to assign the theory a meaning independently of empirical test procedures; in this way they slide, unwittingly or joyfully, into subjectivism. And some objectivists, anxious to show that the theory has an objective reference, tend to neglect its test to the point of introducing concepts designating entities and properties that could not show up in experiment. The current debate might be cleared up and redirected to more fertile issues by showing that a semantical question has been misinterpreted as a methodological problem, and a methodological problem has been neglected for a semantical question. Semanticists might be helpful in this debate if they elaborated on the difference between empirical and objective meaning in connection with scientific issues, instead of secluding themselves in toy languages.*

Further aspects of the problem will be taken up in the next sections.

Problems

3.5.1. Are facts meaningful? In particular, is it enlightening to talk about the meaning of cultural events, of historical happenings, and of dreams? If so, in what sense?

3.5.2. Propose an ordinary language, an epistemological, an ontological and a psychological interpretation of '$p \rightarrow q$'. *Alternate Problem*: Find out whether the formula "$(\exists x)\, Px$", in addition to being formally meaningful, can also be given an empirical and an objective meaning.

3.5.3. Comment of the Stoic doctrine of signs, which distinguished the sound, the meaning, the object denoted, and the subjective image elicited by a word.

3.5.4. Define the concepts of interpretation of a fact, of a formal theory, of a factual theory, and philosophical interpretation.

3.5.5. Suppose a correct quantitative theory of empirical validation were built which permitted us to measure the degree of confirmation of synthetic (nonanalytic) formulas. Suppose further that 'meaningful' = 'verified or verifiable'. We might then switch from degree of confirmation and degree of testability to degree of meaningfulness. What would be the meaning of, say, 'two-thirds meaningful'?

3.5.6. Propose criteria of empirical meaningfulness and objective meaningfulness.

3.5.7. Are there tests for empirical and for objective meaningfulness?

3.5.8. Propose a distinction between "meaning" (a semantical concept) and "understanding of meaning" (a psychological concept).

3.5.9. Examine the thesis that the meaning (not just the test) of a probability statement is that a certain relative frequency close to the numerical value of the probability can be observed. *Alternate Problem*: Determine the meaning and the test of the statement "The rocks on the top of that hill have a positive potential energy". Does the test involve a situation referred to by the statement?

3.5.10. Examine the following expressions common among physicists:

1. 'Mass is a scalar'.
2. 'The electromagnetic field is a hexavector'.
3. 'The gravitational field is a tensor field with 10 components'.
4. 'The spin is a pseudovector'.

Do these expressions take care of the difference between a predicate and its referent? *Alternate Problem*: Discuss the relation between the concept of well-formed formula (or syntactically meaningful expression) and the semantical concept of meaning. Is either necessary for the other? Take into account that "$(\exists x)(x=0)$" is a wff although it seems to make no sense or, rather, it is pointless.

3.6. Interpretation Procedures

A sign, like a puzzle piece, can make sense only in a context, i.e. in relation with other objects. The sign's designata and the latter's referents, if any, are among those other objects which jointly confer a meaning upon the sign. When the symbol has a nonconceptual referent and its interpretation is at least partially determined by a sign-referent relation, we call this relation *referition* and we take care not to mistake it for a definition, which is a sign-sign correspondence (see Sec. 3.3). In the present section we shall examine the following kinds of interpretation procedure: ostensive, coordinative, and operational referition, and semantical rules.

When teaching vocabularies, whether ordinary of technical, we are forced to resort to *ostensive referitions* (usually called ostensive definitions) such as the utterance of the phrase 'This is a so and so' accompanied by some gesture. By itself the verbal expression is meaningless: it is a sentential function (see Sec. 2.1) of the form '... is and so and so', where the blank is not to be filled by a proper name but by a combination of the sign 'this', which is dispensable, with an appropriate bodily movement. Ostensive referitions, then, are not purely conceptual operations but, rather, bridges between rough experience and language.

The didactic virtue of ostensive referitions, namely their closeness to ordinary experience, makes them unfit for introducing high-level terms typical of science, such as 'temperature': it is impossible to point to this property with the finger. Besides, ostensive referitions are much too closely related to the cognitive subject: in fact, not to speak of the gesture — which may be neither unequivocal nor universal — 'this' is a subject-centered (egocentric) word incapable of conveying an objective and universal information. Furthermore, ostensive referitions can rarely yield unambiguous specifications of meaning. Thus, if we say 'This is white' and at the same time point to a sheet of paper, a foreigner might not understand whether we mean the color, the shape, the texture, or the entire sheet. Only some singular names — that is, names of perceptible individuals — can be introduced unambiguously in an ostensive way. Universals, on the other hand, cannot so be introduced: we can point to a perceptible individual and, within bounds, to a collection of individuals but not to classes composed of both actual and potential members; nor can we point to nonsensible properties such as "viscous", or to nonempirical relations, such as "better adapted than". Only phenomenal things, events and properties can be pointed to. Concepts which are assigned an objective meaning but not an empirical one (see Sec. 3.5), like "atom", cannot be introduced by ostensive referition. Ostensive referition has, for this reason, no place in scientific theory although it is indispensable for learning and expanding vocabularies. If science tried to do with ostensive terms, and even if it tried to distil transpenomenal concepts out of facts of experience (phenomena), ostensive referition would be the gateway to science. (The fact that it is not suggests that phenomenalistic philosophy cannot cope with scientific knowledge.) Ostensive referition has been included here only because it is usually confused with a kind of definition and because it is wrongly assumed to occur in science.

A second kind of sign-object interpretation procedure is *coordinative referition* (usually called coordinative definition): it consists in linking a symbol to a particular physical thing or property taken as a standard or base line. In classical biological systematics the first-found complete specimen of a given taxonomic order is taken as a standard, and what the species' name is supposed to name is precisely a set of individuals more or less closely resembling that standard specimen, which is kept in a museum. And physical standards, that is, materializations of magnitude units (kilogram, second, volt, etc.), serve all as referents of symbols introduced by coordinative referition.

Coordinative referitions are all chosen on the ground of practical convenience and, very often, on the human scale. The first international length standard (the standard meter) was built in 1799 and bore the proud inscription *Pour tous les temps, pour les peuples*. Its universality did not reach the Anglo-Saxon countries and its eternity leasted until atomic physics found finer, more stable and more easily reproducible standards. The one adopted in 1960 (which need not be definitive) is a certain orange-red spectral line of Krypton 86. The meter has accordingly become a derived unit, and is no longer introduced by a referition but by an explicit definition, namely thus:

1 meter $=_{df}$ 1,650,763.73 wavelengths of orange-red light from Kr^{86}. (Notice the theoretical terms in the definiens: 'wavelengths' and '86', a

value of the atomic mass. Theoretical terms alone occur in the definiens of the conventional, more exact definition.) Coordinative referitions do not belong to scientific theory but occur in the information-gathering and test phases of factual science. In particular they do not replace the corresponding definitions, if any, but materialize them. Thus the Weston electric cell, used as a voltage standard, does not replace the theoretical definition of "volt" as unit work done along a circuit.

Operational referitions (usually called operational definitions) establish correspondences between symbols, on the one hand, and controlled operations or their results on the other: they are of the sign-experient kind of referition (see Sec. 3.2) and accordingly they supply empirical meaning (see Sec. 3.5). For example, the various known and possible

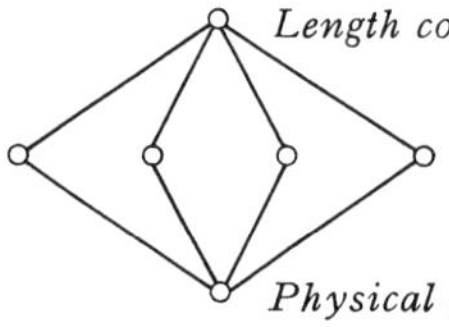

Fig. 3.5. A single concept, denoting a given physical property, is partially elucidated by a number of operational referitions

kinds of temperature measurement (by means of the thermal dilatation of gases, liquids, and solids, or through the thermoelectric effect, and so on) supply as many operational referitions of the temperature concept. This variety contrasts with the singleness of the temperature concept as introduced by thermodynamics. Similarly there is not a unique operational referition of length: the builder, the astronomer and the nuclear physicist employ altogether different procedures for measuring distances and, in the case of the scientists, theories are used as well: still, the same length concept is involved in all of them. This ambiguity is a merit of operational referition, since it shows a good portion of the range of application or extension of certain concepts. By the same token it cannot specify meanings unambiguously (see Fig. 3.5).

A still influential philosophical school, *operationalism*, claims that measurement operations *alone* can endow scientific terms with meaning, and that what we have been calling operational referitions are definitions proper. Operationalism rests on the following misunderstandings. (i) The confusion of *defined* with *definite* (determinate). By means of an apparatus and a sequence of operations the velocity of a molecular beam can be made definite (preparation of an homogeneous beam) or it can be assigned a definite value (measurement). But the velocity concept remains unaffected by such operations; moreover, in planning the operations we make use of a full-fledged velocity concept. (ii) The confusion between *definition* (an equivalence between signs or between concepts) and *referition* (a correspondence between signs and their referents).

(iii) The identification of reference with *empirical* reference, i.e. of meaning with empirical meaning. This has as a consequence the refusal to admit concepts like that of propagation of light in a vacuum just because they have no experiential counterpart (see the next Section). (iv) The confusion between *meaning* and *testability* — hence between semantics and methodology (see Sec. 3.5). If stripped of these misunderstandings operationalism boils down to a few sound requirements: (i) the avoidance of the so-called *verbal definitions* (e.g., PLATO's "Time is the divine image of eternity"); (ii) the empirical *interpretability* of (certain, not all) scientific terms, in order to ensure (iii) the empirical *testability* of (most, not all) scientific hypotheses. But then these are not distinctive of operationalism: they constitute the sound core of empiricism.

Measurements are performed in order to test certain statements, not to discover meanings. True, by delimiting extensions they may, as a byproduct, help to interpret certain signs. When we say 'Temperature is what thermometers measure' we add a dimension of human experience to the objectively meaningful sign 'T' handled by thermodynamics (which is unconcerned with temperature measurements) and obtain consequently a fuller grasp of its meaning, but that does not constitute a definition of "temperature"; it just helps us understand the meaning of 'T' — which is something for psychology rather than for semantics. Suppose it follows from some theory that the temperature of a certain body we may call c is π degrees Celsius, i.e. $T(c)=\pi°$ C. An effective measurement of the temperature of c will support or undermine the preceding statement and it may contribute to understanding it but it will not supply the meaning of the temperature concept. Unless we had this concept before and knew, if only superficially, how it is related to other physical concepts, we would be unable to set up the measurement operation. Furthermore, such an operation cannot even determine (let alone define) the numerical value of the temperature concept in an exact way. In fact any measurement, no matter how precise, will yield a fractionary (rational) number, say 22/7, which is an *approximate* value of the magnitude concerned: irrational numbers, like π, are empirically inacessible. (Usually the *theoretical range* of magnitudes is the set of real numbers, whereas their *empirical range* is a subset of rational numbers.) In short, measurement provides (i) an estimate of the numerical value of a quantitative property, and consequently (ii) a test of statements involving that magnitude; (iii) an illustration of its use, and therefore (iv) a psychological understanding of the meaning of expressions containing the term concerned.

The partiality and ambiguity of operational referition is not limited to quantitative concepts. Take, for instance, the qualitative version of the concept of electric charge (a quantitative concept). An operationalist would propose the following "definition" of that concept: "For every x, if a light body y is placed near x at time t, then: x is electrically charged at time t if and only if y is *observed* to move toward x or away from x at t". The structure of this statement is, basically, that of the conditional definition, i.e. "$C \rightarrow (A \leftrightarrow B)$" (see Sec. 3.3, Formula [3.8]). In this formula 'A' is the attribute to be introduced (in our case 'electrically charged'), 'C' describes some experimental condition (the situation or setting) and 'B' the observed behavior. No doubt such a statement does offer a partial elucidation of the meaning of 'A' — but a highly vague and ambiguous one, because one and the same experimental situation (the one described by 'C') is consistent with alternative outcomes as described by 'B'. Thus, for instance, exactly the same formula holds for *any* other force with macroscopic effects, such as gravity and magnetism; on the other hand it is useless for intermolecular, interatomic, and nuclear forces. Furthermore, the sentence refers to the conditions under which the electric (or magnetic, or gravitational) condition of the body is tested but remains silent about the body's behavior when no tests are made. This is due, of course, to the declared aim of avoiding, with this kind of referition, any reference to independent (real) physical objects — suspect to subjectivism — and to the wish of reducing theoretical to observational terms. And these are among the reasons for regarding operational referitions as supplying only *incomplete interpretations*.

If the meanings of scientific terms were specified in terms of human operations alone objective science would be nonexistent: factual science is objective to the extent to which it truthfully accounts for the external world, and this involves interpretation procedures that do not point to human operations alone but to objective facts. Thus, the theory of electricity has no use for the above-mentioned "operational definition" of electric charge because it simply does not deal with electrically charged bodies in terms of human operations: it describes those systems and explains their behavior in function of theoretical concepts such as "electric charge" and "electric field", which are supposed to have objective referents even though they are not observational concepts. Moreover, the theory gives us a (largely nonpictorial) "picture" of charges and fields while they are not being explored by means of test bodies; and even such test operations are accounted for by the theory as purely physical processes rather than in psychological terms such as the phrase 'is observed to move'. The concept of observation belongs to psychology and to the test stage of factual theories. In other words, man and his operations — the center of operationalism — do not occur in the conceptual mapping of physical reality.

In conclusion, referitions occur in the test stage of science but do not belong to its conceptual framework. The test stage can come only after certain concepts have been elucidated to some extent — if possible with the help of a theory. Referition contributes to concept elucidation by supplying part of the extension of certain (not all) concepts. Exaggerating the role of referition and, in particular, of ostensive and operational referition, leads to mistrust theory and to revert to anthropocentrism, whence it is dangerous for the growth of science. In addition it is philosophically confusing because it drags interpretation, understanding, definition, and test to a single muddle.

Let us finally deal with *semantical assumptions* or rules of meaning. We shall distinguish two kinds of semantical rules: nominal referitions and postulates of interpretation. A *nominal referition* or *rule of designation* is a purely linguistic convention whereby a name is assigned to a thing, as in the case of " '*C*' stands for carbon". In factual science this is not a proposition but a convention or *proposal*: it is not true or false and is therefore not tested for truth. Therefore it would be better to phrase it explicitly as a linguistic convention, namely thus: "Let '*C*' stand for carbon". (Outside the body of science the same expression does mean a proposition, namely an idea about a normal use of the sign '*C*' in certain human groups: it describes a piece of linguistic behavior. But this is not the sense of interest to science and philosophy.) Rules of designation introduce symbols by convention. In contrast to other kinds of referition, they are entirely arbitrary. And contrarily to definitions they do not constitute analyses of the defined concept: nominal referitions merely stipulate a one-to-one correspondence between a symbol and a class of objects. In short, they stipulate names.

(Originally a rule of designation may be suggested and is always preceded by a body of knowledge. Thus, e.g., the name *mercury* to designate a chemical element was originally linked to an alchemical idea concerning astral influences; and the name *organism* for living beings suggests that what differentiates a living being from the juxtaposition of its parts is a certain organization rather than a special substance. Names may begin by suggesting ideas and may not be neutral. But if they are just names they end by being purely conventional tags or identification marks — sometimes very inadequate ones as in the cases of 'real number', 'imaginary number', and 'irrational number', fossils of a mistaken philosophy of mathematics.)

An *interpretation postulate* is an assumption that confers a meaning upon a symbol, not conventionally but in such a way that the factual truth or falsity of the expressions containing the symbol will depend on the acceptance or rejection of the interpretation postulate. Interpretation postulates take an important part in the interpretation, application, and test of the formalism (symbolic skeleton) of scientific theories (see Secs. 7.4 and 8.4). Thus, elementary physical geometry consists of formal (mathematical) statements such as the Pythagorean theorem, and semantical assumptions that postulate correspondences between certain geometrical objects (e.g., straight lines) and certain physical entities (e.g., light rays). Such postulates function as rules of translation from the formal into the physical language and vice versa. In the absence of such postulates the formulas either have no factual and no empirical reference or they are assigned an interpretation in an intuitive, hence uncontrollable way. With the addition of interpretation postulates the mathematical theory becomes a factual theory: it acquires an objective reference and, at the same time, it may become empirically testable. As soon as this enrichment of the original formal theory with a factual meaning is performed, the paradise of formal truth is lost and the theory becomes the prey of experimenters intent on examining its claims to factual truth.

Interpretation postulates are not proposals but full-fledged propositions of the form "The interpretation of *s* is *P*", where '*s*' designates a sign or a concept and '*P*' the property correlated with it. Still, interpretation postulates may be used as semantical rules because they function both as postulates of scientific theories and as rules of meaning. (Recall that formulas are nothing in themselves, apart from the functions they fulfil. The arithmetical formula "$x+y=y+x$" may be regarded either as a law or as a rule for handling the signs occurring in it, depending on the aspect we are interested in.) In any case they are not purely conventional rules like nominal referitions. If the symbol for carbon were changed from '*C*' into '*K*' nothing but practical inconveniences (even for the Germans) would result. But if '*g*' in GALILEI's law of falling bodies were reinterpreted ,say, as the Earth's libido or as the experimenter's visual acuity, a statement which would be physically meaningless (hence untestable) and psychologically false would ensue.

Or take the much-discussed ψ-function of quantum mechanics: it may be regarded as a mathematical symbol without a physical referent, i.e. it may be assigned a formal meaning alone (e.g., through the rule of designation " 'ψ' designates a vector in Hilbert space"). Or it may be regarded as representing the state of a physical system or, alternatively, as a matter wave amplitude, as a field strength, as a probability amplitude, and so on. And some of these postulates may be mutually consistent; i.e. several meanings and roles may consistently be assigned to the symbol 'ψ': it may at the same time belong to an abstract function space, it may represent the physical state of a system as a whole, the DE BROGLIE wave associated with the latter, the amplitude of the probability of position, and so on. For the latter interpretation M. BORN was awarded the Nobel prize — which is given to the fathers of remarkable ideas, not to their baptizers unless the name happens to convey the meaning.

No theory can be regarded as more than a symbolic skeleton unless referitions and/or semantical rules for the interpretation of its basic symbols are added. (Recall that definitions can be helpful in attaching meanings to nonprimitive signs alone.) Every such interpretation will yield an objectively and/or empirically meaningful theory; and one among such alternative interpretations will normally yield a theory (model) that is factually true to some extent — even though a clear-cut choice among such alternative models may not be possible at a given moment. Interpretation postulates and rules of designation — in short, semantical rules — are the chief yet not the sole devices by which the basic (primitive) symbols which make up the substantive assumptions of a scientific theory are endowed with a meaning. The remaining symbols of a theory are each given a context — as long as they are not formal — by means of definitions and/or referitions.

But the whole meaning of a complex scientific theory is never given once and for all by a set of referitions and definitions. The discussion of examples, even so simple as to be of purely academic interest, and the discussion of actual and possible experimental results, usually help get an insight into the meaning of theoretical concepts. Moreover, interpretations need not be final. Thus Newtonian mechanics — not to speak of quantum mechanics — is still in the process of being correctly interpreted although its basic equations are not touched: we can still disagree on the meanings of "mass", "fictitious force", and "inertial systems". There is probably only one secure means for settling once and for all the questions of interpretation of scientific symbols and symbol system, viz., to forget about them. This radical solution to the problem of the meaning of scientific constructs has in fact been proposed and is still being defended, as we shall see in the next Section.

Problems

3.6.1. Identify the following statements:

1. That is a telescope.
2. Philosophers are people like CARNAP and POPPER.
3. The standard of weight is the *kilogramme des archives* at SÈVRES.
4. Length is what yardsticks measure.
5. 'm_i' designates the numerical value of the i-th particle.
6. '$P_{mn}=0$ unless $n=m-1$' means that there is no possibility for the system other than decaying into the next lower state.

3.6.2. Offer examples of coordinative and operational referitions, of rules of designation, and of rules of interpretation. *Alternate Problem*: Study ostensive referitions.

3.6.3. Examine either of the following theses. (i) A concept is synonymous with the corresponding set of operations: P. W. BRIDGMAN, *The Logic of Modern Physics* (New York: Macmillan, 1927), p. 5. (ii) Operational definitions are circular. Thus, e.g., the operational definition of length involves temperature corrections, and in turn the operational definition of temperature involves length measurement: K. R. POPPER, *The Logic of Scientific Discovery* (London: Hutchinson, 1959), p. 440. (iii) Operational definitions are not full equivalences but conditional statements of the form "$C \rightarrow (A \leftrightarrow B)$": R. CARNAP, "Testability and Meaning", *Philosophy of Science*, **3**, 419 (1936) and **4**, 1 (1937). (iv) The so-called operational definitions are criteria of application of the terms under consideration: C. G. HEMPEL, "Introduction to Problems of Taxonomy", in J. ZUBIN, Ed., *Field Studies in the Mental Disorders* (New York: Grune & Stratton, 1961).

3.6.4. Operationalism holds that different kinds of operation define different concepts even when they are designated by the same name and even if theory does not distinguish among them. For example, different kinds of ammeters would "define" different concepts of electric current. What would the unity of every branch of science come to if operationalism were adopted? In particular, what would become of scientific theories? *Alternate Problem*: Study coordinative referitions. See H. REICHENBACH, *The Philosophy of Space and Time* (1927; New York: Dover, 1957), Ch. I, § 4.

3.6.5. Magicians, alchemists and parapsychologists might be able to contrive operational referitions of the peculiar terms they employ, since correspondences between these terms and certain rites and operations do exist. Would this render magic and parapsychology scientific?

3.6.6. Draw a list of the semantical rules (rules of designation and postulates of interpretation) of a scientific theory of your choice.

3.6.7. Examine the following referition of the concept of elementary particle: "An elementary particle is a system that so far has not been decomposed". Would we be justified in adopting this statement as a definition?

3.6.8. Conditional definitions — statements of the form [3.8] in Sec. 3.3 — have been called bilateral *reduction* sentences when they introduce a physical predicate A in terms of an observable predicate B (referring to observable behavior) and another observable predicate C (referring to observable test conditions). The name comes from the hope that physical (objective) properties might be *reduced* to phenomenal (subjective) properties, i.e. to sensible qualities. Has this program been carried out? If not, is it realizable? And is it consistent with the aim of objectivity?

3.6.9. Examine whether, and if so how, is the temperature concept related to the heat feeling. Is it a logical relation? If not, is it purely conventional? Discuss the general problem whether sense-concept correspondence rules might be grounded on psychophysical laws.

3.6.10. It is usually asserted that scientific concepts can be either theoretical or empirical but not both. Does "weight" confirm that this is a dichotomy? And what are the relations between the extensions of the observational and the theoretical concepts of weight?

3.7. Concept "Validity"

It may be granted that "ghost", "lovely" and "big" are not scientific concepts: the first belongs to folklore, the second is much too subjective, the third is exaggerately vague. We can use them in the course of research but we shall exclude them from the results of research. To discard such typically nonscientific concepts is easy, but what about the concepts of gravity, pre-Adamite or neutrino when they were introduced, i.e. at a time when they had no empirical backing? Is there a foolproof criterion for discriminating among scientifically valid and invalid concepts? As will be shown in the sequel there is a simple criterion but it is as fallible as science itself.

First the question of intensional accuracy may be treated. A scientifically valid concept must have a definite intension or connotation. In other words, the intensional vagueness of scientific concepts should be minimal. This excludes from science scandalously vague concepts such as "small", "high" and "possible" when used without qualification or relativization. (Such a relativization may be tacit as in the case of "high energy physics", which refers by convention to the physics dealing with events involving energies larger than one million electron volts.) Now, a precise intension cannot be assigned out of some context. And the proper context of a scientific concept is a scientific system. Our rule can therefore be reformulated as follows: A necessary condition for concept validity in science is the possession of a fairly definite intension in some scientific system. This rule is vague itself and it must be kept vague in order to make room for embryonic concepts that would never be refined unless they were embedded in a scientific system.

A second necessary condition for the scientific validity of a concept is that its extensional vagueness be small. In other words, scientific concepts must have a fairly definite extension, i.e. they must be applicable in a fairly unequivocal way. And this, again, requires us to summon the system in which the concept is embedded, because the empirical data that enable us to fix the extension — or rather the core extension — of a concept are significant only in the same context to which the concept belongs.

The above rules can be stated in a more definite way with the help of the concepts of core intension and core extension introduced in Sec. 2.3. We may say that it is necessary for a concept to be scientifically valid to have both a definite core intension and a definite core extension. Since these in turn determine the core meaning of the sign that names the concept, our rule can be applied to signs by using formula [2.24] in Sec. 2.3: If a sign is scientifically valid it has a definite core meaning in some scientific system.

A fairly precise intension and a fairly precise extension are necessary for accepting a concept as scientifically valid but they are not sufficient. Division by zero complies with both requisites yet it is not valid in arithmetics: it has an empty connotation and also a void denotation, because it does not satisfy the definition of number division. Another way of putting this is to say that "$n/0$", where 'n' designates a number, does not exist in arithmetic. Now existence, in the case of a formal concept, can be secured either by positing it as a primitive in some theory or by defining it in terms of the primitives of a theory (see Sec. 3.4, [3.13]). In short, the criterion of scientific validity for pure concepts is membership in a formal theory. Consequently, logical and mathematical theories are not judged by the concepts they contain but, on the contrary, concepts are judged by the company they keep.

Let us now turn to concrete concepts, i.e. concepts with a spatio-temporal reference. "Adam" has both an exact connotation (provided by scriptural descriptions) and an exact denotation — namely, the empty set. It does not enter scientific history precisely for having no objective referent: why should historians bother to account for a nonentity? This suggests trying the following *Definition* 1: "An intensionally and extensionally precise nonformal concept is valid if and only if it has a *real referent*, i.e. if it denotes a class of spatio-temporal objects." This rule disqualifies ghosts and consecrates atoms. But an examination of not-so-clear cases will show that pointing to a real referent should not be regarded as necessary. In the first place we need the so-called limiting concepts, such as "rigid body", "isolated system" and "rational player", which we know correspond only roughly with their intended referents: after all, scientific theorizing starts by idealizing real objects and rarely, if ever, accounts for anything else than schematic idealizations. In the second place, the reference of many scientific concepts is a disputed question. For instance, are we sure that the uniform time set by our clocks exists in the same sense as clocks exist, or would continue to exist even if there were no clocks? Thirdly, we must remain free to invent new concepts in the building of factual theories that may not immediately be known to be true, but may have to wait for their test and may even so be at most partially true. In view of these objections Definition 1, inspired in naive realism, seems too restrictive. Accordingly we shall regard real reference (nonempty objective extension) as only a sufficient condition for scientific validity.

A slightly more sophisticated criterion is provided by the following *Definition* 2: "An intensionally and extensionally precise nonformal concept is valif if and only if it denotes possible *operations* of some kind". According to this (operationalist) criterion of concept validity, "mass" will be a valid concept if and when masses can be measured. Since the mass of a nonaccelerated body cannot be measured, the concept "mass of a nonaccelerated body" would be invalid according to the preceding definition. Accordingly all theories involving free bodies and particles would have to be banned. Operationalism would likewise request to eliminate the concepts of absolute zero of temperature, quantum number, wave phase, state function, spin of a free particle, adaptation, natural selection, phylogeney, mental state, culture, and nation, none of which denote possible operations. It would lead to sweeping all nonobservational concepts, keeping the observational ones alone. Consequently the requirement of operational reference would lead to beheading science. Since we do not want this result we shall regard the existence of operations as a sufficient criterion for the applicability of a concept, on condition that the operations be scientific — otherwise any action whatsoever might consecrate the corresponding concept as scientific.

What is seemingly a less restrictive version of the operationalist criterion of concept validity is embodied in *Definition* 3: "An intensionally and extensionally precise nonformal concept functioning as a *primitive* in a theory is valid if and only if it denotes *observable* entities or properties". The validity of defined concepts is then automatically warranted. The difference from the previous definition is this: now just the primitive or undefined concepts of a theory are required to correspond to observables. This apparently does not condemn transempirical concepts: these are accepted on condition that they are explicitly definable in terms of the observational primitives. But this is impossible: it is precisely the nonobservables that must occur as primitives of a theory, both because no theory refers directly to observable situations and because a goal of any theory is to explain what is observed rather than to summarize it; and to this end scientific theories are built with nonobservational theoretical concepts, precisely because they are stronger than the observational ones. Definition 3 would permit us to keep transempirical concepts if we were to define them all in terms of observational concepts. But, by definition of transempirical concept, such a reduction is impossible. Consequently Definition 3 is as lethal to scientific theorizing as Definition 2.

A basic defect of the views underlying the above definitions is that they handle concepts in isolation, as if they were self-contained units rather than the products of the analysis of propositions and theories. In the case of naive realism the reason is the belief that every concept reflects a trait of reality. But this is not even true of all extralogical concepts: suffice to recall that a vector quantity representing a directional property, such as a force, can be conceptually decomposed in infinitely many ways depending on convenience. Even though the whole magnitude may have an objective referent none of its components reflects independently a trait of reality. In the case of empiricism the rationale is the thesis that the concepts that count for knowledge (the so-called cognitively meaningful concepts) are just those which name bundles of percepts, the latter being in turn isolable atoms of experience. (Recall HUME's precept, "No idea without an antecedent impression".) The mere existence of fertile transempirical concepts such as those of genus and field refutes this doctrine. Nor is it possible to "infer" transempirical concepts from observational ones — as the mistaken name *inferred concept* suggests — because only statements can be inferred. If a link can be established among concepts it is only via propositions — and such a link is set up in the bosom of testable theories. Once more we see that concepts cannot be judged apart from the systems in which they occur.

There is, of course, a sound motive behind both the naive realist and the empiricist strictures on scientific concepts: namely, the wish to avoid ghost-like concepts such as those of ultimate essence and life force. Only, the proper way of preventing them from creeping into the body of science is not by appointing philosophical prejudices as janitors but by taking into consideration the entire context in which the concept concerned occurs. Such a context is a scientific system — at the very least a scientific hypothesis, at best a system of scientific hypothesis (a theory). Thus, when we say that "ghost" is not a scientifically valid concept we do not merely mean that ghosts do not exist: many concepts in science, such as "average deviation from the mean", have no independent referent either; and many others which were formerly supposed to be denotative, such as "absolute time" and "caloric", were later on found fictitious. Nor do we mean that 'ghost' is meaningless: its denotation is fairly definite from folklore. Or that it covers no experience: some people claim to have experience of ghosts and anyway empirical data are significant in the light of hypotheses only. "Ghost" is unacceptable in science for the simple reason that there is and there can be no scientific theory of ghosts, i.e. no testable system concerning the laws of behavior of what, by hypothesis, are subject to no law. In short, "ghost" is not a scientifically valid concept because it is not a member of a scientific system. The latter statement becomes a trivial tautology if the following convention is adopted: *Definition* 4: "An intensionally and extensionally precise concept is scientifically valid if and only if it is *systematic* (i.e., it occurs in a scientific system)."

The last definition is assumed to apply to all concepts, whether pure or not; in fact, it covers the criterion of concept validity we had laid down for formal concepts. The whole problem of concept validity has thereby been shifted from the territory of semantics to the field of science: it is up to the scientist, not to the philosopher, to decide whether a given concept is scientifically valid or not. What the philosopher can do is to criticize the scientist when he breaks that rule outside his own field of specialization and takes the defence of ghost-like concepts. In the second place, Definition 4 does not bear primarily on the meaning of signs: it automatically warrants the possession of meaning to all signs that name concepts belonging to some systematization of facts — preferably a theory. Moreover, it does not legislate on the meaning of signs which, like 'God', do not belong to science but are perfectly meaningful in other contexts: in contrast to the previous definitions, ours is not directed at obtaining a cheap (linguistic) victory over theology. In the third place our criterion does not require and does not warrant the truth of the propositions and theories built with scientifically valid concepts: we may well play with conjectures the truth value of which we still ignore — as long as we can show that it is possible to make inquiries into their truth value. The refutation of a scientific theory, then, may but need not deprive its specific concepts of validity: some or even all of them may be salvaged and utilized in building a truer theory. Those concepts which occur in moderately true or verisimilar scientific theories may be called *adequate*.

The question of the scientific validity of concepts has now become philosophically trivial although it is a serious concern of scientists. In addition we have come to realize that, although logically speaking concepts are the units of scientific thought, they are not evaluated in isolation from whole systems: their validity, if any, derives from their systemicity — from their occurrence in a system that can be tested for adequacy to fact and for coherence with previously tested systems. Finally, we have come to suspect that the efforts of empiricist philosophy to reduce bold hypothetical concepts to banal observational ones is misguided, not only because it runs counter the history of scientific growth, but because it is a logically unfeasible programme. But this will concern us in Part II.

In Part I, which finishes here, we have assembled some tools of analysis necessary for a methodical examination of the strategy of scientific research and the philosophy behind and ahead it. Let us now proceed to this task.

Problems

3.7.1. Examine the theses of *terminological physicalism* ("All terms should denote spatio-temporal objects") and *terminological empiricism* ("All terms should be observation terms or connectible with them"). See O. Neurath, *Foundations of the Social Sciences*, in *Encyclopedia of Unified Science*, Vol. II, No. 1 (Chicago: University of Chicago Press, 1944), Ch. I.

3.7.2. The complete isolation of a material system in all respects is impossible. Should we therefore ban the concept of isolated system from science? *Alternate Problem*: Study the concept of ideal type such as it occurs in the sociological literature. See R. Rudner, *Philosophy of Social Science* (Englewood Cliffs, N. J.: Prentice-Hall, 1966), Sec. 11.

3.7.3. There are no average men: all averages are logical constructions out of data concerning individuals, and the probability that any given individual will fit exactly all averages is practically zero. Shall we discard the concepts of average and typical man? *Alternate Problem*: No specimen of Pithecanthropus has ever been seen or will probably be seen. Shall we accept the concept "Pithecanthropus" as scientific?

3.7.4. Light cannot be seen to propagate in a vacuum: all we can assert in terms of experience is that a given light beam has been either absorbed or emitted by a certain body; the rest is hypothesis. Is then the concept of light propagation in a vacuum (a primitive concept of optics) scientifically valid? *Alternate Problem*: Behaviorist psychologists have always objected to the consciousness concept because it does not denote any trait of observable behavior. Physiologists, on the other hand, employ the term 'consciousness' (or rather 'conscious state') and are busy trying to discover the neuronal activity likely to correspond to it: see, e.g., J. F. Delafresnaye, Ed., *Brain Mechanisms and Consciousness* (Springfield, Ill.: Charles C. Thomas, 1954). What is to be done: to censure physiologists for ignoring operationalism or to criticize behaviorists for mistaking a good rule of method ("Do not start research by studying consciousness") for a rule of meaning ("The word 'consciousness' is empirically meaningless")?

3.7.5. The concept of field (gravitational, electric, etc.) has been regarded as superfluous because what is observed is never a field intensity but the accelerated motion of a test body (e.g. the displacement of a piece of an electrometer). Comment on this. *Alternate Problem:* Operationalism and mathematical intuitionism have required that all mathematical concepts be constructible: they condemn all concepts which, like that of infinity, are not effectively constructible, i.e. nonoperational. Study the consequences of this prohibition. See M. BUNGE, *Intuition and Science* (Englewood Cliffs, N. J.: Prentice-Hall, 1962), Ch. 2.

3.7.6. "Electric charge of an isolated body" has no meaning in terms of operations. Should it be discarded? *Alternate Problem:* The general concept of energy, in contrast with some of the specific concepts that fall under it (e.g. "thermal energy") can be given no operational reference. Shall we retain it?

3.7.7. The concept of progress is often regarded as nonscientific because it involves a valuation. Examine this contention. Hint: begin by finding out whether every valuation must be subjective. *Alternate Problem:* The concepts of reality and existence have been branded as metaphysical. Are they superfluous in science and in philosophy?

3.7.8. While no scientific theories of goodness, love, and happiness are available, the concepts "goodness", "love" and "happiness" remain nonscientific according to Df. 4. Should they remain in this state or is it possible and desirable to render them scientific? And would goodness, love and happiness cease to be if the corresponding concepts were subjected to the rule of science?

3.7.9. Do the egocentric concepts "I", "now" and "appears" and the anthropocentric concepts "sweet", "rough" and "observable" qualify as physical concepts? And would they qualify as scientifically valid in some other context? Which, if any?

3.7.10. Recall the division of nonobservational concepts into intervening variables and hypothetical constructs (see Sec. 2.6). Pragmatists, phenomenalists and conventionalists accept intervening variables as useful tools but refuse scientific validity to concepts referring to hypothesized transempirical entities, properties, or relations: at most they will accept hypothetical constructs as temporary expedients to be eventually replaced by intervening variables and, even better, by observational concepts. What scientific theories would have to be proscribed if this prescription were adopted? What does the above show concerning the place of philosophy in the construction and acceptance of scientific theories? And how should a decision concerning those two kinds of variable be adopted: in the interest of some philosophy or of the advancement of knowledge?

Bibliography

AJDUKIEWICZ, K.: Die Definition. Actes du congr. Internat. de Philosophie Scientifique, V. Paris: Hermann 1936.

— Le problème du fondement des propositions analytiques. Studia logica **8**, 259 (1958).

BJERSTEDT, A.: Some examples of the possibility of using structural formalizations in sociometric analysis. Sociometry **17**, 68 (1954).

BUNGE, M.: The myth of simplicity, Ch. 1. Englewood Cliffs (N.J.): Prentice-Hall 1963.

CARNAP, R.: Testability and meaning. Philosophy of Sci. **3**, 419 (1936); **4**, 1 (1937).

— Logical foundations of probability, Ch. 1. Chicago: University of Chicago Press 1950.

CHURCHMAN, C. W., and P. RATOOSH (eds.): Measurement: Definitions and theories, Chs. 1 and 9. New York: John Wiley & Sons 1959.

COHEN, M. R., and E. NAGEL: An introduction to logic and scientific method, Ch. XII. New York: Harcourt, Brace & Co. 1934.

COOLEY, J. C.: A primer of formal logic, Sec. 27. New York: Macmillan 1942.

CRONBACH, L. J., and P. E. MEEHL: Construct validity in psychological tests. In: H. FEIGL and M. SCRIVEN (eds.), Minnesota studies in the philosophy of science, I. Minneapolis: University of Minnesota Press 1956.

GOODMAN, N.: The structure of appearance Ch. 1. Cambridge (Mass.): Harvard University Press 1951.

HEMPEL, C. G.: Foundations of Concept formation in empirical science, Vol. II, No. 7 of the International encyclopedia of unified science. Chicago: University of Chicago Press 1952.

LEWIS, C. I.: An analysis of knowledge and valuation, Ch. V. La Salle (Ill.): Open Court 1946.

MARGENAU, H.: The nature of physical reality, Sec. 4.4. New York: McGraw-Hill Book Co. 1950.

PAP, A.: An introduction to the philosophy of science, Part I. New York: The Free Press of Glencoe 1962.

QUINE, W. V.: Word and object, Ch. iv. New York and London: The Technology Press of the M.I.T. and John Wiley 1960.

ROBINSON, R.: Definition. Oxford: Clarendon Press 1950.

RUSSELL, B.: Vagueness. Austral. J. Psychol. Philos. **1**, 84 (1923).

SCHLICK, M.: Allgemeine Erkenntnislehre, 2nd ed., Secs. 6 and 7. Berlin: Springer 1925.

SUPPES, P.: Introduction to logic, Ch. 8. Princeton (N. J.): Van Nostrand 1957.

TARSKI, A.: Logic, semantics, metamathematics, Ch. X. Oxford: Clarendon Press 1956.

WOODGER, J. H.: Taxonomy and evolution. Nuova Crítica, No. 12 (1960/61).

Part II

Scientific Ideas

A piece of scientific research consists in handling a set of problems suggested either by a critical analysis of a fragment of knowledge or by an examination of fresh experience in the light of what is known or surmised. Problems are solved by applying or inventing conjectures which, if testable and grounded, are called scientific hypotheses. In turn, some scientific hypotheses are eventually upgraded to laws supposed to reproduce objective patterns, and laws are systematized into theories. The creative process in science, then, starts with the recognition of problems and culminates with the construction of theories — which pose in turn further problems, among others those of testing the theories. All else is theory application — to explanation, prediction, or action — and theory testing. Let us investigate the members of the central sequence: Problem — Hypothesis — Law — Theory. Once we have found system (theory) we shall turn to its application and test for truth — but this will concern us in Volume 2.

Chapter 4

Problem

Scientific knowledge is, by definition, the outcome of scientific research, i.e. of investigation conducted with the method and the aim of science. And investigation, whether scientific or not, consists in finding, stating, and wrestling with problems. It is not just that research begins with problems: research consists in dealing with problems all the way long. To stop handling problems is to stop research and even routine scientific work. The difference between original research and routine work lies only in that the former deals with original problems or with original approaches to old problems, whereas routine work is concerned with routine problems, i.e. with problems of a known kind approached in a known way.

4.1. The Spring of Science

Apparently all vertebrates have a capacity for taking notice of problems of some kind and for investigating them to some extent. Animal psychology studies the investigatory reflex or exploratory drive, that pattern of behavior — partly innate and partly learned — whereby certain changes in the environment are sensed and examined by the animal with the aim of maximizing their usefulness or minimizing their harm for the organism. All animals search for things and modify the patterns of their own behavior in order to elude or to solve the problems set by new situations, that is, by states of affair that cannot be readily be faced with the stock of reflexes they have already accumulated. Even machines can be designed and built to "sense" and "solve" certain problems — or, rather, to perform operations that are made to correspond to such processes. Yet man alone invents new problems: he is the only problemizing being, the only one which can feel a need to and a pleasure in adding difficulties to those posed by the natural and social environment. Moreover, the capacity for "perceiving" novelty, for "seeing" new problems and for inventing them is an indicator of scientific talent and, consequently, an index of the place occupied by the organism in the evolutionary scale. The more rewarding the problems found, posed, and solved by an investigator the greater his value. He need not nearly solve all of them: he may just feed — directly or indirectly — other investigators with problems the solutions to which are likely to constitute significant advances of knowledge. This must be emphasized at a time when problem finding is neglected to the benefit of problem solving. NEWTON'S *Opticks*, with its 31 profound "Queries" which occupied nearly 70 pages and supplied research problems during one century, might not be regarded as an important scientific work by those who focus on problem solving and on the set of "conclusions" obtained by investigating the sources of research.

The problemizing attitude, characteristic of every rational activity, is the more marked in science and in rationalistic (i.e. critical) philosophy — which is another way of saying that science and rationalistic philosophy consist in a critical investigation of problems. Take, for instance, an archaic thing just discovered in an archaeological site: it may constitute a merchandise for the antiquarian, a stimulus of aesthetic sensations for the art connoisseur, and something to fill a box with for the collector. To the archaeologist, on the other hand, it may become the source of a circle of problems. The object will be "meaningful" for him to the extent to which it is a testimonial of a dead culture some of whose traits he may infer by performing a comparative examination of the object. In fact its form, constitution and function may in principle be explained with the help of conjectures (hypotheses) concerning the mode of life and the mind of the people who made and used the object. To the archaeologist, in sum, the object will not be a mere thing: it will pose him a set of problems, just as finding the object may have been a solution to a previous problem. The solution to any such problems may become in turn the starting point of a new research. Such solutions are often called *conclusions*, which is unfortunate because it suggests that they are conclusive whereas, in fact, they are usually provisional. At other times problem solutions are called *data*, another misnomer because they are never given to the scientist but are extracted and often produced by him in the course of research: what is given to us hardly poses any problem and is therefore of little scientific significance.

To take notice of problems which others may have overlooked; to pose them clearly; to fit them in a body of knowledge and to attempt to solve them with the utmost rigor and primarily in order to enrich our knowledge: such are the tasks of the research worker, a problemizer — though not a mystery-monger — par excellence. The progress of knowledge consists in posing, clarifying, and solving new problems — but not problems of any kind. The schoolmen missed the invention of modern science because their entire approach was wrong: most of them were afraid of new problems and, in general, of novelty to the point of neophobia; the few problems they did formulate were mostly of the wrong kind, i.e. either too trivial or beyond their forces, and in any case too loosely stated; not being interested in the world they could hardly ask any questions concerning its workings; and being dogmatic they did not test their conjectures. In short, although a few schoolmen did produce some acceptable data and hypotheses in the service of a handful of genuine scientific problems — mainly in statics and optics — as a rule they either produced no scientific problems proper or they approached only a few isolated and rather trivial problems with a primarily practical goal in view, such as time keeping and pharmacy. A misselection of problems, resulting in turn from a wrong approach to the world and to research, is likewise the chief reason for the failure of certain schools of thought such as vitalist biology, which has been playing with vague notions about life, purpose and wholeness instead of asking specific questions about the constitution and behavior of organisms.

The choice of problem coincides with the choice of research line, since research is just the investigation of problems. In modern science the choice of problem clusters or research lines is, in turn, determined by various factors, such as the intrinsic interest of the problem as determined by the state of knowledge at the time, the bent and capabilities of the workers involved, the possibility of applications, and instrumental and pecuniary facilities. Practical needs are a source of scientific problems but exaggerated insistence on practical application (e.g. to industry or to politics) at the expense of intrinsic scientific value is sterilizing in the long run — which is all that matters in a collective enterprise such as science. First, because scientific problems are not primarily problems of doing but problems of knowing; second, because creative work can be done only with enthusiasm, and this is apt to be absent if the line of research is not freely chosen out of curiosity. The first consideration in choosing lines of research should therefore be the interest of the problem itself. The second consideration, the possibility of solving the problem — or of showing that it is unsolvable — with the means on hand.

In science, just as in ordinary life, great rewards and great failures alike come from setting great tasks. Nobody should expect deep and sweeping answers to superficial and small questions. The safe course is certainly the choice of trivial problems. Those who seek safety first should select small problems: only venturesome thinkers will risk wasting years in wrestling with big problems that will not guarantee them continuity and promotions in their careers. The big revolutions in pure science have all been made by such people rather than by chance discoveries of prolix and unimaginative workers devoted to small and isolated problems; and what chance discoveries there have been (like the discovery of the white dwarfs in the course of a routine examination of star spectra), they were made by people who expected the possibility of newness and were prepared to recognized it as such: others would see the same things without properly interpreting what they saw.

A few instances of big unfinished tasks are: the building of a general theory of nonlinear oscillations; the *n*-body problem; the investigation of the inners of the fundamental particles; a detailed theory of the origin of life and the attempt to synthesize the grossest components of the protoplasm and eventually a whole unit of living matter; to set up neurological theories of the thought processes; to build mathematical theories of basic social processes in large communities, enabling us to make accurate sociological forecasts. Such problems are ambitious and, at the same time, they seem within the reach of our century, as suggested by the fact that preliminary results are already being harvested.

There are no techniques for generating problems which are at once deep, fruitful and soluble with prescribed means. Yet the following injunctions may be helpful. (i) *Criticize* known solutions, i.e. look for flaws in them — there must be some even if no one has discovered them so far. (ii) *Apply* known solutions to new situations and see whether they still hold: if they do you will have extended the domain of those solutions, if they do not you may have discovered a whole new problem system. (iii) *Generalize* old problems: try new variables and/or new domains of them. (iv) *Look outside:* search for relationships to problems belonging to different fields; thus, in investigating deductive inference as a psychological process ask what its neurophysiological substratum might be like.

Once a problem is proposed for research its worth must be weighed. Again, no hard and fast rules for weighing problems beforehand are known. Only experienced investigators with a wide outlook and far-reaching aims can do good problem-weighing — good, not infallible. (HILBERT'S 1900 list of unsolved problems, which nourished several generations of mathematicians, was as exceptional as NEWTON'S optical queries.) But in addition to choosing the right problem success requires choosing or devising the appropriate means for solving it. That is, wisdom in the choice of research lines is shown in the selection of problems both fertile and of possible solution within our lifetime. And this requires a sound judgment or flair that can be improved but not acquired with experience alone. At this point only one general recommendation can be made: Begin by asking clear-cut and restricted questions: adopt the piecemeal approach to problems instead of starting with sweeping questions like "What is the world made of?", "What is entity?", "What is motion?", "What is man?" or "What is mind?". Universal theories will come out, if at all, as syntheses of partial theories built as an answer to modest — but not trivial — problem systems.

In summary, problems are the spring of scientific activity and the level of research is measured by the size of the problems it handles.

Problems

4.1.1. Comment on the following fragment from I. P. PAVLOV's *Conditioned Reflexes* (1927; New York: Dover, 1960), p. 12: One of the reflexes "may be called the *investigatory reflex*. I call it the 'What-is-it?' reflex. It is this reflex which brings about the immediate response in man and animals to the slightest changes in the world around them, so that they immediately orientate their appropriate receptor organ in accordance with the perceptible quality in the agent bringing about the change, making full investigation of it. The biological significance of this reflex is obvious. If the animal were not provided with such a reflex its life would hang at every moment by a thread". For further information regarding the exploratory instinct see S. A. BARNETT, "Exploratory Behaviour", *British Journal of Psychology*, **49**, 289 (1958), and D. E. BERLYNE, "Curiosity and Exploration", *Science*, **153**, 25 (1966).

4.1.2. Are dogmatists an exception to the law that all vertebrates have a capacity for taking notice of problems? Or are they characterized by deliberately eluding or eliminating problems (and occasionally their finders)? *Alternate Problem:* What are the differences between a social problem and a sociological problem? And could the solution to a sociological problem be helpful in solving the corresponding social problem (in case there is such a problem)?

4.1.3. Why do most philosophical accounts of scientific research begin with data gathering (by means of, e.g., measurement), or by explanation, or by inference? *Alternate Problem:* Did the medieval schoolmen err in dealing with being, potency, act, becoming, and cause? Or rather did they err in not searching for specific variables (such as mass) and their mutual relations (laws)?

4.1.4. Does problem invention illustrate the pragmatist thesis that economy of work and, in general, simplification is the aim of science? *Alternate Problem:* Does the root function of problem in science confirm the empiricist thesis that experience originates all knowledge?

4.1.5. Comment on W. H. WHYTE, Jr.'s *The Organization Man* (1956; Garden City, N.Y., s.d.), Ch. 16, "The Fight Against Genius", and particularly on the effects of short-sighted scientific planning on research, for precluding the free choice of problems as dictated by "idle" curiosity

4.1.6. Give examples of actual research projects that have failed owing to either a wrong choice of problems or to their defective (e.g. loose) statement. *Alternate Problem:* Contrast the views on scientific creativity expressed by A. SZENT-GYÖRGY, in *Perspectives in Biology and Medicine*, V (1962), 173, with those of Lord ADRIAN, ibid., 269.

4.1.7. What is more important in determining the worth of an experimental inquiry: the size and accuracy of the experimental equipment or the size and statement of the problem? *Alternate Problem:* Examine and illustrate the role of phronesis (sound judgment) in the selection of problems. See M. BUNGE, *Intuition and Science* (Englewood Cliffs, N. J.: Prentice-Hall, 1962), pp. 89—90 and 102—104.

4.1.8. W. K. ROENTGEN (1895) noticed the fogging of photographic plates in the vicinity of cathode rays tubes. Many before him had noticed the same fact: in other words, Roentgen had no special set of new data to start with. Why did he rather than others discover X rays? And what does it mean to have 'discovered' them? *Alternate Problem:* State and compare the ideals of the investigator and of the erudite.

4.1.9. Comment on the following articles: A. WEINBERG, *Science,* **134**, 161 (1961) on the problem of choosing research lines that are likely to be fruitful, and P. WEISS, *Science,* **136**, 468 (1962) on trivial research and the bulk/quality alternative. *Alternate Problem:* Mention contemporary research lines that are comparatively trivial (shallow problems) and others that are deep.

4.1.10. Examine the following sentence by Nobel Prize winner I. I. RABI in E. FERMI *et al., Nuclear Physics* (Philadelphia: University of Pennsylvania Press, 1941), p. 25: "The lively state of nuclear physics is indicated by the fact that the newer experimental results seem to raise a new problem for every one which is resolved". Try to set up a measure of the vitality of a science as a function of the rate of emergence of new problems in it. *Alternate Problem:* "Intelligence" has been defined as ability to process information. If this definition is chosen, automatic computers can be said to be intelligent. Is this conception of intelligence more intelligent than the traditional one that equates it with the capacity for "seeing", explicitly posing, and solving new problems?

4.2. *Logic of Problems

The term 'problem' designates a difficulty that cannot be solved automatically but requires a search, whether conceptual or empirical. A problem, then, is the first link of a chain: Problem — Search — Solution. Human problems are problems of doing, knowing, valuing, or saying; in factual science all these kinds of problem are found, but of course the central ones are problems of knowledge. Whatever the nature of a human problem the following aspects of it may be distinguished: (i) the problem itself regarded as a conceptual object different from a statement but epistemologically on a par with it; (ii) the act of questioning (psychological aspect) and (iii) the expression of the problem by a set of interrogative or imperative sentences in some language (linguistic aspect). The study of questioning is taken up by psychology (including the psychology of science) whereas the study of questions regarded as linguistic objects (namely sentences ending with a question mark) belongs to linguistics. We are here interested in problems as a sadly neglected (by far the most neglected) kind of ideas analyzable with the help of other ideas.

In every problem ideas of three kinds are involved: the background and the generator of the problem, and the solution to it in case it exists. Consider the problem "Who is the culprit?". It presupposes the existence of a culprit; it is generated by the propositional function "x is the culprit", where x is the unknown to be uncovered; and it induces a solution of the form "c is the culprit", where 'c' names a definite individual. In other words, our problem is "Which is the x such that x is the culprit?", or "$(?x)C(x)$" for short. The generator of this question is $C(x)$ and the presupposition $(\exists x)[C(x)]$, whereas the solution is $C(c)$. Logically, then, we have the following sequence: (1) *Generator:* $C(x)$; (2) *Presupposition:* $(\exists x)[C(x)]$; (3) *Problem:* $(?x)C(x)$; (4) *Solution:* $C(c)$.

In general, every problem is posed against a certain *background* constituted by the antecedent knowledge and, in particular, by the specific presuppositions of the problem. The *presuppositions* of the problem are the statements that are somehow involved but not questioned in the statement of the problem and in the inquiry prompted by it. Further, every problem may be regarded as generated by a definite set of formulas. The *generator* of a problem is the propositional function which yields the problem upon application of the operator "?" one or more times. Finally, every problem induces a set of formulas — the *solution* to the problem — which, when inserted into the problem's generator, convert the latter into a set of statements with a definite truth value.

At first sight a question such as "Is p true?" does not fit the above schema: it is generated by p itself, which is supposed to be a proposition and not a propositional function. Yet clearly "$?p$" may be reworded as "What is the truth value of p?", which may in turn be restated as the problem: "Which is the value of the function V at p?". In symbols: $(?v)[V(p)=v]$, where V maps propositions p into their truth values v. If these values are just truth (or $+1$) and falsity (or -1) — i.e. if V degenerates into the ordinary valuation function — the original question will not have been modified by the preceding restatement. But if v is allowed to take further values within those bounds, then clearly the new formulation of the problem is more general than the unsophisticated

"Is p true?", which presupposes that a proposition can be just true or not true. In any case truth-value questions presuppose some theory of truth or other and they are questions about the value of an individual variable.

The blanks occurring in a problem may be individual variables or predicate variables. In the question "Who discovered America?", generated by the function $D(x, a)$, the only unknown is an individual variable. Similarly with the question "Where is c?", which involves the quantitative concept of place. On the other hand a problem such as "What does c look like?", directs us to search for the cluster of properties P, so far unknown, that make up the appearance of the individual c. We shall symbolize this question form thus: '$(?P)P(c)$'. Questions asking for the value(s) of one or more individual variables may be called *individual problems;* and questions asking for the value(s) of one or more predicate variables may be called *functional problems.* Every elementary problem is of either of these kinds.

What about the problem form "Does c have the property A?", in which no variable is in sight? The variable is hidden: it is the truth value of the proposition "c has the property A". In fact, the given problem is generated by the function "The truth value of the statement 'c has the property A' is v", and the solution to the problem consists in finding out the precise value of v. Therefore the explicit formulation of the problem is: "What is the truth value of the statement 'c has the property A'?".

Similarly, all the questions concerning existence or universality can be regarded as problem about the truth values of the corresponding existential or universal statements. Thus, "Are there gravitons?" can be restated as "Is it true that there are gravitons?" and, more precisely, "What is the truth value of the proposition 'There are gravitons'?". In the same way, "Is everything changing?" is equivalent to "Is it true that everything is changing?", which may be restated as "What is the truth value of the proposition 'Everything is changing'?".

And what about "What is A?", where 'A' denotes a predicate constant? Here too a variable is hidden and must be dug up in order to complete the question: in fact, what is being asked is "What are the properties P of A?", or $(?P)P(A)$ for short, where 'P' designates a predicate (cluster) of order higher than A. And this is a functional problem, the answer to which is constituted by a set of statements predicating definite properties of A.

The moral we extract from the above cases, all of them characterized by the seeming absence of variables in the questions, is clear: Do not let yourself be misled by language — *cherchez la variable.* Note, in addition, that the question mark — which we are handling as a primitive concept — bears always on an unknown or variable of the individual or of the predicate type. Furthermore, "?" does not bind the variable on which it acts: merely posing a question does not solve it. Only the answer, i.e. the solution, will be free from unbound variables.

In Table 4.1. some typical elementary question forms are listed and their logical form is exhibited. In it c and d are individual constants,

Table 4.1. *Elementary problem forms*

Problem kind	*Question*	*Form*	*Solution form*
Individual			
Which-problem	Which is (are) the x such that $A(x)$?	$(?x)\,A(x)$	$x=c, d, \ldots$
Where-problem	Where (at what place) is the c such that $A(c)$?	$(?x)\,[A(c)\rightarrow B(c)=x]$	$x=d$
Why-problem	Which is the p such that q because p (i.e., p entails q)?	$(?p)\,(p\vdash q)$	$p\equiv c$
Whether-problems	What is the truth value of p?	$(?v)\,[V(p)=v]$	$v=a$
	Does c confirm d?	$(?v)\,[V[C(c, d)=v]]$	$v=a$
Functional			
How-problem	How does c, which is an A, happen?	$(?P)\,[A(c)\rightarrow P(c)]$	$P(c)\equiv B(c)$
What-problems	What are the properties of c?	$(?P)\,P(c)$	$P(c)\equiv A(c)$
	What are the properties of the property A?	$(?P)\,P(A)$	$P(A)\equiv B(A)$

x and v individual variables; A, B and C are predicate constants and P is a predicate variable. The table is meant to be illustrative, not exhaustive.

A single unknown occurs in each of the problem forms listed in the table, none in their answers. This characterizes well-defined, definite, or *determinate problems* in contrast to ill-defined, indefinite, or indeterminate problems. The latter have indeterminate answers, i.e. answers in which free variables occur. A determinate question has a single answer with no unknowns, but the answer may consist of a conjunction of statements. For example, the question "$(?x)(x^2-x=0)$" has a single answer consisting of two members, namely the roots 0 and 1 of the equation. On the other hand the problem "$(?x)(x^2-x+y=0)$" is indeterminate because the unknown y remains free even after fixing x. It can be rendered determinate either by assigning y a definite value or by prefixing to it a question mark bearing on the second variable. In fact, "$(?x)(?y)(x^2-x+y=0)$" is a determinate question with a single answer consisting of an infinity of pairs $\langle x, y\rangle$. In short, a determinate question has a single answer with no free variables, and this answer can be either *single-membered* (as in the case of "What is the value of $y=x^3$ for $x=0.001$?") or *many-membered* (as in the case of "What are the human races?").

Definite answers can be got only if definite questions are asked. The question "What is the length of this rod?" will have a single answer on condition that 'this' is an unambiguous name in the given context, and that the reference system, the length unit, the temperature and the pressure are specified. Likewise "Where is c?" is not quite determinate: a name (in this case 'c') does not individualize anything save in a context; we must specify the cluster of properties A that individuate c and ask accordingly "Where is the c such that c is an A?", or "Given that c is an A, where is c?". Supposing the position of c could be fixed by a single number (the value of a coordinate), the form of this question, once completed, would be "$(?x)[A(c)\rightarrow B(c)=x]$, where '$B$' designates the quantitative concept of position. In short, all the relevant variables occurring in a problem should be displayed explicitly in order to ensure its determinateness unless the context makes it clear what the values of these variables are.

The preceding informal remarks may be summarized in the following *Definition:* A problem is *well-formed* if and only if it satisfies all the following *formation rules:*

Rule 1. The generator of a well-formed problem contains as many variables as unknowns.

Rule 2. As many question marks as variables are prefixed to the generator of a well-formed problem.

Rule 3. Every elementary well-formed problem has either of the following forms:

$$(?x)(\ldots x \ldots), \quad (?P)(\ldots P \ldots), \qquad [4.1]$$

where x is the individual variable occurring in the generator $(\ldots x \ldots)$ and P is the predicate variable occurring in the generator $(\ldots P \ldots)$.

Rule 4. Every nonelementary well-formed problem is a combination of elementary well-formed problems.

Rule 3 amounts to the stipulation that every problem, if well-formed and involving a single unknown, shall be either individual or functional. Rule 4 makes room for problems involving several unknowns, either individual or functional. The word 'combination', occurring in the statement of this rule, is vague and we cannot hope to sharpen it unless we build a full-fledged logic of problems; the following remarks may help elucidate the meaning of 'problem combination'.

Let us call $\Pi(x)$ an elementary individual problem and $\Pi(P)$ an elementary functional problem. The two problem forms can be subsumed under a single one by abstracting from the variable type, i.e. by introducing the concept of variable *v tout court:* $\Pi(v)$. Every elementary problem $\Pi(v)$ can be analyzed as $\Pi(v)=(?v)G(v)$, where $G(v)$ is the problem generator. Let now $\Pi(v_1)=(?v_1)G_1(v_1)$ and $\Pi(v_2)=(?v_2)G_2(v_2)$ be two elementary problems which we wish to combine. We build a third problem $\Pi(v_1, v_2)$ involving the variables v_1 and v_2 if we set up the task of solving $\Pi(v_1)$ *or* $\Pi(v_2)$ or both; this compound problem may be called a *disjunctive* problem and may be symbolized as: '$\Pi(v_1)$ *vel* $\Pi(v_2)$'. Similarly we state a fourth, quite different problem $\Pi(v_1, v_2)$ if we assign the task of solving $\Pi(v_1)$ *and* $\Pi(v_2)$; this may be called a *conjunctive* problem and symbolyzed as: '$\Pi(v_1)$ *et* $\Pi(v_2)$'. In the former case the solution to the compound problem will be the disjunction of the solutions to the elementary problems. Calling a and b the values obtained for the variables v_1 and v_2 respectively, the solution $S(a, b)$ to a binary disjunctive problem will be: $G_1(a) \vee G_2(b)$. Similarly, the solution to a conjunctive problem will be the conjunction of the solutions to the elementary problems; in the case of a binary compound, $S(a, b)=G_1(a) \,\&\, G_2(b)$. In short,

Disjunctive problem

$$\Pi(v_1, v_2)=[\Pi(v_1)\ \mathit{vel}\ \Pi(v_2)] \leftrightarrow \{S(a, b)=[G_1(a) \vee G_2(b)]\}. \qquad [4.2]$$

Conjunctive problem

$$\Pi(v_1, v_2)=[\Pi(v_1)\ \mathit{et}\ \Pi(v_2)] \leftrightarrow \{S(a, b)=[G_1(a) \,\&\, G_2(b)]\}. \qquad [4.3]$$

On the basis of these binary operations more complex problem forms can be built. Conversely, any given problem can be analyzed into simpler problems related by '*vel*' and/or '*et*'. Thus, a three-variable problem (or triatomic problem) may be analyzed in one of the following ways:

$$\Pi_1\ \mathit{vel}\ \Pi_2\ \mathit{vel}\ \Pi_3, \qquad \Pi_1\ \mathit{et}\ \Pi_2\ \mathit{et}\ \Pi_3$$

$$\Pi_1\ \mathit{vel}\ (\Pi_2\ \mathit{et}\ \Pi_3), \qquad \Pi_1\ \mathit{et}\ (\Pi_2\ \mathit{vel}\ \Pi_3)$$

It is clear that the functors *vel* and *et* obey the associative and the commutative laws. Also, the last two formulas can be expanded in the following way:

$$\Pi_1\ \mathit{vel}\ (\Pi_2\ \mathit{et}\ \Pi_3)=(\Pi_1\ \mathit{vel}\ \Pi_2)\ \mathit{et}\ (\Pi_1\ \mathit{vel}\ \Pi_3), \qquad [4.4]$$

$$\Pi_1\ \mathit{et}\ (\Pi_2\ \mathit{vel}\ \Pi_3)=(\Pi_1\ \mathit{et}\ \Pi_2)\ \mathit{vel}\ (\Pi_1\ \mathit{et}\ \Pi_3), \qquad [4.5]$$

which can in turn be generalized to any number of elementary constituents.

So far the analogies between the (future) calculus of problems and the calculus of statements are apparent. They can be stretched further by introducing the concept of *negate* of a problem (expressed by a negative question). We shall do this through the *Definition:* If $G(v)$ is the generator of $\Pi(v)$, then *non* $\Pi(v)=(?v)[-G(v)]$. It is often advantageous to switch from the given problem to its negate. For instance, "Which chemical elements are noble?" may be replaced by "Which chemical elements enter in compounds?", which is the negate of the foregoing. With the help of the concept of negate of a problem and with the assistance of the above formulas any problem can be analyzed into the conjunction (disjunction) of disjunctions (conjunctions) of simpler problems.

In addition to compounds of elementary problems, brought about by the *operations vel* and *et,* we recognize the binary *relations* of problem implication and equivalence. We shall say that Π_1 *implies* Π_2 just in case the generator of Π_1 implies the generator of Π_2; and we shall call the problems *equivalent* if their respective generators are equivalent. In symbols:

Problem implication $$(\Pi_1 \; seq \; \Pi_2) \leftrightarrow (G_1 \rightarrow G_2), \qquad [4.6]$$

Problem equivalence $$(\Pi_1 \; aeq \; \Pi_2) \leftrightarrow (G_1 \leftrightarrow G_2). \qquad [4.7]$$

In the simplest case $v_1=v_2$; in more complex cases v_1 will be a set of variables including v_2 as a subset. *Example* of problem implication: The problem of finding the truth value of a statement implies the problem of finding out whether the same proposition is true. *Example* of problem equivalence: the problem of deducing the conditional $A\rightarrow C$ from the premise P is equivalent to the problem of deriving the consequent C from the enriched premise $P\&A$. In fact, $\Pi_1=(?v_1)[V(P\vdash A\rightarrow C)=v_1]$ and $\Pi_2=(?v_2)[V(P\&A\vdash C)=v_2]$, whence $G_1(v_1)=[V(P\vdash A\rightarrow C)=v_1]$ and $G_2(v_2)=[V(P\&A\vdash C)=v_2]$. By definition of entailment, G_1 is equivalent to the statement that $P\rightarrow(A\rightarrow C)$ is logically true (tautologous); but, by the law of exportation, $P\rightarrow(A\rightarrow C)$ is equivalent to $P\&A\rightarrow C$. Hence, to say that $P\rightarrow(A\rightarrow C)$ is tautologous amounts to saying that $P\&A\rightarrow C$ is tautologous. But the latter is precisely G_2. Q.E.D.

Finally, if G_2 can be deduced from G_1, i.e. if G_1 *entails* G_2, we shall say that Π_1 is *more general* (or *stronger*) than or as general (or as strong as) Π_2. In symbols:

$$(\Pi_1 \geqq \Pi_2) \leftrightarrow (G_1 \vdash G_2). \qquad [4.8]$$

For example, dynamical problems are stronger than the corresponding kinematical problems, since the generators (hence the solutions) of the latter are derivable from the corresponding generators of the former. It is obvious that we must prefer the stronger questions, since these will induce the stronger solutions.

The partition of problems into individual and functional applies only to elementary or *atomic* problems. Rule 4 allows for the statement of non-elementary or *molecular* problems, which may be both individual (with regard to a certain set of variables) and functional (as regards another group of variables). Our classification of elementary problem forms cuts across alternative groupings proposed in the course of history. The best known among these are ARISTOTLE's and PAPPUS'. ARISTOTLE distinguished between *what*-problems, or questions of fact, and *whether*-problems, or dialectical questions. But from a logical point of view there is no difference between the problem "What is the distance between a and b?", a question of fact, and the problem "Does a imply b?", a dialectical question. Both are individual problems: the former asks what the value of the function $D(a, b)$ is, the latter what the value of the function $V(a \rightarrow b)$ is, and in both cases an individual variable is concerned. The difference between the two problems is not logical but methodological: the answer to each problem calls for a method of its own. PAPPUS distinguished between problems of construction (e.g., "Find the average of a set of given numbers") and problems concerning the logical consequences of assumptions. This distinction has recently been worked out and popularized by G. POLYA, who calls them *problems to find* and *problems to prove* respectively. The former are a subset of ARISTOTLE's what-problems, whereas the problems to prove are included in what he called whether-questions. Moreover, problems to prove are a subset of problems to find, since to prove a theorem consists in finding a set of assumptions such that they entail the given theorem; and this, in our classification, is an individual problem. The difference between the two is neither logical nor methodological but rather ontological: the solution to a "problem to find" consists in exhibiting an object other than a statement, whereas "problems to prove" are concerned with statements and their logical relations.

Nor does our classification of elementary problems make a special place for *decision questions*, i.e. problems the answer to which is either a straight "Yes" or a straight "No". Decision questions are special cases of individual problems and particularly of those calling for the determination of truth values. The question 'Is p true?' is of this kind and so is 'Does t belong to the set T?', which is just an instance of the former, namely when p takes the particular form "$t \in T$". Whether or not a given problem is a yes-or-no one is not a logical but a methodological question: the means at hand and the goal in view will enable us to make the decision to regard certain problems as yes-or-no questions. Take, for instance, the problem "How tall is c?", where c names a definite man.

No matter how sophisticated a measurement technique we may choose, the original question can be broken down into a finite sequence every member of which is a yes-or-no problem of the form "Does the top of c fall between the scratches n and $n+1$ of our yardstick?". An improvement in accuracy will enable us to ask more questions of this kind and therefore to come closer to the (supposedly unique) truth, but the non-vanishing error of every measurement procedure will ensure that there will be a finite number of decision questions to ask. And this finiteness is, of course, necessary for the procedure to be *effective*, i.e. performable in a finite number of steps. An exact solution to the given problem would require infinitely many such unit steps and is therefore unattainable.

The hope of the factual scientist (and of the applied mathematician) is that, no matter how complex his problem may be, it can eventually be reduced to a finite sequence of yes-or-no problems. Every achievement of such a methodological triumph hides an epistemological defeat: a strong problem, such as asking to identify one member of a nondenumerably infinite set, has been replaced by a finite set of weak problems, such as asking to decide whether a given individual belongs to a given set. But there is no choice: either we take the solution to the weaker problem or we are left with the unsolved stronger problem.

Let us now turn to a semantical aspect of problems. Abiding by Rules 1—4 is necessary but not sufficient for securing determinate questions, i.e. questions with a unique (though possibly many-membered) answer. In fact, a problem may be well-formed but its background may be defective or just vaguely indicated. For example, the question "Is p true?" is well-formed but in asking it we presuppose that p has or can be assigned a single truth value, which is by no means obvious since p might be true in one system and false in another. Likewise, the problem "What is the melting point of sulphur?" presupposes that sulphur, no matter what its crystalline form may be, has a single melting point — which is false. No question is ever posed without presupposing something. To raise any question at all presupposes our own existence and to ask about the ways of things presupposes at least the possibility of their existence and the possibility of our knowing them to some extent.

Since there is no question without a background, and since the background may be constituted by falsities or just by controvertible ideas, the naive acceptance of a question without examining its background is no better than the naive acceptance of an answer without examining its ground. The defective statement of a question — i.e. the formulation of an ill-formed question — may prevent proper inquiry or perhaps inquiry at all, as is the case of "What is being?" which, far from being similar to "What is motion?", is of the kind of "What barks barking?". Yet the defective conception of a question — that is, the raising of a question with a wrong or indefinite background — can be even more misleading because it may launch research on a fruitless path. Thus, the question "What is the guarantee of truth?" has originated endless and fruitless speculation by presupposing that there is a truth warrant that must only be uncovered.

Let us lay down the following conventions regarding the background of problems. First the following *Definition:* A problem is *well-conceived* if and only if none of its presuppositions is a manifestly false or an undecided formula in the same context. Second *Definition:* A problem is *well-formulated* (proper, sound) if and only if it is both *well-formed* (well-stated, i.e. complying with Rules 1—4) and *well-conceived* (well-backed). With these definitions in mind we formulate our last prescription:

Rule 5. Every problem shall be well-formulated.

A well-formulated problem will be a determinate (well-defined) one: it will have a unique solution and, by displaying all the relevant items, it will at least suggest what further searches may help to solve it. Yet it would be naive to suppose that merely abiding by Rules 1—5 will warrant our asking only well-formulated questions. This, if only because it is always difficult to unearth and examine all the relevant presuppositions of a problem. Even in a formalized theory only those presuppositions which have been realized by its inventor to be relevant are listed; except in trivial cases the list is almost surely incomplete: many advances are made by discovering that a certain formula is either needed or dispensable in the background of a theory. Consequently a question that has *bona fide* been accepted as well-conceived or meaningful may turn out, upon closer examination, to be ill-conceived.

Rigor would seem to require an examination of the presuppositions of every presupposition, and so on until ultimate presuppositions are reached — or not. This is possible, at least in principle, in formal science: here we can dig until we reach set theory and the initial assumptions of set theory. In factual science it is still unknown whether there is any basic theory, although it is usually believed that it must be mechanics — only nobody knows which theory of mechanics: is it classical mechanics, which is known to be only partially true, or relativistic mechanics, which is not independent of electromagnetic theory, or finally quantum mechanics, which includes classical mechanics? In factual science the pattern may be a net rather than a line, and in any case it is premature to legislate in this respect. What we must secure is the *right* to proceed as far as is deemed necessary in each case. Stated negatively: we must refuse to acknowledge unmovable ultimate factual axioms. Presuppositions must be regarded as relative: what is an unquestioned statement in a given context may be the subject of inquiry — and consequently of correction or even rejection — in an alternative context or in a new stage of growth.

The realization of the changeable condition of presuppositions must help us understand the relativity of the well-conceivedness or meaningfulness of questions. There are no *inherently* well-conceived or meaningful questions: a question is well-conceived, it will be recalled, if and only if its background in the same context is sound. Consequently, if we accept the presuppositions of a question we must accept the question itself as well-conceived, otherwise not. Thus, for instance, within the bounds of strict operationalism (see Sec. 3.6) the question about the temperature of the interstellar space is meaningless because there is no operation by which an answer could be given to it: the placement of a measuring instrument and a recording device (e.g., a man) would put an end to the void. The same question is meaningful for the realist (and important for the astronaut) because he does not presuppose that to exist is to measure or to be measured.

Change the context, and the meaningfulness of a question may change accordingly. Now, since the context — i.e. the set of scientific and philosophic theories relevant to the problem — is constantly changing, it would be unwise to dismiss certain questions — e.g., those we dislike — as inherently, hence everlastingly meaningless. It is wiser to adopt a humbler attitude and acknowledge either that the problem does not interest us or that it is interesting but *premature*, i.e. that the adequate tools for handling it are yet to be built.

Limitationist research policies as well as sterile fights over *isms* can be avoided by adopting this more modest strategy which does not kill interesting problems but defers their treatment until the proper theoretical tools are available. The early reflexologists and behaviorists would not have deserved the reproaches of traditionally oriented psychologists if, instead of dismissing as meaningless ("metaphysical") all questions regarding the higher functions of the nervous system, such as consciousness and cognition, had made it clear that theirs was not an ontological but a methodological behaviorism: that they started by studying the simplest problems of animal psychology, not because there were no other problems but because those had to be solved before the much more complex problems of the human mind could even be posed. Pseudoscience and antiscience thrive not only on ignorance and on the deliberate attempt to supress enlightenment but also on the deliberate refusal, by scientists, to even consider perfectly legitimate though perhaps premature problems.

In any case there are formally simple questions, in the sense that they are elementary or atomic (i.e. involve a single unknown) but there are no semantically simple questions: every question has some body of presuppositions. The vexing question "Have you ceased beating your wife?" is not to be laughed away because it is a complex question but because it has a presupposition that may be false. The important thing is not to dispense with presuppositions — an impossible task — but to keep them under control, i.e. to subject them to critical scrutiny as soon as wrong solutions appear. Research being elicited by questioning, we conclude that no research can start from scratch: there is no presuppositionless research. But we are already on the threshold of the methodology of problem stating.

Problems

4.2.1. Symbolize the following questions and identify them.

1. What is the square of x?
2. Which is the number such that, added to any given number, yields this same number?
3. How is a related to b?
4. What is c for?
5. Where are you going?
6. What are you doing?
7. Which is the function f such that, for any x and y in a number set, $f(x \cdot y) = f(x) + f(y)$?
8. Does the theorem t follow from the assumptions A?
9. Does the evidence e confirm the hypothesis h?
10. Does the solution $G(c)$ solve the problem $\Pi(v)$?

4.2.2. Attempt to symbolize the following questions and find out whether they are well-formed.

1. What does doing do?
2. How does becoming become?
3. Am I?
4. Where is nowhere?
5. What is the being of nothingness?

Alternate Problem: Analyze the problem form expressed by 'If p, why q?', or equivalently 'Since p is the case why should q be the case?'. Hint: supply the missing premise.

4.2.3. State three questions that have no unique answer and then complete them by supplying the missing variables and constants. *Alternate Problem:* The statement "Every x satisfies the condition C" is methodologically equivalent to the problem "Which x's fail to satisfy condition C?". Work out the concept of methodological equivalence between statements and problems.

4.2.4. Many criminologists have asked "Why is criminality higher among the lower classes than among the higher classes?" What does this question presuppose? And what should be done before attempting to answer it? See R. K. MERTON, *Social Theory and Social Structure*, 2nd ed. (Glencoe, Ill.: The Free Press, 1957), p. 90. *Alternate Problem:* What, if any, are the differences between problems, questions, and tasks?

4.2.5. Point out the presuppositions of the following questions.

1. Where is c?
2. When did c occur?
3. What is c?
4. What is new?
5. What are the ultimate constituents of matter?

Alternate Problem: Examine the way K. R. POPPER, in *Conjectures and Refutations* (New York: Basic Books, 1963), pp. 21—27, handles the question 'How do you know?'

4.2.6. What, if anything, is wrong with asking "Why is there something rather than nothing?" or, equivalently, "Why being rather than nothingness?". Hint: find out whether this is a root question or whether it does presuppose something. *Alternate Problem:* Prove the following theorem: $(\Pi_1 \geqq \Pi_2) \rightarrow (G_1 \rightarrow G_2)$.

4.2.7. Examine the criterion of meaningfulness (well-conceivedness) of questions proposed by P. W. BRIDGMAN in *The Logic of Modern Physics* (New York: Macmillan, 1927), p. 28: "If a specific question has meaning, it must be possible to find operations by which an answer may be given to it". Some of the questions regarded as operationally meaningless are (op. cit., pp. 30—31): "Was there ever a time when matter did not exist?", "May space be bounded?", "Why does nature obey laws?". Are these questions absolutely meaningless, i.e. are they ill-conceived in every possible context? *Alternate Problem:* Perform a linguistic analysis of questions. See N. CHOMSKY, *Syntactic Structures* ('S-Gravenhage: Mouton, 1962), Ch. 7.

4.2.8. Consider the question: 'What would the world be like if the universe as a whole and everything in it, including our measuring sticks, doubled its size overnight?' Do you regard it as a well-formed and well-conceived problem? If so, do you think one of the following is the proper answer? (i) We could not know how the world would be like because the question involves an untestable assumption. (ii) The world would be exactly as it is now. (iii) There is no reason to assume such an event is possible — quite apart from the possibility of testing the assumption. See N. L. Wilson, "Substances without Substrata", *Review of Metaphysics*, **12**, 521 (1959). *Alternate Problem:* Elucidate the concepts of direct and inverse problem. In particular, find out whether being direct (or inverse) is a logical property of a problem. i.e. one independent of its genesis. Hint: Begin by examining simple cases like the pair $(?x)R(x, b)$, $(?y)R(a, y)$. See G. Boole, *A Treatise on Differential Equations*, 3rd ed. (London: Macmillan, 1872), pp. 385—388.

4.2.9. Consider the class of questions exemplified by: 'What would happen if men managed to become immortal?' and 'What would happen if matter were destroyed (or created) at a given rate?'. Such questions, involving the assumption that one or more fundamental laws of nature might cease to operate, or that the corresponding law statements might eventually be found altogether false, may be called *counterlegal questions*. What is the function, if any, of counterlegal questions: (i) to keep philosophers busy and amuse people?; (ii) to clarify hypotheses and theories?; (iii) to test hypotheses and theories? *Alternate Problem:* Consider the question: 'How does an atom behave when left on its own, i.e. when not subject to observation?'. Does it make sense to a subjectivist (e.g., an operationalist)? And is this question actually being asked in physics?

4.2.10. If we want to follow an electron in its flight we must interfere with it, e.g. by lighting it with gamma rays or by interrupting its motion with screens with tiny holes in them. But then we disturb the electron's motion and consequently cannot answer the original question, namely 'What is the undisturbed electron trajectory?'. The usual solution to this puzzle is the following: "The original question is meaningless (ill-conceived) because it presupposes that the electron has a definite position at each time — an unwarranted hypothesis that leads to trouble". Could the original question become meaningful in a different theoretical context and might it receive an experimental treatment with the help of finer, less brutal means than those now available? Or should the problem be banned forever? *Alternate Problem:* Investigate the mathematics of yes-or-no question sets which have no simultaneous 'yes' answers. See G. W. Mackey, The *Mathematical Foundations of Quantum Mechanics* (New York: W. A. Benjamin, 1963), pp. 64ff.

4.3. Scientific Problems

Obviously not every problem is a scientific problem: *scientific problems* are just those which are posed against a scientific background and are investigated with scientific means in order primarily to increase our knowledge. On the other hand, if the goal of research is practical rather than cognitive, but the background and the tools are scientific, then the problem is one in applied science or technology rather than in pure science (see Sec. 1.5). Yet there is no hard and fast line dividing scientific from technological problems since one and the same problem, approached and solved with whatever aim, may end up in a solution having both a cognitive and a practical value. Thus, studies in the ecology and ethology of rodents may have both a scientific value and a value for agriculture and medicine.

The class of scientific problems — itself a subclass of the problems of knowledge — may be analyzed in various ways. The following dichotomy will be adopted here:

Scientific problems <
- *Substantive or object problems* (Ex.: "How many A's are there?")
- *Strategy or procedure problems* (Ex.: "How shall we count A's?")

Whereas object problems concern the ways of things, procedure problems concern our ways of getting to know about things and about our knowledge. Substantive problems, in turn, can be subdivived into empirical and conceptual, and strategy problems into methodological and valuational. The solution to empirical problems requires empirical operations in addition to thinking, whereas conceptual problems are handled with the brain alone although they may call for conceptualizations of empirical operations, such as data. Methodological and valuational problems are both conceptual as regards the way they are approached and solved; they differ in that whereas the solutions to valuational problems are value judgments, the solutions to methodological problems are value-free. Tables 4.2 and 4.3 display some of the most important species of the four genera of problems.

By definition, empirical problems do not occur in formal science; and when a problem in formal science is translated into a factual analog the solution to the latter must be retranslated into the original terms. Empirical problems intermingle with conceptual ones; they are not characterized by the absence of theoretical considerations in their statement and handling but by the occurrence of empirical operations in the course of their solution. On the other hand conceptual problems require no empirical operations but at most ideas suggested by the latter. As to methodological problems, they are particularly conspicuous in the younger sciences; for instance, the interest in such problems in contemporary sociology is comparable with the interest in methodological problems that accompanied the birth of modern physics toward the end of the 16th century. In either case the traditional approach was found wrong and wholly new methods were sought. Finally, the inclusion of valuation problems in science is apt to raise philosophical eyebrows in view of the deep-seated fact-value dichotomy. Moreover, is it not true that modern science could not begin until nature was freed from values and other anthropomorphic attributes? True but irrelevant: nature is value-free yet natural science is not concerned with substantive problems alone but also with inventing and analyzing ways of handling such problems — and value judgments will be made in the course of

Table 4.2. *Substantive problems*

1. *Empirical problems*
   1.1. *Data finding:* characterizing objects of experience
      1.1.1. Observing
      1.1.2. Counting
      1.1.3. Measuring
   1.2. *Making:* constructing and calibrating instruments, preparing drugs, etc.
2. *Conceptual problems*
   2.1. *Describing:* characterizing individuals and classes
   2.2. *Arranging:* classing and ordering sets
   2.3. *Elucidating:* interpreting signs and refining concepts
   2.4. *Deducing*
      2.4.1. Computing (e.g., finding the value of a variable)
      2.4.2. Proving a theorem
      2.4.3. Checking a solution
      2.4.4. Explaining: accounting for facts and empirical generalizations in terms of theories
      2.4.5. Projecting: predicting or retrodicting facts
   2.5. *Building:* inventing ideas
      2.5.1. Introducing a new concept
      2.5.2. Introducing an empirical generalization
      2.5.3. Introducing a high level hypothesis subsuming empirical generalizations
      2.5.4. Building a system of high level hypotheses (a theory)
      2.5.5. Theory reconstruction (foundational research)
   2.6. *Metalogical:* uncovering and removal of inconsistencies, proving consistency and independence, etc.

Table 4.3. *Strategy problems*

1. *Methodological*
   1.1. *Conventions:* setting up designation rules, measurement scales and units, levels of significance, etc.
   1.2. *Techniques:* devising tactics for solving problems, observing, measuring, etc.
   1.3. *Experiment design:* planning experiments
   1.4. *Theory design:* planning the building of theories
   1.5. *Method examination:* analysis and criticism of any of the above
2. *Valuational*
   2.1. *Weighing* data, hypotheses, theories, techniques and material equipment in terms of given desiderata
   2.2. *Foundational valuation:* examination of the desiderata themselves

this work. Whenever an experimenter is faced with the problem of choosing among different equipments for the same purpose he will somehow weigh factors such as range, accuracy, versatility, reliability, and cost, in attempting to frame an all-round value judgment; similarly the theoretician will compare competing hypotheses and theories as to coverage, depth, support from other fields, and even formal elegance. Every decision rests on a set of value judgments and decisions are made in scientific research all the time even though they do not occur explicitly in the resulting body of substantive knowledge.

The grouping of problems sketched in Tables 4.2 and 4.3 is not entirely adequate as a partition because most "whole" scientific problems are rich enough to fall simultaneously in all four categories. "Empirical", "conceptual", "methodological" and "valuational" should not be regarded as mutually exclusive characteristics but rather as properties that are alternatively stressed in the course of research. Thus, e.g., the problem of finding out the effect of a given drug on the nervous system can in turn be decomposed into the following tasks: (i) the methodological problems of designing the appropriate experiments and choosing the level of significance of the correlations found with the help of experiments; (ii) the empirical problems of making the drug or purifying it, of administering it, and of recording its effects; (iii) the conceptual problems of interpreting the data and hypothesizing the drug's mode of action (e.g., reaction mechanisms in the organism); and (iv) the valuational problem of ascertaining whether the given drug is better or worse, in relation to certain goals, than rival drugs.

Our list of problem kinds does not exhaust the problems that arise in scientific research, many of which are not scientific problems. Problems of budget, supply, division of labor, team training and integration, can all be approached scientifically, thereby becoming scientific problems: as research becomes a major branch of production, the problems of research management tend to be approached with the help of operations research, social psychology, and so on. But usually such problems are still handled on a prescientific level, both because of the weight of tradition and because the theories of action are still underdeveloped.

Having dealt with the taxonomy of scientific problems we may now approach their phylogeny. Scientific problems do not arise in a vacuum but in the midst of an existing body of knowledge constituted by data, empirical generalizations, theories, and techniques. If someone wants to find out, say, the exact chemical formula of platinum oxide, it is because he knows or suspects the existence of platinum oxide, and knows in addition (i) some of the properties of platinum oxide (data), (ii) something about the laws of chemical bonds (theory), and (iii) certain empirical procedures, such as X ray analysis (techniques). On the other hand if somebody asks 'What is the meaning of life?', or 'What is the meaning of history?', he may get along without data, theories, and techniques, because he is asking indeterminate questions, if only on account of the ambiguity of the terms 'meaning', 'life' and 'history'.

The very selection of problems is determined by the state of knowledge — particularly, by the holes in our knowledge — by our aims, and by methodological possibilities. Big problems can only be vaguely stated, hence hardly solved, when the background knowledge is weak. (Consequently we should not be surprised if the sciences of man are still tackling comparatively modest problems leaving momentarily the deeper questions in the hands of pseudoscientists: scientists have not yet built the adequate framework — the theories — where such deeper problems can be stated correctly.) Take, for instance, the question whether a horse might evolve into a tree-climbing animal: it cannot even be posed outside the context of an evolution theory. Or take the question EINSTEIN asked himself as a youngster and which gave rise to his general theory of relativity: Why does the acceleration of a body immersed in a gravitational field not depend on the body's mass? EINSTEIN's question would have been strictly meaningless to, say, GALILEI: it could not have been asked before the classical theories of gravitation and electrodynamics were born. Every theory delimits the set of problems that can be stated.

Moreover, problems do not "arise", they are not impersonally "given" to the investigator: it is the individual scientist, with his fund of knowledge, his curiosity, his outlook, bent and biases, who notices the problem or even searches for it. Therefore the idea that every branch of science has its own permanent conceptual outfit is mistaken: in science, as in catch-as-catch-can, one uses whatever one can get hold of. If all biologists were taught set theory, the theory of relations, lattice theory, differential equations and integral equations, they would use them just because it would occur to them to pose new biological problems requiring such formal tools, or they would apply them for the accurate formulation and solution of current problems. Similarly, if the psychologist who studies the formation and evolution of basic concepts in children were better acquainted with concepts other than class concepts, he would presumably devote more attention to the ontogeny of relation and quantitative concepts. Even physicists would profit from some logical training: they would not speak of operational definitions, would not try to select the basic (primitive) concepts among those referring to observable traits, and would not take successful predictions for the sole and ultimate test of theories.

Curiosity alone does not beget problems: we rarely pose problems to attack which we lack adequate procedures altogether. And if we do lack such procedures and feel that our problem is important, we pose the further problem — of a methodological and not substantive kind — of designing new methods. This is what PAVLOV did when he faced the problem of founding an objective science of behavior; it is also what ASTON did when confronted with the impossibility of separating isotopes by the available (chemical) techniques of analysis. Obviously neither PAVLOV nor ASTON would have formulated their respective problems had they not known that the procedures on hand were inadequate, and had they not hoped to find new ones.

But having a problem-solving technique is insufficient: we must have also a set of data. In the ideal case this will be a set of necessary and sufficient bits of information. In real research we often meet these other cases: (i) *too few data* — which calls for either the completion of the available information or the search for an approximate solution to the given problem; (ii) *too many data:* a large number of bits of information, partly irrelevant, partly raw or undigested by theory, and only partly adequate — which calls for a preliminary data sifting and condensing in the light of new hypotheses or theories.

The possession of a fund of data, techniques and theories is then necessary to pose and attack a scientific problem. But it is not sufficient. We must be reasonably sure, too, that we will be able to *recognize the solution* once we have found it. Moreover, we must stipulate in advance: (i) what *kind of solution* will be regarded as adequate and (ii) what *kind of check* of the proposed solution will be regarded as satisfactory. Otherwise we may get lost in a fruitless search or in an endless argument. For example, if somebody approaches the problem of disclosing the mechanism of the emergence of living matter with a view to refuting the vitalist, the two contenders must agree beforehand on (i) whether synthesizing a virus or an organism of the order of magnitude of a whale will be regarded as necessary and sufficient, and (ii) what kind of properties must the artificial organism possess in order to count as a living being.

In addition to stipulating in advance what the solution should look like in order to be able to recognize it as such if and when we get it, we should approach the problem of the *existence* and the *uniqueness* of the solution before attempting to effectively solve the problem. In pure mathematics and in the sciences employing mathematics these previous questions are standard: the existence of a solution and its uniqueness are proved, or else it is proved that no solution exists or that, if it exists, it is not unique. (In practice one hopes a unique solution exists and does not bother to prove it until trouble arises, but in any case it is recognized that proofs of existence and uniqueness are logically prior to attempting to solve the problem.) Of course, even proving the existence of a solution does not secure the finding of it: very often, for lack of adequate means we can at most obtain an approximate solution. The importance of securing a unique (though possibly many-membered) solution is as clear as the importance of securing the existence of a solution. It is only unique solutions that can be used to give unambiguous accounts of the behavior of things: just think of a force field described by a function with more than one value at every point in space (a multi-valued function). Existence and uniqueness theorems specify under what conditions a solution exists and/or it is unique; these conditions may not belong to the initially given set of data: they may have to be required from the theory in which the problem is immersed.

We may now sum up the conditions, both necessary and sufficient, for a problem to rank as a *well-formulated scientific problem:* (i) a body of scientific knowledge (data, theories and techniques) must be available in which the problem can be made to fit so that it can be handled at all: stray problems are not scientific; (ii) the problem must be well-stated in the sense that the formal requirements laid down in Sec. 4.2 are met; (iii) the problem must be well-conceived in the sense that its background, and in particular its presuppositions, are neither false nor undecided; (iv) the problem must be circumscribed: an approach which is not piecemeal is not scientific; (v) existence and uniqueness conditions must be found; (vi) stipulations as to the kind of solution and the kind of check that would be acceptable must be laid down in advance. Sticking to these conditions will not warrant success but will avoid waste of time.

The above are necessary and sufficient conditions for a problem to be a well-formulated scientific problem. But problems of this kind are occasionally dull, whereas ill-formulated problems can be exciting. A very important condition of a psychological order, that of being interesting to someone well-equipped to handle the problem, must be added if the corresponding research is to be fruitful. Scientific research, no less than art and politics, requires passion if it is to be fruitful. Of course, there are no recipes for falling in love with a problem, other than wrestling with it. Which in turn requires becoming acquainted with the scientific (cognitive, extrapersonal) motivations of the problem, which motivations will be found on examining the problem setting. Now, becoming familiar with the problem setting and developing a taste for it depends both on the individual's leanings and on the state of the science he is interested in. And this state is characterized not only by the achievements that have already been reaped but also by the trends and fashions of the moment. Because there are fashions in science as in every other branch of culture.

Instinctive behavior, such as bird nesting and migration, the spinning of webs by spiders and the forms of communication among bees, had been favorite subjects of biology (specifically, of ethology) during the second half of the last century but had become almost disreputable toward the end of the 1930's. After World War II they again became fashionable, and for good reasons. The previous work had been exclusively descriptive and alienated from theory: one reason for disregarding it. With the development of the science of control and communication a new, deeper approach became possible; also, the relationships between genotype and behavior can now be better controlled; finally, the interest of ethology for the newer psychology and sociology was clear. There were, then, reasonable motives for the revival of interest in instinctive behavior. Yet a small component of snobbery is detectable in it: most people like to be up-to-date not only in knowledge and approach but also in subject — and this is clearly unreasonable, because subjects are essentially problem systems, and problems should die out to the extent to which they are solved, not to the extent to which they are sidestepped.

The realization that the choice of problems is partly determined by the intellectual climate and that this includes fashion is important to prevent underestimating, and consequently undersupporting, unfashionable but serious research — that which only investigators with an established reputation can afford to pursue. The value of problems does not depend on whether many "wear" them at the moment but on the changes their investigation would conceivably force in our body of knowledge.

Suppose, finally, we have hit on a well-formulated scientific problem that happens to intrigue us: can we ascertain whether it will be a *fertile* problem or just an entertaining puzzle? No necessary conditions are known which ensure the fertility of a problem, hence the fruitfulness of research on it. Yet every scientific problem, if investigated seriously, is bound to bear some fruit sooner or later, because scientific problems are *systemic* by definition: they occur in or can be introduced into a system, and this alone warrants their inquiry to have some effect or other. Stray questions receive stray answers leading nowhere, but if a step is taken somewhere along a research line, the whole line may move ahead: that is, new problems can be posed. Therefore intelligent scientific management, far from requesting immediate results, will encourage research on *any* well-formulated scientific problem that has fired the imagination of a competent investigator. That is, scientific management, if intelligent, will ensure freedom of research — which is largely, as we shall see in a moment, freedom to plan.

Problems

4.3.1. Mention one specific scientific problem of every one of the following kinds: empirical, conceptual, and methodological.

4.3.2. Identify the kind of problem to which each member of the following sequence of problems belongs. (i) How is it possible to exterminate a given insect species? (ii) What toxic substances affect most the given species? (iii) How can chemicals be mass-produced at low prices, and how can they be handled without danger, in order to exterminate the species concerned? (iv) How can the destruction of useful species, upon the application of the insecticide, be prevented? (v) How will the ecological equilibria be disrupted by the destruction of the given species?

4.3.3. Propose examples of basic problems, i.e. those which require the critical examination or even the introduction of basic assumptions rather than their application.

4.3.4. Suppose you are assigned the task of finding the Abominable Snowman, that hairy human-like giant said to walk barefooted on the snowy Himalayan hills. Would you regard this as a problem meeting all the conditions laid down at the end of this section?

4.3.5. Consider the problem of searching for signals coming from extraterrestrial intelligent beings. What kind of signal would you regard as conveying meaningful information — long before you had been able to find their code? See S. von Hoerner, *Science*, **134**, 1839 (1961), or A. G. W. Cameron, Ed., *Interstellar Communication* (New York: W. S. Benjamin, 1963).

4.3.6. Could you reasonably question the existence of the whole world? Would this constitute a proper scientific problem — namely, whether the universe exists or not? See S. Hook, "Pragmatism and Existenz-Philosophie", in S. Uyeda, Ed., *Basis of the Contemporary Philosophy* (Tokyo: Waseda University, 1960), p. 401.

4.3.7. Decide whether the following combinations of predicates referring to scientific problems are possible: ill-formulated and important, well-formulated and trivial, isolated and fruitful, fashionable and deep.

4.3.8. Give a few instances of once-popular and now unduly neglected problem systems. *Alternate Problem:* Geometry was so fashionable until mid-19th century that the French for 'mathematician' was *geomètre*; but during our century it went out of fashion. Why? Solid state physics was neglected until about 1950 and is now fashionable. Why? Gravitation theory was forgotten between 1930 and 1960. Why? Evolution theory was displaced by genetics until interest revied in the late 1930's. Why?

4.3.9. Given that scientific problems are formulated against some background of scientific knowledge and that the latter is increasing exponentially, what is your guess concerning the number of scientific problems that will be faced in the future: is it likely to increase or to decrease with increasing knowledge?

4.3.10. "What is x?" has been called a question stated in the *material mode of speaking*, whereas questions of the type of "What is meant by 'x'?" have been christened questions in the *formal mode of speaking*. Members of the Vienna Circle thought the former were metaphysical and eluded them by restating them as language questions. Thus, e.g., instead of asking what a physical object such as an electron really is, we should ask what the syntax and the semantics of the term 'electron' are. What, if anything, is gained by restating a problem of knowledge as a problem of language? See R. CARNAP, *The Logical Syntax of Language* (London: Routledge & Kegan Paul, 1937), Secs. 75 and 80, and P. FRANK *Foundations of Physics*, in *International Encyclopedia of Unified Science*, Vol. I, No. 7 (Chicago: University of Chicago Press, 1946), Sec. 49. *Alternate Problem:* Study the effect, on the choice of scientific problems, of the following factors: (i) preference of governments and private enterprises for safe research; (ii) pressure to produce practical results (e.g., of military or commercial value); (iii) competition for jobs and grants, hence haste in obtaining "concrete" results; (iv) wholesale use of computers. Speculate on the future of science if these factors are not checked on time.

4.4. A Paradigm, a Framework, and a Simile

In contrast to nonscientific problems, scientific problems are members of *problem systems*, i.e. they constitute sets of logically interrelated problems. A problem system is a *partially ordered* set of problems, i.e. a branching sequence of problems arranged in order of logical priority. Discovering and modifying such a partial ordering of problems is part of the *strategy of research:* it must be sketched, if only in outline, if the investigation is not to be haphazard, hence fruitless or nearly so.

Routine problems are those which can be handled with fixed strategies, because no unexpected big novelty arises in the course of their investigation. *Research* problems, on the other hand, call for variable strategies: the (partial) ordering of problems may have to be altered in the course of research more than once as the results of the investigation throw new light on the original problems and as new problems emerge that had not been foreseen when the original strategy was planned.

The need for changing plans corroborates rather than refutes the thesis that scientific research is planned search, if only partially and on a small scale: it could not be otherwise since research consists in handling partially ordered sets (systems) of problems. The freedom of scientific research does not consist in a lack of orientation or program, but in the freedom to choose problem systems, approaches, methods, and solutions with no aim other than truth. Research is not free when planless but when the investigators themselves program their work and change their program in response to internal needs.

Let us illustrate the systemicity of scientific problems with a case of interest in social science: the question of power — which, of course, is not a single problem but a complex problem system. It may be analyzed — though not uniquely — into the following ordered steps.

1. *How is power described?*

1.1. *What are typical instances of power situations?* That is, what cases that are intuitively (presystematically) recognized as involving a power relation shall we keep in mind as standards?

1.2. *What factors are relevant to power:* what are the variables on which power depends? Natural resources? Labor force? Technical level? Repression forces? Ideas? And what factors are concomitant with power? Hierarchical organization? Privilege? Right? Violence? Indoctrination? Corruption?

1.3. *Where does the power relation hold:* in nature of just in society? If in the latter: at the individual level, at the molar level, or at both? That is, what are the relata of the power relation: individuals, groups, or both?

1.4. *What is the taxonomy of power:* what are the kinds of power and the kinds of power situations and how are these kinds related?

2. *How is power analyzed?*

2.1. *How is power to be approached:* what point of view shall we adopt? Shall we select a special kind of power (economic, political, ideological) or shall we study power in general? Shall we study the psychological or the social aspects of power, or both? And are we going to take an externalist (phenomenological) point of view or shall we meddle with the mechanisms of power? *In the former case we might choose as our basic variables the probabilities of the various means that the unit y may employ to achieve a given end, and investigate the way those probabilities alter when y falls under the power of x. In a second stage we might wish to adopt a deeper approach, trying to analyze those probability changes in terms of the resources that x and y can mobilize in the attempt to achieve their aims, and of the ability with which they handle their resources.*

2.2. *How is power defined:* what properties are necessary and sufficient to characterize the power relation? It is clearly an order relation —

but what else? If the definition is to serve as an operational criterion for recognizing the exercise of power — if it is to answer the question 'How is power recognized?' — the definiens concepts must be accessible to observation, whether directly or not; otherwise no such restriction is necessary. *We might, for instance, try the following definition: "x exerts power over y in the respect z if and only if the behavior of y in the respect z in the presence of x differs markedly from the behavior of y in the respect z when x is removed".* Any definition will in turn pose further problems: is it formally correct and does it cover the typical cases of power we have in mind?

2.3. *How is power measured*? Should we rest content with a comparative concept of power or can we analyze it in terms of objective quantitative features? In case we try latter course, what unit of power shall we adopt?

3. *How is power interpreted*?

3.1. *What is the statics of power:* what are the power relations among the members of a set when in equilibrium? (Search for the laws of power equilibrium.)

3.2. *What is the kinematics of power:* how do power relations emerge and how do they change in the course of time? What configurations are unstable and what are the most probable directions of change: towards equilibrium or away from it? (Search for the laws of power evolution.)

3.3. *What is the dynamics of power:* what forces can alter the balance of power and what forces can restore equilibrium? (Search for the laws of power mechanisms.)

The above paradigm of research strategy consists of a sequence with three main steps: *description, analysis,* and *interpretation.* The solution to the description problems requires a survey of the relevant sociological and historical data and uses rather elementary analytic tools. Yet the performance of the descriptive tasks will depend on the analytic ability of the investigator as well as on his background knowledge. To begin with he has to recognize that power is not a thing or a substance secreted by powerful entities but a relation; then, the taxonomy of power can be clumsy or subtle according as set-theoretical ideas are ignored or used. The second problem cluster, namely analysis, is conceptual and methodological. Once a refined concept of power has been elaborated at this stage, the investigator may go back to stage one in order to improve on his previous description. The last stage — interpretation — consists in hypothesizing law statements regarding power and in establishing logical relations among such statements: they are problems of building. Once a reasonably satisfactory theory of power has been built, the number of empirical and methodological problems will increase: in fact, the theory will have to be tested, perhaps not just with the evidence at hand but with new evidence, the search for which may be suggested by the theory itself — in case the theory is not merely a phenomenological account. In the context of such a theory more ambitious problems may be posed — such as "Why is power sought?", and "When and how is such and such a power configuration going to change and in what direction?". Finally, answering the valuation problem "To what extent is the theory true?" will involve checking the adequacy of the answers provided by the theory to the foregoing questions.

The above paradigm illustrates the following theses. (i) Scientific problems come in clusters or systems. (ii) These systems must be analyzed into unit problems. (iii) These individual problems must be ordered at least temporarily. (iv) This ordering, i.e. the strategy of problem-solving, must be set up according to the nature of the problems themselves rather than in response to extrascientific pressures. (v) Every such strategy of research, even on a modest scale, cannot restrict itself to data gathering, but will have to deal with conceptual, methodological, and eventually valuational problems as well.

We may now attempt to disclose the over-all pattern of problem-solving in factual science. Handling a problem or, rather, a problem system, does not begin with the actual solving and does not end when a solution has been found. Five main stages in the process can be distinguished: formulation, preliminary exploration, description, interpretation, and solution control. Each stage can in turn be divided into a number of individual problems; in the following an illustrative rather than an exhaustive list is given.

1. *Formulation.*

1.1. What is the problem? (Problem recognition.)

1.2. What are the data? (Fund of information.)

1.3. What are the assumptions? (Fund of ideas.)

1.4. What are the means, e.g. the techniques? (Fund of procedures.)

1.5. What are the logical relations involved, e.g. between the data and the unknown? (Conditions relating the problem's constituents.)

1.6. What kind of solution is wanted? (Profile.)

1.7. What kind of check is requested? (Solution recognition.)

1.8. Why is a solution sought? (Aim.)

2. *Preliminary exploration.*

2.1. What is it like? (Search for similarities with the known.)

2.2. Is it defined, and if so how? (In the case of concepts.)

2.3. Is it presupposed, and if so on what grounds? (In the case of assumptions.)

2.4. Is it hypothesized, and if so on what evidence? (In the case of assumptions.)

2.5. Is it observable? (In the case of physical objects.)

2.6. Is it countable of measurable? (Idem.)

2.7. How can it be counted or measured? (Idem.)

3. *Description.*

3.1. What is it? (Reterent.)

3.2. How is it? (Properties.)

3.3. Where is it? (Place.)

3.4. When does it happen? (Time.)

3.5. What is it made of? (Composition.)

3.6. How are its parts, if any, interrelated? (Configuration.)

3.7. How much? (Quantity.)

4. *Interpretation.*

4.1. What are the relevant variables? (Factors.)

4.2. What are the determining factors? (Causes.)

4.3. How are the relevant variables related? (Laws.)

4.4. How does it work? (Mechanism.)

4.5. Where or whence does it come from? (Origin.)

4.6. Into what does it get transformed? (Prediction.)

5. *Solution control.*

5.1. What is the validity domain of the solution? (Limits.)

5.2. Can the same solution be obtained by other means? (Possible independent check.)

5.3. Was the solution known? (Originality.)

5.4. Is the solution consistent with the accepted body of knowledge? (Matching.)

5.5. What difference, if any, does the solution make to the available knowledge? (Effect.)

Questions of formulation, preliminary exploration and control occur in formal science as well as in factual science (as shown by G. POLYA). The first three questions of preliminary exploration are also common to all sciences, whether formal or factual, and the same is true of the first two description problems. Problems of interpretation of fact are peculiar to factual science.

Let us finally examine the similarities and differences between scientific problems and crossword puzzles: this will shed additional light on our problem. The following common traits can be noticed.

(i) *A body of knowledge is presupposed* in both cases. Just as an illiterate person could not attack a crossword puzzle so an untrained amateur can seldom approach a scientific problem. The rare cases of recent distinguished contributions by amateurs (animal behavior and radio astronomy) have been in new fields and have involved some previous specialized knowledge.

(ii) Fairly *well-formulated problems* are at stake in either case. In the case of the game the unknown is a set of interrelated words; in the case of science the unknown may be an object (e.g., a source of radio waves), a property (e.g., a wavelength), a proposition (e.g., a law statement), or any other cognitively valuable item. The constituents of the problem are known in both cases as well, and so are the means in most scientific problems.

(iii) The operator *advances by making conjectures* in both cases. In the case of the crossword puzzle the conjectures consist in assuming that certain words fitting the description in the instructions will adequately combine with the remaining words. In the case of the scientific problem, too, the hypotheses are to satisfy compatibility conditions: they are to fit data and they are to be consistent with one another and with the bulk of knowledge. A double coherence is, then, required in both cases.

(iv) The conjectures are *tested* in both cases: the operator checks whether they correspond to the data and the conditions of the problem and whether they match with the remaining hypotheses.

(v) The *solution is controlled* in either case. In the case of the puzzle the solution is compared with the one published by the periodical. Measurements are repeated or taken with different instruments, and ideas are evaluated with the help of other ideas. Also, in either case the control is open to the public.

These similarities should not blind us to the contrasts between crossword puzzles and scientific problems. In the first place, the test of both the component assumptions and the final solution is never final in factual science: disconfirming evidence or unfavorable arguments must always be expected even in the case of the best established ideas. Consequently there are no final solutions to scientific problems concerning facts: unlike puzzle-solving, scientific problem-solving is open-ended. In the second place, the primary aim of research is not entertainment but the growth of knowledge. Unlike games, which are artificial obstacles set up with a short run and personal aim, scientific problems are "natural" obstacles in the sense that they are tied to the evolution of the modern culture and their solution may be socially valuable. The entertainment value of research is a bonus

Problems

4.4.1. Pick up a scientific problem and distinguish its constituents. For a clarification and illustrations of the concept of problem constituent, see G. Polya, *How to Solve It* (New York: Doubleday Anchor Books, 1957).

4.4.2. Select a scientific problem from *Scientific American*, *Science*, *Discovery*, *Endeavour*, or better any specialized publication, and dissect it along the lines sketched in the framework in the text (formulation-preliminary exploration-description-interpretation-solution control).

4.4.3. Ask a scientist what problem system is he dealing with at the moment. Take note of his answers and check whether the analysis in the text applies to the case or needs mending — or whether the scientist in question could use some research strategy instead of groping in the dark.

4.4.4. Try to find an empirical research that raises problems of description alone. Should you find it, make sure it is scientific. *Alternate Problem:* Discuss the motivations of scientific research.

4.4.5. Compare research in history, archaeology, or palaeobiology with puzzle solving. Point out the similarities and the contrasts. *Alternate Problem:* Find out whether scientific problems, just as some puzzles, may not be partially indeterminate and consequenty with no unique solution. What kind of conditions are missing in such a problem? And how can the indeterminateness be decreased or removed?

4.4.6. Comment on I. KANT's statement that what is a datum for the senses is a task for the intellect. (KANT used the words *Gabe* and *Aufgabe.*)

4.4.7. Suppose you are assigned the task of finding out whether intelligence is inborn or the product of education. What presuppositions would you point out to, what problems would you state and how would you order them?

4.4.8. Suppose you were asked to set up a hypothesis by means of which the degree of corruption accompanying the uncontrolled exercise of power could be estimated. Would you try several conjectures right away or would you rather precede this search by posing and solving certain logically prior problems? If so, which ones?

4.4.9. Problem-solving is a rule-directed activity in the case of routine problems: in this case at least it is an activity that may be made to conform to a set of rules or prescriptions. What about problem-finding?

4.4.10. Computers are designed, programmed and operated to solve problems: they are "fed" with instructions and data to handle them. Could a computer, present or future, be designed and programmed to *find* fresh problems? Supposing the whole of human culture could be "fed" into a computer, could the machine be instructed to find flaws (contradictions and holes) in that cultural heritage? Could it be made to state the corresponding problems? Could it be instructed to elaborate its own instructions in the case of problems of an unforeseen kind? And would this exhaust the possible problems?

4.5. Heuristics

No infallible recipes are known to prepare correct solutions to research problems by handling the problem ingredients: only routine problem-solving is — by definition — largely a rule-directed activity (see Sec. 4.4). Yet some advice concerning the handling of research problems may increase the probability of success. For instance, the following dozen rules.

1. *State the problem clearly.*

Minimize the vagueness of concepts and the ambiguity of signs.
Select adequate symbols, as simple and suggestive as possible.
Avoid logically defective forms.

2. *Identify the constituents.*

Spot the premises and the unknowns, and display the generator.

3. *Unearth the presuppositions.*

Point out the most important relevant presuppositions.

4. *Locate the problem.*

Determine whether the problem is substantive or strategical; if the former, whether it is empirical or conceptual, if the latter whether it is methodological or valuational.

Insert the problem in a discipline (unidisciplinary problem) or in a group of disciplines (interdisciplinary problem).

Find out the recent history of the problem, if any.

5. *Select the method.*

Choose the method adequate to the nature of the problem and to the kind of solution wanted.

Estimate beforehand the possible advantages and disadvantages of the verious methods, if any.

In case no method is at hand state the strategical problem of devising a method and attack this first.

6. *Simplify.*

Eliminate redundant information.
Compress and simplify data.
Introduce simplifying assumptions.

7. *Analyze the problem.*

Divide et impera: break the problem down into simpler units, i.e. shorter steps (subproblems).

8. *Plan.*

Program the strategy: order the unit problems in order of logical priority; if this is not possible, order them according to their degree of difficulty.

9. *Look for similar solved problems.*

Try to include the given problem in a known class of problems, thereby routinizing your task.

10. *Transform the problem.*

Vary the constituents and/or the formulation, trying to convert the given problem into another, more tractable problem in the same field. Whenever possible, switch to an equivalent problem.

11. *Export the problem.*

Failing the above attempts try to change the given problem into a homologous problem in a different field, as when a problem in human physiology is tranferred to the physiology of frogs.

12. *Control the solution.*

Check whether the solution is right or at least reasonable.

Recall the simplifying assumptions and, if necessary, relax some of these restrictions, attacking the new, more complex problem.

Repeat the whole process and, if possible, try another technique.

Estimate the precision attained.

Indicate possible ways of improving the solution.

The first operation, that of *stating the problem,* is often the most difficult of all, as known to the mathematician who is asked to formulate a mathematical model (a theory) on the basis of a messy bundle of more or less foggy conjectures and data concerning certain social facts. A solution, if only a roughly approximate one, can in most cases be obtained by making simplifying assumptions and/or gathering additional data: what is seldom given in the beginning, particularly along the research frontier, is a clear statement of the problem. Usually the problem statement develops into a well-formulated and clear question as work on the problem itself progresses; many problems begin in an obscure embryonic way and end up into a question that only resembles the original half-baked question. Some of the remaining operations listed above — particularly those of identifying the constituents, unearthing the presuppositions, simplifying, and analyzing — are aimed not only at solving the problem but also at restating it in a viable form. "A good statement is half the solution" is one of the few true popular sayings.

The second operation — *identification of constituents* — looks trivial but may be difficult to perform, particularly if the problem has not been well-stated. To ascertain that the given conditions tying the data to the unknowns (e.g., the equations containing both) are all necessary may be easy; to make sure they are also sufficient, so that the problem is a determinate one, is not so easy.

The third operation — *bringing the presuppositions to light* — will involve as deep an analysis as required. It may lead to a restatement of the problem or even to its dissolution.

The fourth operation — *problem location* — is performed automatically in the evolved sciences but is far from obvious in the younger disciplines. For example, problems of perception, of empirical semantics, and even of political doctrines, are often still classed as philosophical. A consequence of such a mislocation is that the wrong background knowledge and the wrong method are chosen, and so the whole problem is spoiled. A right location of problems, particularly in the younger sciences, requires a broad and up to date scientific outlook.

The fifth operation — *method selection* — is, of course, trivial when a single method is known; but this is not always the case: often several methods exist or can be evolved either for obtaining equivalent solutions or for obtaining solutions of different kinds (e.g., of various degrees of approximation). What kind of solution is wanted should be made clear in the problem statement. Thus, e.g., certain equations may be solved to yield compact analytic solutions if enough labor and ingeuity are spent; but for certain purposes (e.g., theory interpretation) an approximate solution may be enough or even preferable, and for other purposes (e.g., theory test) a numerical solution in a certain domain may be sufficient. Finally, if no known technique works or if no known method can yield the kind of solution wanted, the problem solver has been honored with a problem of an entirely new class and his attention will be redirected from substantive to strategic questions.

The sixth operation — *simplification* — is crucial since it may lead to the restatement of a complex rebellious problem into a simpler tractable question or set of questions. Problem simplification may go as far as performing brutal amputations leaving a core slightly similar to the given problem; this is usually the case of theory construction, which starts by focusing on what is believed to be essential — although a closer look may show it to be secondary. Simplifying assumptions may be grotesque in a first stage; thus a real finite elastic beam may be simplified, in the theoretical representation, to an imaginary infinitely long beam. The elimination of irrelevant information ("noise") is part of this stage. Sometimes the information may be relevant but, due to the great variety and amount of data, a small number of data sets must be chosen, i.e. a few variables must be seized upon if work is to start at all — and this involves making definite assumptions as to which variables are primary, which of secondary importance.

The seventh operation — *analysis* — consists in the atomization of the given problem, i.e. in its resolution into simpler, not further reducible problems. Analysis is necessary but not sufficient for obtaining a solution: there are problems with an elementary statement that have so far defied all efforts, such as proving that four colors are sufficient for coloring a map in such a way that no two contiguous countries be assigned the same color. What is wanted in such cases is not a clearer statement or a set of simpler problems equivalent to the given problem but a powerful enough theory or a powerful enough technique.

The eighth operation — *planning* — was analyzed and exemplified in Sec. 4.4.

The ninth operation — *looking for similar solved problems* — is related to problem location. It usually involves scanning the relevant bibliography, a task which is becoming more and more difficult owing to the exponential growth of the volume of scientific literature. In the case of difficult or just time-consuming problems it will be worth while to assign this task to machines capable of recognizing problem similarity and sorting out and abstracting the relevant literature. As long as such machines are not available the existing literature is of limited use; what is worse, when the investigator becomes aware of it his dare may be drowned in paper.

The tenth operation — *problem transformation* — may be called for whether the preceding move has been successful or not. Changes of variables can perform such a problem restatement once the problem has been stated in a mathematical form. For example, the problem "$(?x)(ax^2+bx+c=0)$" is transformed into the atomic problem "$(?y)[y^2=(b^2-4ac)/4a^2]$" through the change of variable $x=y-(b/2a)$; in fact, the second problem is equivalent to the first and is solved by just extracting a square root. A problem restatement, by definition, will not affect the problem. Sometimes, however, a nonequivalent problem may have to be attacked; for instance, a nonlinear term in an equation may have to be neglected for want of a theory capable of handling the complete equation.

The eleventh operation — *problem exportation* — is becoming increasingly frequent as the integration of the sciences advances. For example, the distinction among animal groups, often difficult on the basis of morphological, ethological and other superficially observable characters, can be made at the molecular level, e.g. by studying the proteins and their ratios: in this way a problem in zoological systematics — wrongly supposed to be simple — is exported to biochemistry and the results obtained in this field are finally reexported to the original field. This move goes back to the origins of arithmetics and geometry, which were introduced as tools for converting empirical operations of counting and measuring into conceptual operations.

The twelfth and last operation — *solution control* — was commented on in Sec. 4.4 but bears an additional remark. The solution may be controlled in either of the following ways: by repeating the same operations, by trying an alternative approach (e.g., a different technique), and by seeing whether it is "reasonable". The reasonableness will often be judged intuitively but in rigor only a theory and/or a set of data can determine whether a solution is reasonable because 'reasonable', in science, means nothing but compatibility with what is known, and the body of knowledge contains highly unintuitive theories and data.

This is about all heuristics — the art of expediting problem-solving — can say at the moment without going into the specific differences among the several fields of science. Let us next inquire into the fate of scientific problems.

Problems

4.5.1. Illustrate the analysis of molecular into atomic problems with examples drawn from factual science.

4.5.2. If we are assigned the task of multiplying 1,378,901 by 78,000,671, we shall do well to start by making an approximate computation, replacing the given numbers by one million and one hundred million; the product, one hundred million, will be the order of magnitude of the wanted result. Why should we perform this approximate preliminary computation? *Alternate Problem:* Why is it desirable to precede every quantum-mechanical computation, whenever possible, by a rough classical or semiclassical calculation?

4.5.3. Some philosophers have imagined that the ultimate desideratum of factual science is solving the problem: "Give a complete description of the state of the universe at any specified instant of time". Is this a reasonable problem both in the sense that it is solvable and that its solution would be cognitively valuable? *Alternate Problem:* Given a problem, should the first move be to search for its solution?

4.5.4. What kind of operation, in terms of problem manipulation, do geologists and archaeologists perform when they resort to radioactivity measurements in order to date their findings?

4.5.5. Suppose seven apparently identical things are weighed with a scale to within 1 milligram and the average value 100 plus or minus 1 mg is obtained for the set. What is the average weight per piece? Solve and discuss this problem as a prolegomenon to the following *Alternate Problem:* Shall an exact solution be always sought?

4.5.6. Suppose you have found the solution to a problem. What will you do next (apart from rewarding yourself)? *Alternate Problem:* Analyze examples of problem mislocation, like the treatment of certain social problems, such as juvenile delinquency, as psychiatric problems.

4.5.7. A *decision procedure* is a "mechanical" or "thought-less" procedure of testing an expression for either validity (in the case of arguments) or truth (in the case of statements). Decision procedures are employed for checking the answers to decision problems. Give an example of decision procedures. *Alternate Problem:* Work out and exemplify the distinction between algorithmic (routine) problem solving and heuristic (creative) problem solving.

4.5.8. The problem of finding the intersections of two or more lines can be solved if the equations of the lines are set up; the points are then determined by the solutions of the simultaneous system of resulting equations. Exemplify this procedure and analyze it in terms of the operations described in the text. *Alternate Problem:* Analyze the demonstration of any given theorem in terms of the heuristic hints given in the text.

4.5.9. When little knowledge is available to prove a theorem by some direct method, the negation of the statement to be proved can be used as a premise; if a contradiction is derived, the premise must be false, too, and the original statement true. Analyze this technique (indirect proof) with the help of the concepts introduced in the text. *Alternate Problem:* Elucidate the notion of subproblem relating it to the concepts of atomic problem and of problem strength or generality (see Sec. 4.2).

4.5.10. Study the capabilities and limitations of computers for solving problems. See, e.g., M. MINSKY, "Artificial Intelligence", *Scientific American* **215**, No 3 (1966) and the bibliography cited therein. *Alternate Problem:* Report on the present state of the psychology of problem-solving. See, en particular, J.W. GYR, "An Investigation into, and Speculations about, the Formal Nature of a Problem-Solving Process", *Behavioral Science,* **5**, 39 (1960) and the bibliography cited therein.

4.6. The End of Scientific Problems

Scientific problems can be forgotten, eliminated, clarified, solved, shown to be unsolvable, or left for the entertainment of future generations. In no case is a scientific problem declared to be a mystery beyond human reason. Let us deal separately with the various ways of bringing a problem to an end.

1. *Oblivion.*

Some problems pass from the scientific scene without having been either explicitly eliminated or solved: for some reason, which is often extrascientific, they no longer attract the attention of the new generations of scientists. Thus, e.g., most problems concerning the origin of social activities and institutions remain unsolved, yet they are not dealt with nowadays with the enthusiasm they were attacked in the early days of expansion of the evolution theory. Likewise botany and invertebrate zoology are at present out of fashion. The professionalization of science has engendered careerism, and the careerist wishes to show off that he is up to date: this is one illegitimate motive for the fading away of genuine problems in our time. But there are also legitimate motives for this. One is that new, more pressing tasks may displace old unsolved problems. A second legitimate cause of problem oblivion is that the invention of a new theory will shift attention from isolated problems to whole new problem systems that can be attacked with the new theory: in this way a whole class of loosely related problems is displaced by a tightly knit problem system. A third legitimate rationale can be that the adequate theory for answering the given questions is not yet at hand. In this case the problem is not entirely forgotten but rather postponed in favor of more tractable problems, as is currently happening with the deepest (and often also the cloudiest) problems of human behavior. In any case a good problem, if forgotten, is bound to be eventually rediscovered, perhaps in an entirely unsuspected context.

2. *Elimination.*

Science progresses not by solving problems but also by eliminating questions, i.e. by showing that questions of certain kinds are pointless. Legitimately eliminable problems are of the following kinds.

2*a*. *Trivia:* problems that, even if well-formulated, are shallow and ill-integrated, i.e. they do not fit rich theories or promising programmes. Examples of trivial problems are counting the number of pebbles on a beach and tabulating unusual functions to no ulterior purpose. Many problems in the descriptive stage of science (e.g., in pre-evolutionary biology and in pretheoretical behavioral science) are trivial; a way of progress has been to elude them, i.e. to abstain from asking such trivial questions demanding no theories, and concentrating efforts on more difficult and rewarding tasks. The degree of advancement of a science is not measured by the number of problems it poses but by the depth and complexity of its problems. The sciences of man are still posing problems by far simpler than those of physics, and they ask simpler problems because our systematic knowledge of man is much poorer than our systematic knowledge of atoms. The more we know the deeper questions can we ask and the deeper questions we answer the more we know. The objects of a science are not inherently simple or complex: our approach can be simple-minded or deep.

2*b*. *Pseudoproblems:* ill-stated or ill-conceived questions. Traditional philosophy is full of then, but once in a while they creep into scientific minds. For example, "When was the universe created?", which presupposes that the universe has been created by a supernatural act — an untestable assumption. An analysis of presuppositions is necessary and sufficient to eliminate pseudoproblems of this kind.

2*c*. *Chimeras:* fanciful questions arising outside scientific contexts. Examples: finding the philosopher's stone and the fountain of youth; establishing communication with the dead and communicating with other people without employing the senses. Chimeras, like pseudoproblems, have untrue presuppositions. Thus, the chimera of trying to communicate with the dead pressuposes the existence of an immortal or at any rate a body-independent soul. What distinguishes chimeras from pseudoproblems is their practical aim: just as some pseudoproblems belong to pseudoscience, so chimeras might give rise to the technology of pseudoscience.

3. *Clarification*

In many cases research may not solve but does clarify certain problems. For instance, the question "What is life?" has been clarified in the course of the last century to the point that a special kind of organization or structure rather than a special substance or agent is now sought as the distinctive property of living matter; such a clarification amounts to a fresh approach to the problem, i.e. to its restatement and to a change in the problem's background. Once a problem has been clarified by stating it in a more exact way and by making clearer its relation to its background the probability of its solution is enhanced.

Sometimes the converse happens, i.e. the nature of the problem becomes clearer in the light of its solution. Thus, e.g., if an absurd solution is obtained the problem is seen to have been ill-formulated, and if several mutually inconsistent solutions are obtained the problem is seen to be indeterminate. *The latter is the case with the several possible interpretations of the mathematical formalism of quantum mechanics:

the mere multiplicity of mutually incompatible solutions to this problem should make it clear that the problem of interpreting a formal structure in physical terms is indeterminate unless its philosophical presuppositions are made clear; in this case the presuppositions concern the existence of the referents of hypothetical constructs, and the nature of knowledge. Consequently, a discussion of such interpretations should be preceded by a general philosophical discussion leading to the explicit statement and foundation of such philosophical presuppositions.* In any case, problem clarification is one of the ways of scientific progress. The sooner science managers become clear of the importance of problem clarification, the better.

4. *Solution*

As regards their truth value, solutions are of three kinds: true, approximately true, and false. In factual science most solutions are either outright false or at best they have an element of truth: this contrasts sharply with formal science, where most propositions are (formally) true. Most problems in factual science have only approximate solutions; only comparatively trivial problems have exact solutions. The important thing is not to find entirely true solutions to problems of all sorts: this is a chimera, i.e. a problem that must be eliminated. What is important is to possess means for spotting errors and inaccuracies, and for improving on the available solutions.

*In formal science it is often possible to find standardized techniques for obtained solutions and/or for checking them: the former are called *algorithms*, the latter *decision procedures*. If both an algorithm and a decision procedure are available for a certain class of problems, these are called *effectively solvable*. If no algorithm exists, so that the investigator must zig-zag by trial and error using his intuition and heuristic hints, but it still is possible to recognize the solution as such in a mechanical way, the problem is *well-defined*. If algorithms and/or decision procedures are not available for a class of problems, the following situations are possible: (i) proofs exist that either or both of them are possible, but such techniques are unknown; (ii) no proof exists that such techniques are either possible or impossible; (iii) proofs exist that either or both such standardized techniques cannot possibly exist. If the latter is the case the problem may still be solvable, but in an artisanal or artistic way rather than in the fashion of mass production made possible by automatic computation. (Incidentally, the present interest in effectively solvable problems stems partly from the computer boom and in turn contributes to it. Which in a way is a pity, because effectively computable problems are routine problems; on the other hand research on the conditions for effective computability is not a routine problem.)

*Finally, a problem is declared *unsolvable* with certain means if and only if it is *proved* that no solution to it can be attained with the given means: a verdict of unsolvability of Π is not a dogmatic dictum but a solution to the metaproblem "Is Π solvable?". For example, it is possible to prove (K. GÖDEL) that certain arithmetical statements are not provable with finite methods, and even the consistency of elementary arithmetics cannot be so proved. This result, important as it is for the philosophy of formal science, is comparatively unimportant for the philosophy of factual science, since not a single factual statement can be *proved:* at most it can be satisfactorily (but just temporarily) substantiated. And it is no motive for despair because anyhow the set of problems is potentially infinite.

*In short, formal science problems fall into one of the classes displayed in the following classification:

- *Problems*
  - *Solvable*
    - *Effectively Solvable* (algorithm and decision procedure)
    - *Well-defined* (decision procedure)
    - *Ill-defined* (neither algorithm nor decision procedure)
  - *Unsolvable*

The decision to place any given problem in one of the preceding categories is preceded by a rigorous proof; no such proofs are possible in factual science.*

If no solution to a given factual problem has been found after the recipes of heuristics (see Sec. 4.5) have been tried, the following moves should be attempted in succession. (i) *Reexamine the problem statement:* make sure you are dealing with a well-formulated problem and have all the premises (data and hypotheses) as well as the appropriate techniques; (ii) *try alternative known means* (theories and techniques); (iii) *invent new, more powerful means.* Even after reexamining the problem statement and trying alternative theories and techniques the problem may be intractable — as is the case with the three-body problem in physics. One may then suspect that it is unsolvable — or, rather, unsolvable with the specified means, because solvability is contextual rather than absolute.

Are there unsolvable problems in science? There certainly are: it is enough to limit the data, the theory, or the technique, in order to generate an unending sequence of unsolvable problems. Since data, theories, and techniques are of limited precision anyhow, an artificial limitation is unnecessary. Let us mention a few typical unsolvable problems. (i) Trace the history of any given pebble on a beach; (ii) supposing that our present cosmic era began with a "big bang" which destroyed the traces of previous eras, reconstruct the latter; (iii) find evidences of life in the Precambrian period. These problems are unsolvable because of lack or incompleteness of evidence. Moreover, their unsolvability is not circumstantial, i.e. such that it might be made for in the future: the evidences that would be needed in order to make the desired reconstructions are no longer extant.

The realization that there are everlastingly unsolvable problems should not blind us to the relative or contextual nature of solvability. Every problem is set against a background of data, theories and techniques, hence the solvability of problems is not absolute but is *relative* to its background and to the body of fresh knowledge that is gained in the course of its investigation. If, in order to transmute atoms, we are allowed to employ chemical methods alone, we will fail as the alchemists did; but if more powerful means are introduced, such as bombarding them with subatomic projectiles, we may succeed. Likewise, trisecting an angle is impossible with ruler and compass alone, as required by the Greek constructivist philosophy of mathematics, but it becomes trivial if the given geometrical problem is translated into an arithmetical one. Consequently, unless we prove that a given problem is unsolvable let us abstain from prohesying that no attempt to solve it can possibly succeed; and if we prove unsolvability let us not forget to mention the means relative to which the problem is unsolvable. Nothing more than this *relative* or *conditional unsolvability* can be asserted, both in formal science and in factual science. But the statement that a problem is unsolvable with the given means is equivalent to the statement that the given means are insufficient to solve the given problem. This way of putting it is preferable because it shifts the focus from problems to means: instead of driving us away from unsolvable problems it directs our attention to the search for more powerful means.

According to obscurantism, not only are there inherently unsolvable problems (mysteries) but the deeper a problem the more mysterious it is. The illuministic tradition, on the other hand, asserts the solvability of every problem subject to the following conditions: (i) that the problem be well-formulated; (ii) that the necessary means be available; (iii) that what is sought be not physically impossible (i.e., incompatible with natural laws), and (iv) that the term be not limited. In short, *qualified* solvability is asserted by illuminism: it does not say that every problem is solvable but rather that every *well-formulated* scientific problem is solvable *in principle*, i.e. if the adequate means (data, theories and techniques) are found. Clearly, this assertion is in keeping with the history of science, whereas the obscurantist thesis is refuted by the latter. Yet it cannot be proved, just as it cannot be disproved. A refutation of the illuminist thesis would amount to proving the existence of at least one problem absolutely unsolvable and consequently adamant to every possible strategy. But the latter would involve the disclosure of all the time to come — an impossible task because it falsely presupposes that the present sows all the seeds of the future. We adopt therefore the thesis that every well-stated and well-conceived scientific problem is solvable in principle: this is a fertile programmatic conjecture.

What comes after the solution? Once an animal or a computer has solved a problem it comes to rest. Once man has solved a problem of knowledge a new sequence of related questions are likely to be raised by the investigator or by somebody else, immediately or after a while. Research is a self-sustained chain reaction by virtue of the imperfect and systemic character of scientific knowledge: the newly conquered solution can surely be improved on and may induce corrections in the antecedent knowledge, which corrections may call for whole new research lines; or perhaps the solution can be generalized, or else specialized, or it may suggest a fruitful analogy for solving another problem, or even initiate a pathway in a new territory. Solving scientific problems is neither like filling a finite bucket nor like working on PENELOPE'S loom; it is not even like exploring an infinite universe which is already there. The march of science consists in *building* new problem systems rather than in uncovering existing problems; these new problem systems overlap partially with the older ones and, far from being smaller than they, are of a bigger size: not just more numerous, but deeper and more general.

The best criterion for finding out whether a discipline is still alive is to ascertain whether it still poses new problems the solutions to which would presumably add to our knowledge. (ARISTOTLE'S mechanics is in this sense dead whereas NEWTON'S mechanics is still alive, not only as regards routine problems but also as regards research problems.) Similarly, the peculiar mark that distinguishes original investigators from everyone else is the capacity for finding, inventing, and handling new problems. No new problems can be raised within the framework of a dead doctrine: this requires commentators but no research workers, i.e. investigators who can advance the subject. There can be nothing to advance if the problems concerned are pronounced solved. Untouchable solutions and unapproachable problems set a limit to scientific inquiry. Yet this limit, however voluntarily accepted by the dogmatic, is not set by science itself, which is essentially problemizing. Living science recognizes no inherent and everlasting limitations since it consists in handling problem systems, and problems come in clusters and do not go without giving birth to further problems.

Problem-seekers are often regarded as trouble-seekers. Rightly so, because a questioning mind is a critical mind, one which cannot conform to any achievement, however great, because it senses that it must be imperfect, hence perfectible. Human progress depends more and more on problem-seekers and a free society is one in which the posing of problems of every kind and research into them find no limits other than those of the state of knowledge. We owe more to those who, ignorant, ask and seek an answer, than to the wise who are in the possession of truth and can answer any question without erring — i.e. without contradicting received opinion. The wisest among the wise is no longer the one who knows more solutions — the erudite — but the one who knows just enough to pose and attack entirely new and fertile problem systems and who has the moral (and if necessary the civil) courage to do so.

The recognition that it is problems which, by sparking ideas, keep the engine of science running, should modify the standard philosophy of science, centered as it is on data rather than on what is sought. Moreover, it should eventually lead to deep changes in the current psychology of intelligence, committed as it is to the definition of "intelligence" as the ability to answer questions (too often asked by unintelligent testers) rather than as the ability to both answer and ask original questions. In turn such a shift from problem-solving to problem-posing should affect our present educational policy and techniques, which aim at the most efficient learning of problem-solving recipes and ready-made solutions. A nondogmatic educational policy, one in agreement with the spirit of science and which did not try to make us race against computers, would stimulate the maturation of an inquisitive attitude and of the most powerful methods and theories: it would be a problem-centered pedagogy, and would thereby train men to face the rapid obsolescence of received opinion.

Problems

4.6.1. Illustrate the concepts of trivial problem, pseudoproblem, and chimera.

4.6.2. Consider the following ways of handling contradictions. (i) Elimination of one of the conflicting propositions or theories by showing it to be false. (ii) Conciliation or harmony of the two. (iii) Synthesis or building of a third system encompassing the valid and not mutually inconsistent elements of the given pair. (iv) Construction of an altogether new proposition or theory.

4.6.3. Mention some problems that were regarded as unsolvable but were finally solved. Examples: (i) Do atoms exist or are they just fictions? (ii) What is the nature of chemical affinity? (iii) Is it possible to control the mind by material agents (e.g., drugs)?

4.6.4. Mention a handful of presently unsolved problems. Examples valid at the time of writing: (i) What forces hold the nuclear particles together? (ii) Do gravitons (gravitation quanta) exist? (iii) Is the universe spatially finite or infinite? (iv) What produces gene mutations: external agents (e.g., cosmic rays) or internal changes (such as changes in cell metabolism)? (v) What is the detailed mechanism of photosynthesis? (vi) Why are hybrids usually more vigorous than their immediate ancestors? (vii) What are the functional properties of the brain that account for voluntary muscular movement? (viii) What causes childbirth? (ix) Why do we age? (x) What is the neural mechanism of hypnosis?

4.6.5. Comment on the following fragment from S. ALEXANDER's "Natural Piety" (1922), repr. in D. J. BRONSTEIN, Y. H. KRIKORIAN, and P. P. WIENER, Eds., *Basic Problems of Philosophy* (New York: Prentice-Hall, 1955), pp. 517ff.: "The natural piety I am going to speak of is that of the scientific investigator, by which he accepts with loyalty the mysteries which he cannot explain in nature and has no right to try to explain. I may describe it as the habit of knowing when to stop in asking questions of nature [...] There is a mental disease known as the questioning or metaphysical mania, which cannot accept anything, even the most trivial, without demanding explanation". *Alternate Problem:* Distinguish between *Ignoramus* (We do not know) and *Ignorabimus* (We shall never know), between modesty and humility.

4.6.6. A technological version of the law of conservation of energy is this: "It is impossible to build perpetual movers, i.e. engines running without energy supply". Does this statement refute the programmatic conjecture "Every well-formulated problem is in principle solvable"? *Alternate Problem:* Does science advance by amassing observations, by building theories, by refuting them, or by posing and solving consequential problems?

4.6.7. Kant claimed that there will always be insoluble contradictions or antinomies, i.e. pairs of mutually contradictory and equally likely propositions regarding fundamental matters. Comment on this thesis and, if possible, examine the present status of the antinomies proposed by KANT in his *Kritik der reinen Vernunft*, 2nd ed. (1787), p. B 454. *Alternate Problem:* Discuss the oblivion of problems caused by the loss of prestige of the theories in which they were born. Example: important problems of biological origin and evolution, as well as anthropological and sociological problems of the same kind, were forgotten during the eclipse of the evolution theory in the 1920's and 1930's.

4.6.8. Comment on the following conflicting prognoses concerning the future of physics. (i) "Future developments in theoretical physics will consist in applications of the present theories, which are essentially correct". (ii) "The next revolution in theoretical physics will be the last and it will consist in introducing one single master idea that will permit the solution of all the present difficulties". (iii) "The next revolutions in theoretical physics will consist each in the introduction of a new strong idea which will help solve the present difficulties and raise a host of presently unconceivable new problems". *Alternate Problem:* Comment on the usual phrase 'Your solution raises more problems than it solves'. Why is it so often used in a derogatory vein? Where might it be in point: in science or in practical affairs?

4.6.9. Build a theory concerning the rates of problem-solving and problem-finding in science. Hint: refine and mathematize the following hypotheses. *Axiom* 1: The rate at which new problems are found in a field is proportional to the number of solved problems in the same field. *Axiom* 2: The problem-solving rate is proportional to the number of fresh problems in the same field. From both assumptions we derive the *Theorem:* The number of fresh problems in a field is proportional to the number of available solutions in the field, i.e. $F = kS$. For a dying science $0 \leq k < 1$, for an emerging science $k > 1$, and for a stationary science $k \cong 1$. *Alternate Problem:* Speculate on the possible evolution of a culture which paid attention to problem-solving alone and which neglected every problem that could not be handled with the help of computers of the presently available kind.

4.6.10. Elucidate the concepts of *problem complexity* and *problem depth*, such as they occur in the following examples. Solving the equations of motion for a number of bodies interacting in a known way is extraordinarily complex but not deep: the forces are known and the laws are known, hence it is only a question of mathematical ingenuity to invent techniques for finding approximate solutions to the problem, or even of patience and skill in programming computers to do the work. On the other hand, the problem of the equilibrium and the transformations of a bunch of nuclear particles is both complex and deep, because neither the forces nor the laws of motion are accurately known in this case. *Alternate Problem:* Much work is being done nowadays on information-retrieval (e.g., recovery of information buried in old or obscure scientific journals). What is being done on *problem-retrieval* (recovery of unjustly shelved problems)?

4.7. Philosophical Problems

First-hand philosophers are those who deal with philosophical problems; second-hand philosophers, those who deal with what first-hand philosophers have said or failed to say. The former concentrate on problems, the latter on solutions. The former are primarily interested in ideas, the latter in the expression of ideas and the circumstances accompanying their birth and spread. Just as some writers deal with life and others with books dealing with life, so first-hand philosophers do the primary philosophizing and second-hand philosophers record, comment on, explain, develop or criticize the formers' doings.

The preceding statements are not valuational but descriptive: the two kinds of philosopher exist and, moreover, n-th hand is not identical with n-th rate: first-hand (original) philosophers may be second-rate thinkers and even quacks, whereas second-hand philosophers may be first-rate thinkers. Both kinds of philosopher are needed to keep philosophy going, but the progress of philosophy, just as the advancement of science, requires a clear understanding that (i) original research consists in disclosing, inventing, dissolving and solving problems — if possible deep and fruitful ones, and (ii) original research is indispensable to keep a discipline alive.

The above, plain as it is, bears repetition because of the popularity of the view that philosophy is nothing but a set of subjects and opinions to be taught — i.e. a set of doctrines — rather than a set of problems to wrestle with. When the upholders of the doctrinal view refer to philosophical problems at all, they do not mean problems proper but rather whole subjects such as "the problem of knowledge". And, if invited to exhibit a specific member of such a problem system, they are apt to misunderstand the question and offer some historical problem — e.g., "Who may have influenced *A*'s doctrine?" — or some linguistic problem — e.g. "What do people around here mean when they say that they mean what they say?" — or perhaps some psychological problem — e.g., "If a distraction makes me forget my headache, does it make my head stop aching or does it only stop me feeling it aching?" (an actual competition problem in *Analysis* for January MCMLIII). Historical, linguistic, psychological and many other problem kinds occupy both first-hand and second-hand philosophers, just as cytologists must often struggle with their electron microscopes, archaeologists with jeeps, and prehistorians with geological evidence. Research into historical, linguistic, psychological and other problems may shed light on philosophical problems and are often propeadeutic to them — but they are not philosophical problems.

What is a philosophical problem? This is a problem for the theory of philosophy — and there are as many metaphilosophies as there are philosophies. If a somewhat traditional position is taken, the answer may be given in the form of a simple denotative *Definition:* "A philosophical problem is a problem in logic, or in epistemology, or in ontology". If asked to spell out this definition, one may add that a philosophical problem is either a problem of form, or a problem of knowing, or a problem concerning being. But this is both unclear and insufficient: inquiring whether two conceptual systems, such as two theories, are isomorphous or not, is a problem about form yet it may be a strictly mathematical problem; finding out how we know about unexperienced things is a problem of knowledge but not an epistemological problem (at least it is no longer since psychology took it up); and inquiring into the nature of enzymes is a problem of being but not an ontological problem. Logical problems are included in the wider set of formal problems: they are generic problems concerning form and they may arise in any research. In any field whatsoever we may have to tackle problems such as "Is p equivalent to q?", "Is q deducible from p?", "Is p, which contains the concept C, translatable into some equivalent proposition not containing C?". Epistemological problems are not problems of knowledge proper but certain nonempirical problems concerning knowledge — such as "What are the criteria of factual truth?", "What is the truth value of the conjunction of two partially true statements?", "How are theories tested?", or "What is the role of analogy in scientific inference?" And ontological problems are not specific problems of being but generic, equally nonempirical problems concerning pervasive traits of reality, such as "What is the relation of time to change?", "Are there natural kinds?", "Is chance irreducible?", "Is freedom consistent with lawfulness?", or "How are the various levels related to one another?" With these qualifications we may retain the above definition of philosophical problem, even realizing that every denotative definition is elusive.

A peculiarity of philosophical problems is that no empirical data (such as, e.g., nuclear moments or historical dates) occur in their statement. Empirical data may, though, be relevant to philosophizing: they may give rise to philosophical problems and they may refute solutions to philosophical problems; but they cannot occur in their statement, for if they did then philosophical problems would be investigated with empirical means, i.e. they would belong to some empirical science. Secondly, philosophical problems belong to no particular science either by subject or by method, although scientific research — as we shall see in Sec. 5.9 — both presupposes and suggests philosophical theses (e.g., the reality of the external world) and theories (e.g., ordinary logic). Thirdly, philosophical problems are all conceptual, but many of them — say, the problem system of scientific laws — presuppose a body of factual science. As a consequence, they are often solved (or dissolved) with the help of science or even in science. Philosophers may have done more by asking intelligent questions that were eventually taken up by science, than by proposing crazy solutions to quaint problems. Fourthly, philosophical problems of the nonlogical kinds cannot be solved in a fully exact way, particularly if they are tied to science, which is never final. Hence epistemological and ontological problems are, like the fundamental problems of factual science, eternal in the sense that they have no final solution. They may receive better and better solutions and they may eventually cease to appeal to inquisitive minds but they will always remain, in the best of cases, half-solved. This, of course, does not exempt us from being careful in the statement and conception of philosophical problems: the solution will be the truer the better stated and conceived the problem has been.

A fifth, most unfortunate peculiarity of nonlogical philosophical problems is that no criteria are usually available in order to recognize the solutions, let alone to decide whether a given solution is correct. That some philosophical questions are inherently undecidable, is known: they are not proper problems but pseudoproblems such as "How much more being has man than the lower animals?" (asked at the XIIth International Congress of Philosophy, 1958). But that so many genuine philosophical problems are the subject of long and unfinished controversy has troubled many thinkers. Is there anything wrong with philosophical problems *per se*, or does the fault lie with our sloppiness in stating them and in stipulating the techniques whereby philosophical solutions (i.e., philosophical hypotheses and theories) should be tested? Before jumping to a pessimistic answer we should recall that all of formal logic and most of semantics have been turned into rigorous disciplines, to the point that they are often regarded as independent sciences. These successes suggest adopting a definite philosophical methodology, and more precisely one inspired in the method of science.

We propose the following rules as a basis for a philosophical methodology. First, the treatment of nonlogical philosophical problems should conform to ordinary logic. Logical mistakes will therefore invalidate philosophical discourse wholly or in part; they will not disqualify every philosophical problem and every philosophical program, but they will certainly weed out much philosophical argument. Second, the treatment of nonlogical philosophical problems should not clash with the main body of scientific knowledge and, moreover, should be scientifically up to date; this will not condemn unorthodoxies as long as they are in the spirit of science, but it will chop off much nonsense. Third, the statement and working out of philosophical problems, as well as the checking of their proposed solutions, should parallel the corresponding operations in science — that is, the method of philosophizing should be scientific. Fourth, the proposed solutions to philosophical problems should be judged as to their truth value alone, in independence from noncognitive, e.g. political, considerations. These four rules for philosophizing and for evaluating philosophical work should guide the very choice of philosophical problems. If logic is not respected any absurdity can be inquired into, from Hegelianism to existentialism; if the bulk of science is not respected any superficial or even silly question will be approached, such as whether there might not be traces of the future; if the method of science is not imitated, the benefit of the most successful experience of mankind is renounced; and if the aim of philosophizing is not the search for truth (the free search for perfectible truth), a handmaid of some fossil doctrine is readily obtained.

The problem of choosing the right problem and the right approach in the philosophy of science is as important as in any other branch of learning. Here, as in the rest of philosophy, it is tempting to tread only on paths open by authority, no matter what relevance the problem may have to actual scientific research. Characteristic recent instances of this kind of problem are the following: (i) the question of contrary-to-fact conditionals, the solution of which is said to be a prerequisite for the theory of scientific law; (ii) the question of finding logically satisfactory definitions of qualitative dispositional concepts such as "soluble", which is thought to be indispensable for approaching the problem of theoretical concepts; and (iii) "the" problem of induction, which is held to exhaust the problem of scientific inference. Unfortunately the problem of counterfactuals is so far obscurely stated, hence unsolved, whereas the theory of scientific law is going ahead — as could not be otherwise, because what is interesting about counterfactuals is that they occur in inferring rather than in stating premises of factual theories. As to dispositionals, scientists usually prefer to derive dispositional qualitative or comparative concepts from quantitative ones, and this they do in the midst of theories rather than outside them (see Sec. 3.3). And the role played by induction in scientific inference is much more modest than usually supposed (see Sec. 15.4). A problem inflation has taken place for lack of acquaintance with real science, and an artificial theory of science has been evolved which is not really about science but about certain ideas that have occurred to distinguished philosophers in relation to problems that are of little interest, and sometimes of no interest, for the advancement of knowledge: they are often handled with an enormous expenditure of rigour and ingenuity only because they are wrongly assumed to be vital for science or for the philosophical explanation of science.

The theory of science need not deal exclusively with problems that might attract the attention of scientists — who usually overlook the philosophical theses they take for granted — but it certainly must deal with real science rather than with a simplistic picture of science. And if a fertile exchange between philosophers and scientists is found desirable, both for the enrichment of philosophy and for the cleansing of science, then it is mandatory to deal with the philosophical problems that arise in current research. Nowadays physicists are faced with the need for building theories of elementary particles and they would be helped by competent discussion on the general problem of the possible approaches to physical theory construction. Cosmologists are faced with insecure evidence in favor of highly speculative theories: they should welcome a competent discussion on the testability and accuracy to be required of theories. Chemists are feeling uneasy with much *ad hoc* hypothesizing of wave functions and much blind computation: they would benefit from a discussion of ad hocness as well as from an examination of the status of models. Biologists are faced with a widening gap between observational and experimental research, as well as between molar and cellular biology: they should be helped by a discussion on the value and interrelations of these various approaches. Psychologists are learning chemistry and might welcome a discussion on whether psychical events are nothing but chemical reactions and on whether there are specifically psychological laws in addition to the chemical ones. And so on. A choice of living problems will enliven the philosophy of science and will render it helpful for the advancement of science.

In conclusion, the right approach to philosophical problems — their choice and handling — does not or should not differ too much from the right approach to scientific problems, however much the subjects and techniques may differ. But this is just an ambiguous way of saying that there is only one way of approaching problems involving knowledge, be it in pure science, in applied science, or in philosophy: scientifically. Which may be dogmatic but should be tried for a change.

Problems

4.7.1. Draw a list of philosophical problems raised by scientific research. Hint: scan the basic assumptions of any scientific discipline and examine whether they either presuppose or suggest philosophical ideas.

4.7.2. Show why, or why not, the following are philosophical problems — according to your own conception of philosophy, that is. (i) How is the perceived image related to external objects? (ii) Can we perceive social relations? (iii) Is the mind in the brain, and if so how: as a separate substance or as a system of functions? (iv) What is pain? (v) What is evil? (vi) What are obligations? (vii) What are excuses? (viii) What is work (or action)? (ix) What is wrong with society? (x) What are the attributes of the deity?

4.7.3. Does philosophy grow by learning to abstain from asking certain questions — or rather by learning to formulate correctly deep problems? In particular, are there "dangerous" questions? For the latter question see O. NEURATH, *Foundations of the Social Sciences, Encyclopedia of Unified Science*, vol. II, No. 1 (Chicago: University of Chicago Press, 1944), p. 5: we should avoid "dangerous" questions "such as how 'observation' and statement are connected; or, further, how 'sense data' and 'mind', the 'external world' and the 'internal world', are connected. In our physicalist language the expressions just mentioned do not appear". *Alternate Problem:* Discuss the opinion of F. P. RAMSEY, in *The Foundations of Mathematics* (London: Routledge & Kegan Paul, 1931), p. 268, that "We are driven to philosophize because we do not know clearly what we mean; the question is always 'What do I mean by *x*?,"

4.7.4. Comment on the following statements by L. WITTGENSTEIN — one of the founders of the Oxford school of language philosophy — in his *Philosophical Investigations* (New York: Macmillan, 1953): (i) ... "philosophical problems arise when language *goes on holiday*" (p. 19). (ii) "A philosophical problem has the form: 'I don't know my way about,'" (p. 49). (iii) ... "the clarity that we are aiming at is indeed *complete* clarity. But this simply means that the philosophical problems should *completely* disappear" (p. 51). (iv) "The philosopher's treatment of a question is like the treatment of an illness" (p. 91). *Alternate Problem:* Examine the proposal that the task of philosophy is to find, or to stipulate, what can and what cannot be said with propriety ("felicitously", as J. L. AUSTIN and his followers say).

4.7.5. The proper subject of philosophy, according to R. G. COLLINGWOOD, is the study of ultimate or absolute presuppositions. Comment on this proposal. (Incidentally, Collingwood did not study the logic of presuppositions, nor did he show that there are absolute presuppositions.)

4.7.6. Many histories of philosophy are catalogues of opinions and data; some go as far as proposing and substantiating hypotheses on the influences of some philosophers on others. How would you try to write a history of philosophy: by starting with the outcomes (doctrines) or by focusing on the problems and the means used to solve them? Hints: begin by drawing a distinction between the historical and the systematic (or theoretical) approach to philosophical problems and select a case history.

4.7.7. Metaphysical questions are usually supposed to be independent of fact and, accordingly, unanswerable with the help of experience. Does this hold for questions such as "Are there objective connections?", "Is symmetry prior in some sense to asymmetry or conversely?", "Does change require something unchanging?", "Is progress objective?" "Do properties tend to cluster?", "Is randomness a swerving from law or is it an outcome of laws operating on a different level?", "Are there individuals in reality?", and "Are genera real?" *Alternate Problem:* Examine the opinion of C. S. PEIRCE on the relation of metaphysics to science. See his *Collected Papers*, Ed. by C. HARTSHORNE and P. WEISS (Cambridge: Harvard University Press, 1935), Vol. VI: metaphysics is an observational science and "the only reason that this is not universally recognized is that it rests upon kinds of phenomena with which every man's experience is so saturated that he usually pays no particular attention to them" (p. 2).

4.7.8. Does philosophy have a fixed object? *Alternate Problem:* Examine the list of philosophical problems proposed by K. R. POPPER, *Conjectures and Refutations* (New York and London: Basic Books, 1962), pp. 59ff.

4.7.9. Does philosophy have a method of its own? Recall and examine the main answers to this question: the method of philosophy consists in (i) tracing the psychological genesis and development of ideas (British empiricists); (ii) discovering the social determination of ideas (sociologism); (iii) a linguistic description and analysis (Oxford philosophy); (iv) the logical analysis of scientific discourse (M. SCHLICK and R. CARNAP); (v) the clear statement of problems and the critical examination of its various proposed solutions (K. R. POPPER); (vi) the general method of science and the techniques of formal logic and semantics.

4.7.10. Consider the following antinomy of the scientific basis of philosophy. Suppose scientific philosophy employs not only the method of science but also some of its results — e.g., that space and time are mutually dependent. Suppose further that a desideratum of philosophizing were the construction of lasting theories. Now, among the results of science the particular ones have usually a longer life span than the general results. But if we select particular results — such as, e.g., the molecular weight of water or the duration of the Tertiaty period — we shall be unable to construct philosophical theories. And if we select the general results (such as a theory of gravity) we will get no stability.

Bibliography

ACKOFF, R. L.: Scientific method: Optimizing applied research decisions, Chs. 2 and 3. New York: John Wiley & Sons 1962.

AGASSI, J.: The Nature of scientific problems and their roots on metaphysics. In: M. BUNGE (ed.), The critical approach. New York: Free Press 1964.

ARISTOTLE: De interpretatione, 20b, 27—31.

BUNGE, M.: Qué es un problema científico? Holmbergia **6**, No 15, 47 (1959).

FREUDENTHAL, H.: Analyse mathématique de certaines structures linguistiques, Folia biotheoretica **5**, 81 (1960).

HILBERT, D.: Mathematische Probleme. In: Archiv der Mathematik und Physik, vol. I, p. 44 and 213 (1901), repr. in R. BELLMAN (ed.), A collection of modern mathematical classics. New York: Dover 1961.

HIŻ, H.: Questions and answers. J. Philosophy **59**, 253 (1962).

KOLMOGOROFF, A.: Zur Deutung der intuitionistischen Logik. Math. Z. **35**, 58 (1932).

KUBINSKI, T.: An essay in logic of questions. Proceedings of the XIIth Internat. Congr. of Philosophy. Firenze: Sansoni 1960, vol. V.

LINDLEY, T. F.: Indeterminate and conditional Truth-values. J. Philosophy **59**, 449 (1962).

MEHLBERG, H.: The reach of science, Ch. 3. Toronto: University of Toronto Press 1958.

MORRIS, R. T., and M. SEEMAN: The problem of leadership: an interdisciplinary approach. Amer. J. Sociol. **56**, 149 (1950).

PÓLYA, G.: How to Solve it? New York: Doubleday Anchor Books 1957.

POPPER, K. R.: The nature of philosophical problems and their roots in science. In: Conjectures and refutations. New York and London: Basic Books 1963.

STAHL, G.: La lógica de las preguntas. Anales de la Universidad de Chile, No 102, p. 71 (1956).

— Preguntas y premises. Revista de Filosofía de la Universidad de Chile **8**, 3 (1961).

Chapter 5

Hypothesis

Once a problem, or a problem system, has been set up and examined, its solution will be sought — unless it can be shown that it has no answer. The procedure for finding the solution will depend on the nature of the problem. Some problems in factual science are solved by asking the world, i.e. by setting up scientific experiences (observations, measurements, experiments); others are solved by working out testable theories about the world. In other words, a factual problem will generate an experience, a conjecture, or both. But in science experiences are not made in a vacuum: they are designed with definite ideas in mind and they are interpreted with the help of theories — for example, theories concerning the probable behavior of the observation means. Even data-gathering experiences involve theories, the more involved the deeper and more accurate such data are: just think of the amount of theorizing required by an experiment in genetics or in physics. In short, no scientific problem is ever answered by rushing to the laboratory. Therefore it will be best if, before dealing with scientific experience, we examine the scientific ideas tested by experience — namely the guesses called hypotheses, the hypotheses upgraded to laws, and the system of laws called theories. Let us begin with hypothesis.

5.1. Meanings

That the earth is round is a fact but not an observable fact: nobody has seen the whole planet and even astronauts can see only a part of it at a time. The proposition "The earth is round" was first hypothesized in order to explain certain observed facts such as the disappearance of the hull of a distant ship from the visual field; and it was subsequently corroborated by independent findings such as the earth circumnavigation and the measurement of the earth's figure. "The earth is round" is, in short, a surmise or conjecture about certain facts, i.e. a factual hypothesis — or a hypothesis in the epistemological sense. This naming rests on the following convention inherent in contemporary metascientific usage: *Definition:* A formula is a *factual hypothesis* if and only if (i) it refers, immediately or ultimately, to facts that are as yet *unexperienced* or in principle *unexperientiable,* and (ii) it is *corrigible* in view of fresh knowledge.

Hypothesis is not to be equated with fiction and contrasted with fact except in so far as both hypotheses and fictions are mental creations whereas facts other than mental facts are out there or may be made to happen in the external world. Factual hypotheses, being propositions, can be opposed to propositions of another kind, namely particular empirical propositions or *data*, i.e. bits of information. A datum is not a hypothesis: every hypothesis goes beyond the evidences (data) it purports to account for. That is, hypotheses have a greater content than the empirical propositions they cover. The information that the needle of a given meter is pointing to the 110 volt mark is a singular empirical datum: it may be tested by mere ocular inspection. (In general experiences, either single or in bunches, are necessary to corroborate singular empirical data. They are not sufficient, though: some theoretical element will always be needed.) That this datum refers to an electric current in the meter is no longer a datum but a hypothesis. In fact (i) electric currents are inferable but not observable, and (ii) the hypothesis may turn out to be false, as the meter may be out of order, so that its indications may be wrong.

Note that singular data are in principle as corrigible as hypotheses: they do not differ from hypotheses with regard to condition (ii) (corrigibility) but in relation to condition (i): in fact, data refer to *actual* experiences, whether objectifiable and intersubjective (e.g., observations) or not (*Erlebnisse*). Note also that the fact that hypotheses do not express single experiences does not preclude them from summarizing experiences in certain cases, namely when they are just generalizations of singular experiences, i.e. empirical generalizations. If "$P(a)$", "$P(b)$", ..., "$P(n)$" are singular propositions each expressing an experience of a given kind, their conjunction, namely the bounded universal proposition "$(x)\, P(x)$" is such a package of singular experiences (data). Yet it is not a datum itself but a logical construction out of data. (The most interesting hypotheses in science contain nonobservational predicates and are not analyzable into conjunctions of data.) A third characteristic worth mentioning is that, by not referring to singular experiences, hypotheses cannot be established by any single experience: single data can only refute hypotheses. Fourth, the condition of corrigibility in our definition is necessary to sort out hypotheses from propositions of other kinds. Thus, e.g., "God is omnipotent" does not refer to an experienced or experientiable fact: it complies with condition (i), but it violates condition (ii), since it is not regarded as corrigible in the context in which it is meaningful. (In the context of atheism it is not even a proposition, since neither it nor its negation are true.)

Hypotheses, rather than data, are the center of cognitive activity among humans. Data are gathered in order to serve as evidence in favor of or against hypotheses; and even data-gathering presupposes a host of hypotheses (e.g., that there is something to be observed, that the observation means are in order or can be corrected, and so on). Consider the procedure of a medical practicioner when presented with a case. He does not begin by observing his patient in a haphazard and unprejudiced way in order to obtain raw data of any kind which he might finally interpret. The gathering of data is itself guided and justified by certain hypotheses underlying his empirical procedures. Thus, e.g., his auscultation, palpation, and use of instrnments each presuppose a body of anatomical, physiological, and even physical hypotheses. The data he obtains with the aid of such procedures and on the basis of such hypotheses are in turn used by him to formulate diagnostic hypotheses which he may wish to check with finer procedures, such as biochemical analyses. In short, when presented with a diagnostic problem the physician does not start from scratch but from a body of presuppositions some of which function as guiding hypotheses, others as interpreters of his data. It is against this background that data will be produced by him; and such data will be interpreted and used to generate and test further (diagnostic) hypotheses.

The central role of hypothesis in science is often overlooked because 'hypothesis' is still employed, in popular parlance, in a pejorative sense, namely as designating a groundless and untested assumption, a doubtful and probably false conjecture that has no place in science — in turn supposed to be infallible. Yet many statements that pass for more or less direct reports on facts of experience are elaborate constructions, hence hypotheses even if they are true. A historian might feel offended if told that his stories are hypothetical reconstructions rather than bare sequences of data: yet that is what they are even when true, since what a historian narrates is his interpretation of certain documents purportedly concerning facts which he has not witnessed and which, even if witnessed, need be interpreted in the light of a body of ideas regarding human behavior and social institutions. Likewise when a physicist announces the increase of the decrease in fallout level he is interpreting certain instrument readings with the help of scientific laws and he is consequently stating a hypothesis concerning something that is no less real for being intangible.

In ordinary life we build hypotheses the whole day long: even when we act automatically we act on certainly tacitly accepted hypotheses, i.e. presuppositions. Thus, when we take the subway to school we assume it is in working condition (which may prove false); we also assume that the school building is open (a riot may falsify our assumption), that the students are interested in our lecture (which may be sheer conceit), and so on. Every activity involves assumptions stretching beyond our information to the extent to which it is a rational activity, i.e. one conducted with the aid of knowledge or made up of learned reflexes conducive to consciously preestablished goals. At any given moment only a small portion of the field where our activities are displayed is perceptible to us: most of it, though existing on its own, must be hypothetically reconstructed by us, if only in outline, to the extent to which we have got to understand it or to master it. In short, since the world is never *given* to us in its entirety, we must hypothesize it to some extent.

Hypotheses, indispensable in rational action, are even more central in the rational conception of the world (science) and in its rational modification (technology). Sensing is the animal, prescientific requisite for thinking the world; and to conceive the world is nothing but to frame hypotheses about the world. The fact that most scientific hypotheses are stated in a categorical mode should not mislead us. When the biologist says that life emerged 2 billion years ago, that the first terrestrial organisms were lichens, that plants synthesize carbohydrates out of dioxide carbon and water, that oxygen is indispensable for animal life, or that all mammals are homeothermal, he is not conveying information about experience but is stating hypotheses by means of which certain chunks of experience can be interpreted: his assumptions, being hypotheses, are not about experience but about nonexperientiable facts, and he will employ them in order to explain his biological experience.

Sometimes the hypothetical character of a proposition is made apparent by its logical form. Every *hypothetical*, i.e. every proposition of the form "If p then q", is a hypothesis, since it is a logical construction out of two propositions that may, but need not, refer each to a fact. Thus, e.g., "If the dog is angry then the dog barks" is a hypothetical that links two categoricals, namely "The dog is angry" and "The dog barks". The first categorical is inferable from data concerning the dog's behavior (in analogy with human behavior) and is therefore a hypothesis itself; the consequent, on the other hand, may be a datum. But there is no state of affairs, no event or process, whether experientiable or conceivable, corresponding to the *whole* conditional (or hypothetical). Having the form of a conditional, i.e. being a hypothetical is, then, sufficient for being a hypothesis. But it is clearly not necessary, as the case of the dog's anger shows, and the following existential hypotheses will confirm: "There are many planetary systems", "There is probably life on Mars" and "There is radiocarbon in every living being". That a categorical sentence may express a hypothesis is no paradox; the air of paradox vanishes as soon as the traditional name 'hypothetical' for propositions of the "if-then" form is replaced by 'conditional'. In general, logical form alone is not a reliable indicator of epistemological and methodological status.

So far we have alluded to the epistemological and methodological sense of 'hypothesis'. The logical sense of the word is that of *assumption*, premise, or starting-point of an argument (e.g., a proof). This is one of the original meanings of 'hypothesis' and the one retained in formal science. In this context a premise is either a previously accepted formula (axiom, theorem, or convention such as "T is an Euclidean triangle"), or a tentatively introduced formula making some deduction possible (by an argument *ex hypothesi*) and retained or rejected on the strength of its consequences. In either case a hypothesis is a premise used in a reasoning and is therefore as good as an assumption or a supposition.

In the logical sense of 'hypothesis' all the initial assumptions (axioms) of a theory, whether formal or factual, are hypotheses; they are distinguished from the remaining hypotheses of a theory by being called *fundamental* or *basic* hypotheses (or assumptions). The procedure consisting in developing a theory by first stating its starting-points or basic hypotheses and then deducing their consequences with the help of the underlying formal theories is adequately called *hypothetico-deductive method*. The axioms of a formal theory are, then, hypotheses in the logical sense, whereas the axioms of a factual theory are hypotheses both in the logical and in the epistemological and methodological senses: they go beyond experience and they are corrigible. And all theories, whether formal or factual, are *hypothetico-deductive systems*.

The background of a problem, a hypothesis, or a theory, is not usually exhibited in its entirety. The tacit and unquestioned assumptions of an idea are its presuppositions (see 4.2). A somewhat loose but handy way of characterizing this important though neglected concept is by means of the following *Definition*: *B presupposes A* if and only if (i) A is a necessary condition for the meaning or the verisimilitude of B, and (ii) A is not questioned while B is used or tested. A possible symbolization of "B presupposes A" is '$A \dashv B$', not to be confused with '$A \vdash B$', meaning "A entails B".

The presuppositions occurring in any scientific investigation are hypotheses in the logical sense of the word, i.e. basic assumptions. They are consequently assayed just as any other hypotheses, i.e. by evaluating their consequences. *And presuppositions may be recognized as such much as axioms can be shown to be independent of one another. Let B be a body of premises entailing the consequences C. In order to find out whether $A \dashv B$, i.e. whether B presupposes A, we negate A and conjoin it with B, and finally check whether $-A \& B$ entails C. If the new set of premises entails C the suspect formula is not a presupposition of B; if $-A$ introduces a change in the consequences C, then obviously A backs B up.*

Presuppositions may be divided into generic and specific. *Generic presuppositions* are those formulas which are not peculiar to the special field of inquiry. For instance, the laws of ordinary logic and the laws of physics are generic presuppositions of biological research; and certain philosophical hypotheses to be studied in 5.9, such as the principle of lawfulness, are generic presuppositions of factual science. *Specific presuppositions*, on the other hand, are those formulas in the same field that constitute the immediate and peculiar background of the formulas under consideration. For example, the existence of light is a specific presupposition of optics, not however of mechanics. (On the other hand the test of the hypotheses of mechanics presupposes the existence of light: it takes it for granted.)

Let us now take a look at the ways in which scientific hypotheses are formulated.

Problems

5.1.1. Comment on the role of "preconceived ideas" (hypotheses) according to CLAUDE BERNARD. See his *Introduction to the Study of Experimental Medicine*, Part I, Ch. 2, Sec. 1. *Alternate Problem:* Study the nature and function of the clues that suggest to the archaeologist where to dig.

5.1.2. State whether, and if so in what sense and why, the following formulas are hypotheses.

1. She loves me.
2. Copper is a good conductor.
3. In all organisms mutations occur.
4. Some of my colleagues will achieve distinction.
5. Every psychical process has a physiological correlate.
6. Man emerged ca. 1 million years ago.
7. Political preferences are largely determined by social status.
8. The clash of economic interests generates social readjustments,
9. *Beowulf* was written by Vikings.
10. There is no scientific research without hypotheses.

5.1.3. Did the geographic expeditions of COLUMBUS, MAGELLAN, and LIVINGSTONE involve hypotheses? If so, which were they in each case?

5.1.4. Does invention involve hypothesizing? Consider both scientific invention (e.g., of a new theory or of a new instrument) and technological invention (e.g. of a new processing technique or of a new machine).

5.1.5. State one lowest-level hypothesis (of the data-summarizing kind) and one going beyond every conceivable set of data.

5.1.6. State a few conditionals, either in the context of ordinary knowledge or in a scientific context, and search for their referents if any. *Alternate Problem:* Report on I. BERLIN's "Empirical Propositions and Hypothetical Statements", *Mind* (N. S.), LIX, 289 (1950).

5.1.7. Think of a scientific or technological investigation and point out (i) the generic and specific presuppositions of it, and (ii) the main hypotheses formulated in the course of it. Hint: do not try to be exhaustive because you will never finish the assignment.

5.1.8. Examine the following chain:
Logic ⊣ Mathematics ⊣ Physics ⊣ Chemistry ⊣ Biology ⊣ Psychology ⊣ Sociology ⊣ History.

Alternate Problem: Does the historical development of the sciences parallel the preceding chain?

5.1.9. What are the formal (logical) properties of the relation of presupposition? See N. RESCHER, "On the Logic of Presupposition", *Philosophy and Phenomenological Research,* **21**, 521 (1961).

5.1.10. H. DINGLER, the German operationalist, wished to build science on a set of "unaccompanied assertions", i.e. statements of experience and action presupposing no other propositions and providing therefore a basic and immutable foundation for science, both factual and formal. And E. HUSSERL, the founder of phenomenology, wanted to build a philosophical system free from presuppositions (whether logical, epistemological, ontological, or scientific). Both DINGLER and HUSSERL wanted to start from zero. Is this possible at all? If not, why? In case no such program is realizable, are we committed to espousing presuppositions as incorrigible beliefs? *Alternate Problem:* How are infallibilism and fundamentalism related? See M. BUNGE, *Intuition and Science* (Englewood Cliffs, N. J.: Prentice-Hall, 1962).

5.2. Formulation

Factual hypotheses are surmises formulated to account for facts, whether experienced or no. Now, a great many hypotheses can be conceived to cover any given set of data referring to a bunch of facts; that is, data, do not uniquely determine hypotheses. In order to sort out the most verisimilar among all such empirically backed conjectures some restrictions must be imposed. In nonscience, where truth may not be the ultimate desideratum, criteria such as conformity with established authority, simplicity, or practicality are employed. In science three main requisites for the *formulation* (not yet for the acceptance) of hypotheses are imposed: (i) the hypothesis must be *well-formed* (formally correct) and *meaningful* (semantically nonempty) in some scientific context; (ii) the hypothesis must be *grounded* to some extent on previous knowledge; if entirely novel it must be *compatible* with the bulk of scientific knowledge; (iii) the hypothesis must be *empirically testable* by the objective procedures of science, i.e. by confrontation with empirical data controlled in turn by scientific techniques and theories.

The above are requisites necessary and sufficient for regarding a hypothesis as a *scientific hypothesis* regardless of whether the conjecture is actually true or not, that is, the above are conditions that the formulation of scientific hypotheses must abide by. Moreover, the three requisites are not mutually independent. Being well-formed is a necessary condition for having a definite meaning in some language. (Imagine, on the other hand, the task of interpreting HEIDEGGER'S "Temporaneity times itself"). Having a definite meaning is in turn necessary both for receiving support from available knowledge (i.e., for being reasonable) and for facing fresh experience. (Just imagine trying to find support for, or to test, FREUD'S hypothesis that "Sleep represents a return to the maternal womb".) And being grounded in at least a part of available knowledge is the only warrant (even though it is a faint one) that the empirical test is worth trying — and possible, since scientific experience presupposes some scientific knowledge. Let us now make some exercises in hypothesis formulation: this will render the above three requirements plausible and their ulterior examination (in 5.5 to 5.7) possible.

If we insert a stick in a pool of clear water we may observe that the stick looks as if it were broken at the place where water and air join. If we are not interested in knowing we may simply wonder at this phenomenon. If we are pseudoscientists we may venture some more or less fantastic conjecture without caring whether it meets the above-mentioned requisites. If we are fact collectors we will look carefully, make drawings, perhaps take some pictures and measurements, and will end up by embodying these data in a careful but superficial description of the phenomenon. If we are scientists we will attempt to explain this very description by trying hypotheses that are logically sound, scientifically grounded and empirically testable. Such hypotheses will in turn help us look at the same phenomenon in a new light: they will prompt a deeper description, couched in theoretical terms and not simply in ordinary language.

Now, two classes of scientific hypotheses are possible in the case of any given observable fact (phenomenon). Type I *(physical hypotheses):*

the phenomenon is an objective fact, i.e. it is independent of the observer. Type II *(psychological hypotheses):* the phenomenon is subjective, i.e. it depends on the observer. In our case we have, in particular, the following possibilities, which are not exhaustive and which must have occurred to myriads of people:

h_1=The broken stick appearance is a delusion;

h_2=The broken stick appearance is due to the bending of the stick;

h_3=The broken stick appearance is due to the bending (refraction) of the light beams at the air-water interface.

The three statements are hypotheses proper: they do not describe appearances but purport to explain them in nonobservational terms and they are capable of being corrected. Moreover, they satisfy the conditions for being called scientific hypotheses. In fact, they are all logically (formally and semantically) sound; they are grounded: we know there are delusions and we know or suspect that both sticks and light beams can bend; and they are testable: delusions can be conjured off by changing the subject or the conditions of observation, the bending of sticks can be checked by touching them, and the bending of light can be tested in the absence of sticks, i.e. independently. All three conjectures count, then, as scientific hypotheses. In order to know which is the truest we test them. And this we can do only by first *deriving some consequences* from them, in conjunction with our background knowledge, and by *facing these logical consequences with empirical information,* whether available or new. Let us proceed to this test.

Empirical test of h_1. (i) *Derivation of a consequence:* From what we know about delusions, if the effect is subjective then it will disappear upon summoning another observer or changing the conditions of observation, such as the color. (ii) *Confrontation with experience:* We do not attempt to reinfore the hypothesis by choosing the most favorable conditions but try to shake it by alternatively changing both variables, the subject and the conditions of observation. Result: the phenomenon is observed by different subjects under widely varied circumstances. (iii) *Inference:* Assumption h_1 is false. The logic of this inference is the following: the testable consequence t_1 had been derived from the hypothesis h_1 and a certain body A of antecedent knowledge: $\mathsf{A} \& h_1 \vdash t_1$. Experience showed t_1 to be false, i.e. $-t_1$ to be true. Applying the *modus tollens* inference schema we infer that the premise $\mathsf{A} \& h_1$ is false. But in this experience A, far from being questioned, was presupposed, i.e. antecedently asserted (mostly in a tacit way). Hence the falsity of the logical consequence t_1 bears only on the specific assumption h_1: it is enough for $\mathsf{A} \& h_1$ to be false that h_1 be false. In short formal logic, with the assistance of an empirical datum ($-t_1$), has enabled us to refute h_1.

Empirical test of h_2. (i) *Derivation of a consequence:* If the stick is in fact bent or broken, then we shall be able to feel this with our hand. This consequence t_2 follows from h_2 in conjunction with our knowledge concerning our experience with broken sticks. (ii) *Confrontation with experience:* Again, we do not attempt to protect the hypothesis by abstaining from touching the stick but touch it and, of course, feel no difference. That is, we are able to assert that t_2 is false. (iii) *Inference:* Applying once more the *modus tollens* we infer that $A \& h_2$ is false; and, since we

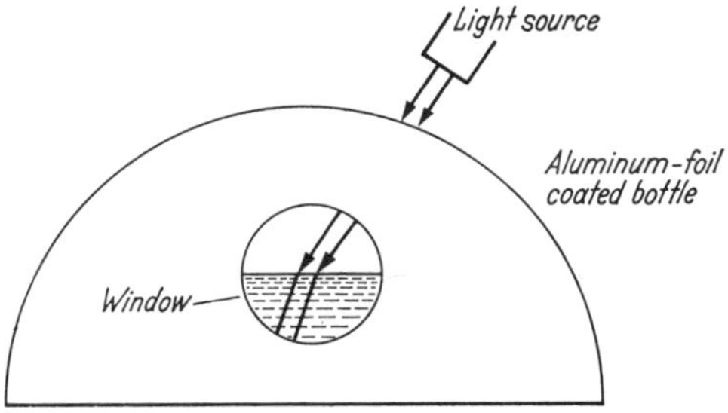

Fig. 5.1. Empirical test of the light refraction hypothesis

had used our background knowledge for deriving the testable consequence and are anyhow not investigating A but rather h_2, we conclude that h_2 is the sole culprit.

Empirical test of h_3. (i) *Derivation of a consequence:* If the broken stick appearance is an optical fact, namely light refraction (h_3), then the stick is irrelevant. Consequently we better test the logically prior hypothesis h'_3= If a beam of light falls on the air-water interface then it is refracted. That is, we strip the phenomenon of one of its ingredients thereby subjecting h_3 to a tough test: in fact, the presence of the stick might muddle things up. The given hypothesis, h'_3, is universal: it covers all possible angles. Consequently it entails the statement that if a beam of light falls on the air-water interface at a given angle, say 45°, then it is refracted. (ii) *Confrontation with experience:* In order to better control the variables we use the device shown in Fig. 5.1. We then make the antecedent of h'_3 true by illuminating the interface with light beams falling at various angles, and test for the consequent of the hypothesis by watching the light bending in water. The outcome is that the consequent is true save for light that strikes the interface at right angles. (iii) *Inferences:* (*a*) h'_3 must be changed — so as to allow for the exception just noted — into "If a light beam strikes the air-water surface at an angle other than a right angle, then it is refracted"; let us call this h''_3. (*b*) Since both the antecedent and the consequent of this conditional may be regarded as satisfactorily corroborated for a large variety of angles, the compound statement h_3'' is pronounced *confirmed* in its universality, even though only a finite number of cases have been investigated. A deeper search will show that h_3'' is only partially true. In fact, if we illuminate the interface from below we shall see that null refraction, i.e. total reflection occurs for a certain angle span above a critical value; and we shall also find that refraction depends on the light color. Such corrections are taken into account in the statement of the law of refraction — which, incidentally, contains theoretical predicates such as 'light ray' (instead of 'light beam') and 'refraction index' (instead of the qualitative 'refraction'). Finally, this refined law statement is given support from other fields by being deduced from much stronger hypotheses (namely, those of the electromagnetic theory of light). In any case it will be noted that, whereas h_1 and h_2 had been *refuted*, h_3'' was not *verified* but just *confirmed* by empirical evidence and by being incorporated in a theory: formal logic enables us to refute hypotheses, not however to establish them, and there is no *logic* of confirmation (see Vol. II, Ch. 15).

Let us now take up a case of a different kind: the test of a statistical hypothesis. It is a "fact" that smokers are more liable to get lunger cancer than nonsmokers. More exactly, extensive observation on the smoking-cancer relation has established the following low-level *statistical hypothesis:* "The frequency of lung cancer among people who smoke around 30 cigarettes a day is about 30 times the frequency among nonsmokers". Our problem now is not to explain a singular fact but to explain the preceding hypothesis. To this end we must frame some stronger hypothesis which, by itself or in conjunction with some body of knowledge, will entail the given statistical generalization. Two scientific hypotheses have so far been proposed to this effect:

h_1 = Cigarette smoking causes lung cancer.

h_2 = Lung cancer and cigarette smoking are both determined by an unknown third factor.

Both hypotheses account for the generalization to be explained and both are compatible with the bulk of knowledge: we know, indeed, that smoking is harmful in other respects and we would not be surprised if it did stimulate the formation of tumors. Further, we know that there are often spurious correlations, i.e. that a close association between two variables, A and B, may result from their linking to a common source or intervening variable C (in the statistical sense of 'intervening'); this third variable, C, can be related to A and B in either of the following ways: $A \rightarrow C \rightarrow B$, $B \rightarrow C \rightarrow A$, or $C \begin{smallmatrix} \nearrow A \\ \searrow B \end{smallmatrix}$. In our case a certain genetic factor C might mediate between cancer and cigarette smoking.

As to empirical testability, h_1 clearly satisfies this condition since a variation in the number of cigarettes would, according to h_1, make a difference in the cancer frequency. On the other hand h_2 is much too vague, as it stands, to be testable: if we are to search for something we must have at least a hint of what that something should look like: so far almost anything, hence nothing in particular, might support h_2 or undermine it. Hence although h_1 is a scientific hypothesis h_2 is, strictly speaking, a propositional function in which a predicate variable (the unknown third factor) occurs. Yet it is possible to choose, among the class of conjectures h_2 actually covers, a rather less indeterminate hypothesis, namely

h_2' = Lung cancer and cigarette smoking are both favored by a genetic factor.

Although h_2' does not specify what the determiner may be, it is a testable hypothesis to the extent to which it asserts the existence of a factor of a determinate kind. Furthermore, it is related to genetics, and genetics will direct us to investigate the possible association of lung cancer with a number of characters known to be inheritable. We now have two scientific hypotheses, h_1 and h_2', that must be tested.

But before rushing to gather more data in order to decide between h_1 and h_2' we should be clear on the *kind* of data we want. We certainly do not want more data on the lung cancer-smoking correlation because this correlation is what we wish to explain. Hence we shall not insist on making further observations on experimental groups (smoking) and control groups (nonsmoking). The nature of h_1 and h_2' themselves will direct the new observations: h_1 is clearly asking us to experimentally produce lung cancer in animals with cigarette smoke, whereas h_2' suggests examining identical twins and looking for correlations among lung cancer, age, sex, ethnic group, eating and drinking habits, personality traits, family background, etc. Hypotheses may not only explain but may also *orient research* and, in particular, the research undertaken to test them. Consequently different bodies of empirical data may be gathered by investigating different hypotheses, so that the data may be relevant to some of the competing hypotheses but not to all. In short, rival hypotheses may have to be weighed on the basis of non-comparable bodies of data. But let us return to the cancer problem.

The empirical results at the time of writing are these. Although smoke does not seem to produce cancer in laboratory animals, tobacco tar does produce it; hence there is no conclusive corroboration of h_1 — much as we would like to confirm it for aesthetic, moral and economic reasons. Also, there is a definite correlation among cigarette smoking and other characteristics such as alcohol and coffee consuming, parents with hypertension or coronary disease, etc.; but, again, the correlation is not significant enough. In short neither h_1 nor h_2' is conclusively established for the time being. What should the next move be: an increase of observations and experiments or an increased theoretical activity? Apparently the latter, because h_1 and h_2' are too *weak:* we need stronger hypotheses, suggested by theoretical considerations regarding the probable detailed *mechanism* of both the action of smoke on cells and of cigarette-appetite. The former requires a more intense cooperation of cytologists, the latter of physiologists and geneticists.

To summarize, the situation in lung cancer research at the time of writing is this: (i) we have a huge mass of empirical information and a few weak rival hypotheses (in particular h_1 and h_2') that are almost equally well confirmed by empirical findings; (ii) bolder hypotheses are needed, sensitive to the available empirical evidence and eventually suggesting the gathering of evidence of a new kind (perhaps at the molecular level). It would be a waste of time to accumulate more data of the known kinds without first refining the available hypotheses or conceiving radically new ones. And this is no exception in science: much too often stalemates are produced by want of strong hypotheses rather than by scarcity of evidence. And a wrong philosophy of science — Dataism — contributes to such a stagnation.

Let us now undertake a methodical analysis of the kinds of scientific hypothesis: it is needed in view of the widespread belief that they are all universal empirical generalizations.

Problems

5.2.1. Consider the following hypotheses explaining the gold, bronze and blood of autumnal foliage.

h_1 = Cold induces in plant the segregation of substances that replace chlorophyll or subdue it.

h_2 = Cold produces the disintegration of chlorophyll: only the more resistent pigments remain.

h_3 = Cold produces a change of color in chlorophyll.

h_4 = Waste products accumulate in leaves shortly before they fall.

Are they all scientific?

5.2.2. A coroner is faced with a death with no know cause. What hypotheses is he likely to frame? Can he assign a priori weights (i.e., weights prior to empirical tests) to his hypotheses? If so, what relation(s) should the various weights satisfy? And how would the coroner test the various conjectures? *Alternate Problem:* Pick up a page of a scientific paper and underline the sentences expressing hypotheses.

5.2.3. It is a "fact", i.e. a well-corroborated hypothesis, that in the U.S.A. most Northern Negroes are mentally superior to the Southern Negroes. Essentially two hypotheses have been conceived to account for this statistical generalization.

h_1 = The superiority has a genetic (e.g., ethnic) origin: the Negroes who emigrated to the North were already the ablest.

h_2 = The superiority is due to environmental influences: the North is economically, socially and culturally more favorable to the mental development of Negro children and youngsters.

Draw testable consequences, suggest empirical tests, and if necessary propose alternative hypotheses. See O. KLINEBERG, *Negro Intelligence and Selective Migration* (New York: Columbia University Press, 1935), or the excerpts of it in P. LAZARSFELD and M. ROSENBERG, Eds., *The Language of Social Research* (Glencoe, Ill.: Free Press, 1955), pp. 175 ff.

5.2.4. It is a "fact", i.e. a well-corroborated hypothesis, that cancer incidence has steadily increased during our century. Discuss the following and, if possible, alternative hypotheses to account for that fact.

h_1 = The increase in cancer is not real: the number of correct cancer diagnoses has increased owing to refinement in histological techniques.

h_2 = The increase in cancer follows the increase in life expectancy, because cancer is an old-age degeneracy.

h_3 = The increase in cancer is due to the increase of soot (known to be cancer-producing) and the latter is in turn due to industrialization.

Are they all testable hypotheses? And are they mutually incompatible?

5.2.5. Draw some consequences from the controversy over the cancer-smoking correlation — metascientific consequences, of course. *Alternate Problem:* Pick any other currently controverted subject and examine the hypotheses involved.

5.2.6. An experimenter, on measuring a certain magnitude, finds the following values in succession: 1, 3, 5, 7, 9. Being a generalizer, he imagines some hypotheses that condense and generalize these data:

h_1 $y = 2x + 1$, with $x = 0, 1, 2, \ldots$

h_2 $y = 2x + 1 + x(x-1)(x-2)(x-3)(x-4)$, with $x = 0, 1, 2, \ldots$

h_3 $y = (2x+1)(-1)^{x+1} \cos(x-1)\pi$, with $x = 0, 1, 2, \ldots$

h_4 $y = 2x + 1 + f(x)$, where $f(x) = 0$ for integral values of x but is otherwise arbitrary.

Can he decide among these conjectures without further data or without theoretical considerations about the nature of the x-y relation?

5.2.7. A tourist has been told to take the bus number 100. He watches five buses in succession: they bear the numbers 1, 2, 3, 4, and 5 in this order. What conjectures may the tourist conceive and how can he test them? And are these conjectures hypotheses?

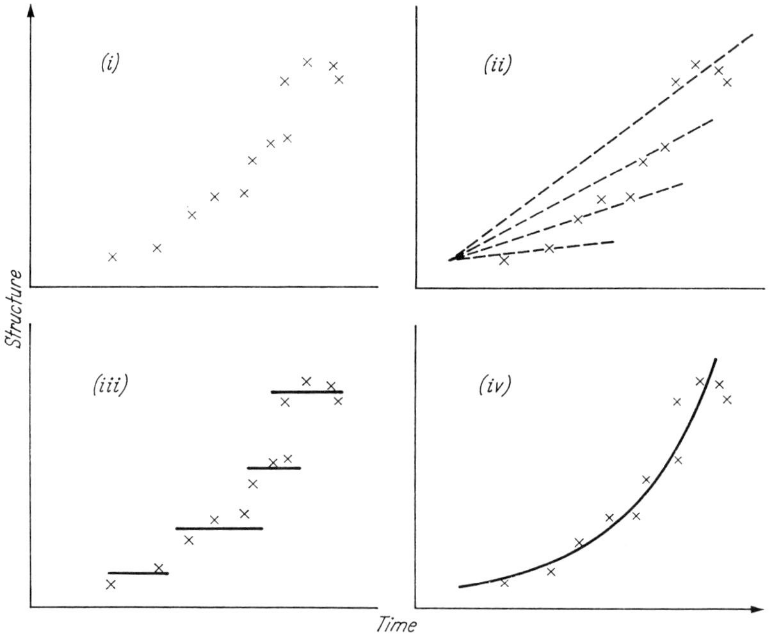

Fig. 5.2. Three different hypotheses — (ii), (iii) and (iv) — for the interpretation of the palaeontological data (i)

5.2.8. Does every hypothesis prescribe the kind of data that may serve to test it? If so how? If not why?

5.2.9. Examine NEWTON'S metascientific hypothesis: "I frame no hypothesis". See his *Principia*, Book III, General Scholium. Is it true that Newton feigned no hypotheses? If he did invent hypotheses, why did he not realize it? Was he perhaps reacting against the speculative procedure of DESCARTES? Was he under the spell of BACON'S empiricist philosophy? And if he did not formulate hypotheses, how did he manage to explain so many facts and to build theoretical mechanics? Hint: begin by enquiring whether NEWTON had our contemporary meaning of 'hypothesis' in mind.

5.2.10. The evidences in favor of evolutionary hypotheses regarding the remote past are incomplete: they consist of random samples the members of which are scattered in space and time. Every line of descent connecting fossil evidences is a hypothesis and, since the evidences are sparse, there is room for diverging interpretations (hypotheses). G. G. SIMPSON, in *The Meaning of Evolution* (1949; N. Haven: Yale University Press, 1960), p. 138, offers the following example. In Fig. 5.2. (i) a set of data are represented; (ii) represents the orthogenetic hypothesis (bundle of straight lines representing each an evolution in a definite direction); (iii) represents the macromutation hypothesis (sequence of stages emerging in leaps from one another); (iv) represents the hypothesis that the remains are members of a single evolutionary line changing in direction. Every one of the three mutually conflicting hypotheses fits the data at hand, but (iv) is the most likely because it also agrees with independent (extrapalaeontological) evidence concerning genetic changes, adaptive changes, etc. What does this suggest concerning (*a*) the determination of hypotheses by evidence and (*b*) the freedom to invent hypotheses?

5.3. *Kinds: Form and Content

Scientific hypotheses may be classed in a number of respects. Classifications as to form (syntactical), reference (semantical) and cognitive status (epistemological) will be particularly useful to our purpose.

Formally (syntactically), hypotheses may be arranged with regard to various characteristics, chiefly structure of predicates, range, systemicity, and inferential power. As regards predicate structure a first trait to be considered is *predicate number:* a hypothesis may contain a single predicate (e.g., "There are neutrinos") or, more commonly, several predicates (e.g., "All modern societies are stratified"). In the second place, the *predicate degree* or number of argument places of the predicates occurring in a hypothesis are important in a logical analysis: thus, e.g., "valence" is a unary or monadic predicate whereas "descends" is at least binary. The degree of a predicate is, though, a function of analysis depth; thus, "x is observable" is just a first approximation to "x is observable to y under conditions z with means w". In short the degree of predicates, hence the logical structure of hypotheses, is contextual rather than absolute: it depends on the state of the body of knowledge in which they occur and on the analytic fineness that is required or possible. (In general, logical form is contextual rather than absolute.) Thirdly, the *metrical character* of predicates is of interest: a necessary condition for a hypothesis to be vague is to contain qualitative (non-metrical) predicates only, as in the case of "Heat passes from the hotter to the cooler bodies". But this condition is far from being sufficient:

many accurate hypotheses contain dichotomic (presence-absence) predicates only, as, e.g., "The transitions between states of different symmetry properties are discontinuous".

As regards *range*, scientific hypotheses are of all possible extensions, not just universal:

1. *Singular* hypotheses, like "That was an intelligent act".

2. *Pseudosingular* hypotheses, like "The solar system is dynamically stable", have a hidden quantifier, usually with respect to time and/or space. In our example the ostensive form is "$S(s)$", but what we mean is that the solar system is stable at every time t in a time interval T, i.e. $(t)_{t\in T}\, S(s, t)$. Many sociological and historical hypotheses could qualify as laws if their hidden universality were made apparent.

3. *Indefinite existential* hypotheses, like "There are indivisible (elementary) particles", which specify neither place nor time and are therefore hard to disclaim.

4. *Localizing existential* hypotheses, like "There is plenty of iron in the Earth's core". Localization may be spatial, temporal, or spatiotemporal.

5. *Quasi-general* hypotheses, such as "If a system is isolated then in most cases it will go over to states of higher entropy". Such hypotheses make place to exceptions, whether in a specified or in an unspecified number.

6. *Statistical* hypotheses, like "Ectomorphic persons tend to be cerebrotonic". They state correlations, trends, modes, averages, scatters, or other global (collective) properties.

7. *Bounded universal* hypotheses, such as "*Laissez-faire, laissez-passer* is the counsel given by industrialists and merchants to statesmen from the 18th century until recently": it refers to a limited interval. A bounded universal quantifier, '$(x)_{x\in S}$', meaning "for every x in S", where S is a bounded set, occurs in these hypotheses.

8. *Unbounded universal* hypotheses, like the laws of physical optics, which are supposed to apply to all instances of a certain kind, in all places, and at all times. An important scientific task is to establish in each case the limits of such a claim to unrestricted universality, and an interesting scientifico-philosophical investigation is to speculate on the possibility that such a universality be nothing but a delusion of short-lived and short-sighted beings.

With respect to systemicity or systematic character a hypothesis may be either *isolated* or *systemic* (i.e. belonging to some system). All empirical generalizations are isolated during their infancy: it is only when they grow into laws that they acquire a systemic status. But, of course, no statement is entirely isolated: if it were it would be unintelligible. 'Isolated' does not mean in this case self-contained or cut out from the body of knowledge but just presystematic and, in particular, pretheoretical. Systemic hypotheses, on the other hand, are entrenched in some system either as axioms (starting-points) or as theorems (logical consequences). Thus, e.g., NEWTON'S laws of motion are axioms in elementary mechanics and theorems in general analytical mechanics.

Inferential power is the last formal trait to be considered here: it is the capacity formulas have of giving rise to further formulas — with the assistance of a logical or mathematical forceps. Actually no proposition is sterile, i.e. every proposition has some inferential power. Even a singular proposition e entails infinitely many conditionals with e as a consequent: $e \vdash h \rightarrow e$. (Proof: Suppose the inference invalid, i.e. $h \rightarrow e$ false. This requires h to be true and e false. But this contradicts the assumption that e is true.) Under 'inferential power' we shall subsume in this context specifiability (possibility of instantiation) and counterfactual force (possibility of deriving contrary-to-fact conditionals). As regards *specifiability*, general hypotheses may be specifiable, conditionally specifiable, or unspecifiable:

1. *Specifiable* hypotheses are those from which singular propositions can be derived by merely substituting variables by constants, in order to account for (describe or explain) singular facts. Low-level empirical generalizations and lowest-level theorems of factual theories comply with this condition.

2. *Conditionally specifiable* hypotheses are those which can be applied to individual cases only after convenient formal or semantical operations. Thus, an equation referring to an individual (e.g., a cell or a cell system) must first be solved, then be interpreted in empirical terms if it is to account for (describe or explain) a fact involving the individual it refers to. Likewise, law statements involving theoretical probabilities will have to be transformed into statements involving frequencies if they are to be interpreted as describing collective properties since "frequency", not "probability", is a descriptive concept. And the converse substitution will be performed in statements of empirical statistical uniformities if anything is to be concluded about the individuals of a collection. Thus, e.g., from "The frequency of the property B in the class A is f" we cannot conclude that every A or any given A is a B or is not a B: what can be concluded is that the probability that an A be a B is close to f — if we are willing to regard probabilities not only as collective properties but also as properties of individuals *qua* members of certain sets.

3. *Unspecifiable* hypotheses do not allow the drawing of singular propositions by specification even after syntactical or semantical transformations. Examples of this kind are existential and quasi-general propositions such as "Most salts of the alkali metals have a high solubi-

lity in water", and statistical hypotheses containing nondistributive (global) predicates, such as "The less homogeneous a population the more widely its quantitative properties are scattered around the respective averages".

As regards the power to infer subjunctive conditionals, hypotheses may be divided into *counterfactually powerful* and *counterfactually powerless.* Most singular and general hypotheses are counterfactually powerful. Thus, from "Uranus rotates around the Sun" we can infer that, if that tiny light spot in the sky were Uranus, it would rotate around the Sun. And from "Mesons are short-lived" we can infer that, if this particle were a meson, it would be short-lived. On the other hand, from "Everyone in this room is a scientist" we cannot infer that if the janitor came into the room he, too, would be a scientist: the association between the predicates 'scientist' and 'being in this room' is accidental rather than systematic. Existential hypotheses, like "There are many planetary systems", seem to be counterfactually powerless — but this is an unsettled question.

In some cases the possibility of counterfactual inference is not apparent. For instance, at first blush the statistical law "The mode for family size in the Western hemisphere is two children" is counterfactually powerless. On closer inspection this conclusion is seen to be mistaken. In fact, the hypothesis can be spelt out to read: "For every x, if x is a random sample of the population of Western families, then the mode of the number of children of x is 2". Now if on inspecting demographic data concerning any given community of unknown location we find that the typical family size is significantly different from two children, we may conclude that the community does not belong to the Western world. And if someone were to doubt our conclusion we might rejoin that, if the community in question were in fact located in the Western world, then its most frequent family size would be two children. In many cases, then, counterfactual force depends on the depth of analysis.

We now leave the syntactical point of view and concentrate on a few semantical properties of hypotheses, namely some properties of the concepts occurring in them and their reference. First of all, predicates may be either *distributive* (hereditary) or *global* (nonhereditary). Thus in "Physical space is homogeneous" the concept "homogeneous" is a distributive or hereditary concept because homogeneity is supposed (perhaps wrongly) to be true of every part of physical space. On the other hand "composition", "average", and "living" cannot be predicated of every part of their referents: they are collective or global concepts. This point is relevant to the discussion whether the properties of wholes are already present in their parts (primitive mechanism) or are genuine novelties emerging from the latter (emergentism); unfortunately the discussion of emergence, so often muddled, have not taken advantage of this distinction.

A second semantical property of predicates that must be considered here is *order* or semantic category. Most predicates refer to properties of individuals (whether simple or complex), but some scientific hypotheses contain higher-order predicates, i.e. they predicate something of properties or relations. Examples of such higher-order predicates are 'symmetric relation' and 'biological property'. Such higher-order predicates occur also in metascientific analysis.

A third semantical respect of interest is *precision*. A first dichotomy in this regard is the division of hypotheses into coarse and refined. *Coarse* hypotheses are imprecise either because of the occurrence of vague predicates in them (as in the case of "A depends on B") or because of indefinite range. Logical disjunction and consequently existential quantification may be responsible for range imprecision, as in "Some substances do not combine with any other substance", and in the many theorems of statistical physics that begin with the phrase 'For almost all points (or trajectories) ...'

Refined hypotheses, on the other hand, are those which are both predicate- and range-precise, such as, e.g., "The oscillation period of a simple ideal pendulum is $T=2\pi(L/g)^{\frac{1}{2}}$". Refined hypotheses are often equalities, such as "$y=kx$", rather than inequalities, such as "$y>x$". At first sight only singular and universal (bounded or unbounded) hypotheses can be given precision, but this is not so: "There are exactly n A's in B", and "The average P of x equals y", are perfectly precise although the former is existential and the latter statistical. There are, of course, degrees of precision: the coarse-refined dichotomy is itself vague. At any rate it is clear that maximum precision is desirable for the sake of strength, testability, and truth.

Let us finally tackle the problem of the referents of scientific hypotheses as wholes rather than of their constituent predicates. Every statement contains what may be called *leading predicates*, the analysis of which will show what its referent is. Thus the referent of "Atoms are restless" is the set of atoms, the referent of "Molecules are composed of atoms" is the set of all pairs of atoms and molecules (the Cartesian product of the set of atoms by the set of molecules), and the referent of "Temperature is a state variable" is the physical property temperature. Temperature, the property, is the referent of the last statement, but this has also an indirect or mediate referent, namely the set of molar physical systems. If we concentrate on immediate referents we shall find that scientific hypotheses purport to refer either to experience (possible rather than actual), or to both experience and objective fact, or to objective fact alone, or to a conceptual model of facts. More in detail, we have the following possible kinds of scientific hypotheses as regards their immediate referent.

1. *Experience-referent* hypotheses, such as "All color sensations can be produced with only two lights of different colors", refer to phenomena or experienced facts; they contain therefore phenomenal predicates, i.e. concepts referring to sensory experience. They are not subjective on this count; only, they are inconceivable apart from some subject. And, of course, in order to be hypotheses and not data they must not refer to an actual experience but to possible experiences; such is the case of a universal conjecture, i.e. an all-statement (if unbounded).

2. *Experience and fact-referent* hypotheses, such as "The probability of obtaining the value a when measuring the property A is p". (Actually this is a hypothesis schema rather than a hypothesis, since it involves the predicate variable A.) Hypotheses of this kind, in which both the object and the subject of knowledge are involved, are often found in the stage of theory testing and in the attempt to interpret factual theories in terms of operations.

3. *Fact-referent* hypotheses, such as "Earthquakes tend to occur near faults", are supposed to refer to objective facts and their properties. Actually, the most refined scientific hypotheses do not refer to whole facts but to selected traits of concrete systems, events, or processes — such as, e.g., "Hydrogen has three isotopes". Whereas experience-referent hypotheses contain only observation concepts, fact-referent hypotheses contain, either in their place or in addition to them, transempirical concepts such as "inheritable". In a sense the progress of knowledge consists in replacing observation concepts by transempirical concepts, and correspondingly experience or subject-centered hypotheses by fact or object-centered hypotheses.

4. *Model-referent* hypotheses have no concrete immediate referent: their referents are theoretical models that, in turn, purport to be approximate reconstructions of real systems (see Sec. 8.4). Thus, the theoretical laws of physics and economics refer immediately to ideal objects (frictionless motions, free enterprise, etc.) which are at best rough approximations. Actually all quantitative and transempirical law statements are model-referent, so that they must be expected to apply to real systems only to within certain errors.

In the next section we shall examine further aspects of the richness of hypotheses.

Problems

5.3.1. State a scientific hypothesis and perform a syntactical analysis of the predicates occurring in it (number, degree, and metrical character).

5.3.2. State one scientific hypothesis of each possible range.

5.3.3. Illustrate the concepts of isolated hypothesis and systemic hypothesis.

5.3.4. Take a quantitative law statement having the form of an equation and incapable of having substitution instances unless it is subjected to some transformation.

5.3.5. Mention a couple of distributive (hereditary) and another couple of global (nonhereditary) predicates.

5.3.6. Illustrate the concepts of coarse hypothesis and refined hypothesis.

5.3.7. Why are equalities preferred to inequalities?

5.3.8. Examine the opinions that (i) scientific hypotheses are universal generalizations concerning experience and (ii) scientific hypotheses express relationships among facts.

5.3.9. Lay down in detail the conditions a conjecture must satisfy in order to refer to human experience and yet be a hypothesis rather than a summary of actual experience. *Alternate Problem:* Examine the view that hypotheses with the form of differential equations cannot be universal. Hint: Distinguish the object variable(s) from the numerical variables.

5.3.10. Make a thorough study of as-if hypotheses, such as "Gravitational forces act on bodies as if the latter were condensed in their center of mass". It is possible to dispense with the expression 'as if'?

5.4. Kinds: Epistemological Viewpoint

Let us now adopt an epistemological point of view and focus on the inception, ostensiveness, and depth of hypotheses. As regards *inception* scientific hypotheses may be originated by analogy, induction, intuition, deduction, or construction. Actually these are just ideal types: every hypothesis proper is a construction framed with the help of all kinds of inference. We should therefore speak of hypotheses found *predominantly*, not exclusively, by analogy, induction, deduction, or construction.

1. *Analogically* found hypotheses are those inferred through arguments from analogy or by the intuitive realization of similarities. Two

kinds of analogical jump may be distinguished: (i) *substantive analogy*, as when the response to a stimulus by one organism suggests the hypothesis that a similar stimulus-response relation will take place in a different organism; (ii) *structural analogy*, as when the law of growth of a population is suspected to have the same form as the law of growth of an individual. Substantive analogy (similarity in kind) regards specific properties and goes from individual to individual; structural analogy, on the other hand, bears on formal similarities of systems, whether physical or conceptual. Such analogical inferences may take place spontaneously: it is only their justification that resorts to more strict patterns of inference.

2. *Inductively* found hypotheses are those framed on the basis of case examination. We may distinguish two kinds of inductive generalization: (i) *first degree induction*, or inference from particular to general statements, as when "Studying French interferes with the simultaneous learning of Italian" is "concluded" from the examination of a number of individual cases; (ii) *second degree induction*, or generalization of first degree generalizations, as when the general conjecture "Learning of any subject interferes with learning of any other contiguous subject" is conceived on the basis of first degree generalizations regarding the learning of particular pairs of subjects. Induction, particularly when it does *not* start from singular empirical statements, has an important place in the construction of science, but a still more important place in the "drawing of conclusions" from the comparison between general theoretical predictions and empirical data (see Vol. II, Sec. 15.4).

[*Empirical* induction, i.e. generalization of observed cases, has been grossly overestimated by philosophers who have concentrated on the early (pretheoretical) stages of research, as well as on the empirical test of noninductive hypotheses. Inductivism has also been stimulated by the behaviorist (Watsonian) and mechanistic (Pavlovian) learning doctrines, according to which not only preconceptual learning (such as the learning of a skill or a language) but learning of *every* kind is done on the basis of cumulative trial-by-trial reinforcement and by generalization of associations. On these doctrines — when extrapolated from maze-racing to theory-building — the growth of scientific knowledge would be just an accumulation of useful (reinforced) behavior patterns initially hit on by blind trial and error. Actually not even laboratory rats conform strictly to this view but have rather definite expectancies; furthermore, they not only strengthen but also modify their behavior with experience. But whatever rats may do, it seems that men learn to pose and solve *conceptual* problems by making conjectures and testing them methodically. Some such conjectures are indeed the result of cumulative experience in one direction (empirical inductions), but it so happens that these conjectures are scientifically uninteresting precisely because they do not go far beyond experience. The most important conjectures are gained on the basis of little or no experience of the preconceptual kind: they are not solutions to recurring empirical problems but to new problems of a conceptual kind. Inductivism, which accounts for certain routine procedures, fails to explain the posing of original problems and their solution by the invention of entirely new hypotheses, and precisely of hypotheses that refer to objective facts or to idealized models of them rather than to immediate experience (see Sec. 5.3).]

3. *Intuitively* found hypotheses are those the introduction of which has not been traced and looks both natural and straightforward: at a first, superficial sight, they look as if born by spontaneous generation, without prior research and with no logical processing. This impression is false, since all hypotheses must at least be "felt" (obscurely suspected) to be logically sound, compatible with the bulk of knowledge, and testable if they are to count as scientific (see Sec. 5.1). Many hypotheses which now seem "natural", "obvious", or "intuitive" are elaborate constructions that could not have been conceived in earlier epochs or in different intellectual climates. For example, the hypothesis there is a fixed relationship between the quantity of heat released by a heater and the quantity of electric energy it consumes seems now obvious because we pay for energy supply; but it was not imagined, let alone tested, before it was suspected that electricity could be converted into heat — a suspicion confirmed by J. P. Joule in 1843. The equally "natural" hypothesis that the biological effect of a drug is related to its chemical constitution was not advanced until about the same time (J. Blake, 1841). The two hypotheses were so unintuitive at the time they were proposed that they had to struggle for recognition. No hypothesis is inherently intuitive, natural or obvious: to overlook this is as foolish as to deny that the invention of hypotheses is not made by accumulating data but requires some insight and occurs often like a flash — though never without some prior knowledge and pondering.

4. *Deductively* obtained hypotheses are those deduced from stronger propositions. Three subclasses may be distinguished: (i) *theorems* or logical consequences of some of the prior assumptions of a theory, such as, e.g., the hypotheses regarding the geographical distribution of a given species when they are derived from general biogeographical postulates; (ii) *derivations from wider scope theories*, as when a thermodynamical relation is deduced from statistical mechanical principles.

5. *Constructs*, i.e. more or less elaborate constructions that are not ostensively derived from anywhere but are guessed with the explicit help of certain conceptual tools. For example, the strongest physical principles (the variational principles) are tailored after given equations of

motion, just as NEWTON tried several functions of the inter-body distance until he hit on the inverse-square law, which alone led to the Keplerian laws via his own laws of motion, other typical constructs. Without constructs of this kind there would be no science proper.

Yet the fact that constructs are not inferred or concluded from other propositions should not lead us to believe that they are *freely* invented guesses: scientific hypotheses are born in response to definite problems that are stated in a given body of knowledge, and they are expected to pass the test of fresh experience. Although any given set of data can be accounted for by a number of hypotheses, the latter grow in well-trained minds and must satisfy certain requirements rather than being arbitrary: this is, precisely, the difference between a wild guess and a scientific hypothesis.

And the fact that there are no infallible techniques of hypothesis formation does not entail that no definite *avenues of approach* exist: in fact, there are as many as modes of thinking. A mode of thinking characteristic of our time is the probabilistic style. Consider, for instance, the transmission of messages along a noisy channel, such as a long distance telephone line. Let the problem be to formulate a hypothesis concerning the intelligibility of messages. An adequate measure of the intelligibility of a word for a receiver is the probability that the word may be correctly identified by the receiver on a first presentation. Once the concept of intelligibility has been thus quantified and elucidated at the same time, it is rather easy to construct a hypothesis concerning the intelligibility of a message after a certain number of repetitions: the mathematical theory of probability acts as a nest for the hatching of the factual hypothesis. Similarly the theoretical biologist will use physical theories for the formation of biological hypotheses and the historian may use sociological theories for the formation of historical hypotheses. This procedure, of letting a theory of a different species do the work of hatching a hypothesis, may be termed the *cuckoo technique*.

In any case scientific hypotheses are born in a number of ways and they have very often spurious origins, in the sense that the arguments leading to them are unsound or that they proceed from false hunches. The paths leading to the formulation of scientific hypotheses are devious and often muddy; this is why, when expounding them in written form, scientists almost always reconstruct these paths entirely, to the despair of the historian and the psychologist of science. [The systematic presentation of a subject coincides only rarely with its historical presentation; quite often the one is the reverse of the other. Thus, a historical presentation of genetics would lay bare the following chain: Individual variations — Mendelian inheritance — Chromosome basis — Genes — *DNA* molecules. On the other hand a systematic presentation at the time of writing could start from *DNA* molecules (high level hypotheses) ending with the observable consequences in phenotypic characters (low level hypotheses).] Scientific hypotheses are not legitimate or illigitimate on account of their origin but on the strength of their tests, both empirical and theoretical: they are given test certificates rather than birth certificates.

Let us now approach the problem of the *degree of ostensiveness* (or conversely of the degree of abstractness) of scientific hypotheses. This characteristic is determined by the degree of ostensiveness of the predicates occurring in them. Consequently we may adopt the observational/nonobservational dichotomy for concepts (see 2.6).

1. *Observational* or low-level hypotheses contain only observational concepts, i.e. concepts referring to observable properties like position, color and texture. Many objects which pass for facts are actually low level hypotheses such as "Birds lay eggs in nests". Strictly speaking no purely observational hypotheses can make part of theories proper, since the latter are made of constructs; observational hypotheses can only generalize observable states of affair, and if they are included in some theory it is only by being first translated into nonobservational statements.

2. *Nonobservational* hypotheses are those containing nonobservational concepts, whether intervening variables (e.g., "average") or hypothetical constructs (e.g., "inertia"). Such concepts are absent from raw empirical data but ordinary knowledge teems with them: "joy", "love", "thought" "purpose" and thousand others are never observed to be true or false of something: this must always be surmised or inferred. Nonobservational *ordinary* concepts may occur in scientific hypotheses in the descriptive stage of a discipline; this is the case of "Suicide is more frequent among Protestants than among Catholics". But in the more advanced stages only *theoretical* nonobservational hypotheses will be found, such as "The inhibition of digestion during states of emotional stress favors the use of blood by effector organs". Intermediate-level hypotheses, i.e. hypotheses containing both ordinary and theoretical concepts also exist and function as bridges between theory and experience. An example of such a *mixed* or intermediate level hypothesis is "Meat is rich in proteins". The occurrence of mixed hypotheses in a theory is sufficient for its empirical testability but is not necessary: in most cases the theory's predictions will be translated into a semiempirical language: e.g., 'light ray' will be translated by 'light beam'. What is important in a scientific hypothesis is not to secure the occurrence of ordinary, let alone observational concepts in it: what must be secured is the absence in the hypothesis of inscrutable concepts such as "libido energy" or "motion from future to past". For a hypothesis to be testable it must contain *scrutable* predicates only, however sophisticated they are.

The third and last epistemological trait of hypotheses to be dealt with here is *depth*. In this regard scientific hypotheses may be divided into phenomenological (not to be confused with phenomenal or experience-referent) and nonphenomenological or representational.

1. *Phenomenological* hypotheses are those which, whether they contain observational concepts or are rather abstract (that is, epistemologically uplift) constructions, they do not meddle with the inner workings of systems but only with their outward behavior. All input-output relations in thermodynamics, electrical engineering and economics are phenomenological to the extent that they do not refer to processes whereby inputs are converted into outputs. Likewise, chemical formulas that specify neither chemical structure nor reaction mechanisms are phenomenological hypotheses. Consider, for example, the formula for the synthesis of glucose in the leaves of green plants:

$$\text{Carbon dioxide} + \text{Water} \xrightarrow[\text{Chlorophyll}]{\text{Light}} \text{Glucose} + \text{Oxygen}$$

This formula says only that, "under the action of light" and "in the presence of chlorophyll", certain substances get transformed into certain other substances. The modes of action of light and chlorophyll are not referred to: the whole thing is treated as a structureless black box that in some mysterious way converts certain inputs into certain outputs. It is only when, with more knowledge, the photochemical and enzymatic mechanisms of photosynthesis are inquired into, that the phenomenological stage is left behind.

2. *Representational* or "mechanistic" hypotheses go beyond input-output balances: they specify some mechanism — which, of course, need not be mechanical in the strict gear-and-pulley sense. The theoretical concepts of representational hypotheses purportedly denote real properties: they are not just intervening variables useful for condensing and computing data. For example, a representational approach to ferromagnetism and ferroelectricity will not restrict itself to stating phenomenological relations among polarization and temperature but will try to explain such relations by deducing them from deeper, representational hypotheses; in particular, it will attempt to explain the sudden drop in polarization at a certain critical temperature as the effect of a discontinuous change in microscopic or semimicroscopic structure. Or take the study of biological growth. Empirical studies on the growth of individuals and populations can be summed up and generalized by growth curves. Since these curves refer to limited time intervals, they can be fitted by infinitely many functions relating the size and age of the biological entity. Every such function will be a phenomenological

hypothesis on growth. In the absence of conjectures concerning the growth mechanism we are unable to decide which among these infinitely many phenomenological hypotheses is the truest. This uncertainty may be decreased by making definite assumptions concerning the growth process — different assumptions for individuals and for populations. For example, in the case of individual growth we may hypothesize that cell expansion is at least as important as cell reproduction, whereas for populations we may hypothesize that reproduction alone is operant. In this way stronger, representational hypotheses are built. Moreover, there will be no end to their correction; for instance competition with other entities may be taken into account. In any case, whereas the phenomenological approach gave an infinity of rival hypotheses the representational approach yields a handful of hypotheses competing for the same data; moreover additional evidence, not directly concerning growth but other processes (e.g., intraspecific competition) may now be adduced in favor or against the representational hypothesis.

Certain philosophical schools, notably positivism and conventionalism, abhor representational hypotheses because these go far beyond summarizing data; they tolerate the use of nonobservables as long as they are regarded as symbolic intermediaries (intervening variables) between observational concepts rather than as representing real though nonobservable traits. The decision between this policy and the one consisting in encouraging the introduction of hypothetical constructs as long as they are scrutable is a philosophical matter, yet it should not be dictated in the interest of a philosophical school because this would be lethal for science. Thus, the decision to regard "drive" as a hypothetical construct rather than as an intervening variable must be made on the finding that every drive so far investigated has a neural correlate. The same holds for all laws of behavior psychology: since we know that in a vertebrate organism a process of excitation of the central nervous system mediates between every stimulus and its response, we are led to hypothesize that every phenomenological (behavioristic) hypothesis concerning behavior has an underlying set of neurophysiological laws (see Fig. 5.3). The refusal to investigate this assumption isolates psychology from biology, deprives it of evidence of a new kind (neurophysiological) and of a foundation capable of explaining the higher in terms of the lower. In other words, the intervening variable-hypothetical construct feud is a metascientific dispute but it should be solved in the interest of the advancement of science rather than by resorting to philosophical dogmas. And scientific progress has largely consisted in hypothesizing, and eventually confirming, the existence of unperceivable things, properties, and mechanisms explaining what can be perceived. Representational hypotheses, by assuming organs and functions beneath behavior, are *deeper* than the corresponding phenomenological hypotheses in the sense that they reach deeper levels of reality. At the same time they are logically *stronger*, since they entail the phenomenological hypotheses; and they are *better testable* because they are sensitive to more minute detail and to more varied evidence. To oppose them is therefore to oppose the maturation of science.

It is now time to investigate the ground of hypotheses.

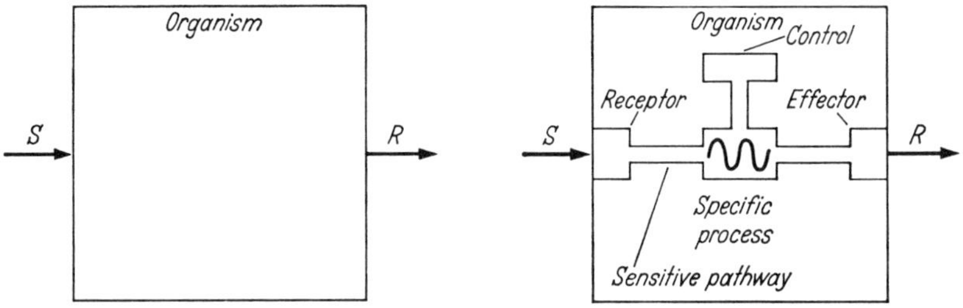

Fig. 5.3. The two approaches to psychology. (i) *Phenomenological* (black box) approach: functional relationships $R = F(S)$ between stimuli S and responses R. (ii) *Representational* (translucid box) approach: a system of conjectures (a theory) regarding the mechanism responsible for overt behavior — i.e. explaining the phenomenological hypotheses $R = F(S)$

Problems

5.4.1. Illustrate the following kinds of scientific hypothesis: found by analogy, by induction, by intuition, by deduction, and constructed.

5.4.2. Examine the opinion of H. Poincaré that some scientific hypotheses are *natural*, in the sense that we could not possibly miss them. Among these hypotheses he mentioned (i) contiguity (action by contact), (ii) continuity of the causal relation, and (iii) linearity (superposability) of small motions. See his *Science and Hypothesis* (London: Walter Scott, 1905), Ch. IX. *Alternate Problem:* Work out the distinction, sketched in the text, between nonempirical induction and empirical induction.

5.4.3. Comment on the following case history of hypothesis invention. J. Dalton invented the law of partial pressures: "The pressure of each constituent in a gas-mixture is independent from the pressures of the other constituents". He found this law on the basis of a false assumption as told by T. G. Cowling in *Molecules in Motion* (1950; New York: Harper & Brothers, 1960), p. 34: "Dalton, believing as he did that gas pressures arise from the mutual repulsions of the molecules, took this law to mean that a molecule was only repelled by like molecules. As a consequence, he said, no amount of air pressure can stop water evaporating if the air is perfectly dry, but the evaporation at once stops when the right amount of water is present in the air — the vapour molecules pushing back any molecules that might otherwise rise from the water. He was quite right in saying that evaporation can only be stopped by the presence of sufficient water vapour; he had the gift, common to all really great men, of being usually correct in his conclusions, even though led to them by unsound arguments". Pay attention to both the inception of the law and to the depth of it as well as the depth of the underlying hypotheses. *Alternate Problem:* Examine GOETHE's inference of his evolutionary hypothesis from the idea that every living being is a copy of an ideal type (archetype).

5.4.4. State one observational hypothesis and one nonobservational hypothesis and perform an epistemological analysis of both. *Alternate Problem:* Frame several hypotheses to explain some recent event in your own experience and class them.

5.4.5. Diagnose the following hypotheses as to observational or nonobservational character. Pay attention to the differences between actually observed properties and observable properties (including those that might have been observed but were not). (i) "The Vikings visited North America during the Middle Ages". (ii) "Every body continues in its state of rest or of uniform motion in a straight line unless it is compelled to change that state by forces impressed upon it" (NEWTON's first law of motion). (iii) "90 per cent of neurotic patients are much improved or cured after five years, whether they are treated or not". *Alternate Problem:* Report and comment on "The Black Box", Ch. 6 of W. R. ASHBY, *An Introduction to Cybernetics* (1956; New York: John Wiley, 1963).

5.4.6. Superstitions are often born in either of the following way: (i) a chance conjunction between A and B is observed a few times or even just once and the conjecture is there upon framed that all A's are B's or conversely; (ii) a conjecture is invented to account for some fact and is accepted either because no alternative hypothesis is available or because it agrees with the dominant body of beliefs. Do scientists proceed in the same way? *Alternate Problem:* According to HUME and his followers every hypothesis (i) is generated by induction — however much it may outstrip its evidences, and (ii) is held by habit. Examine this doctrine in the version due to N. GOODMAN, *Fact, Fiction & Forecast* (London: Athlone Press, 1954), where psychological habit is replaced by "entrenchment" in language of the predicates "that we have habitually projected", i.e. concerning which we have made forecasts.

5.4.7. W. GILBERT (1600) explained the fairly fixed orientation of the compass assuming (h_1) that our planet is a huge magnet interacting with the needle. And in order to explain the magnetic terrestrial field we frame the hypotheses that there are underground magnetic materials (h_2) and subsurface electric currents (h_3). Analyze this situation and determine whether further questioning might not lead to an infinite regress. *Alternate Problem:* Examine the popular beliefs that KEPLER got his laws by looking at TYCHO BRAHE's tables, and that NEWTON derived his inverse square law of gravitation from KEPLER's laws alone. If you believe these stories try the recipe yourself.

5.4.8. It has been learned from experiment that the mobility of electrons inside semiconductors (such as the germanium used in transistors) is fairly independent of the applied electric field (h_1). The first hypothesis advanced to explain this generalization was that the electrons meet a resistance that counteracts the external field (h_2). What may have suggested h_2? And was h_2 phenomenological or representational? The next step was to explain h_2. It was first assumed that the resistance was due to the collisions of the electrons with the atoms (h_3); but this hypothesis had testably false consequences. The hypothesis accepted at the time of writing is that what opposes the electron drift are the elastic (sound) waves generated by the thermal motion of the atoms (h_4). Analyze this situation. *Alternate Problem:* What are the differences between the medicine limited to describing and correlating symptoms and the medicine which, according to C. BERNARD, tries to disclose "*la filiation physiologique des phénomènes*"?

5.4.9. Are hypotheses drawn from somewhere or are they invented? (Psychological aspect.) And are they inferred or posited? (Logical aspect). Hints: (i) consider whether experience can yield or at least suggest propositions concerning unexperienced facts; (ii) consider the "fact" that only problem-oriented (inquisitive) people do not remain satisfied with what they sense; (iii) take a look at W. WHEWELL'S *Novum Organum Renovatum*, 3rd ed. (London, 1858), pp. 64ff. *Alternate Problem:* According to inductivism scientific hypotheses are got by inference from data. If this is so, how come that, by definition, hypotheses go beyond data?

5.4.10. The perceived intensity ψ of a physical stimulus having the intensity S is given by S. S. STEVENS' psychophysical law: $\psi = k \cdot S^p$, where the precise value of p is characteristic of every kind of stimulus and ranges between 0.3 and 3.5. This phenomenological law has been confirmed for more than a dozen kinds of perception (brightness, apparent length, duration, heaviness, velocity, etc.). The constants k and p in it are just numerical constants with no psychological, let alone a physiological interpretation. Examine the derivation of STEVENS' law proposed by D. M. MACKAY, *Science*, **139**, 1213 (1963) by hypothesizing that every subject contains an internal organizer and a comparator tuned to every receptor organ. *Alternate Problem:* Social and economic history investigate the mechanisms responsible for certain historical regularities. Study this as a case of interpretation of phenomenological hypotheses in terms of representational ones.

5.5. Ground

Scientific hypotheses, when true, are lucky hits but they are not born by spontaneous generation and they are not accepted just when they are lucky, i.e. when they fit the facts. Scientific hypotheses are all more or less *grounded* on previous knowledge, i.e. they are advanced, investigated, and entertained on definite grounds other than the data they cover. The fact that such foundations are never regarded as final and must often be replaced is an evidence for the thesis that the best ground is always sought for a scientific hypothesis. The task of founding scientific hypotheses on grounds other than empirical evidence may be called their *theoretical justification* (or validation). The justification for this name is that the best foundation a hypothesis can be given is its embedding in a theory, i.e. a system of mutually supporting and controlling hypotheses. (To regard the relation of foundation as a relation of strict linear order would lead to infinite regress.)

Moreover, scientific hypotheses deserve being put to the test of experience only when there is some reason to suspect that they may pass the test: i.e. they must first be shown to be *reasonable conjectures* rather than wild guesses. No effort is usually wasted on groundless guesses: thus, no grant is normally given to researches into unjustified conjectures, even if testable. A hypothesis may be entirely new and even eccentric, but it must somehow respect the main body of knowledge and the tradition of science, not so much in the letter (results) as in what is called the spirit (method, goals, and great ideas). That is, the raw material to which the method of science is to be applied, i.e. the unbaked or half-baked ideas that are to be inquired into must be conceived in the spirit of science. In short, the prior truth value of the hypothesis, i.e. its truth value with regard to the antecedent knowledge, must not be falsity. (In symbols, $V(h|A) \gg -1$.) But even the best grounded hypothesis should be advanced, as SZENT-GYÖRGI is fond of saying, with a smile.

What is regarded as the spirit of science depends on the state of knowledge and even on scientific fashion: after all, it is but part and parcel of the so-called spirit of the times (*Zeitgeist*), i.e. the set of basic ideas and norms that mold our very selection and approach to problems (see 4.3). This concept of intellectual climate is ill-defined but, once stripped of its spiritualistic overtones, it is seen to be instrumental for the understanding of the gestation and reception of scientific hypotheses: it helps us understand, for instance, why so many ideas that seem now obvious were not "seen" before and why so many false ideas were taken for granted. Hypotheses, whether scientific or not, are never born in a vacuum and are never evaluated in isolation from the general intellectual heritage inherent in the intellectual atmosphere of a circle or period. To put it in a slightly different way: the invention, investigation, acceptance, and rejection of hypotheses is but one aspect of the creation of culture. Consequently, to pay attention to their logical form and empirical support alone is to betray a short-sighted view of culture. A couple of case histories will illustrate this point.

In 1630 J. REY, an obscure French physician, published a work in which he put forward two hypotheses borne out by his own experiments, as well as one rule of method he actually employed in them. The hypotheses were that the weight of metals increased on roasting (owing in turn to the "absorption of air"), and that weight is conserved in all transformations of this kind. The rule of method: The weights of bodies ought to be controlled by means of balances in all chemical reactions. The book was commented on by the efficient MERSENNE (1634) and reprinted in 1777, but REY'S ideas gained no currency. They were independently reinvented and refined by A. L. LAVOISIER from 1772 to 1789; moreover, REY'S above-mentioned contributions were precisely the kernel of the LAVOISIER revolution. Why was REY not a genuine precursor of LAVOISIER: why did he play no role in the development of chemistry? His ideas were neglected because they were not in tune with the prevailing chemical doctrine and because the latter was judged satisfactory. In fact, the very use of scales was at that time reserved to physicists; the conservation of weight could not appeal to people sharing the Aristotelian belief that weight was not a primary but an accidental property; and the increase of weight (explained by REY as due to the absorption of air) during calcination could not be believed by people who held that an earth or "calx" (what we call an oxide) is more elementary, hence simpler, than a metal. In short, REY was unable to give his ideas and procedures a foundation acceptable to his contemporaries. (More on this story in Vol. II, Sec. 14.1).

Our second story is that of I. SEMMELWEIS, the Viennese physician who, in 1847, explained the deadly puerperal fever as a consequence of the transport unawares, by midwifes and physicians, of "cadaveric material" from the dissection room to the maternity. He consequently proposed that all the personal washed and disinfected their hands before passing from one room into the other. This was enough to reduce mortality from 12 per cent to 1 per cent — which gave a strong empirical support to SEMMELWEIS' hypothesis. Yet SEMMELWEIS was not believed; moreover, he was fought and eventually driven mad. It is now easy to indict his contemporary critics, including the great pathologist R. VIRCHOW; yet theirs, although a dogmatic attitude, was not an unfounded one. SEMMELWEIS' hypothesis conflicted with the then prevailing pathological theory according to which sickness develops and resides in our bodies (endogeneous factors theory). This theory had been tremendously fertile because it had directed physicians to investigate the human body instead of leaving them content with pointing to vague exogenous factors, such as demons, miasmata, and germs. The germ theory of disease, which had explained malaria in antiquity and tuberculosis and pest in the 17th and 18th centuries, had become discredited for good reasons: in the first place, it had not been independently corroborated, i.e. the germs it hypothesized had not been identified and isolated; in the second place it counseled resignation in front of the unavoidable rather than research and fight, whereas the half-false theory of the cellular origin of morbidity had given cytology and pathology a powerful impulse. Moreover, SEMMELWEIS had offered no explanation of the contagion mechanism: this had to wait for PASTEUR and his school, who showed that microbes could reproduce at enormous speeds. In short, the SEMMELWEIS hypothesis not only lacked theoretical justification but contradicted a fruitful accepted theory: it was a lucky hit that was not accepted until it was incorporated in the theory of pathogenic germs.

The cases of REY, SEMMELWEIS and countless other unrecognized predecessors teach various morals. Firstly, the requirement of foundation or theoretical validation is *double-edged:* on the one hand it protects us from crackpot ideas; on the other hand, if exaggerated it may kill any number of truths and, in particular, it may delay or even prevent revolutionary changes in science. Secondly, the requirement of agreement with fact (empirical validation) is *double-edged* too: on the one hand it is a necessary condition of truth and a protection against wild speculation; on the other hand it may consecrate groundless and definitely false hypotheses (e.g., long-lived chance correlations) and, in many minds, satisfaction with it may alleviate the itch of theoretical validation. The requirement of foundation and the requirement of empirical test, when met independently from one another, must be handled with care in order to avoid the dogmatic rejection of truth and the dogmatic acceptance of error. The optimum course is to work out the theoretical and the empirical validation hand in hand.

We rarely, if ever, hit right away on fully grounded and corroborated hypotheses that are also interesting: usually we proceed by trial and error guided by more or less obscure intuitions, and we often start from groundless untested guesses. We may, in fact, distinguish the following levels in conjecture making: guesses, empirical hypotheses, plausible hypotheses, and corroborated hypotheses.

1. *Guesses:* unfounded and untested hypotheses. Guesses are certainly suggested in an obscure fashion by previous knowledge and by fresh

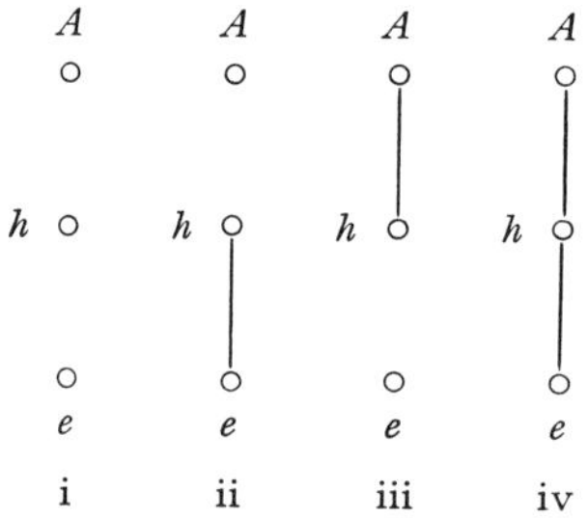

Fig. 5.4. Four levels of validation: antecendent knowledge A and fresh empirical evidence e may combine to yield (i) guesses, (ii) empirical hypotheses, (iii) plausible hypotheses, or (iv) corroborated hypotheses

experience, but they are never adequately justified by them: they hang in the air [see Fig. 5.4. (i)]. The predominance of guesses characterizes speculation, pseudoscience, and the early stage of theoretical work.

2. *Empirical hypotheses:* ungrounded but empirically corroborated conjectures. An empirical hypothesis is a rather isolated conjecture with no support other than the ambiguous one offered by the facts it covers: it lacks theoretical validation [see Fig. 5.4. (ii)]. Empirically found correlations in medicine, the rules of thumb of synoptic meteorology, and those of metallurgy and agriculture, belong to this level. When empirical hypotheses are dominant in a field we can speak of empirical knowledge proper.

3. *Plausible hypotheses:* founded but untested hypotheses. The plausible hypothesis is a reasonable conjecture that has not passed the tests of experience but may, on the other hand, suggest the very observations or experiments that will test it: it lacks an empirical justification but it is testable [see Fig. 5.4. (iii)]. A plausible hypothesis that opened an entire new field of research was J. C. Maxwell's conjecture of the existence of electromagnetic waves.

4. *Corroborated hypotheses:* well-grounded and empirically confirmed hypotheses [see Fig. 5.4. (iv)]. The predominance of hypotheses of this level characterizes theoretical knowledge and is the mark of mature

science. If a corroborated hypothesis is, in addition, general and systemic, we dignify it with the title of law; and if its logical status in a system is that of initial assumption (axiom), we call it a principle. But if we strongly feel that no fresh experience and no fresh reasoning could possibly affect our corroborated hypothesis, then we class it with logical truth — or we found a new religion.

The history of science is largely a sequence of transitions between the above kinds — not only forward from guesses to corroborated hypotheses but also backwards, from "final" truths to groundless speculations. The conversion of a partially supported hypothesis into a corroborated hypothesis is beautifully illustrated by the discovery of the planet Neptune. The original problem was not to gaze at the heavens in search for a new object but to explain the "anomaly" in the apparent motion of Uranus. This "anomaly" was just a discrepancy between the observed and the calculated values of the positions of this planet. Since the computed values were slightly wrong there had to be something wrong with the hypotheses employed in the calculation. These were mainly the following: The solar system is an essentially self-determined object owing to its great distance from the remaining celestial bodies (h_1); Uranus is the outermost planet, hence the Sun and the inner planets alone influence its motion (h_2); NEWTON's laws of motion (h_3, a conjunction of hypotheses); and NEWTON's law of gravity (h_4). The problem of explaining the "anomaly" of Uranus' motion was that of spotting the false conjunct in the compound $h_1 \& h_2 \& h_3 \& h_4$. Owing to the great success of the general hypotheses h_3 and h_4 within and beyond the solar system, the more specific hypotheses h_1 and h_2 were suspect; of the two h_1 could not possibly be guilty in this case: if the "anomaly" were the effect of an influence coming from outside the solar system, why should precisely Uranus, of all the planets, exhibit it? F. W. BESSEL guessed that h_2 might be false and proposed, but did not work out, its negate $-h_2$, i.e. "Uranus is not the outermost planet" or, equivalently, "There is at least one planet beyond Uranus". This was not an ungrounded conjecture: the discovery of Uranus itself had been preceded by the conjecture of the existence of a "wandering star"; furthermore, h_2 had no support aside the weak observational evidence. In short, $-h_2$ was a plausible but untested hypothesis.

Some time later J. C. ADAMS (1843) and U. J. LE VERRIER (1846) independently worked out (theoretically) the plausible hypothesis $-h_2$, to which end they had to introduce several auxiliary hypotheses, the chief one being that the new planet moved on the ecliptic plane. Their problem, then, was to find the orbit, velocity, and mass of the hypothesized new planet in such a way that the bunch of hypotheses accounted for the "observed" motion of Uranus. The sole "evidence" in this case was the discrepancy between the observational data and the predictions made on the basis of h_2: the data by themselves called for no hypothesis. The computations made on the basis of $-h_2$ and with the help of the mathematical theory of perturbations included one testable consequence, namely, the precise direction in which a telescope had to be aimed at on a given night in order to see the hypothesized planet. The night chosen was that of September 23—24, 1846; the astronomer J. GALLE looked at the predicted place and saw the new planet, which was baptized Neptune — but any other astronomer could have found the same confirmation of the theoretical prediction. Thereupon the weakly grounded and entirely untested plausible hypothesis $-h_2$ passed that night into the highest category: it became a corroborated hypothesis. In due time anomalies were found in Neptune's orbit, too, and a new planet, Pluto, was hypothesized and subsequently discovered (1930). Finally, at the time of writing the existence of Pluto is being seriously challenged.

Note the nature of the argument in all three cases (Uranus, Neptune, and Pluto). It does not start with observations but from a discrepancy between theoretical predictions and data, a discrepancy suggesting that something is wrong with at least one of the assumptions. (If $A \vdash t$ and experience falsifies t, then we infer $-A$. But logic alone does not tell which member of the set A of assumptions is wrong.) The record of every assumption is critically examined until the most likely culprit is spotted and eventually replaced by a new plausible hypothesis. (Wild guesses, such as that the anomalies are just planetary caprices, or psychokinetic effects of a powerful magician, are not even considered.) Testable consequences of the new hypothesis are worked out with the available means and if necessary with additional conjectures and techniques. Finally a set of observations is conducted to test it. The whole procedure is *hypothetico-deductive* and the very search for new evidence is suggested by the hypothesis itself — not the other way around.

Let us finally give a couple of words of caution concerning the concept of *rival hypotheses*, inherent in every situation in which hypotheses are involved. Firstly, rival hypotheses are mutually incompatible, i.e. cannot be asserted jointly — but 'incompatibility' does not mean "contradictoriness". In particular, if quantitative concepts are involved in the hypotheses, the possibility of an infinity of mutually incompatible but not contradictory hypotheses is ensured. Just think of the infinite set of hypotheses covered by the formula "$y=x^n$", where n runs over the set of integers: "$y=x$", "$y=x^2$", "$y=x^3$", and so on. 'Contradictory' applies only to pairs of formulas such that one is the logical negation of the other. Thus, the contradictory of "$y=x^n$" is "$y\neq x^n$", and the contradictory of "Every A is a B" is just "Not every A is a B", not "No A is a B" (or "Whatever is an A is not a B") nor "Some A's are B's". Moral: Spare the word 'contradictory'.

Secondly, there is and there will always be a number of rival hypotheses to account for one and the same set of data — but they need not be equally entitled to court truth. Thus, e.g., if a hypothesis h accounts for the evidence e, the mutually inconsistent hypotheses $h\&p$ and $h\&-p$, where 'p' designates an arbitrary assumption, will cope with the same evidence because both entail h which in turn entails e. But if there is no reason to accept p rather than $-p$ we should forget about p: we shall add p, or $-p$, to the picture if and only if it can be subjected to an *independent* test, i.e. if it entails new testable consequences not entailed by h alone. In general, unless they are all groundless the rivals disputing an area of fact will not be equally well grounded and equally well tested, hence they will not be equally verisimilar.

But the subject of testability deserves a section by itself.

Problems

5.5.1. What grounds did COLUMBUS have to suppose that he might find firm ground if sailing westward? Did his conjecture match with prevailing opinion? And would he have received financial support had he offered no arguments in favor of his hypothesis? *Alternate Problem:* Examine COPERNICUS' reasons in favor of the heliostatic hypothesis.

5.5.2. Look at the cartoon in *The New Yorker* for October 20, 1962, p. 53. It illustrates the hypothesis that the megalithic constructions (such as those of Stonehenge and Carnac) might well have been the work of a giant playing with little stones. Is this a worthy alternative to the accepted hypothesis? Consider whether there is independent evidence for it, such as fossil bones of giants found near those megaliths, and take into account that the usual hypothesis is independently based on evidence concerning, e.g., the transportation and use of the said stones. *Alternate Problem:* Can a biologist believe in human immortality? Whichever is the answer, will it be supported by empirical information alone?

5.5.3. What, if any, are the grounds for assuming (i) that all stars are spherical or nearly so; (ii) that there are other minds besides our own; (iii) that there are extraterrestrial organisms.

5.5.4. Education experts usually claim that the essential determiner of good teaching is the mastering of educational techniques (h_1); at least this is the theoretical justification for the existence of schools of education. Science teachers, on the other hand, tend to believe that what essentially determine good teaching, at least good science teaching, is a clear grasp of the subject and love of it as well as interest in the pupils (h_2). Discuss the theoretical and the empirical justifications, if any, of these rival hypotheses. And, if you have power, proceed in consequence. *Alternate Problem:* The hypothesis of continuity of motion, an assumption of mechanics, is empirically untestable in a direct way. Why is it upheld? Hint: Think what discontinuity would commit us to.

5.5.5. In order to explain the precession of Mercury's perihelion, not accounted for by Newtonian astronomy, LE VERRIER hypothesized the existence of a new planet (Vulcanus), which he supposed to be unobservable from the Earth for being constantly hidden by the Sun. What suggested to him the invention of this hypothesis? Was it a testable hypothesis before the advent of astronautics? And how was the problem finally solved?

5.5.6. When they were first proposed the following hypotheses were far from being well grounded: (i) that matter in bulk is made up of invisible atoms (Greek atomists); (ii) that many a star must be the focus of a planetary system (G. BRUNO, end of 16th century); (iii) that chemical binding is basically electrical (J. DAVY, 1807); (iv) that hydrogen is the building stone of all chemical elements (J. PROUT, 1815); (v) that the brain is a sort of electric pile (J. HERSCHEL, 1830). Were they therefore totally useless? *Alternate Problem:* Does every plausible (well-grounded) hypothesis turn out to be true? And is every true hypothesis plausible?

5.5.7. Class the hypotheses involved in the following summary of recent studies on hunger. First a systematic correlation between obesity and certain lesions of the hypothalamus is found in man. Then experiments are made with animals and the correlation is confirmed. This is accounted for by supposing that the lesions eliminate certain inhibitions. This hypothesis is experimentally refuted by the observational finding that rats with the appropriate lesions press the food-providing bar less rapidly than normal rats. Finally the hypothesis is tried, and corroborated at the time of writing, that certain chemicals stimulate appetite when implanted in the appropriate places in the brain.

5.5.8. Examine the following metascientific norms. (i) Scientific hypotheses are to be founded on a priori principles of reason (traditional rationalism). (ii) Scientific hypotheses ought to be based on empirical data alone (traditional empiricism).

5.5.9. Examine the thesis that scientific hypotheses are unjustified and unjustifiable, the only reason for holding them (temporarily) being their survival to severe empirical tests, all attempts to justify them leading to either circles or infinite regress. See K. R. POPPER, *The Logic of Scientific Discovery* (1935; London: Hutchinson; New York: Basic Books, 1959), Ch. x. *Alternate Problem:* Study the nonjustificational approach to philosophy suggested by POPPER and worked out by W. W. BARTLEY III in "Rationality versus the Theory of Rationality", in M. BUNGE, Ed., *The Critical Approach* (New York: The Free Press, 1964).

5.5.10. Examine the following situation presented by N. GOODMAN in *Fact, Fiction, and Forecast* (London: Athlone Press, 1954), Ch. III:

$e =$ All known emeralds are green;
$h_1 =$ All emeralds are and will forever remain green;
$h_2 =$All emeralds will remain green until the year 2000, when they will turn blue — in short: All emeralds are grue.

Clearly, h_1 and h_2 are mutually incompatible and yet they have the same empirical justification or inductive support. R. CARNAP tried to cope with this riddle by stipulating that nontemporal predicates are to be preferred to temporal ones. Queries: (i) Is the above an uncommon situation in science or does it exemplify the metascientific law that no set of data points unequivocally to a given hypothesis? (ii) In which of the four levels discussed in the text (see Fig. 4) does h_2 belong? (iii) What could be the support for CARNAP's stipulation? (iv) How would a scientist react to h_2, i.e. the guess concerning "grue" (green before 2000 A.D. and blue after 2000 A.D.)? (v) Are predicates invented *ad libitum* in science? (vi) Is it legitimate to ban those predicates which lead philosophers into trouble? *Alternate Problem:* Discuss and exemplify (i) the actual "reasons" for proposing a hypothesis, (ii) the plausible or good reasons (grounds) for advancing the hypothesis before it is worked out and tested, and (iii) the actual and the "good" reasons invoked to accept or to reject the hypothesis once it has been tested.

5.6. Testability

Untestable formulas are those that cannot be subjected to test and cannot consequently be assigned a truth value. If they cannot be evaluated they have no truth value: they are neither true nor false. In other words, truth values are not inherent in formulas but are attributed to them by metastatements such as "p is factually true". And attributions of factual truth can only be made on the strength of empirical tests. In fact, we know of no other method but experience to test for factual truth.

Empirical testability, then, is a means for ascertaining factual truth values — not for obtaining truth, because tests do not dictate hypotheses but rather the other way around, and because tests may be unfavorable or inconclusive. A review of the tricks that can be used to elude empirical test will shed light on the concept of testability, which is both tricky and central to the methodology of science.

The best way to dodge empirical testing is to abstain from hypothesizing, restricting ourselves to thinking of our precious private experiences without attempting to understand them or even to act, since explanation and rational action require hypotheses. We shall seldom err if we limit our language to phrases like 'I have now a feeling of roughness'. What is more, we will hardly feel the need to test such egocentric statements by means of further statements of the same kind, since what is in need of test is the uncertain leap beyond immediate experience. It is not that judgments of immediate experience are beyond error, but that — except in the case of psychological investigations of self-delusion — they are seldom worth being checked.

The so-called *protocol statements* couched in a phenomenalistic language — like 'I now see a red patch' — are sometimes supposed to be incorrigible, hence the ideal foundation stones of science regarded as certain (incorrigible) knowledge. However, phenomenalistic statements are both corrigible in principle and useless for building science. In fact, judgments of perception can sometimes be shown to be wrong when judged in the light of physical-object propositions: we may and in fact do correct and refine phenomenalistic statements with the help of scientific instruments and reasonings employing physical-object hypotheses — as when we say that the rays we see in stars are not in the stars but are produced in the refractive media (atmosphere, telescope, eye). In short, perception judgments, though incorrigible by judgments of the same kind (related to the same sense organ), can be corrected by judgments of a higher level. Secondly, phenomenalistic statements are not undistorted expressions of pure, preconceptual, unprejudiced experience, merely because human experience itself is never entirely free from the influence of expectations and opinions: to some extent we see what we are ready to see and fail to see what we do not expect to see. Moreover, the very expression of experiences is made with the of a language, and no language proper can help but teem with universal words, such as 'feel' and 'red'. Thirdly, nothing scientifically interesting can be concluded from protocol statements alone, without the help of some theory. When someone claims he "draws a conclusion" from a set of perception judgments he can rightly mean only that the latter trigger (psychologically speaking) some hypotheses; different individuals, with a different background, might "draw" quite different "conclusions" from similar experiences. Fourthly, crude phenomenalistic statements are not utilizable as evidences for or against hypotheses: they must first be interpreted, i.e. transformed into objective statements couched in the same physical-object language as the hypotheses are. That is, such as they stand protocol statements are neither substitutes for scientific hypotheses nor evidences relevant to them.

We grant, of course, that phenomenalistic propositions are the least uncertain among empirical propositions for being least-committal. Yet they are not immune to criticism and, what is more important, they do not enter science precisely because they are confined to the subject, whereas science aims at objectivity (see Sec. 1.1). Phenomenalistic propositions are in this regard worse off than ordinary knowledge, which expresses itself chiefly in physical-object sentences: usually we do not say 'I see a tablelike brown patch' but rather 'Here is a brown table' or a similar object-centered sentence. The claim that physical objects are *inferred* or even metaphysical and that only the phenomenal object — e.g., the sensed table rather than the physical table — is directly given, hence certain, is controverted by physics, psychology, and anthropology. In fact, while it is becoming increasingly possible to analyze sensations in terms of physical processes, the converse reduction is not possible. Also, children, primitives and philosophically unsophisticated adults are not phenomenalists and do not speak as such: it is only certain philosophers who, in their quest for certainty, invent purely phenomenal objects and phenomenalistic languages. At any rate abiding by immediate experience, in case this were possible, would not deliver us from physicalistic demons, because physical-object judgments are thought spontaneously in relation to phenomena whereas phenomenalistic propositions are most often inferred from them. And even if the phenomenalistic purification rites were effective, science would not be interested in them because scientific research is the risky attempt to leap beyond appearances, into objective facts, which compels it to invent hypotheses. Science is not interested in appearance save as an equivocal clue to reality.

A second trick for eluding empirical test is to maximize *vagueness*. Certain qualifying phrases such as 'under certain circumstances', 'in favorable conditions', and '*mutatis mutandis*', can produce facile truths, i.e. truths so insensitive to empirical details that they come close to logical truths. Thus, that "The condition of the nervous system at a given moment determines behavior at a later moment" will hardly be doubted by a modern psychologist. Yet this proposition is so vague that it is nearly untestable — and untestability is much too high a price to pay for certainty. In fact if the condition, the behavior, and the relation between the two are not specified, any piece of behavior will count as confirming evidence of our hypothesis since, whatever it be, a given behavioral event is certainly preceded by some state of the nervous system. Only if a precise relation between neural states and behavioral states is hypothesized do we get a fully testable hypothesis. In general undetermined functional relations, i.e. functions that are not specified, so that one of the variables cannot be inferred from the remaining variables, are untestable. An expression such as "*y depends on x*", i.e. "$y=f(x)$", is not a proposition but a propositional function if just the variables are specified (interpreted) but the function f is left undetermined. Obviously we cannot test something that has not been stated. In this case we have not a hypothesis proper but a *hypothesis of relatedness*, which is a working or programmatic hypothesis, i.e. a blank to be eventually filled by research. Sometimes we adopt vague formulas either because of ignorance or because we want them to be versatile, just as the legislator often adopts vague expressions in order to leave the details to the judge. But before setting out to test such assumptions we must render them precise: as they stand they are much too noncommittal — and empirical tests are aimed at forcing us to make commitments.

Yet the most widespread and candid — or perverse, as the case may be — way of eluding empirical test and attaining certainty is to hypothesize *inscrutable* objects. DESCARTES' evil spirit, "who is supremely powerful and intelligent, and does his outmost to deceive" him without ever allowing him to realize that he is being deceived, is such an inscrutable object. For, no matter what DESCARTES does, he cannot detect that evil spirit: to detect him would require eluding his vigilance, which is ruled out *ex hypothesi*. Similarly the soul hypothesis is impervious to evidence: even if a person is beheaded, his ceasing to manifest feeling and thought will not falsify the hypothesis since one may retort that, all that happened upon the beheading, was that the soul lost its normal channel of communication with the material world. We do not reject the soul hypothesis because it has brilliantly been refuted by modern experiments but rather because no conceivable experiment could correct it even slightly. A third distinguished member of the class of dodging hypotheses is "Whatever an organism does it does it because some instinct impels the organism to behave in the way it does". If a mother cares for its young the maternal instinct is called to duty and if a man attacks another the aggressive instinct is invoked. Should a mother not care for its young or a man not attack anyone, the instinct is said to be weak, or latent, or repressed, or overpowered by a stronger instinct such as self-preservation. In this way every conceivable datum is reckoned as evidence in favor of the hypothesis. We do not reject such conjectures because they are false but because experience is irrelevant to them and so we cannot assign them a definite truth value. Since we want to attain truth, even if partial, we do not consider dodgers as candidates for scientific hypothesis.

A more sophisticated device for eluding empirical test while feigning to speak truthfully of experience is to contrive conjectures that may or may not be individually testable but in any case are saved from refutation by *ad hoc* assumptions. In order to effectively shield the basic conjecture the protective assumption must not be *independently testable*. Suppose we wish to save Aristotelian dynamics by making it consistent with the principle of inertia, according to which in the absence of external forces a body remains at rest or continues moving with constant velocity. A patching up of this kind is readily performed by adding the *ad hoc* hypothesis that every body is subject to a constant inner force independent of the body's constitution and structure — and therefore inscrutable. With this addition the basic law of ARISTOTLE's dynamics becomes: "The total (internal plus external) force is proportional to the velocity". In the absence of external forces we are left with "The inner force is proportional to the velocity"; and, since this inner force is constant by hypothesis, the velocity is constant too, which agrees with the law of inertia. The case of rest is next accounted for as an equilibrium or "inner strife" between the body's inner force and the inner forces of the environment. But the patched-up theory is inconsistent with other facts, not taken into account in contriving the *ad hoc* hypothesis. Thus, for free fall in a void we would add the inner force to the constant weight of the body and would get a constant velocity in contradiction with experiment. Like all conjectures of this kind our *ad hoc* hypothesis is not consistent with the whole evidence — and this was bound to occur as it was introduced to save the basic Aristotelian law from just one of its unfavorable consequences. Few *ad hoc* hypotheses are all-purpose protective devices, such as FREUD'S repression hypothesis (see Sec. 1.6).

*Empirical testability is not always easy to determine: occasionally a scientific hypothesis is wrongly regarded as empirically testable (or untestable). A famous case in point is the so-called competitive exclusion principle in ecology, which can be stated thus: "Complete competitors do not coexist for long" — meaning that, if two species occupy the same ecological niche (i.e. have the same requirements) and the same territory, then one of them will eventually become extinct. It has been held that this principle, though true, is immune to experience because, if two competing species in fact coexist, we may save the principle by assuming that there is some small as yet undetected difference in their ecology, that will eventually be discovered. Leaving aside the puzzling reference to the truth of an allegedly untestable factual statement, the objection holds. But then it just exemplifies a difficulty common to all scientific hypotheses: assumptions of exact identity can at best be approximately true if they refer to concrete objects (see Sec. 6.1). In fact, we can be sure that no two real species have exactly the same ecology, so that we know beforehand that the ecological exclusion principle cannot be rigorously true. The thing to do is not to reject the principle but to give it a more realistic form by relaxing the clause and requiring only a similar ecology for the competing species: in short, let it refer to *almost* complete competitors. In this form the principle becomes: "For every x and every y, if x and y are almost complete competitors, then either x becomes extinct or y becomes extinct". In order to test this hypothesis we bring together two populations occupying very close ecological niches, and watch the developments. That, is we tentatively assert the hypothesis, firmly assert its antecedent, and test for its consequent. After a lapse we gather data concerning the consequent (i.e., eventual extinction of one species) and perform an inference about the truth value of the hypothesis. If we find out that one of the populations has become extinct or nearly so, we conclude that the principle has been *confirmed in this case*. If, on the other hand, we discover that the two species coexist, we infer that the hypothesis has been *refuted* in this case, hence refuted as a universal hypothesis. In the latter case we might try to save the hypothesis by assuming that the observation time has been insufficient to allow for the advantages of one species over the other to show up. In this way the unfavorable decision may be postponed for a while, but the critic has every right to demand that the principle be then reformulated in a more precise way, including a reference to the expected extinction time of one of the populations, or even to the intensity of the competition. In this respect, the vague formulation of the ecological exclusion principle does not differ from any other loosely stated hypothesis. In conclusion, the principle is *weakly testable* but not altogether untestable. Moral: a comparative concept of testability should be introduced.*

Empirically testable means "sensitive to experience". Now a hypothesis sensitive to empirical data may be supported (confirmed) or undermined (disconfirmed) by them. Testable hypotheses may then be (i) purely *confirmable*, (ii) purely *refutable*, or (iii) both *confirmable and refutable*. Purely confirmable hypotheses are the least testable, yet we may try them if they have some ground. If we did not care for confirmability at all — i.e. if we were extreme refutabilists — we would miss the opportunity to apply our hypotheses to particular cases and watch how they fare. Moreover in that case we would not care for factual truth, since factual truth is tested, in part, by *agreement* with fact, i.e. by confirmation. What is worse, we would have to accept as scientific a crackpot conjecture like HIMMLER'S, that stars are made of ice, for such a crazy conjecture could only be refuted. We shall consequently stipulate that (i) confirmability is *necessary and sufficient* for empirical testability; (ii) refutability is neither necessary nor sufficient for empirical testability but is necessary for *optimal* testability, which is enjoyed by hypotheses that can be *both* confirmed and refuted, i.e. for which favorable and unfavorable evidence can be conceived on the strength of what is known. And we shall keep in mind that testability unaccompanied by foundation or theoretical validation is insufficient to regard a hypothesis as scientific (see the preceding section).

If we were to reject all purely confirmable (irrefutable) hypotheses we would mutilate large sections of science, in which indefinite existential hypotheses ("There are A's") and probability hypotheses ("The probability of an A being a B equals p") play an important part. Let us examine these two kinds of hypothesis. "There are signals faster than light" is an *indefinite existential hypothesis*. It can only be confirmed: no failure to detect or produce such a signal would conclusively refute the possibility of discovering or making it in the future, just as the sad failures of the 1789 device *Liberté, égalité, fraternité* do not overturn the programmatic hypothesis that it is possible to build a human society on the basis of freedom, equality, and brotherhood. Only very good theoretical arguments might (temporarily) dispose of such hypotheses. To deny the existence of faster-than-light signals just because no evidence for them has so far been found, or because no known theory predicts them, would be sheer dogmatism. *It is usually believed that relativistic theories prohibit the existence of faster-than-light signals, but this is mistaken: all such theories say in this respect is that bodies cannot attain the velocity of electromagnetic waves. They do not and could not preclude the existence of other kinds of field, with a different propagation velocity.*

The *heuristic value* of many such irrefutable but confirmable hypotheses is undeniable. Just think of "There are transuranians", "There are antiprotons" and "There are neutrinos", all of them exclusively confirmable and extremely fertile. Moreover, no historical research would be possible without hypotheses of this kind. Consider, for instance, the case of the palaeontologist who, on the basis of his knowlegde of a living species and its extinct ancestor, hypothesizes the existence of an intermediate form. His sole ground so far is the basic hypothesis of the quasicontinuity of biological lineages. The failure to find evidence, whether living or fossil, for his missing link hypothesis will not count against it. In fact, the palaeontologist will protect his grounded conjecture with the *ad hoc* (but in principle testable) hypothesis that no remains of the intermediate link may have reached us either because the fossilization conditions were not realized or because of some physical process, such as a geological cataclism, has obliterated all evidence. Only the effective finding of a specimen resembling the imaginary reconstruction will establish the hypothesis — or, rather, an improved version of it; failing such a finding the hypothesis will have to be reckoned with in order to save an important body of knowledge. Had this policy not been abided by, DARWIN'S hypothesis of the descent of man would have been rejected and, as a consequence, it would not have guided the field research which led to its own confirmation. Similarly, the detective will account for some of his failures with the hypothesis "Some crimes leave no perceptible traces" and the epistemologist will postulate: "Most events are unobserved". There are good grounds for regarding many such hypotheses as true, hence for accepting them, yet there is no way of disproving them.

Probability hypotheses, too, are confirmable but either irrefutable or only weakly refutable. Consider the hypothesis: "The probability of casting an ace with an arbitrary fair die is $\frac{1}{6}$". This is a grounded conjecture rather than just an empirical generalization from observation of past performances. In fact, the dynamical study of the die shows that all its faces are equally probable. The problem, then, is to ask experience whether the grounded hypothesis "$P(\text{ace})=\frac{1}{6}$" is in fact true. Before putting this hypothesis to the test we must restate it in empirical terms, because probability is as unobservable as fairness: we replace "probability" by "long-run relative frequency" and "fairness" by "well-balancedness". Besides, we require that the throwing of the die be random (unbiased). With these transformations our original probability hypothesis is converted into the following conjecture, which is both statistical and physical: "The long-run relative frequency of the 'ace' event in a sequence of random throwings of a well-balanced die is close to $\frac{1}{6}$". Suppose now we throw the die 60 times and observe only 1 ace instead of the 10 apparently predicted by our hypothesis. Shall we reject the latter? Not at all: we know that failures do not count as strong refuters in the case of statistical hypotheses. What we should do is to make sure that the conditions of well-balancedness and random throwing are in fact being satisfied; should our scrutiny fail to exhibit any defect we would conclude that we have witnessed an improbable yet possible streak of "bad luck" (i.e. below-chance) events. We know that in the short run anomalies like this one can happen and that it is only in the long run that statistical laws hold. We accordingly try a longer sequence of, say, 600 throws. If the observed frequency is still far different from the predicted frequency we may still stick to our hypothesis, which we know excellently *confirmed* by past experience, and may demand a still longer sequence. Yet, no matter how long a sequence of throws is taken, the possibility of large fluctuations around the probability will remain; moreover, if such fluctuations from the average are ruled out and a perfectly smooth sequence remains, we may not be dealing with a random phenomenon. But we can at least be sure that the probability of a large deviation from the average relative frequency will uniformly decrease with the length of the sequence: this we can both prove theoretically and confirm observationally (by first translating this second-degree probability statement into the corresponding relative frequency of discrepancy). And this is the closest to certainty we can get. We conclude that statistical hypotheses are hard to falsify empirically but not untestable. (The weak testability of statistical hypotheses is a strong reason for demanding their link to better testable hypotheses. In other words, since no ultimately or irreducible statistical hypothesis can be subjected to tough empirical tests, it is desirable to relate every hypothesis of this kind to more exacting hypotheses, perhaps referring to lower level or to higher level events. This methodological consideration makes irreducible indeterminism suspect.)

The above shows why we regard confirmability as necessary and sufficient for testability. Still, if a hypothesis is irrefutable or nearly so we shall regard it as having a *weaker testability* than a hypothesis which is refutable as well as confirmable. Moreover, we shall attempt to *compensate* for such a weak testability with a strong foundation. For example, if the hypothesis is statistical we shall provide a ground for it by setting up a theoretical model. Only conjectures which are both irrefutable and groundless will be left to speculation or to pseudoscience.

Moreover, we shall not require that every hypothesis be *directly* testable: this demand, made by empiricism, would render scientific theory impossible. In fact a scientific theory, far from being a lump of empirical statements, is a multilevel system the highest formulas of which are not directly comparable with empirical reports. Moreover, some perfectly respectable physical theories contain intermediate-level formulas (theorems) that are empirically untestable — such as the quantum-mechanical formulas regarding the motion of a particle inside a closed box. Yet we keep such theories, which so manifestly violate the requisite of full testability, because they yield other consequences which have been well corroborated. Accordingly we adopt a realistic viewpoint and require of every theoretical formula that it either (i) *entails* directly testable formulas or (ii) is *entailed* by formulas with testable succedents. If a formula is neither directly testable nor has empirically testable succedents yet is a theorem of an otherwise testable theory, we may regard that formula as *indirectly testable:* every evidence relevant to the assumptions that entail the formula may be regarded as an *indirect evidence* relevant to the formula (see Fig. 5.5).

Finally, we shall not demand the testability of every formula occurring in science because certain formulas, although fruitful, are incapable of being even false. These are, in addition to the rules of inference, the formulas of the following kinds: (i) *rules* of method, (ii) *conventions*, such as those regarding units, and (iii) *criteria* for evaluating pieces of science, e.g., testability criteria. None of the formulas of these kinds can be tested for truth, but all of them should be tested for either convenience or fruitfulness. Thus, a rule of method that does not lead to the desired result is pronounced fruitless; and a rule of method that cannot be checked as to fruitfulness is clearly undesirable. We can require the testability of every piece of science on condition that we

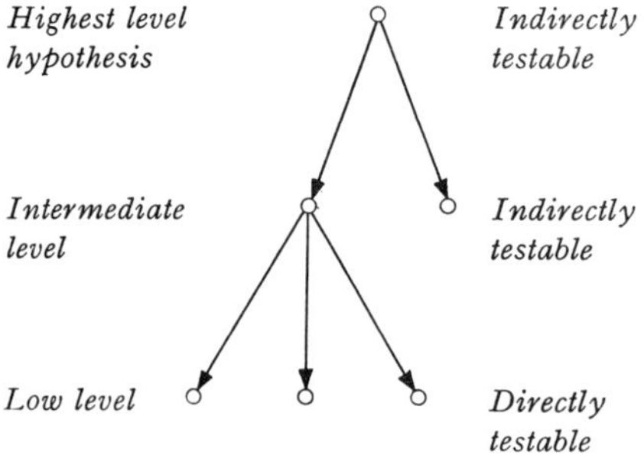

Fig. 5.5. The theorem on the right branch of the tree entails no empirically testable consequence: it can be supported or undermined from above only

widen the concept of testability to include the *pragmatic testability* of a formula, that is, the possibility of showing that the formula is, or is not, convenient or fruitful.

In short, we lay down a liberal criterion of testability, according to which every hypothesis (not every formula) must be at least confirmable — preferably refutable as well — either directly or indirectly.

We are now prepared to collect all the requirements we have so far imposed on scientific hypotheses.

Problems

5.6.1. Examine the singular hypothesis: "It snowed on Manhattan Island on the first of January in the year 1 A.D.", discussed by B. Russell in *An Inquiry into Meaning and Truth* (London: Allen & Unwin, 1940), pp. 277ff. Is it meaningful, empirically testable, grounded, and/or verisimilar? And could you alter it so as to render it more likely?

5.6.2. Examine the testability of the following ideas. (i) "Whatever happens is the work of the Providence". (ii) "Mishaps are either a punishment for sins or a test of faith". (iii) "Whatever happens is historically necessary". (iv) "Whenever an epoch needs a great man it begets him". (v) "Birds can learn to follow the movements of almost any suitable object" (read in a study of animal behavior).

5.6.3. Examine the following conjectures as to testability. (i) "The cause of social change is a certain economic process". (ii) "The ideas of an epoch depend on the mode of production prevailing during that epoch." (iii) "All animal species are the modifications of an original type (*Urtier*), the eternal morphological model copied more or less successfully by the various real species" (GOETHE).

5.6.4. C. G. JUNG stated that outwardly extroverted persons are inwardly introverted and vice versa: i.e., that people who behave as if they were extroverted are unconsciously introverted and conversely. Examine the testability of this conjecture.

5.6.5. Examine the testability of the following psychoanalytic hypotheses. (i) "Infants derive sexual pleasure from performing their excretory functions". (ii) "All males have an Oedipus complex, whether in a manifest or in a repressed form". (iii) "The critics of psychoanalysis illustrate the psychoanalytic hypotheses of the aggressive instinct and of the defense mechanism (unconscious protection of something the subject wishes to conceal)". *Alternate Problem:* Compare the status of ad hocness in factual and in formal science. Recall that most proofs in mathematics are tricky rather than straightforward, in the sense that they require the introduction of *ad hoc* premises, i.e. of assumptions that were not laid down to begin with.

5.6.6. How many data are necessary and sufficient to establish, and how many to refute (i) a singular hypothesis, (ii) an existential hypothesis, (iii) a bounded universal hypothesis, and (iv) an unbounded universal hypothesis? *Alternate Problem:* If a hypothesis is altogether untestable then it is not scientific. Does the converse hold, i.e. is it true that if a hypothesis is testable then it is scientific?

5.6.7. The upholders of the subjectivist interpretation of probability as a measure of a degree of personal belief (and nothing else) claim that beliefs (whether of ordinary knowledge or scientific) are impervious to facts. From this they (correctly) conclude that every probability statement, such as the estimate of the probability of an event, is subjective and incapable of being proved or disproved by experience — in short, untestable. Are they right? If so, must we shun probability theory in science or relinquish the requirement of testability? See K. R. POPPER, *The Logic of Scientific Discovery* (London: Hutchinson, 1959) and "The Propensity Interpretation of Probability", *Brit. Jour. Phil. Sci.*, **10**, 25 (1959).

5.6.8. Examine the differences between hypotheses purporting to represent an aspect of reality and *fictions* serving as auxiliares. According to Kantianism the former should be testable, the latter useful. See H. Vaihinger, *Die Philosophie des Als Ob*, 4th ed. (Leipzig: Meiner, 1920), Ch. XXI, and M. Bunge, *Metascientific Queries* (Springfield, Ill.: Charles C. Thomas, 1959), Ch. 10. See Also Alternate Problem 8.2.10. *Alternate Problem:* Study the relation between inscrutable concepts and untestable assumptions.

5.6.9. Discuss the testability of the following hypotheses. (i) "There are at least two exactly identical things in the world". (ii) "There are things that can interact with no other physical system". (iii) "There are fields that cannot be detected by physical means and which convey ideas". (iv) "Physical space is embedded in a space of a higher dimensionality — only, such additional dimensions do not show up". (v) "The universe is spatially infinite". (vi) "The universe is spatially finite". (vii) "The universe was created several billion years ago". (ix) "There are infinitely many levels below the level of the presently known fundamental particles". (ix) "All organisms, including plants, can feel pain". (x) "A complete central nervous system kept alive in a laboratory could feel and think even if deprived of the organs by which it normally manifests its functioning".

5.6.10. Examine the doctrine that the recognition of phenomenal properties (qualia) is untestable and precisely for this reason qualia should be chosen as the basic units for the systematization of experience. For the most complete and cogent realization of this anti-Democritean programme, see N. Goodman, *The Structure of Appearance* (Cambridge, Mass.: Harvard University Press, 1951). For the untestability of quale recognition, see op. cit. pp. 99ff.; for the reduction of every individual object to sums of one or more qualia, see pp. 175ff. For the unfeasibility of the programme, see any physics textbook. *Alternate Problem:* Elucidate the concepts of *possible supporter* and *possible falsifier* of a hypothesis and build with their help the concept of degree of testability of a hypothesis relative to an empirical procedure and in the light of a set of theories. Take into account that, in the case of scientific hypotheses, their supporters and falsifiers may not have the same referents — i.e. they may not be just either instances or counterexamples; thus, every evidence relevant to a microhypothesis is a statement referring to some macrofact.

5.7. *Requirements

Long before a conjecture is awarded the title of True Hypothesis — which anyhow may not last longer than that of World Champion — it must show that it is a scientific conjecture. In successive preliminary tests of a nonempirical character the hypothesis must exhibit logical fitness, compatibility with the bulk of scientific knowledge, and capability of being checked by experience. That this is the right order of testing was stated in Sec. 5.2 and will be justified in the following.

Logical fitness includes certain syntactical and semantical characters that must be ascertained before empirical tests are envisaged. First among them is logical strength, a syntactical concept that can be introduced through the *Definition:* "A formula A is *logically stronger* than a formula B if and only if A entails B". In obvious symbols:

$$[S(A) \geqq S(B)] =_{df} A \vdash B.$$

Thus p is stronger than $p \vee q$ since, given p, by the principle of addition we can adjoin it an arbitrary q. For this reason $x = y$ is stronger than $x \geqq y$ — i.e., $x = y$ or $x > y$ — because $(x = y) \vdash (x = y \vee x > y)$. Also, $p \& q$ is stronger than either p or q because a conjunction entails its conjuncts. For the same reason $(x) P(x)$ is stronger than $P(c)$. And, by the principle of addition, $P(c)$ is in turn stronger than $(\exists x) P(x)$ — which warns us against the temptation to identify strength with generality.

Every axiom is logically stronger than the theorems derived from it. On the other hand if A and B are interdeducible they are equally strong. And the formulas which are logically valid, i.e. the analytic formulas, are the weakest of all because they follow from no other formulas. Tautologies, then, have minimal strength. Conversely logically false formulas (contradictions) have maximal strength because anything can follow from them. (In symbols: $\emptyset \vdash L$ and $-L \vdash U$, where '$\emptyset$' denotes the empty set, 'U' the set of all formulas, and 'L' the set of logical truths.)

In logic we seek to establish maximally weak formulas, i.e. such that follow from no premises and are therefore self-validating. In mathematics, on the other hand, we prefer the strongest assumptions compatible with logic, i.e. those yielding the richest set of mutually consistent theorems. (It could even be argued that this strength difference is the main difference between logic and mathematics.) Finally in factual science, just as in mathematics, we steer a middle course between the maximal weakness of logical truth and the maximal strength of contradiction.

Logical strength is a source of *definiteness* or nonvagueness as well as of information content. That definiteness, a semantical property, is related to strength — a syntactical property — seems clear yet it has not been proved. Thus, e.g., "All P's are Q's" is both stronger and more definite than either "Most P's are Q's" or "Some P's are Q's". And, whatever the aura of vagueness surrounding "$P(c)$", it is smaller than the mist enveloping its succedent "$(\exists x)\,P(x)$". But, of course, two equally definite formulas may have different strengths, as in the case of "$(x)\,P(x)$" and "$P(c)$". That is, strength is sufficient but not necessary for definiteness. More precisely: $[S(A) > S(B)] \rightarrow [D(A) \geqq D(B)]$, where '$D$' stands for definiteness or nonvagueness (see 3.1).

A second bonus of strength is rich *information content:* of two formulas with unequal logical strength the stronger can store more information. This is obvious in the extreme cases of formally true formulas, which are both weakest and void, and of their contradictories, which have both maximal strength and maximal content. The tautology "Things are either extended or not extended" says nothing about things: the concept of thing occurs vacuously in it, i.e. it may safely be replaced by any other concept of the same type. And the contradiction "Things are both extended and not extended" says too much: it is too accomodating, so much so that it is consistent with conflicting evidence. Formal truths and contradictions are therefore undesirable in science: the former because their content is nil, the latter because their content is universal. For a similar reason we should keep clear of modal propositions — whenever possible! A proposition such as "It is possible to return alive from the Moon" can be conjoined without contradiction with "It is possible to return dead from the Moon", but such a conjunction,obviously stronger than either conjunct, is of no interest to astronauts: although it is not self-contradictory, it does accomodate contradictory evidences: it says too much.

Since we want our scientific hypotheses to be informative and testable we must avoid the two extremes of nil content and universal content: we must seek a *via media* between logical truths, which say nothing about the world and are therefore untestable by experience (there being nothing in experience to compare them with), and logical falsities which, by saying too much, are untestable as well because they agree with anything that may happen. We state our requirement in the following way: Scientific hypotheses shall be self-consistent and shall have *maximal logical strength with respect to the empirical evidence relevant to them.* In symbols: $S(-t) > S(h) > S(e) > S(t)$, where '$t$' designates any tautology and '$-t$' its negate (a contradiction).

Another way of stating the strength condition is this: Scientific hypotheses shall be synthetic (i.e. nonanalytic or factually meaningful) and shall have maximal logical strength with respect to the empirical evidences relevant to them. The condition of syntheticity, understood in a wide sense, eliminate not only contradictions but also the formulas that are true by virtue of the meanings of the concepts occurring in them. In this way the pseudoproblem of testing conventions is eliminated.

Thus, e.g., we shall not inquire whether pure water might not boil at exactly 100° C at normal pressure, or whether the atomic weight of oxygen might not be slightly different from 16.

Now, we want our hypotheses to be synthetic relative to the empirical evidences relevant to them and, at the same time, we want them to be logically as close as possible to the bulk of the body of available knowledge, which alone can provide a ground for them. We are not requiring that *all* our hypotheses be deducible from a certain body of knowledge, for this would render the body of knowledge nonhypothetical and would make progress impossible. But we do wish to make *most* hypotheses derivable from high level hypotheses (postulates), minimizing the number of stray (extratheoretical) hypotheses. This desideratum may be put thus: Scientific hypotheses should be *as nearly analytic as possible relative to the bulk of available knowledge*.

At first blush this desideratum — which blends the condition of foundation with the condition of systemicity — is incompatible with the condition of syntheticity. Yet there is no contradiction: we are requiring syntheticity with respect to the body of empirical information, and maximal analyticity with respect to the main body of prior knowledge. If preferred, our desiderata are: maximal analyticity of hypotheses with respect to accumulated experience, and syntheticity with respect to fresh experience. In terms of strength, any hypothesis h should be midway between any empirical information e relevant to it (weakest) and the body of knowledge A to which it may eventually be incorporated either as a theorem or as a postulate: $S(\mathsf{A}) \geqq S(h) > S(e)$. An alternative way of stating the condition of maximal analyticity is to say that we want the body of antecedent knowledge A to give h a greater truth value than it would have if A were rejected: $V(h|\mathsf{A}) > V(h|-\mathsf{A})$. The larger the value of $V(h|\mathsf{A})$ the better the theoretical validation of h. In the extreme case in which h follows from A we get maximal verisimilitude: $(\mathsf{A} \vdash h) \rightarrow [V(h|\mathsf{A}) = 1]$. If, on the other hand, $-h$ follows from A, then the prior (theoretical) truth value of h will be minimal: $(\mathsf{A} \vdash -h) \rightarrow [V(h|\mathsf{A}) = -1]$.

Strength, by being relevant to definiteness and information content, is relevant to *testability*. The class of possible supporters and the class of possible refuters of a formula are the larger and the better delimited the stronger the formula. Conversely, the weaker a formula the less informative and definite it is apt to be and, consequently, the less sensitive to empirical contingencies. Thus, a disjunction will take less risks than either of its disjuncts: being weaker than them it is also more elusive. By contrast a factual universal proposition (an indefinite, perhaps infinite conjunction), like "The electric charge of electrons is constant", has an unlimited number of opportunities of coming to grips with experience: its strength lends it a high degree of testability. In short, ***the stronger a hypothesis the greater its testability*** (K. R. Popper).

Strength is necessary but not sufficient for testability, and consequently these concepts are not interdefinable. In fact, whereas the strength of a formula can be determined with reference to the context in which it occurs, the testability of the same formula will be judged not only in the light of that body of knowledge but also in the light of the available or conceivable empirical procedures, such as measurement techniques. For example, at the beginning of our century several investigators suggested that the mechanism of nervous impulses was the release of chemicals; at the time (1903) this hypothesis did not seem testable and was abandoned until, later on, adequate techniques were deliberately invented to test it. Moral: testability is not inherent in hypotheses but is a methodological property scientific hypotheses possess to varying degrees relative to a body of empirical and theoretical knowledge. Consequently, let us not write '$T(h)$' for "h is testable" but rather '$T(h|\mathsf{A}\,\mathsf{E})$' for "$h$ is testable relative to the antecedent knowledge A and the empirical procedures E".

A further necessary condition for testability, in addition to technical maturity, is the occurrence of empirical concepts somewhere along the line of the test process: otherwise no experience would be relevant to our hypothesis. We shall not require that the hypothesis itself have an empirical content because this condition would outlaw all the more important scientific hypotheses, none of which refer to experience although they can be tested with the help of experience bearing on certain distant consequences of the hypotheses. All we shall require in this respect is that the hypothesis, in conjunction with some body of knowledge, entail formulas approximately *translatable* into observational propositions. For instance, the hypothesis that the free electron has a spin (a sort of intrinsic rotation) has no observational content whatsoever: not only is the spin itself unobservable, but it is possible to show theoretically that no experiment could possibly measure the spin of the free electron. Yet the spin hypothesis, conjoined with other assumptions, accounts for certain observations (such as the splitting of spectral lines by magnetic fields) concerning more complex systems (e.g., atoms) composed of spinning electrons. (Moreover, no theory other than the theory of spinning electrons can turn data of this kind into evidences relevant to itself: in a sense, then, empirical tests are a family affair.) Or take psychological hypotheses: it is not by refraining from hypothesizing mental events that we shall be faithful to facts, but by being able to deduce consequences regarding objectively recordable events, whether behavioral or physiological. The assumptions regarding higher-level activities will not, by definition, be testable by direct inspection, but they will have to entail consequences containing behavioral and/or physiological concepts alone if they are to count as testable. In short, testability does not require empirical reference but rather the possibility of becoming part of a net connected with observational formulas.

Furthermore, no hypothesis is *independently testable*, and this because no premise is by itself sufficient for the derivation of testable consequences. In the case of stray conjectures we shall need a part of the body of antecedent knowledge (see Sec. 5.2). In the case of a theory we will need, in addition to the hypothesis itself, other assumptions of the theory and/or empirical data (since no hypothesis yields by itself empirical information). The question, then, is not whether a theory contains some assumption which is not independently testable and must therefore be regarded as suspect, but rather whether (i) every one of the assumptions of the theory is actually *necessary* for the derivation of testable consequences, (ii) the axioms make up a consistent whole having at least *some* empirically testable consequences (lowest-level theorems), and (iii) there are no *ad hoc* hypotheses serving just to prop up some of the hypotheses and incapable of being tested independently of the hypotheses they protect.

The requirement that a system of hypotheses (theory) shall not contain assumptions unnecessary for the deduction of testable propositions is aimed at excluding parasitic untestable hypotheses clinging to an otherwise testable system and obtaining apparent support from the confirmation of the latter's low-level theorems. And the relaxation of the condition that all theories shall have nothing but empirically testable consequences is aimed at retaining theories which, like quantum mechanics, contains assumptions without testable consequences, often because they refer to systems that are not subjected to the disturbances necessary to extract information from them. We regard such hypotheses as indirectly testable (see Sec. 5.6), i.e. as testable via other formulas of the theory, which do enjoy empirical support. Finally, the condition concerning the protecting hypotheses is laid to avoid situations common in pseudoscience, where every single hypothesis may be testable but the whole doctrine is so contrived as to escape test: i.e. the various conjectures shield one another so that they come to hold in all possible worlds.

We can now tackle a thorny problem. On the one hand we have throughout endorsed a restricted version of the principle of empiricism: namely, that whatever the origin and the reference of scientific hypotheses *experience must test them*. On the other hand we have agreed to call hypotheses proper certain formulas about *unexperienced* facts (see Sec. 5.1). But how can experience test what our hypotheses assert if they say nothing about experience? Have we not indulged in a contradiction? The answer is this: strictly speaking, experience can only test things (e.g., cars) and empirical (descriptive) propositions (e.g., "It rains"). But experience cannot test interpretative formulas, i.e. hypotheses. When we say that experience must test hypotheses of fact we mean that in the last resort *reports* on experience (i.e. certain propositions purporting to describe experience) are employed to support or undermine hypotheses. In other words no scientific hypothesis is, strictly speaking, ever subjected to test. Only some "translations" of the lowest level consequences of scientific hypotheses are testable, namely their translations into the language of experience.

Now this translation of theoretical into empirical statements is not a purely linguistic affair: it consists in establishing certain correspondences between conceptual objects (e.g., "mass-points") and empirical objects (e.g., small bodies). Strictly speaking any statement about mass-points we encounter in mechanics is empirically untestable, because there are no mass-points in reality: what can be observed is bodies that look small and that can be regarded as realizations or concrete models of mass-points. These concrete models are pieces of machinery in some cases, stars in others, but in any case our hypotheses caricature rather than portray them. That is, what we test are some propositions referring to these empirical objects: these are the propositions we confront with the theoretical propositions. In summary, the standard statement that experience tests hypotheses and theories is elliptical: propositions containing nonobservational theoretical concepts entail no empirical statements. What we do is to establish certain empirical models that can be more or less accurately matched with some low-level theoretical statements. This is why we do not say that a hypothesis h entails its evidence e, but rather that h entails a testable consequence t that, when suitably translated, can be compared with an evidence e.

To sum up, the requirements an assumption must satisfy in order to pass for a *scientific hypothesis* are: (i) it must be well-formed, self-consistent, and have maximal strength with respect to the empirical evidences relevant to it; (ii) it must fit the bulk of relevant available knowledge; and (iii) it must, in conjunction with other formulas, entail consequences translatable into observational propositions.

We shall now turn to the function of hypothesis in science.

Problems

5.7.1. Discuss POPPER's argument that confirmation has no value because any number of hypotheses (or theories, as the case may be) can be constructed to cope with a given set of empirical data. Does the argument retain its force if the requirement of foundation is added to empirical testability?

5.7.2. Does one exception suffice to ruin a universal hypothesis if it is otherwise well-grounded? Hint: consider the cases of (i) observations rejected on theoretical grounds, and (ii) empirical generalizations with exceptions that are eventually explained by some theory (e.g., the law of DULONG and PETIT).

5.7.3. Is strength necessary, sufficient, or necessary and sufficient for information content?

5.7.4. Examine POPPER's thesis that the probability of any statement is the greater the less the statement says; and that, accordingly, we should seek highly improbable hypotheses rather than highly probable ones. Begin by establishing what interpretation of probability POPPER has in mind. See, e.g., his *Conjectures and Refutations* (New York: Basic Books, 1963), Addenda.

5.7.5. Examine the following proposals to quantify the concept of logical strength or content in terms of probability. (i) $C(h)=1-P(h)$ (K. R. POPPER); (ii) $C(h)=-\log_2 P(h)$ (J. G. KEMENY). See POPPER, op. cit. in the preceding Problem, and J. G. KEMENY, "A Logical Measure Function", *Journal of Symbolic Logic*, **18**, 289 (1953).

5.7.6. Of two rival hypotheses the more general one will usually be preferred if both are consistent with the same data, if only because the more general conjecture entails a more varied set of consequences and is therefore apt to be tested more fully than the less general hypothesis. Now, the less general implies the more general; thus, e.g., $(y=\text{const.})\rightarrow[y=f(x)]$ but not conversely. Discuss this point in connection with the strength of hypotheses.

5.7.7. Would you rate as scientific the hypothesis of spontaneous generation before the time of PASTEUR? Take into consideration the following versions of the hypothesis: (i) "Living beings are formed out of inorganic matter in a short time (e.g., frogs from lime)". (ii) "The original ancestors of the present living beings evolved out of complex nonliving systems in a long process". *Alternate Problem:* What, if anything, is wrong with the "partly-baked ideas" found in I. J. GOOD, Ed., *The Scientist Speculates* (London: Heinemann, 1962)?

5.7.8. Do we test entirely isolated hypotheses? Hint: formulate a simple hypothesis and see whether, in the process of testing it, you need not make use of other hypotheses. *Alternate Problem:* Examine the verifiability doctrine of meaning, according to which the meaning of a proposition consists in the method of its verification and, accordingly, a propositions is meaningful to the extent to which it is verifiable. See H. REICHENBACH, *Elements of Symbolic Logic* (New York: Mac Millan, 1947), p. 7. Hint: Discuss a hypothesis which is manifestly untestable yet meaningful in the context in which it occurs, such as the hypothesis of reincarnation.

5.7.9. The enemies of theorizing reject the learning theories of C. L. HULL and his followers alleging that these theories are too high in the scale of abstraction, i.e. too far from experience, since they include nonobservational concepts. Make a case for a criticism on the opposite count, namely that the basic hypotheses of those theories are not strong enough, that they should come out as solutions of more basic equations regarding the dynamics rather than the kinematics of behavior.

5.7.10. Discuss the following rule: "The hypothesis to be tested shall not belong to the very body of formulas involved in the design and interpretation of the test: otherwise the chance of accomodating the evidence to the hypothesis is enhanced to the point of making the test superfluous." Clearly this rule is logically sound since it aims at preventing circularities. Is it feasible? *Alternate Problem:* The finding of a red raven would refute the well-known universal statement about ravens; the red raven could not be counted as an experimental error. Does this suggest that qualitative hypotheses — clearly weaker than the corresponding quantitative ones — are better refutable?

5.8. Functions

Hypotheses overreach experience and therefore cannot be certain. Why do we frame them if we cannot be sure that they are true? Could we not stick to facts and dispense with hypotheses? Infallibilists, i.e. believers in and seekers after indubitable knowledge, have always distrusted hypotheses and have proposed regarding them as, at best, temporary and purely instrumental and heuristic devices. They have proposed, in their stead, trusting one of the following allegedly unshakable varieties of knowledge: (i) singular empirical propositions and, at most, inductions from them (*empiricism*); (ii) propositions derived from the "eternal principles of human reason" (*rationalism*); or (iii) propositions derived by immediate, total, and infallible insight (*intuitionism*).

Unfortunately for infallibilism, experience is not secure and, above all, it is not self-explanatory but is an object of scientific explanation; reason is not invariable in time and, although it does organize and process empirical knowledge, it can yield no empirical information; and intuition is hazy and undependable. We cannot help making hypotheses in everyday life, in work, in technology, in science, in philosophy, and even in art. We frame hypotheses to the extent to which we think and act rationally and effectively. Hence, instead of trying to avoid hypotheses we should try to control them.

Hypotheses are present in all steps of research, both in pure and in applied science, but they are particularly conspicuous on the following occasions: (i) when we attempt to sum up and generalize the results of our observations, (ii) when we try to interpret the foregoing generalizations, (iii) when we attempt to justify (ground) our opinions, and (iv) when we design an experiment or plan a course of action in order to either gather further data or test an assumption. Scientific knowledge is hypothetical to such an extent that some researchers do not realize it (just as a fish may not realize it is immersed in water) and think that there can be a self-contained piece of research that neither presupposes nor involves hypotheses. But this is a mistake: research consists in handling problems and, as we have seen in 4.2, every problem (i) is posed within a body of knowledge involving hypotheses, and (ii) is generated by a schema (the problem generator) that, when filled, becomes either a datum or a hypothesis. If we ask the individual question "Which is the x such that x has the property A?", i.e. $(?x)A(x)$, we presuppose that there exists at least one x with the property A and we tacitly assert the schema $A(x)$ which is the generator of our problem; the solution to the latter will consist in filling the blank 'x', i.e. in converting the generator into a datum or a hypothesis, as the case may be. Similarly, if we ask the functional question "What are the properties of the individual c?", i.e. $(?P)P(c)$, we introduce the schema $P(c)$ which, once the unknown P is assigned a definite value, becomes either a datum or a hypothesis. In any case the generator of a problem is a schema or propositional function and the solution to it is a statement that is a datum or a hypothesis, depending on whether it overreaches the experience at hand and is corrigible, or not. The flow diagram of research is, then, the following:

Hypothesis schema → *Hypothesis* ↘
Antecedent knowledge → *Problem* ⟨   *New problem*
Datum schema → *Datum* ↗

Eliminate hypotheses and you are left with data of a comparatively uninteresting kind — superficial, stray, unexplained — as well as with minor problems, generated by data schemata.

The main functions of hypotheses in science are the following.

1. *Generalizing experience:* summing up and extending available empirical data. A conspicuous subclass of this kind of hypotheses is the generalization, to an entire population, of "conclusions" (hypotheses)

"drawn" from particular samples, such as those made by agriculturists on the effects of fertilizers on the crop yield of a given species. Another conspicuous member of the same class is the so-called *empirical curve*, i.e. the continuous line that is made to join a set of points on a plane representing each an empirical datum. (As a matter of fact quantitative data are nearly always subject to error, so that they are represented by segments, or even by parallelograms, rather than by points. Moreover, empirical curves are not made to pass through the given "points" but rather close to them.) The extension beyond the set of data may be done,

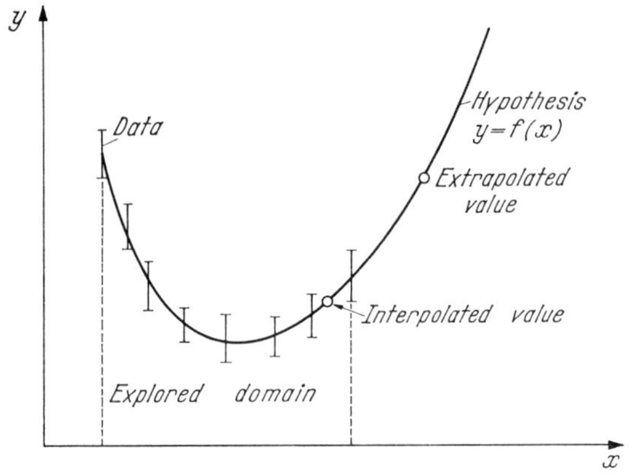

Fig. 5.6. An "empirical curve" fitting a set of data and representing the hypothesized $y=f(x)$

in this case, either by *interpolation* (assumption of values intermediate between those observed) or by *extrapolation* (assumption of values beyond the explored range, as in prediction). In Fig. 5.6 one of the infinitely many hypotheses (continuous curves) fitting a set of imaginary data is shown: it is a smooth curve [representing a continuous function $y=f(x)$] passing near the centers of the segments representing the data. Note that an empirical curve (or the corresponding function) is not a summary of data, if only because (i) the curve consists of an infinite set of points whereas data are always in finite number, and (ii) on the basis of the hypothesis (curve) we can anticipate experience in as yet unexplored domains.

2. *Inference starters:* initial posits or wagers, trial hypotheses, working hypotheses, or simplifying assumptions serving as premises in an argument even if suspected to be false. Examples: (i) in an indirect demonstration we try the negate $-t$ of the thesis of the theorem "$h \rightarrow t$" to be demonstrated and find out whether such a conjecture leads to a contradiction; (ii) the initial value assumed in the computation of a function by a method of successive approximations, or in the measurement of a magnitude, is likewise hypothetical; (iii) a grossly simplifying assumption that renders the application of a theory possible — such as, e.g., that

the earth surface is flat, or perfectly spherical over a certain region. Such hypotheses, when they are known to be strictly false, can be called *pretences*. No quantitative theory can be developed without pretences.

3. *Research guides:* exploratory guesses, i.e. more or less reasonable (founded) conjectures that are both the object of investigation and its guide. They range from precisely stated working hypotheses to rather vague conjectures of a programmatic kind. Examples: "Electrically neutral particles are composed of pairs of oppositely charged particles", "Mental processes consist of physiological processes in the brain", "Living beings can be synthesized by reproducing the physical conditions reigning on our planet two billion years ago".

4. *Interpretation:* explanatory hypotheses, or conjectures supplying an interpretation of a set of data, or of another hypothesis. Representational hypotheses are all interpretive since they enable us to interpret, rather than just generalize, data in terms of theoretical concepts. By contrast phenomenological hypotheses are of the generalizing type. For example, the electromagnetic field hypotheses (organized into a theory) explain the behavior of perceptible bodies of a kind.

5. *Protection of other hypotheses: ad hoc* conjectures the sole *initial* function of which is to protect or save other hypotheses from either contradiction with accepted theories or refutation by available data. For example, W. HARVEY (1628) hypothesized the circulation of the blood, which is not an observable process and did not account for the difference between venous and arterial blood; to save the hypothesis he introduced the *ad hoc* hypothesis that the artery-vein circuit is closed by invisible capillary vessels, which were in fact discovered after him.

Let us concentrate on empirical curves (included in type 1 hypotheses) and on *ad hoc* hypotheses, since they seem to be the least understood. When a scientist draws a curve passing near a set of empirically found dots (or rather segments), or when he applies an interpolation formula to construct a polynomial representing such a curve, he may not realize the jump he is making and the risk he is assuming. In fact, he is betting that the next observed value will fall very near the hypothesized curve — which may not be the case. He is adopting the principle of continuity— an ontological hypothesis. Pure logic does not dictate him which among an infinity of possible curves, all compatible with the same handful of data, he should prefer (see Fig. 5.7). He may, of course, argue that he will prefer *the simplest* hypothesis, and in fact the interpolation formulas are so built as to yield the formally simplest expressions (least power polynomials, representing least-wavy lines). But this preference has neither a logical nor an empirical ground. One might try to render it plausible by postulating the ontological hypothesis that reality is simple. But this conjecture is refuted by the history of science, which shows that progress is, to a large extent, the discovery of complexities behind simple appearances. There *are* reasons for preferring the simplest empirical generalization compatible with the data at hand and *as long as no theory is available* to provide additional suggestions — only, these reasons are methodological in character.

One reason for preferring formal (e.g., mathematical) simplicity in the case of empirical generalizations is that nothing so far warrants the assumption of a more complex pattern: in other words, given empirical

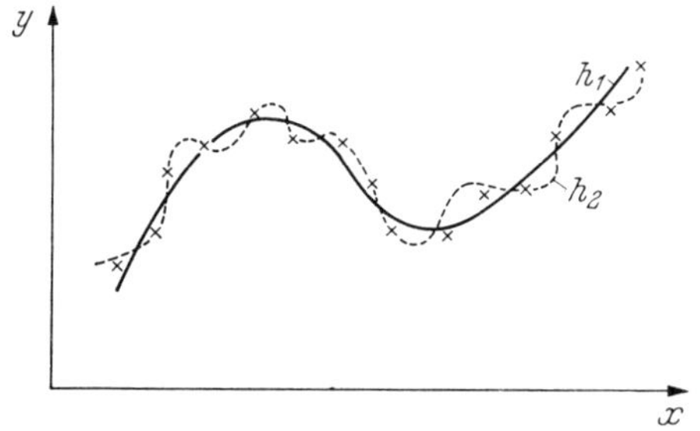

Fig. 5.7. Any set of data is consistent with an infinity of mutually incompatible hypotheses

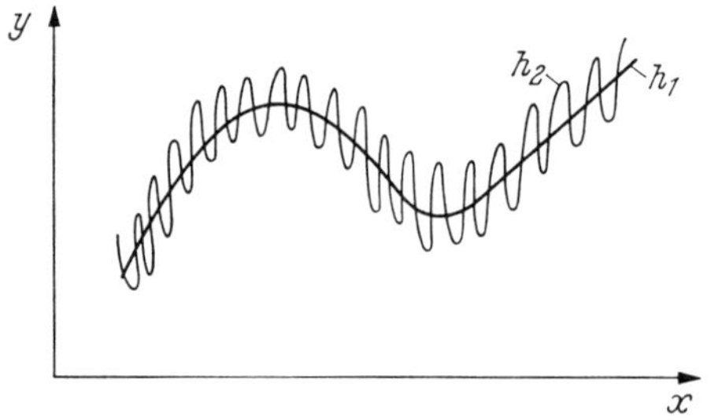

Fig. 5.8. The more complex hypothesis (h_2) is the safest, hence the least desirable in the absence of additional evidence and theory

evidence alone the more complex hypotheses are *groundless:* they assume too much. (This situation may be altered radically as soon as theoretical considerations are available, as will be seen shortly.) A second reason is that, so far as possible, irrefutable hypotheses should be avoided in science and, the more wavy an empirical curve, the nearer any future datum it will lie, i.e. the less it will learn from experience if from the start it anticipates every possible experience (see Fig. 5.8). In other words, the more complex empirical curves will be less liable to be refuted and more liable to be confirmed by any new evidence, however anomalus this evidence would be with regard to simpler, riskier hypotheses. This has not been realized by the partisans of confirmation as the supreme test of hypotheses; yet since they seek maximum confirmability they ought to preach maximum complexity instead of arbitrarily (i.e. groundlessly) positing the nonempirical principle of simplicity.

The argument from simplicity is powerful in the pretheoretical stage, which is the one envisaged by inductivist philosophers. But as soon as an attempt is made to include an empirical generalization in a body of theory, the more powerful requisite of compatibility with the rest of knowledge will function. Take, for instance, GALILEI'S semiempirical law of falling bodies, namely "$s=5t^2$", where 's' designates the height (in meters) of a body that has been freely falling during t seconds. This may be regarded as the simplest among infinitely many possible relations between observed values, but it is true near the surface of the earth only and it contains a number, "5", for which no explanation is given. The corresponding theoretical law (unknown to GALILEI) is "$s=\frac{1}{2}gt^2$". This formula is deduced from NEWTON'S axioms and has a wider domain of validity than GALILEI'S semiempirical law, since it applies, within limits, to any homogeneous gravitational field of intensity g. The theoretical law could not have been obtained by induction because it contains a nonobservational theoretical concept, "acceleration of gravity", designated by 'g', which is a construct that makes sense in a theory of gravitation. Considerations of simplicity are likewise alien to this law — which, incidentally, has to be considerably complicated in reference to very strong fields and high speeds. The final word, then, is given neither by induction nor by simplicity but rather by the continuity with the bulk of theoretical knowledge. Simplicity seems important only in the early, simple stage.

Let us now attack the problem of ad hocness. Of all kinds of hypotheses, protective conjectures are the most tempting and the most distrusted by honest thinkers of all philosophical affiliations save conventionalists, who regard all hypotheses as being on the same footing and as merely useful tools for the condensation and processing of data: as a consequence of this view the conventionalist feels no repugnance to the propping of one fiction with the help of another fiction. We reject conventionalism because it is inconsistent with both the goal of science (building the truest possible models of things) and the method of science (which includes testing the hypothetical models for truth). But even rejecting conventionalism we shall argue that *ad hoc* hypotheses are unavoidable and welcome in science as long as they contribute to enriching the foundation of important hypotheses and to ensure their consistency with other hypotheses. *Ad hoc* hypotheses are acceptable when they protect important ideas from *rash* criticism (such as the one based on the latest measurement), just as they are unacceptable if they prevent *every* criticism. What must be demanded before admitting (temporarily) a protective *ad hoc* hypothesis is that it be *independently testable*. According to our stipulations in Secs. 5.2 and 5.7, if a conjecture is not testable then it is not scientific; and once a protective hypothesis has been tested, it becomes either a false proposition or a likely normal hypothesis. There is, then, no need to impose the extra requirement of non-ad hocness on scientific hypotheses: testability will suffice to screen out the undesirable protectors. The analysis of a few examples will illustrate this thesis.

The hypothesis that atomic elements may have different isotopes (i.e., physically different subspecies of chemically homogeneous species) was initially framed in order to save the hypothesis that atoms were composed of a discrete number of particles. In fact, the isotope hypothesis (F. SODDY, 1913) saved the atomic theory in the face of the finding that most atomic weights are not exact integral multiples of a basic unit. This anomaly was explained by assuming that the naturally occurring samples of chemical elements might be mixtures of several isotopes, so that the measured atomic weight was a weighted average of the atomic weights of the various isotopes occurring in the samples. This protective hypothesis was subsequently confirmed in an independent way: in fact, the isotopes of a number of elements were separated and weighed with the help of the mass spectrometer (F. W. ASTON, 1919), a device invented in order to test the isotope hypothesis. The ad hocness of the isotope hypothesis was, then, a negligible birth accident.

*It is not always apparent whether a hypothesis introduced to protect another hypothesis, or a whole theory, is testable: after all testability is not inherent in hypotheses but is relative to the means (theories and empirical techniques) available at a given moment. This was the case of the contraction hypothesis, that all bodies contract in the direction of their motion. With this hypothesis G. F. FITZGERALD and H. A. LORENTZ attempted to rescue classical mechanics from the ruinous „conclusions" that would otherwise derive from the Michelson-Morley experiment. It is often maintained that the contraction hypothesis had to be rejected because it is not independently testable since it specifically states that the alleged contraction is absolute or, if preferred, is relative to the (nonexistent) immobile ether. Yet the contraction hypothesis was eventually tested; only, it took some time to realize its measurable consequences and to understand that it *presupposed* untestable assumptions such as the existence of absolute space. One of the testable consequences of the contraction hypothesis is that, since the electrical resistance of a wire is proportional to its length, if the contraction were real then the resistance during motion should decrease — which was disproved in 1908 by TROUTON and RANKINE. Another testable consequence is that, since the frequencies of the proper vibrations of a cube depend on its side, they ought to change with motion, which again was found not to be the case (WOOD, TOMLINSON, and ESSEX, 1937).

*The contraction hypothesis was rejected or, rather, reinterpreted by EINSTEIN not because it was protective but because it presupposed the untestable assumption of an absolute space and consequently an absolute motion. Were it not for this and for the factual falsity of the contraction hypothesis physicists would have hailed it as the savior of something worth saving indeed, namely, classical mechanics. When a big theory that has rendered distinguished service is endangered by an exception, a first and legitimate move is to try to prop it by some lesser but testable *ad hoc* hypothesis.*

The tendency to frame protective hypotheses is psychologically understandable: on the one side we normally resist changes in our system of beliefs, and on the other the inscrutable hypothesis has the advantage, from the point of view of belief conservation, that it cannot be independently inquired into, so that it not only provides protection but remains itself immune to attack from experience. This is the case, e.g., with the claim of the mind dowser that any failure to perceive thought signals in the correct time order is due to the subject's precognitive faculty: in this way any evidence unfavorable to the telepathy hypothesis is reckoned as a confirmation of the precognition hypothesis. Similarly, according to psychoanalysis certain infantile experiences originate aggressiveness but, if a case is found of timorous behavior when aggressive behavior ought to occur, the finding is not counted as a counterexample: the *ad hoc* hypothesis is introduced that the subject has formed a reaction against his natural bent. In this way unfavorable evidences cannot occur and the innocent is persuaded by a gang of accomplices that are never caught because they supply each other with alibis. Moral: Although scientific theories are tested as wholes because most of their testable consequences are derived from a number of their basic assumptions, the *ad hoc* conjectures among these assumptions must be independently testable.

To conclude. We must agree with the infallibilist that hypotheses are risky. On the other hand to take such risks is all we can do in the field of science, since scientific research is essentially handling problems that demand the conception, working out, and testing of hypotheses. Moreover, the riskier hypotheses are, within limits, the better because they say more and they are consequently more sensitive to experience. There are, to be sure, some dangerous hypotheses that should be avoided: the a priori (ungrounded) limits set on human ingenuity, the naive conjectures commended on no ground other than their simplicity ("natural", "obvious", "intuitive"), the sophisticated conjectures that cannot be checked with the help of experience, and the *ad hoc* hypotheses that dodge independent test. But the joint requirements of *foundation and testability* will check any excess of this sort. The question is not to minimize hypotheses but to maximize their control, because rational beings face experience, multiply it, and transcend it by inventing hypotheses. The slogan is not *Be blank minded* but rather *Be open minded.*

In the next section we shall argue that, upon embarking on any scientific research, we carry not only a body of factual hypotheses and data, but also a bundle of philosophical hypotheses.

Problems

5.8.1. Examine the following opinions on hypothesis. (i) SEXTUS EMPIRICUS, *Against the Professors*, Bk. III, 9—10, in Works, transl. R. G. BURY (Cambridge, Mass.: Loeb Classical Library, 1949), IV, p. 249: ... "if the thing is true, let us not postulate it as though it were not true. But if it is not true but is false, no help will emerge from the hypothesis" ... Hint: Does the objection hold in the framework of an epistemology which acknowledges neither complete factual truth nor complete certainty concerning it? (ii) F. BACON, *Novum Organum*, Aphorism I, in *Philosophical Works*, ed. J. M. ROBERTSON (London: Routledge, 1905), p. 259: "Man, being the servant and interpreter of Nature, can do and understand so much and so much only as he has observed in fact or in thought of the course of nature: beyond this he neither knows anything nor can do anything". The task is then not to frame "anticipations of Nature" (hypotheses) but "interpretations of Nature" (inductions). The latter procedure, according to Aphorism XIX, "derives axioms from the senses and particulars, rising by a gradual and unbroken ascent, so that it arrives at the most general axioms last of all. This is the true way, but as yet untried". Hint: find out whether it is logically possible to ascend from the weaker to the stronger formulas. (iii) J. TOLAND, *Christianity not Mysterious* (London, 1702), p. 15: ... "since Probability is not Knowledg, I banish all Hypotheses from my Philosophy; because if I admit never so many, yet my Knowledg is not a jot increas'ed: for no evident Connection appearing between my Ideas, I may possibly take the wrong side of the Question to be right, which is equal to knowing nothing of the Matter. When I have arriv'd at Knowledg, I enjoy all the Satisfaction that attends it; where I have only Probability, there I suspend my Judgment, or, if it be worth the Pains, I search after Certainty".

5.8.2. Examine the following statements. (i) L. PASTEUR, in R. DUBOS, *Louis Pasteur* (Boston: Little, Brown & Co., 1950), p. 376: "Preconceived ideas are like searchlights which illumine the path of the experimenter and serve him as a guide to interrogate nature. They become a danger only if he transforms them into fixed ideas — this is why I should like to see the profound words inscribed on the threshold of all the temples of science: 'The greatest derangement of the mind is to believe in something because one wishes it to be so' ". (ii) T. H. HUXLEY, *Hume* (London: Macmillan, 1894), p. 65: "All science starts with hypotheses — in other words, with assumptions that are unproved, while they may be, and often are, erroneous; but which are better than nothing to the seeker after order in the maze of phenomena. And the historical progress of every science depends on the criticism of hypotheses — on the gradual stripping off, that is, of their untrue or superfluous parts — until there remains only that exact verbal expression of as much as we know of the fact, and no more, which constitutes a perfect scientific theory". (iii) M. SCHLICK, *Sur le fondement de la connaissance* (Paris: Hermann, 1935), p. 33: "all the propositions of science, all without exception are seen to be *hypotheses* when their worth, i.e. their truth value, is examined". Purely empirical statements (protocol statements), on the other hand, are certain but are not propositions belonging to science and are not understood unless accompanied with gestures. Thus, e.g., 'Here now two yellow lines' is not a scientific sentence. On the other hand, 'Sodium presents a double line in the yellow part of its spectrum' is a genuine scientific sentence (p. 47).

5.8.3. Expand DARWIN's remark that "all observation must be for or against some view if it is to be of any service". *Alternate Problem:* Discuss the nature of "the" principle of simplicity and its role in science. In particular, examine whether it is a principle proper or can be derived from stronger assumptions, and whether it has ontological commitments.

5.8.4. Extreme behaviorists are suspicious of hypotheses and advocate experiments which are not aimed at testing hypotheses but at answering questions of the type "I wonder what will happen if ...", because such experiments never produce negative results, whereas other experiments may give negative results, i.e. may result in the refutation of a hypothesis, which event they regard as wasteful. See M. SIDMAN, *Tactics of Scientific Research* (N. York: Basic Books, 1960), or A. J. BACHRACH, *Psychological Research* (N. York: Random House, 1962). Granting that behaviorism has been theoretically barren, is it true that it contains and presupposes no hypotheses?

5.8.5. Are the so-called empirical curves strictly empirical? And are such graphs hypotheses proper or rather nonverbal (geometrical) symbols of hypotheses? *Alternate Problem:* Determine whether the proposition "87 per cent of big business people in the U.S.A. are Conservative" is a hypothesis, and if so of what kind.

5.8.6. For purposes of mathematization, population can be regarded as a continuous variable. What kind of assumption is this?

5.8.7. According to W. D. MATTHEW, all terrestrial animals originated in the Holarctic region (North America, Europe, northern Asia, northern Africa, and the Arctic). This hypothesis can be held only if it is further assumed that transatlantic and transpacific bridges, now sunken, existed in the past — although geology provides no evidence for this hypothesis. Discuss this case.

5.8.8. H. BONDI and T. GOLD (1948) postulated (i) that the universe at large is everywhere and always the same (the "Perfect Cosmological Principle") and (ii) that the galaxies are constantly receding from one another (expansion of the universe). These two postulates are mutually incompatible (since expansion leads to the thinning out of matter, which contradicts the hypothesis of over-all homogeneity) — unless the hypothesis is added that matter is continually being created out of nothing at exactly the rate required to compensate for the expansion of the universe. Examine this latter hypothesis as to foundation and testability. See M. BUNGE, "Cosmology and Magic", *The Monist*, **44**, 116 (1962).

5.8.9. J. C. MAXWELL (1864) postulated that every electric current is closed, which was apparently refuted by the existence of condensers. In order to save that basic hypothesis he assumed that a variable current did not end up at the plates of a condenser but is propagated in the intervening dielectric (or in the vacuum) as a "displacement current". This hypothesis was severely criticized on methodological grounds both because it introduced an unobservable and because it was *ad hoc*. The hypothesis was independently confirmed by H. HERTZ (1885) after MAXWELL'S death. Discuss this case. *Alternate Problem*: Discuss any other case of *ad hoc* hypothesis.

5.8.10. According to E. MACH, *History and Root of the Principle of Conservation of Energy* (1872; Chicago: Open Court, 1911), p. 49, "in the investigation of nature, we have to deal only with knowledge of the connexion of appearances with one another. What we represent to ourselves behind the appearances exists *only* in our understanding, and has for us only the value of a *memoria technica* or formula, whose form, because it is arbitrary and irrelevant, varies very easily with the standpoint of our culture". If this is true, science abounds with untestable hypotheses and, being untestable, there is no point in choosing among them. On the other hand the methodological rationale of MACH'S — and DUHEM'S — phenomenalism is the elimination of untestable assumptions. Discuss this paradox. *Alternate Problem:* In Sec. 5.7 it was stated that no hypothesis is tested in isolation from other hypotheses, and in the present section it was stated that the independent testability of *ad hoc* hypotheses must be required. Discuss this apparent contradiction.

5.9. Philosophical Hypotheses in Science

Scientific knowledge contains no philosophical assumptions. From this it is often concluded that scientific research has neither philosophical presuppositions nor a philosophical import, whence science and philosophy would be water-tight compartments. But this is a hurried conclusion. Philosophy may not be found in the finished scientific buildings (although this is controversial) but it is part of the scaffolding employed in their construction. And conversely philosophy can and should be built with the method of science and on the basis of the achievements and failures of scientific research (see Sec. 4.7). This latter point cannot be made here: what concerns us at this point is to substantiate the thesis that scientific research does *presuppose and control* certain important philosophical hypotheses. Among these the following stand out: the reality of the external world, the multilevel structure of reality, determinism in an ample sense, the knowability of the world, and the autonomy of logic and mathematics.

1. *Realism: The Reality of the External World.* Some philosophers hold that factual science neither presupposes nor employs nor confirms the philosophical hypothesis that there are real objects, i.e. that there is anything that exists independently from the cognitive subject. But this is mistaken. First, the very notion of factual truth, or adequacy of a proposition to a fact, involves the notion of objective fact; formal truth alone, being a syntactical property, is independent of fact, and this is why it can be complete and therefore final. Second, when a factual hypothesis is framed to cover a set of facts, the presumption is that the facts are real (actual or possible); no time is wasted in science to account for nought. Third, the very tests for the factual truth of a hypothesis presuppose that there is something outside the subject's inner world which will either agree or disagree to some extent with the proposition in question. If that something were to depend entirely upon the subject we would not speak of objective tests and of objective truth. Fourth, every empirical procedure in science begins by drawing a line separating the inquiring subject from his object of research: if no such line is drawn, so that any other operator can have access to the same object, then the procedure should not be acceptable to scientists. Fifth, natural science, in contrast with prescientific views such as animism and anthropomorphism, does not account for nature in terms of typically human attributes, as it should if nature somehow depended on the subject. Thus, we do not account for the behavior of an object in terms of our own expectations or other subjective variables but, on the contrary, base our rational expectations on the objectively ascertainable properties of the object as known to us. Sixth, there would be no need to experiment and theorize

about the world unless the world existed by itself; a factual theory refers to something which is not the subject (although it can be a person regarded as an object) and the empirical test of the theory involves handling and eventually modifying (by way of experiment) the theory's referent. Seventh, factual science contains interpretive rules that presuppose the real existence of the referents involve. Thus, e.g. the semantical rule "'*Z*' designates the atomic number of an element" is not invented for fun or for correlating certain perceptions but is supposed to establish a relation between the sign '*Z*' and an objective (but nonobservable) physical property, namely the number of electrons in an atom. Eighth, no successive corrections of factual theories would be necessary if they were mere conventional constructions that did not attempt to depict reality in a symbolic way. If we had less faith in the existence of atoms than in our atomic theories we would not be willing to correct the latter as soon as they exhibit their defects but would give up the hypothesis of the existence of atoms altogether. Ninth, the axioms of a factual theory are affirmative rather than negative statements, not only because negative propositions are rather indefinite and therefore scarcely fruitful but because an affirmative proposition suggests the search for some existent entity or property, as only the existence of that referent will make it true; negative propositions, on the other hand, are true if nothing exists that falsifies them. Tenth, law statements presuppose the objective existence of the objects the properties of which they refer to; otherwise they would be vacuously true. In short, factual science does not *prove* the existence of the external world but it definitely *presupposes* this philosophical hypothesis. Those who would like to disprove this hypothesis would accordingly have to dispense with science.

It is often claimed, though, that contemporary atomic and nuclear physics questions or even refutes the hypothesis of the reality of the external world — an impression certain authors give indeed. Yet a semantical examination of the fundamental statements of the quantum theory shows that they fall into the following classes: (i) statements referring to autonomous objects unperturbed by measurement, such as an atom in a stationary state (which neither absorbs nor radiates energy) or a photon travelling in empty space, where no absorbers can detect it; (ii) statements referring to objects under observation, measurement, or generally in interaction with macroscopic systems, such as an electron beam passing through a slit system; (iii) statements referring to (possible) results of observation or measurement, and (iv) statements referring to properties of the basic laws themselves. Class (i) and class (ii) statements point to physical objects assumed to have an independent existence: the former to nonobservable microobjects, the latter to complex systems involving both a microscopic entity (the object of research) and a macroscopic entity (a means of research). Class (iii) and class (iv) statements do not refer immediately to physical objects existing on their own: the former refer to results of physical operations done on them, the latter to other statements. But physical operations are here regarded as purely physical processes (even though they have certainly been planned by some mind or other), and the statements referred to by class (iv) statements are statements of either class (i) or class (ii). In no case do quantum-theoretical formulas deal with the mental states of the observer, which are something for the psychologist to inquire into. The most that certain interpretations of the theory claim is that it contains no class (i) statements. But this — which is false — does not touch the hypothesis of the reality of the external world: it just converts the quantum theory into a theory of objects under experimental control — which, if true, would render it inapplicable to astrophysics.

2. *Pluralism: The Multilevel Structure of Reality.* An ontological hypothesis involved in and encouraged by modern science is that reality, such as known to us today, is not a solid homogeneous block but is divided into several *levels*, or sectors, each characterized by a set of properties and laws of its own. The main levels recognized at present seem to be the physical, the biological, the psychological, and the sociocultural ones. Every one of these may in turn be divided into sublevels. For instance, the main sublevels of the physical level are the physical proper and the chemical sublevels, and the main sublevels of the sociocultural level are the economic, the social proper, and the cultural sublevels. Finer divisions can be introduced and none are clear-cut and rigid.

A second, related presupposition is that *the higher levels are rooted in the lower ones*, both historically and contemporaneously: that is, the higher levels are not autonomous but depend for their existence on the subsistence of the lower levels, and they have emerged in the course of time from the lower in a number of evolutionary processes. This rooting of the higher in the lower is the objective basis of the possibility of partially explaining the higher in terms of the lower or conversely.

The two basic ontological hypotheses just referred to are built into the contemporary outlook to the point that they underly the usual classification of the sciences and they dominate more and more our system of higher education. Thus, the scientific psychologist is required to learn more and more biology, and even chemistry and physics, because it is increasingly being recognized that psychical events are rooted in these lower levels; but the psychologist is also required to become conversant with sociology, because we have come to realize the reaction of the sociocultural level on the immediately lower ones: thus, we recognize the influence of religion on food habits and the reaction of the latter on food production. Only physicists are entitled to ignore the upper levels — and sometimes they ignore them to the point that they talk of direct mental influences on physical events, thereby skipping all the intermediate levels.

Moreover, the above mentioned level hypotheses underlie several important principles of scientific *methodology*: we shall call them the principles of level parsimony, level transcendence, source level, and level contiguity. (According to some philosophers levels are just a methodological affair with no ontological import. But this is another ontological hypothesis, and one that severs methodology from the rest and is thereby incapable of explaining why methods work or fail to work.) The principle of *level parsimony* is this: "Begin by investigating your class of facts on their own level: introduce further levels only as they become indispensable". For example, do not introduce psychology and psychiatry in the study of international politics since you can go along quite a stretch without their company. The principle of *level transcendence*: "If one level is insufficient for the truthful account of a set of facts, scratch its surface in search for contiguous levels". For example, in order to explain chemical binding do not stop at the particular laws of chemical reactions or the accompanying thermodynamics but look underneath the molecular level, into the atomic level, in search for their mechanism. Principle of *source level:* "Try to explain the higher by the lower and reverse the process only in the last resort". For example, try to explain problem-solving among animals in terms of trial and error and learning; introduce insight and intelligence only when the previous approach is insufficient and where the complexity of the nervous system of the organism concerned make insight and intelligence possible. This principle may be called also the principle of *methodological reductionism* — not to be confused with ontological reductionism or the denial of levels. Principle of *level contiguity:* "Do not skip levels — that is, do not miss the intermediate levels when establishing interlevel relations". For example, do not regard as adequate an explanation of a social behavior pattern in physical terms, because physical stimuli can reach the social level only through organisms endowed with certain psychical abilities. Level skipping may, however, be unavoidable when little is known; and it may be desirable when the intermediate processes are of no interest in the research under consideration. These, however, are pragmatic considerations of no avail when the aim is a faithful reproduction of reality.

3. *Ontological Determinism: Lawfulness and Non-Magic.* The philosophical doctrine of determinism has two aspects, one ontological and the other epistemological, and the two are often merrily confused. *Ontological determinism* holds the determinateness of things and events; *epistemological determinism* holds the possibility of conceptually determining (knowing) facts and their patterns in their entirety. In a narrow sense

ontological determinism equals mechanical or Laplacian determinism, a component of the Newtonian world view according to which the cosmos is a set of particles interacting and moving in accordance with a handful of mechanical laws. The ample version of determinism involves only (i) the hypothesis that all events are lawful (*principle of lawfulness:* see Sec. 6.8) and (ii) the hypothesis that nothing comes out of nothing or goes into nothing (*principle of non-magic*). This lax determinism does not restrict the kind of law: it makes room for stochastic laws and recognizes objective chance. It just denies the existence of events either lawless or unproduced by prior events.

Various shades of strict determinism, none of which recognized objective chance, were clung to until the 1920's. It was not realized that, even assuming that every one of a set of entities did behave in a perfectly determinate (nonrandom) way, some amount of play or randomness would result from the comparative mutual independence of those entities (complete rigidity being nonexistent). Narrow ontological determinism was finally beaten by the quantum theory, which acknowledges objective chance not only as a trait of complex systems but even at the level of "elementary" particles, which obey stochastic laws. *Whether such a randomness is final or will eventually be analyzed as the outcome of either complex inner processes or interactions with as yet unknown, lower-level fields, it is premature to decide. Moreover, it is important to realize that both the quantum theory and its philosophy are still in the making, whence no detailed conclusions should be drawn with an air of finality about the behavior of microsystems. Yet the kind of chance and the exact levels on which it occurs is of secondary importance compared with the recognition that randomness is a mode of becoming, and indeed one which fits laws. Also important to us at this place is the realization that the quantum theory abides by the principles of lawfulness and non-magic: it states laws covering most known patterns on the atomic level, and among these laws some are conservation laws, i.e. statements about the non-creation *ex nihilo* and the non-traceless annihilation of material systems (particles or fields) — however much particles "annihilate" (i.e. get transformed into photons) and conversely. The quantum theory, in short, respects general determinism as much as any other scientific theory. And how could it be otherwise if that theory purports to share the goal of science — the conceptual reconstruction of the patterns (laws) of being and becoming?* To imagine lawless events would be to recognize that no science can account for them, which would be to prejudge the issue. And to imagine events that are lawful but indeterminate (such as, e.g., the creation of atoms out of nothing) would be to acknowledge that no laws are really necessary since anything, even magic, is possible: in fact, if an atom can pop up from no antecedent condition why not a molecule, if a molecule why not a chromosome, if a chromosome why not a cell, if a cell why not a full dinosaur? In short, general determinism is entrenched in science *qua* science, insofar as scientific research is the search for and application of laws — which laws in turn set limits upon logical possibilities such as the emergence of something out of the blue or the disappearance of something into thin air.

4. *Epistemological Determinism: Knowability.* Strict epistemological determinism is the programmatic hypothesis that anything can be known if we only care to: that in principle it is possible to exhaustively know the present, past and future states of any object in such a way that no uncertainty about it remains. This narrow form of determinism was abandoned *de facto,* if not *de jure,* in the second half of the 19th century with the advent of field physics and statistical physics. The former showed that it is in principle impossible to get to know every portion of a field because a field is a system with infinitely many degrees of freedom. And statistical physics showed that the state of every particle in a system cannot be known exhaustively if only because of the minuteness and large number, not to speak of the involved motions of the particles. But this was regarded as a practical limitation on knowledge, whereas the limitations imposed by continuous media, such as fields, are limits *de jure* and therefore insurmountable. Yet they are limits of *experience,* and such that theory can transcend. In fact, although we cannot hope to measure the value of the field strength at every point in a region, we can *compute* it with the help of theory and of a few well-chosen data. It is only by keeping in mind that empirical knowledge does not exhaust scientific knowledge that we can avoid being the victims of either complete skepticism or irrationalism.

Be that as it may, just as we have given up strict ontological determinism and adopted instead a richer doctrine, we must relax strict epistemological determinism and adopt the (philosophical) hypothesis of *limited knowability.* This lax version of epistemological determinism will allow for both the uncertainties rooted in objective chance and those inherent in our knowledge. All this version of epistemological determinism commits us to is the hope that chance effects (or rather their probabilities) will eventually be computed and that the latitudes, whether objective (indeterminacies) or subjective (uncertainties) may gradually be analyzed, computed, and reduced to some extent. Lax or general determinism incorporates the valuable contributions of indeterminism, namely the recognition that there is objective chance on any level (hence that there are stochastic laws), and the acknowledgment that no final certainty is possible. This metaphysical doctrine is not to be found in scientific contexts for the mere reason that it is presupposed by scientific research: drop the hypothesis of (limited) knowability and all research aiming at obtaining scientific knowledge stops; drop the restriction marked by the word 'limited' and absurd inquiries are allowed — such as, e.g., trying to communicate with the past or with the future. The genuine epistemological problem, then, is not *whether* we can know but to what extent we do in fact know and how far can we stretch the present frontiers of the known — recalling that scientific knowledge, far from being indubitable, is fallible.

According to phenomenalism we can know appearances only: we ignore what things in themselves, apart from our transactions with them, can be; moreover, there is no point in trying to get at them because the assumption of their independent existence is an unwarranted metaphysical conjecture. Phenomenalism covers a part of ordinary knowledge: the one dealing with appearance. But science goes beyond phenomena: otherwise it might be intersubjective (interpersonal) but not objective. In fact scientific theories, far from asserting relations among phenomenal predicates, contain nonphenomenal ones; moreover, science explains appearance in terms of (hypothesized) objective facts rather than the other way around. Whereas the phenomenalist and the radical empiricist must accept "Tonight I can see more stars than last night" as an ultimate, since it expresses an experience, the scientist will attempt to explain such an experience, e.g. in terms of atmospheric conditions. And he will assume that, whether someone sees one thousand stars or none, the stars are all there — the visible ones and those which we may never see.

Granting that science does get hold of the thing in itself and not just of its appearance to us, how far does it get? According to traditional positivism what science yields is just the external behavior of the object and its relations to other objects. This thesis, though false and crippling, has a sound methodological root, namely the rule of method: "Test hypotheses concerning the composition and internal structure of systems by their external manifestations". The ground of this rule is, in turn, the generalization that externalization is a necessary, though insufficient, condition of observability: in other words, that we do not grasp a thing's inners unless these manifest themselves, however indirectly, to our senses. But this should not lead us to confuse the reference of scientific hypotheses with their test: external behavior is not the thing but a portion of the thing. Behavior and internal structure are just two aspects of real sytems; and we explain behavior by inner structure and test structure hypotheses by observable behavior. As to the relationalist prescription, it should be clear that (i) except in pure logic we never state just relations, let alone relations among relations, but rather relations among variables representing each a supposedly objective trait, and (ii) a set of interrelated systems is a higher-order system, so that the relations among the members of the latter produce the structure of the whole. In short, a study of relations can be profound — if we only care to.

Yet obviously if just a description of external behavior is sought no more than this will be obtained. But then the limitation of our approach should not be attributed to the object of research, or even to every possible research. A deeper approach — a representational rather than a phenomenological one — can be taken up to inquire into the inner sources of behavior. Such an approach will set itself the goal of finding (i) the source properties and relations of the object, and (ii) the fundamental relations among such essential variables, i.e. the essential laws of the object, those which account for the inner mechanisms ultimately responsible for its external (partly observable) behavior. Such source variables and invariant relations among them, rather than some special core substance, is what is nowadays meant by the *essence* of a thing. Science, then, does attempt to discover the essence of things — only, in this more sophisticated sense of 'essence'. Also, we would be unwise to proclaim at any one moment that the essence of something has been caught once and for all: what we obtain are less and less blurred glimpses of essential laws on different levels.

In sum, science presupposes that its objects are knowable to some extent, and it acknowledges that some of the limits to knowledge are set by the objects themselves, whereas other are temporary. In turn, the possibility of knowing anything at all, held by epistemological determinism, is grounded on the assumed determinateness of the world: if events were all unpatterned and were neither produced by other events nor left traces, only stray and fleeting impressions would be possible. The empirical fact that scientific research manages to get hold of some patterns of determinateness beneath the chaotic flux of appearances suggests and confirms ontological determinism. The two branches of a lax determinism — ontological and epistemological neodeterminism — support each other.

5. *Formalism: The Autonomy of Logic and Mathematics.* A good tool should not change with use: otherwise no task could be finished with its assistance. Logic is such a tool for science: however much the science of logic may change, it does so internally or in response to purely rational problems rather than in an effort to adapt itself to reality. Logic is self-sufficient as regards both its object and its methods: it has no object other than its own concepts and its proofs owe nothing to the peculiarities of the world. Not that logic is other-worldly, e.g., that logical formulas dwell in a realm of Platonic Ideas: logic is made by rational beings and will disappear with the last logician (it has happened before); only, it is unconcerned with reality. No matter how the world may look to the successive generations of scientists, logical truths, like "$p \to (q \to p)$", and logical arguments, like "$\{p \vee q, -q\} \vdash p$", remain unchanged because they have no factual content. That they have not always been recognized

as self-sufficient is a different story, one of interest to the psychologist and the historian of science: what is claimed is that logical formulas and logical arguments stand or fall by themselves. The same is true of mathematics. In short, the validity of formal science is independent of the world because it is not concerned with it. That factual science, when employing mathematics, poses mathematical *problems*, and that mathematical research poses logical *problems*, is again a story for the psychologist and the historian of science as well as for the methodologist: it does not prove that formal science deals with the external world or that it is logically dependent on it but rather that formal science does not live a secluded life: that it is often thought by people interested both in ideas and in the world. It would be very different if the world had formal properties; but only ideas, whether pure or about the world, can have formal properties.

All logical theories contain, and all mathematical and factual theories presuppose in a way or other, the laws of identity and noncontradiction, and the rule of detachment or *modus ponens*. (The excluded middle law is absent from intuitionistic logic, which is not employed in science partly for this reason. In ordinary logic this law, "$p \vee -p$", is equivalent to the principle of noncontradiction, i.e., "$-(p \;\&\; -p)$", so that it need not be mentioned apart in our discussion.) Suppose, for the sake of argument, that factual science did not presuppose these logical principles. Then it would either presuppose alternative logical principles or none at all. If the former, this would be discoverable upon analysis — just as the analysis of ordinary discourse led to classical logic and the analysis of mathematical discourse led to symbolic logic. If, on the other hand, factual science presupposed no logical principles at all, it would either remain so or make an empirical search for logical principles of its own. If the former, there would be no limitation upon either logical form or inference: anything could be stated (every possible string of symbols could be taken to represent a well-formed formula) and anything could be concluded (any sequence of statements, however logically incoherent, would be acceptable as a valid argument.) Since this is not the case, let us suppose the second alternative: namely, that factual science would take up the task of logic and search for its own principles of reasoning. How could it go about this? Concepts, propositional functions, propositions and the like — that is, logical objects — have no material existence and consequently cannot be experimented with: only their symbols have a material existence, but they are inessential, i.e. they can be changed for any other symbols without changing what they denote. Factual science would then have to turn inwards, that is, it would have to make its own analysis in order to discover such logical principles as might be built into itself. But with what tools would it proceed to perform such an analysis, if not with the logical tools it had initially refused to countenance? Factual science must therefore presuppose some logic.

The logic presupposed by factual science is just one among infinitely many logically possible (consistent) logical theories: it is the so-called *ordinary* or two-valued logic. The remaining logical theories are interesting in themselves but have no application to the analysis of scientific discourse. Yet all of them, applicable and nonapplicable (or, if preferred, applied and as yet not applied) logical theories, either contain the above mentioned logical principles or are built in such a way that these are not violated. Suppose for a moment that science rejected such logical principles. If the logical principle of identity were given up we would admit the miracle that a statement could change of itself and would be unable to deal with the same proposition twice, e.g., along an argument. If the principle of noncontradiction were relinquished we would be unable to make definite assumptions, since we might be asserting their negates at the same time. Furthermore, contradictory hypotheses and evidences would be assigned the same worth and consequently the whole concept of test would become pointless. Finally, without the rule of detachment or some stronger principle of inference no assumption would be fertile: we would be unable to conclude, or at least to validate our conclusions. The above logical principles must therefore be accepted by science somehow and somewhere. It matters little if they are stated as axioms or as theorems, or even as rules; if they are placed in logic, in metalogic, or even in mathematics: we must have them if correct and incorrect formulas and inferences are to be discriminated. In short all science, whether factual or formal, presupposes a modicum of logical principles and all of formal science is logically (not psychologically or historically) independent of factual science.

There are other philosophical hypotheses relevant to factual science but it is not our (impossible) task to examine them all. The aim of the preceding study has been to show that scientific research logically presupposes certain wide scope philosophical hypotheses: that science is not philosophically neutral but partisan. It should not be concluded that science needs a solid philosophical *basis*, in the sense that a philosophy is needed to *validate* scientific hypotheses: it would be disastrous if, once again, the philosopher were given the final word concerning matters of fact. It is not a question of *basing* science on philosophy or conversely but rather of recognizing that the one does not *exist* without the other and that the one is not likely to *advance* without the other's support and criticism.

No philosophical principle provides a conclusive justification for a scientific hypothesis: a factual hypothesis is merely nonscientific if it is handled as true on a priori grounds or as incorrigible by fresh experience.

In particular, it is a waste of time to search for philosophical principles that could validate nondeductive scientific inference — metaphysical hypotheses such as "The future resembles the past", "Nature is uniform", or "Every event has a cause". It is not possible to validate and it is not wise to give rigidity to essentially insecure heuristic arguments such as the inductive ones; and it is not worth while to try it either, because inductively found hypotheses are shallow and are best (yet imperfectly) validated by tying them logically with other hypotheses. The various unstated philosophical presuppositions of scientific research provide no *ultimate* foundation for science but are themselves in need of support — and how else could they be justified if not for their guiding a successful (yet not infallible) search for truth?

While philosophy cannot claim to validate scientific ideas and procedures it can and must examine them: criticize and refine them as well as speculate on possible alternatives. And, while it takes a scientific mind to realize that most of philosophy is still in a protoscientific stage — to say the least — and to lay down the desiderata of scientific philosophizing, it takes a philosophical mind to realize the inescapable infirmities and some of the unexplored possibilities of science at every one of its stages. Needless to say, such a philosophical mind is not the exclusive property of philosophers: as a matter of fact every great scientist has had a philosophical outlook or other, however incoherent, and he had suffered philosophical pangs upon planning research lines and evaluating their results— which is not surprising since a great scientist is one who grapples with some deep problems, and deep problems call for deep hypotheses, that is, hypotheses that are somehow linked to philosophical views about the world and our knowledge of it.

Let us now turn to those hypotheses that are supposed, rightly or wrongly, to represent general patterns: namely, law statements.

Problems

5.9.1. Most scientists have had no formal training in logic: they reason (often incorrectly) in an intuitive way unless they cast their thoughts in mathematical forms, in which case mathematics takes care of the logic. Does this prove that science is independent of logic? And does this prove that scientists need no formal training in logic? *Alternate Problem:* Traditional empiricists and materialists have held that logic, far from being presuppositionless, presupposes a number of principles taken over from metaphysics, science, etc., such as the hypothesis of the independent existence of the world and the hypothesis of lawfulness. Study any doctrine of this kind, e.g. the system of "material or objective logic" proposed by J. VENN in *The Principles of Empirical or Inductive Logic*, 2nd ed. (London: Macmillan, 1907), Ch. I.

5.9.2. Some philosophers, notably HEGEL and his followers, have rejected the logical laws of identity and noncontradiction arguing that they do not allow for change. Examine this contention. For a criticism of the belief that logic has ontological commitments, see E. NAGEL, *Logic Without Metaphysics* (Glencoe, Ill.: The Free Press, 1956), Ch. 1. *Alternate Problem:* Discuss the philosophical and heuristic principles dealt with by physicist J. A. WHEELER in "A Septet of Sibyls: Aids in the Search for Truth", *American Scientist,* **44**, 360 (1956).

5.9.3. Is any of the five philosophical hypotheses dealt with in the text not only presupposed but also corroborated by scientific research? *Alternate Problem:* Are philosophical hypotheses testable? If so, how? In particular, how could we test ontological hypotheses concerning change?

5.9.4. Search for further philosophical presuppositions of science. *Alternate Problem*: Is it true that modern physics forces us to regard 'reality' and its relatives as dirty words?

5.9.5. T. GOUDGE, in *The Ascent of Life* (Toronto: University of Toronto Press, 1961), pp. 155ff., cites the following as metaphycisal presuppositions of the theory of evolution. (i) "There is an actual evolutionary past which can be scientifically known, but which can never be observed". (ii) "'Objects called *fossils* are evolutionary remains' is a true assertion". (iii) "Factors and laws now discovered to be operative in the biological domain were operative throughout all or most of the history of life". In what sense are these metaphysical hypotheses? *Alternate Problem:* Discuss the impact of the mechanistic ontology on biology and psychology — on the very kind of problems these disciplines dealt with — from the 17th century onwards.

5.9.6. G. SCHLESINGER, in *Method in the Physical Sciences* (London: Routledge & Kegan Paul, 1963), p. 46, states that the "principle of micro-reduction" ("The properties of physical systems should be explained in terms of the properties of its parts and not vice versa") is unjustifiable, being nothing but a prejudice involving "a partiality toward a method that is not objectively superior to its alternative in any way". Discuss this contention.

5.9.7. Extreme propounders of the so-called Copenhagen (or orthodox) interpretation of quantum mechanics hold that the latter establishes the impossibility of neatly separating the object of research from the observer; some go as far as asserting that the theory establishes the primacy of the mind over matter. If this were true, would physical research be distinguishable from psychological research? For a sample see E. P. WIGNER, "Remarks on the Mind-Body Question", in I. J. GOOD, Ed., *The Scientist Speculates* (London: Heinemann, 1962), p. 285: "it was not possible to formulate the laws of quantum mechanics in a fully consistent way without reference to the consciousness. All that quantum mechanics purports to provide are probability connections between subsequent impressions (also called 'apperceptions') of the consciousness, and even though the dividing line between the observer, whose consciousness is being affected, and the observed physical object can be shifted towards one or the other to a considerable degree, it cannot be eliminated". For criticism see M. BUNGE, *Metascientific Queries* (Springfield, Ill.: Charles C. Thomas, 1959), Chs. 8 and 9, and *Foundations of Physics* (Berlin: Springer-Verlag, 1967), Ch. 5.

5.9.8. Mechanism can be ontological and/or methodological. *Ontological mechanism* claims that all reality is just (or at least at bottom) physical. *Methodological mechanism* (or rather *physicalism*) is the strategy of applying the methods and theories of physics and chemistry to biology as far as possible, without making ontological commitments. Which, if any, of the two kinds of mechanism is favored by biology? And can either kind of mechanism be stretched over to psychology and sociology? *Alternate Problem:* Does scientific research leave irrational residues as claimed, e.g., by E. MEYERSON?

5.9.9. Philosophical individualism, like mechanism, can be ontological and/or methodological. *Ontological individualism*, or nominalism, holds that there are only individuals, not wholes, let alone classes. *Methodological individualism* says only that wholes can be understood if analyzed into their parts; e.g., that social trends must be analyzed as the outcome of the actions of groups, and in turn the actions of groups must be analyzed into individual actions. Is individualism of either kind employed in science? And does methodological individualism oblige us to give up the search for, say, societal laws on their own level? *Alternate Problem*: Contrast the contemporary search for essential properties and essential relations to Aristotelian essentialism.

5.9.10. One of the philosophical assumptions entrenched in science is that material objects on the macroscopic scale persist between observations. There is no possible proof of this principle by experience: only indirect evidence will support it, as well as the lack of ground of the alternative hypothesis. Does it follow that the hypothesis is not truer than the hypothesis that material objects cease to exist while unobserved, or that they turn into objects of an altogether different kind unless we keep an eye on them? If this were true we might stick to the persistence hypothesis only by stipulating that the simplest hypothesis is to be chosen — a prescription as unempirical as the hypothesis it is supposed to salvage. Thus, for example, B. RUSSELL wrote in *Human Knowledge: Its Scope and Limits* (London: George Allen & Unwin, 1948), p. 343: "Suppose I were to set up the hypothesis that tables, whenever no one is looking, turn into kangaroos; this would make the laws of physics very complicated, but no observation could refute it. The laws of physics, in the form in which we accept them, must not only be in agreement with observation, but must, as regards what is not observed, have certain characteristics of simplicity and continuity which are not empirically demonstrable". Is there no way out? Does the kangaroo-table conjecture satisfy the condition of foundation? And does it entail testable consequences different from the table-table hypothesis?

Bibliography

BERNARD, C.: Introduction to the study of experimental medicine, transl. H. D. GREENE. New York: Macmillan 1927.

BEVERIDGE, W. I. B.: The art of scientific investigation, Ch. IV. New York: W. W. Norton & Co. 1950.

BUNGE, M.: Intuition and Science, Ch. 3. Englewood Cliffs, N. J.: Prentice-Hall. 1962.

— The myth of simplicity, chs. 4—8, 9 and 10. Englewood Cliffs, N. J.: Prentice-Hall 1963.

DUHEM, P.: The aim and structure of physical theory (1914). New York: Atheneum 1962, part II, ch. VII.

HANSON, N. R.: The logic of discovery. J. Philosophy, **55**, 1073 (1958). Criticism by D. SCHON, J. Philosophy **56**, 500 (1959) and rejoinder by HANSON, J. Philosophy **57**, 182 (1960).

NAVILLE, E.: La logique de l'hypothèse. Paris: Alcan 1880.

POINCARÉ, H.: Science and hypothesis 1905. New York: Dover 1952.

POPPER, K. R.: The logic of scientific discovery 1935, chs. IV and X. London: Hutchinson 1959.

— Conjectures and refutations, chs. 1 and 8. New York: Basic Books 1962.

SCHILLER, F. C. S.: Hypothesis. In: C. SINGER (ed.), Studies in the history and method of science, vol. II. Oxford: Clarendon Press 1921.

WILSON, E. BRIGHT: An introduction to scientific research, Sec. 7.5. New York: McGraw-Hill 1952.

WISDOM, J. O.: Foundations of inference in natural science, chs. III and IV. London: Methuen 1952.

WOODGER, J. H.: Biology and language, Lect. I, Sec. 1 and Lect. II, Sec. 1. Cambridge: Cambridge University Press 1952.

Chapter 6

Law

A scientific law is a hypothesis of a kind: a confirmed hypothesis that is supposed to depict an objective pattern. The centrality of laws in science is recognized upon recalling that the chief goal of scientific research is the discovery of patterns. Laws summarize our knowledge of actuals and possibles; if deep, they will go as far as nibbling essences. In any case it is laws that theories unify and it is in turn by means of theories — webs of laws — that we understand and forecast events.

6.1. Variables and Invariants

Variety and change are all-pervasive in the world; moreover, change arises out of variety and variety is in turn but the outcome of change. Probably no two things and not two events are identical or remain self-identical in all respects, in every detail, and forever. Strict identity may not be of the real world: identity in all respects, whether of coexistents or of successives, is a simplifying hypothesis — one without which no science would be possible. If two real objects (things or events) seem exactly alike or do not seem to be changing or about to change in at least one respect, we may assume that our appearances are wrong.

More specifically, we may hypothesize: (i) that experiential identity results from a failure to discriminate tiny but real differences among coexistents or successives, and (ii) that our mistake may eventually be corrected by finer observation and deeper analysis. These two assumptions are methodological rather than either scientific or metaphysical, and they are meant to apply not only to large-scale (e.g., historical) happenings but also to atomic objects. The latter differ from one another if only as regards position in the various fields in which they are embedded — which is an indirect way of saying that they differ at least in the interactions they hold with the rest of the universe.

*That every real object is unique and unrepeatable in all respects, is not a scientific but a metaphysical (ontological) hypothesis. But it is a grounded rather than a wild conjecture. To be sure, it is not attested by scientific practice, which includes purposely blurring minor distinctions in order to bring to the fore the essential sameness of all the members of a given natural class. But it can be justified by a scientific (e.g., physical) analysis of real systems, which shows that even the so-called indistinguishable particles, such as the electrons in an electron gas, are different in some respects: otherwise we would not be able to ascertain that they are distinct objects and consequently we would be unable to count them (in an indirect way). In this context 'indistinguishability' is not synonymous with 'identity' but amounts to lack of individuality. (For example, two electrons in a system can be exchanged without there occurring a change of system or even a change in the state of the system: the electrons are exchangeable although they are not identical.)

*We accordingly reject Leibniz' principle of the *identity of indiscernibles* for being near-sighted: from our inability to discriminate between two objects — an inability which may be temporary — we cannot conclude to their identity. On the other hand we accept the reciprocal principle of the *indiscernibility of identicals:* if two objects are identical then they are indiscernible. (Indiscernible $=_{df}$ empirically indistinguishable.) This principle is pointless for conceptual objects; and it is vacuously true for material objects because the condition is never exactly realized though often assumed. Identity is always partial in the real world and our principle of the uniqueness of every existent does not preclude *partial identity*, identity in at least one respect, or even *near identity* — identity in every respect save one. Partial identity is the basis for classifications, generalizations, and laws expressing the patterns or invariants of things and events amidst variety and change. Strict identity is an indispensable pretence.*

Consider a system of atoms of the same chemical species, e.g. helium, all of them in the same state, e.g. the ground state of energy. These objects will then be identical as regards both chemical species and state: these two properties will be *constant* rather than variable in the present context. Yet there will be differences among these otherwise identical atoms; for example, no two atoms will have exactly the same position in space. In other words, position is a *variable* that may take on a number of values — in fact, a nondenumerable infinity of values. In principle every one such object — and in general every thing and every event — can be exhaustively characterized by specifying the values of some of the variables representing its properties; in the first place, but not exclusively, position in spacetime relative to some reference frame. Such an exhaustive characterization or *identification* of (not with) a real object is far from exhausting the object's properties, just as having the identification card of a person does not afford a knowledge of his personality. *Also, the possibility of identifying and naming real objects by specifying the values of some of their variables does not mean that the real objects *are* nothing but bundles of properties. In fact, every property in the real world is the property *of* something. Thus, e.g., if we write merely 'M' for mass, the context makes it clear that we mean the mass of a thing of some kind, so that our first task in a logical analysis will be to disclose the object variable, i.e. to write '$M(x)$' instead of 'M' (syntactical analysis) and to indicate what is the domain of individuals over which the object variable x ranges (semantical analysis). The elimination of physical objects in favor of bundles of properties — as proposed by some philosophers — results from a failure to logically analyze the properties occurring in science, all of which contain object variables although they are not normally mentioned explicitly.*

The hypothesis that no two real objects in the world are identical in all respects, in every detail, and forever, can be rephrased in the following way: "Given any two real objects, there exists at least one variable which has not exactly the same values for both". Admittedly, this principle is irrefutable. We state it because it is grounded and because of its fertility: it challenges the scientist to seek for diversity beneath the apparent identity. But we also postulate this other ontological hypothesis: "Given any two real objects, there is at least one variable with a value shared by both." If every real object were entirely different from every other real object, i.e. if all classes were unit sets, science would be impossible and the concepts of variable would be next to useless: proper names would suffice for identification.

The concept of variable enables us to accurately discriminate diversity and uncover partial identity: it serves both the purpose of accounting for variety and change and of accounting for the patterns of variety and change. The further metaphysical hypothesis, that variety and change are neither boundless nor chaotic, is the assumption that there exist constant relations among certain variables — i.e. that there are laws. But before approaching the concept of law it will be convenient to further examine the concept of variable. The term 'variable' covers a family of concepts. What is common to all members of this family is that a variable may take on at least one definite (fixed, particular) value.

In logic we are interested essentially in three kinds of variable: propositional variables, individual variables, and predicate variables. *Propositional variables* are symbols denoting indeterminate propositions, or schemata the values of which are determinate propositions. Thus, in "$p \rightarrow q$" the propositional variables p and q do not stand for given propositions but for any propositions whatsoever: every propositional formula covers an infinity of propositions. (Accordingly, 'propositional calculus' is a convenient misnomer: it is convenient because it is short, inadequate because it does not deal with propositions but with propositional variables. But this is just a pedantic remark: every scientific theory is general to some extent, and generality requires the introduction of variables.) *Individual or object variables* are symbols denoting indeterminate individuals, such as 'x' in the formulas "x is long" and "the length of x is y cm". These variables are called individual because they range over individuals: they stand for unspecified individuals in a set. Numerical variables, a subclass of individual variables, are symbols which designate any elements of a set of numbers. Finally, *predicate variables* are symbols designating indeterminate or unspecified properties, whether of individuals — as in "$P(137)$" — or of further properties — as in "P(mechanical properties)". In order to avoid nonsense and paradox (e.g., "How hard is hardness?") it is ruled that whatever is predicated must be predicated of a *lower-order* variable: we thus have a whole hierarchy of variables: first order predicates or predicates designating properties of individuals; second order predicates or predicates designating properties of first order predicates, and so on. Thus, 'harder' is a first order predicate, whereas 'asymmetrical', which applies to 'harder', is a second order predicate.

Every scientific formula is or can be analyzed into a propositional function, i.e. a certain combination of variables of various orders. Thus, "Lethal genes are manifested in the phenotype" is a first order formula, i.e. one predicating something about individuals; on the other hand, "Center of mass is a nondistributive (nonhereditary) property" is a second order formula. In short, as regards its logical structure every scientific formula is *a formula of the predicate calculus*. It is as if this logical theory had been "there" all the time waiting to be filled with a factual content. But this is just a fiction: there is no "there" for ideas; besides, logical formulas are invariant under changes in interpretation. In factual science, furthermore, we are not interested in variables in general but in specific variables and specific relations among them. In scientific contexts 'variable' designates "property", "trait" or "aspect". We shall distinguish the following kinds of extralogical (factual) variables:

1. Qualitative variables or dichotomic predicates, such as "solid". Anything is at a given instant of time in the solid state or it is not in it, whence the name 'dichotomic variable'. But, of course, if we are investigating just solids then non-solids are not taken into account and „solid" becomes a constant. Qualitative variables occur not only in factual science but also in mathematics. Thus, when we think of the set of all plane triangles we actually use the notion of a variable ranging over such a set: this, in effect, is the case of any phrase beginning with 'Consider an arbitrary plane triangle ...'

2. Ordinal variables, such as "hardness" and "cohesiveness of a social group". The values of ordinal variables can be ordered; but the variables themselves cannot be subjected to arithmetical operations such as addition. Thus, e.g., if we have rated the pleasure value of three candies by assigning each a number between 1 and 3, we may sum up the result of this ordering operation in the obvious statement: "$3>2>1$", which we regard as an abbreviation for "Candy number 3 is better than candy number 2, which is better than candy number 1". But this does not entitle us to infer that candy number 3 is thrice as good as candy number 1 or one and a half times as good as candy number 2: this would attest to our confusing an ordinal variable with a cardinal variable. (We shall come back to this in Vol. II, Sec. 13.1.)

3. Cardinal variables or *magnitudes* or simply *quantities*, like the size (numerosity, cardinality) of a population and the habit strength. Magnitudes are also called numerical variables because they range over a set of numbers, i.e. their range of variation is a set of numbers; but this name is misleading because numerical variables are a component of magnitudes. In fact, the structure of the simplest magnitude ist: "$P(x) = y$", where 'x' designates the object variable and 'y' the numerical variable. The numerical variables of magnitudes can be subjected to mathematical operations, though with restrictions: thus, we may the add populations of two cities but not their population densities.

Three further concepts we need in order to characterize the concept of law are those of independent variable, dependent variable, and parameter. In an expression like

$$y = mx + n \qquad [6.1]$$

'x' and 'y' are usually called the *independent variable* and the *dependent variable* respectively, whereas 'm' and 'n' are *parameters*. The distinction between independent and dependent variable is contextual and, more precisely, it is relative to the formula in which the variables occur. Every explicit function sending x's to y's can, under certain conditions, be inverted to yield x in terms of y; thus, [6.1] is equivalent to "$x = y/m - n/m$". In science the independent variable is often (not always) the *control variable*, i.e. the variable whose values can be assigned or changed at will (within bounds). This pragmatic distinction has an ontological root: changes in the values of the control variable are usually called *causes*, whereas the resulting changes in the values of the dependent variable are called *effects*. Example: by varying the volume of our radio set (cause) we may bother our neighbors to any desired extent (effect). Finally, the name of parameter is given to an arbitrary constant. In the example above m and n are parameters because their values are assigned independently from those of x and y. In other words, parameters are variables that are kept frozen in a given context.

For every pair of values of m and n, save the trivial case $m = n = 0$, the function $y = mx + n$ may be regarded as the analytic representation of an infinite straight line on the coordinate plane (x, y). If m and n are allowed to take different values we get an infinite set of straight lines; each member of this set may be visualized as the law of an individual. (See Fig. 6.1.) The linear relation [6.1] can then be interpreted in the following way: "For every pair $\langle m, n\rangle$ any given y-value is related to the corresponding x-value in the way: $y = mx + n$". This is a propositional function with the numerical variables m, n, x and y, where m and n are bound by universal quantifiers while x and y are free, i.e. can be specified in any way.

So far [6.1] is not a scientific law but just a blank or *schema* for a scientific law, because the variables occurring in it have only an arithmetical interpretation. It is only when at least the variables proper (x and y) are *interpreted*, not just as unspecified numbers (which they are already) but as the numerical variables of *properties* of some real system, that [6.1] may become a scientific law. An unlimited number of interpretations of

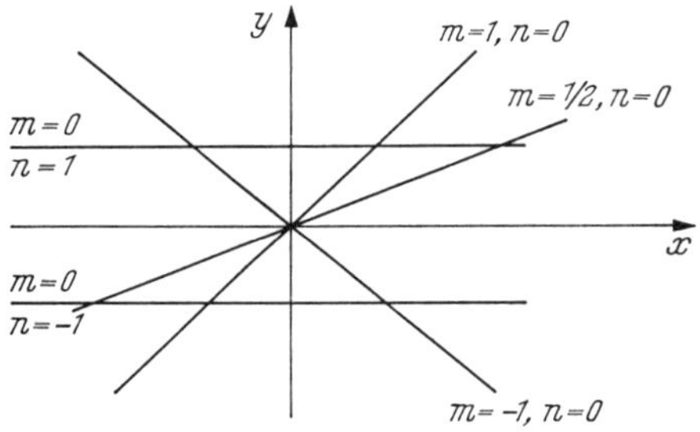

Fig. 6.1. The linear function [6.1] represents an infinite set of straight lines on a plane

any given law schema are possible; some will be true, others false, others just meaningless in a given context. In other words, every law schema can be assigned a potential infinity of interpretations in factual terms.

A possible interpretation of [6.1] is the one determined by the following semantical rules: 'y' designates the numerical values characterizing the successive positions of a freely moving mass point, 'x' the duration of the motion from a conventional beginning (zero time, i.e. $x = 0$), 'm' the velocity and 'n' the position at the initial time. Adopting the usual symbols, suggestive of this particular interpretation of the law-schema [6.1], we have

$$s(t) = vt + s_0. \qquad [6.2]$$

(We have written '$s(t)$' for the distance to indicate that the latter is a function of the duration t.) This, which is among the simplest quantitative laws, clearly exhibits the chief trait of every law, namely, that of being a *constant relation* among two or more variables referring in turn (at least partly and indirectly) to *properties of real objects*. The constancy consists in that the particular (linear) relation between distance and duration does not change either in time or for different individuals (specified by particular values of the parameters v and s_0).

Formula [6.2] is a *general law* since the values of the parameters v and s_0 occurring in it are not specified. We can form infinitely many couples of numerical values $\langle v, s_0 \rangle$ of the parameters, one for each *possible* freely moving mass point. Hence [6.2] subsumes an *infinity of special laws*. Moreover, since the above mentioned parameters range over real numbers, and these make up a continuum, the set of special laws condensed by [6.2] is a *nondenumerably infinite set of special laws*. This is characteristic of all general quantitative laws: they not only encompass an infinity of individuals but also an infinity of circumstances.

These infinities would be superfluous in the case of an empirical generalization of the kind envisaged in inductive logic, since experience can give only a finite number of data. Laws are not summaries of experience: laws are supposed to reconstruct objective patterns, and this objective reference, this pointing out to a reality beyond experience, requires the

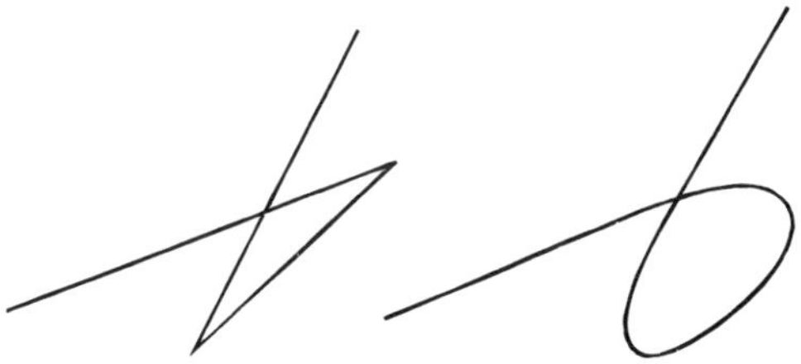

Fig. 6.2. Laws as restrictions on logical possibilities: (i) an impossible motion for an aircraft flying at constant velocity; (ii) a possible motion for the same object

introduction of infinities. In fact, a law such as [6.2] specifies the physically *possible* motions of a class; by the same token it rejects as physically *impossible* any motion of the same kind that would take, say, less time than the one it allows. Generally, every law statement specifies a class of *possible facts*; the complement of this set is the class of logically possible but physically impossible facts (see Fig. 6.2). Both sets, the one of possible facts and the one of impossible facts, are conceivably infinite. Any law involving numerical variables excludes or "prohibits" many more facts than it "permits"; the stronger the "prohibition" the narrower the class of possible facts. But such "prohibitions" must, of course, be understood in a metaphorical way: laws do not tower above facts.

The values of the properties related by a law may change from individual to individual and from moment to moment. Thus, e.g., a developing embryo — a unique process that will never recur in exactly the same way — has a given size for each age, yet for all members of a given species the size-age *relationship* is assumed to be the same at least on the average even if the specific function relating both variables is not accurately known. That is, we state the universal law schema: "For every x, if x is a developing embryo of a given species, then the volume of x at any time t is a definite function of t". (Symbolically: $(x)\,[E(x)\,\&\,S(x) \rightarrow (V(x,t) = F(t))]$.) In contrast with the previous case we have now taken care to indicate the object variable; on the other hand we have left the function

F unspecified. Embryology writings are unlikely to contain full formulations of law schemas such as the preceding: usually they will write only the *consequent* of the whole conditional and will disregard the object variable x, i.e. they will write: "$v=F(t)$" for the relation between the numerical values of the volume and the age, and they will state in words that this law schema is supposed to hold at any time for every member of the set of developing embryos of any given species. The neglect of the antecedent and of the object variable is justifiable for practical reasons but it can be misleading.

Notice that our law schema does not assert that all embryos have the same initial size and consequently the same size at any age: for all we know no two fertilized eggs may have exactly the same number of molecules. That is, our law schema is not the type "Whenever A is the case B is the case", in which 'A' and 'B' designate particular facts. Such generalizations belong to ordinary knowledge rather than to science. Scientific laws do not state conjunctions of facts but relations among selected traits (variables); and they do not assert the sameness of individuals but the invariance of certain relationships irrespective of changes in the values of the object variables. In particular, a law statement involving time need not be a law of recurrence: recurrent pattern is just a proper subclass of pattern. All that a law of science asserts is that individual differences in certain respects fit certain patterns. In short, a law is a pattern of variety and change.

We close this section with a preliminary characterization of the concept of scientific law, which will be refined later on: A scientific law is a confirmed scientific hypothesis stating a constant relation among two or more variables each representing (at least partly and indirectly) a property of concrete systems.

Problems

6.1.1. What difference is there between exhaustive characterization and exhaustive knowledge? *Alternate Problem:* A point in spacetime is characterized or identified by an ordered quadruple of real numbers. Does this entail confusing concrete objects with abstract objects?

6.1.2. Examine the sentence: 'A change Δy in the dependent variable corresponds to a change Δx in the independent variable'. Are changes in the variables meant or rather differences in the values of the variables? See W. V. Quine, "Variables Explained Away", *Proceedings of the American Philosophical Society*, **104**, 343 (1960). *Alternate Problem:* Examine Kant's contention that all the properties of things belong, not to the things in themselves, but to their appearances to us. See his *Prole-*

gomena to any Future Metaphysics (1783; Indianapolis: Bobbs-Merrill, 1950), espec. Sec. 12, Remark II.

6.1.3. Comment on the following fragment from A. RAPOPORT, in L. GROSS, Ed., *Symposium on Sociological Theory* (Evanston, Ill.: Row, Peterson & Co., 1959), p. 351: in sociology the process of selection of variables "is so laborious and involved that it often constitutes the bulk of the social scientist's effort, and so he hardly ever gets around to stating 'postulates'. He must first relate his terms to referents. These referents cannot be simply exhibited; they must themselves be abstracted from a rich variety of events, generalizations, and relations. By the time a number of these referents have been so abstracted and christened, one already has a bulky 'system' before the work of seeking out 'laws' has ever begun. Such 'systems', particularly in sociology, are sometimes taken to be 'theories' ".

6.1.4. Take any mathematical form (function, equation, etc.) other than [6.1] and interpret the symbols occurring in it in two alternative ways so as to obtain two candidates to law statement. *Alternate Problem:* Analyze the law statement: "The total momentum of a system of particles subjected to nonfrictional forces is conserved (constant in time)." Symbolize it taking care of identifying the object variable (in this case a numeral) and of stating the antecedent of the conditional.

6.1.5. Examine the way B. RUSSELL, in *An Inquiry into Meaning and Truth* (London: George Allen and Unwin, 1940), Ch. VI, "abolishes particulars" replacing them by Platonic universals. In particular, examine the contention that "wherever there is, for common sense, a 'thing' having the quality *C*, we should say, instead, that *C* itself exists in that place, and that the 'thing' is to be replaced by the collection of qualities existing in the place in question. Thus '*C*' becomes a name, not a predicate" (p. 98).

6.1.6. Most generalizations in sociology and history concern unanalyzed facts: they express relations among events rather than relations among properties referred to by more or less sophisticated (nonobservational) variables. Could this account for the immaturity of these disciplines? *Alternate Problem:* Many philosophers of science think that there is no essential difference between a common sense generalization of the form "Whenever *A* is the case *B* is the case" and a scientific law. Could this account in part for the immaturity of the philosophy of science?

6.1.7. When physicists maintain that the so called fundamental particles (e.g., electrons) are identical or indistinguishable, do they mean (i) that they are not objectively distinct objects, or (ii) that although they may be objectively distinct we have no means for discriminating among them, or (iii) that they are distinct but can be exchanged without the whole system (of which they form part) suffering any change? *Alternate Problem:* Examine K. POPPER's view that the more a law statement prohibits the greater its content is. Study the case of stochastic (probability) laws and the case of an imaginary region where the law "Nothing changes" held, i.e. where all change were "prohibited". And distinguish laws from constraints (e.g., "Pressure= const").

6.1.8. Elucidate the concept of near identity such as it occurs in the proposition "*A*ny two complete helium atoms are almost identical". *Alternate Problem:* Study the biochemical and genetical explanation of the uniqueness of the individual.

6.1.9. What would be the point of searching for laws (i) if there were neither variety nor change, as PARMENIDES thought?; (ii) if we did not assume the existence of constant relations among variable relata?; (iii) if we had not somehow formed the concept of constant relation or the concept of invariant of a transformation?, or (iv) if individuality were incompatible with membership in classes?

6.1.10. The idea of natural law was conceived by a few thinkers in Antiquity and in the Middle Ages but it did not become widespread until about DESCARTES' time. The evolution of the concept of law was apparently correlated with the concept of social constraint — but little is known in this regard. Trace the probable evolution of the concept of law of nature from Antiquity to our time. For interesting glimpses see E. ZILSEL, "The Genesis of the Concept of Physical Law", *Philosophical Review*, **51**, 245 (1942) and J. NEEDHAM, *Science and Civilization in China* (Cambridge: University Press, 1956), vol. II, Ch. 18. *Alternate Problem:* Examine the doctrine that the goal of science is the reduction of apparent change and diversity to essential identity and permanence. What does it presuppose concerning the relation of diversity to oneness? And what view of explanation does it lead to? See E. MEYERSON, *Identity and Reality* (1908; New York: Dover, 1962).

6.2. Searching for Law

Scientific laws do not relate all possible aspects but just a finite number of selected variables at a time. (But a law may contain an infinity of variables of a certain kind.) What traits or variables should be selected in the search for lawfulness is determined primarily by our outlook, both generic and specific. The Democritean world outlook suggests that, except in the sciences of man, secondary properties (sensible qualities) do not really matter: that in the natural sciences we should accordingly select primary properties such as wavelength in preference to secondary properties such as color. It took physics a long time to discover a bunch of fundamental, hence transphenomenal variables such as "mass", "electric charge" and "field strength". Why this is so should be clear: physics attempts to account for observable properties in terms of objective and fundamental variables, which rarely represent observable traits. No wonder psychologists and sociologists, who deal mostly with non-observable aspects of human behavior, are barely beginning to discover fundamental variables for the explanation of mind and society. Unfortunately one does not really know that a variable is fundamental until after it is found to occur in a set of high level (strong) law statements on the basis of which derived variables can be expressed as functions of fundamental variables. The search for fundamental variables goes hand in hand with the search for high-level constant relations among them, i.e. rich laws. Whence the efforts of so many behavioral scientists to discover the basic variables through observation and without hypothesizing lawful relations is a waste of time.

The simplest relation among two variables is, of course, the relation of *independence*, i.e. the lack of systematic relation. An example of a statement of mutual irrelevance among two variables is this: "The acceleration of a freely falling body does not depend on its mass". Would this statement qualify for the status of scientific law, as characterized in Sec. 6.1? It certainly would, since the relation of independence or irrelevance is a kind of relation not to be confused with unrelatedness. If we say that y is independent of x we mean that the values of y stay the same no matter what values may x take; a number of relations can be set up among mutually independent variables, none among unrelated ones: even the statement that two variables are unrelated establishes a relation among them. *In other words, when we say that the acceleration of a freely falling body is independent of its mass we mean something like "$da/dm=0$", not "$(R)-[R(a, m)]$".* Now, an unlimited number of statements of mutual irrelevance could be set up concerning any given set of variables: for instance, we might truly say that the acceleration of a freely falling body is independent of its color, texture, price, aesthetic value, and so on without end. Consequently a criterion is needed for sorting the laws among all possible statements of mutual irrelevance. No definite criterion is in use save this: A statement of irrelevance, if confirmed, may be upgraded to law if it conflicts with statements of relevance made in a rival theory or proposed intuitively — i.e. when the variables concerned, which to begin with had seemed to be mutually dependent, turn out to be independent.

Next to irrelevance comes, of course, relevance. We say that two variables are *mutually relevant* in a given domain if, and only if, a change in the values of one of the variables makes a difference in the values of the other. The simplest kinds of mutual relevance are those of favorable and unfavorable relevance, but these by no means exhaust the concept of mutual relevance. We can say that armament is favorable to world tension and conversely, and that old age is unfavorably relevant to metabolism. But these would hardly qualify as law statements for being much too vague. Law statements are much stronger, hence less facile truths.

*A quantitative statement concerning the degree of correlation between two variables comes closer to a law. The statistical concept of *correlation*

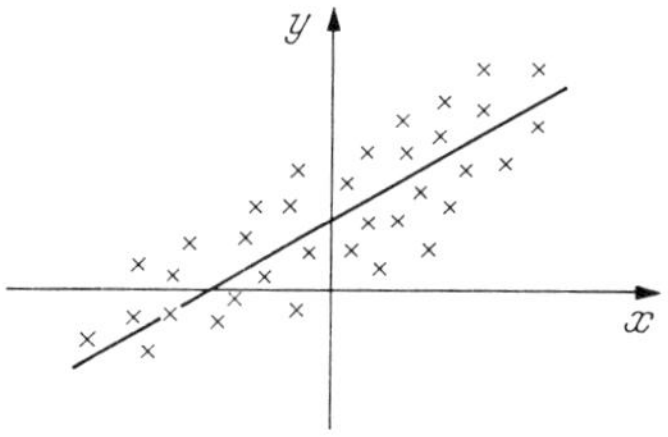

Fig. 6.3. Linearly correlated variables. The straight line (linear regression line) is something like the average around which the data cluster

$$r(x, y) = \frac{(1/N)\sum(x - x_i)(y - y_i)}{\sigma_x \cdot \sigma_y}$$

coefficient $r(x, y)$ between the variables x and y is one of the elucidations of the intuitive concept of correlation. And any statement assigning $r(x, y)$ a definite value is stronger than a qualitative statement of favorable or unfavorable relevance. In fact, if $r(x, y)$ is near $+1$ we say that x is favorably relevant to y and conversely, whereas if $r(x, y)$ is near -1 we say that x and y are unfavorably relevant to one another. If $r(x, y)$ is exactly $+1$ or -1 we obtain, as a special case, the linear relation [6.1] between x and y. (See Fig. 6.3.) Variables that are functionally related among each other are correlated but not conversely: if the coefficient of correlation is near $+1$ or -1 we may suspect the existence of a law but this is all. We cannot expect to find always a law underneath a constant statistical correlation: we must be prepared to find various degrees of tightness in the relations among variables, particularly among observational variables. If the correlation is high, i.e. if the data are near a line such as the one (linear regression line) shown in Fig. 6.3, we are justified in interpreting the line as a *trend* — not however as a law proper.

*We cannot assume laws underneath trends without further ado, but we shall always search for grounds in favor or against the hypothesis that a given trend is actually a law blurred by random effects. In fact, the straight line in Fig. 6.3 can occasionally be interpreted as the law that *would* hold were it not for random disturbances, i.e. as a kind of "message" distorted by a more or less strong random "noise". But in order for this interpretation to be reasonable we need some ground, if possible both theoretical and empirical. More precisely, we may assume that a trend line hides a law only if (i) the data tend in fact to collapse to the line when the disturbances are rendered negligible (e.g., by cooling in the case of physical systems, and by choosing homogeneous groups in the case of social systems); or (ii) a theoretical model is available which accounts for the central line underlying the random process.

*At any rate the computation of correlation coefficients and the fitting of regression lines should not be taken for a *law-finding method*, as is so often the case in the behavioral sciences. When a linear regression model is assumed and the parameters are computed from the data, the central law that is supposed to run through the "noisy" (scattered) information is not found but is assumed beforehand. No amount of statistical data processing produces new hypotheses by itself, let alone laws; in general no amount of technique, whether empirical or mathematical, saves us the labor of inventing new ideas, although it can effectively conceal the lack of ideas.*

The mutual relevance of variables is either hypothesized or discovered by chance before the precise relations (laws) among them can be found. That is, a *hypothesis schema* concerning the relatedness among certain variables is found first, and then filled. Such schemas, when not hit upon by chance, can be suspected by any of the following procedures. One: theoretical considerations may suggest that a given variable is relevant to certain other variables; thus e.g., our coarse sociological background susggests that occupation is relevant to most other variables of interest to sociology. Two: an imaginary experiment is often possible to suggest relatedness: our knowledge is frequently sufficient to imagine what might happen if a given variable were missing, or if its values changed in a prescribed way.

Yet such procedures can no more than suggest the existence of a law or systematic (nonaccidental) relationship among two or more variables. The suspicion must be empirically tested, and this is done either statistically or experimentally, according to the nature of the system and to the possibility of effectively controlling some variables. The statistical test of a hypothesis of relevance may consist in inquiring, on the basis of observational data, whether there is a significant correlation among the suspected variables. And the experimental test will consist in deliberately changing the value of one of the variables and watching whether, and if so how much, the suspect partners are thereby affected.

Note that so far we have not been concerned with law statements but rather with hypotheses of relatedness, those programmatic conjectures we frame before formulating law statements. The coining and checking of such hypotheses, important as it is, does not replace the search for and test of scientific laws. Thus e.g., the mere statement that a guessing score is above chance (e.g., that it is right more than half the time) is not a law statement and accordingly its confirmation establishes no law statement. At most it would support the plan of searching for laws explaining the presumptive anomaly. This is, incidentally, one of the many reasons why parapsychology does not qualify as a science: it contents itself with making vague assertions of relatedness without specifying the relations, i.e. without stating laws — let alone testing them. (See Sec. 1.6.) No laws, no science.

Once a hypothesis of relatedness is established, the task of stating a precise relation is faced, and after this has been completed the test of the hypothesized law statement will be undertaken. Unfortunately there are no recipes for finding precise law formulas save of the lowest level kind. Careful observation, so often commended as the way to law, will not suffice by itself because laws are not observable: we observe at most selected aspects of phenomena which we report in data, but a heap of data is precisely what a law statement is supposed to account for, usually in terms of transempirical variables. Moreover, the flux of anyone's experience is lawless: a sequence of experiential items fits no law. In order to obtain laws we must therefore posit entities behind tangible bodies and properties not accessible to the senses though related in a lawful way to sensible qualities.

Careful *observation* with some hypotheses in mind is a way to low-level laws, i.e. to observational hypotheses — and to such only. Suppose we suspect there is a systematic relation between the percentage of a certain chemical substance C in protoplasm, and biological species — i.e., that the percentage of C depends on the species. A preliminary test of this hypothesis of relatedness may consist in testing for C in a few specimens of distant orders. The next step might be to find out the exact C-content in a random sample of a given species S. In this way we may be able to state a smallish law of this form: "The average C-content in S is s per cent with a standard deviation σ". We may next try to relate the C-content of closely related species in an attempt to discover the philogeny of the given species, or we may investigate possible environmental influences (e.g., the effect of the salinity of water on the salt-content of various fish species). In this way we may establish thousands of more or less insignificant low-level laws. But unless we face the problem of finding the role of C (e.g., in metabolism) and unless we try to account for the differences in C-content among different species we will just add to the boring swelling of protoscientific literature, in which isolated data and isolated empirical generalizations pile up without purpose.

A common technique for finding low-level laws relating a few quantitative variables (magnitudes) is this, which we shall describe with reference to two variables. One first gathers quantitative empirical data and tabulates them. Then one applies a standard *interpolation technique* — with the help of a computer if the variables are many — and obtain the simplest polynomial fitting the data. A geometrical interpretation of the input (data) and the output (polynomial) is this: every datum is a point in a space of as many dimensions as variables, and the polynomial is the smoothest figure (line or surface) passing near the "empirical points". The usual interpolation formulas for two variables yield polynomials of the $(n-1)$th degree for n data. Let us sketch the working out of one example.

Suppose we have found out that the angle of refraction r of a light ray depends on the angle of incidence i (relatedness hypothesis). We wish to

Table 6.1

i	r	i	r
0°	0° 00′	50°	22° 31′
10°	4° 59′	60°	25° 40′
20°	9° 51′	70°	28° 01′
30°	14° 29′	80°	29° 30′
40°	18° 44′	90°	30° 00′

find the precise law relating these two variables in the case of a given transparent medium and a given light color. We may then proceed as follows. We first perform angle measurements at intervals of 10 degrees with, say, 1 minute accuracy, and tabulate the results of our measurements. In this way we get a table like Table 6.1, which displays our findings. But this table, a summary of experimental results, is clearly insufficient: (i) it contains a finite number of data only, and (ii) it does not help to explain the refraction phenomenon. Consequently we search for a formula of the form "$r=f(i)$" covering infinitely many possible pairs $\langle i, r\rangle$ of angles. To this end we first plot the data on the $i-r$ plane and join the "empirical points" with a smooth line: this will give us an intuitive representation of the formula. (See Fig. 6.4.) We see that, up to about 30°, the linear approximation — dotted line representing the function "$r=0.5\,i$" — is quite good. From then on the discrepancy increases noticeably, reaching 40 per cent at 90°. Before the law was found KEPLER had guessed that a linear relation between the angle of incidence and the angle of refraction holds. This is common in the history of science: *first-order approximations*, i.e. the simplest hypotheses, are usually (not always) the first to be conjectured.

To improve on the first-order approximation we may add a quadratic term to the previous expression, i.e. we may put $r=0.5\,i+a\,i^2$, where a

should be a small negative number to curb the curve downwards. But we need not grope at random: GREGORY and NEWTON, among others, bequeathed upon us a mechanical interpolation technique by means of which our ten pairs of numbers in Table 6.1 can be made to fit a 9th degree polynomial. This procedure can be improved without limits other than those set by the sensitivity of our measuring instruments. In fact, we may next take one degree intervals, then one minute intervals, and so on until we hit the afore-mentioned limitations (which are both technical and physical). We accordingly get better and better fits though at the

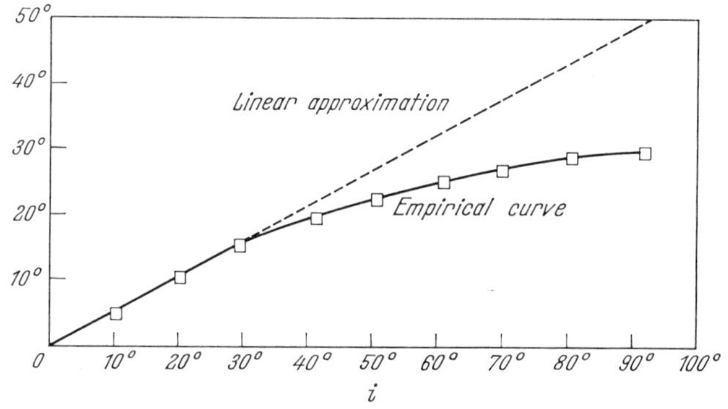

Fig. 6.4. Representation and generalization of Table 6.1 concerning the relation between the angle of incidence i and the angle of refraction r. Each square represents an empirical datum

price of dull complexity. For example, with one reading every one-tenth of a second, and some patience, we could obtain a polynomial with 54,000 terms. But this would take us no single step nearer the true hypothesis, which is SNELL's law.

SNELL's law can be stated thus: "The sine of the angle of incidence divided by the sine of the angle of refraction equals a constant [the refractive index of the pair of substances concerned]". In symbols: $\sin i/\sin r = n = \text{const}$. This law holds, with qualifications, not only for the particular pair of substances dealt with in our example, but for all known pairs of transparent media. It supplies some insight into the refraction phenomenon and it enjoys the support of the wave theory of light, since it is a theorem deduced in it. A law such as SNELL's cannot be obtained by an interpolation technique because it involves a nonalgebraic (transcendental) function, and transcendental functions (like the sine, the logarithm, and the exponential) can be expanded in infinite power series but not in finite polynomials. *Transcendental functions are infinitely more complex than the most complex algebraic function; they are only typographically simpler. It is certainly possible to quantitatively approximate any given transcendental function by means of a polynomial and to improve the approximation to any desired degree so that there remains no noticeable numerical difference between the exact function and its algebraic approximation; but the function itself remains essentially as different from the one-million-order approximation as from the first-order approximation. This difference may not matter for practical purposes; for example, a lens manufacturer can do very well with a second-order approximation to the SNELL law. Moreover, SNELL's law is *empirically indistinguishable* from the corresponding empirical generalization, as long as the interpolation process is carried far enough. But theoretically the difference is abysmal. First: whereas the polynomial covers and general-

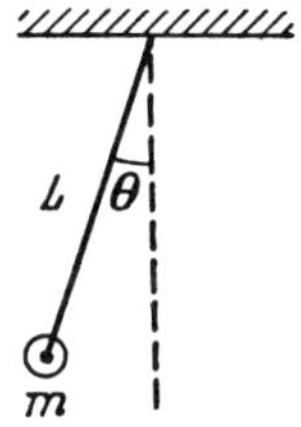

Fig. 6.5. A theoretical model of a pendulum swinging in a vacuum with small oscillations

izes a finite set of data, the exact law covers a potentially infinite set of facts. Second: we cannot systematize and explain any of the algebraic approximations to the Snell law, which are on the other hand entailed by higher-level principles such as FERMAT's principle of extremal (in particular, minimal) duration of light paths, or the even richer wave equations of physical optics; in other words, whereas the exact law is theorifiable the empirical generalization remains outside theory.*

In short, given a set of empirical data, infinitely many functions may be found to fit them, and by sheer arithmetic a good algebraic function (polynomial) can be built to fit the given data. There is no single and simple criterion — such as simplicity of some sort — for choosing among them. The chief selection criteria are: (i) goodness of fit; (ii) possibility of theorification (embedding in or expansion into a theory), and (iii) possibility of interpreting the constants occurring in the function. Polynomials as those an interpolation technique yields satisfy the first condition as well as desired but they do not satisfy the other two requirements: in the first place they are stray formulas rather than members of extended families (e.g., sin and log); in the second place they are ridden with purely numerical, factually meaningless constants.

*A different, but also limited technique helpful in the search for low-level laws is *dimensional analysis*. Suppose we wish to find the law of swing of the simple pendulum (see Fig. 6.5) and for some reason do not wish to use the only reasonable method for discovering low-level laws in explored domains: namely, applying some theory such as Newtonian mechanics. We begin by listing the suspect relevant variables and their corresponding dimensions:

Variable	*Symbol*	*Dimensional formula*	*Unit*
Period of oscillation	T	T	sec
Length of pendulum	L	L	cm
Mass of pendulum	m	M	g
Acceleration of gravity	g	$L T^{-2}$	$\mathrm{cm/sec^2}$
Angle of swing	θ	—	degree

In making this list we had a definite theoretical model in mind even though we did not use a theory. In fact, we dispensed with secondary properties, we eliminated draughts and even air, we assumed that the bob of the pendulum is suspended from a rigid frame by an inextensible thread, and so on; in other words we neglect as secondary the properties of the frame, the bob, the thread, and the environment, with the exception of those listed. Ours is, in short, an *ideal* pendulum and we seek the law of this ideal object. More precisely, we look for a relation $R(T, L, m, g, \theta)$ among the suspect variables, such that it remains *invariant under changes in units*. (The invariance of laws with regard to the choice of units has to be emphasized in view of the widespread mistake that units are essential to science. As a matter of fact the consideration of units does not enter until the test stage: see 13.5.) Suppose we solve this relation for the period of swing: $T=F(L, m, g, \theta)$. Now, a change in the unit of mass cannot be compensated for by a change in any of the remaining units because none of the suspect variables, except mass itself, depends on mass. Hence m cannot be a relevant variable and the former relation reduces to $T=F(L, g, \theta)$. If we now change the length unit, g will be affected, hence L and g will have to be combined in such a way that T does not change; in other words, any change in L must be compensated for by a change in g upon the adoption of a new unit of length. The only combination satisfying this requisite is L/g: this quotient does not depend on length. Accordingly, $T=F(L/g, \theta)$. Now, since θ is dimensionless it may occur in any form whatsoever as far as dimensional analysis is concerned; hence we may separate it as follows: $T=F(L/g) \cdot f(\theta)$. But L/g must occur in such a way that the dimension of the second hand member be the same as the dimension of the first hand member (principle of dimensional homogeneity). Since the dimension of L/g is T^2 (see the foregoing table), we must require that L/g occurs under the square root sign, i.e., that $T=\sqrt{L/g} \cdot f(\theta)$ where $f(\theta)$ remains undetermined. This is as far as the method of dimensional analysis can take us. Experiment teaches us that, for small swing angles, $f(\theta) \cong 6$. And analytical mechanics shows that, under the same conditions, this hitherto empirical constant is exactly 2π, a constant that, for large angles of swing, must be replaced by a function of the angle. It is obvious that no mass of experimental data could ever yield either the square root function or the exact value of $f(\theta)$.*

In conclusion, there are definite techniques for summarizing and generalizing data, i.e. for obtaining *low level law statements*. But these methods (i) presuppose that the relevant concepts (variables) are at hand, (ii) they employ more or less sketchy theoretical models of the object concerned, and (iii) they are of limited scope, if only because they provide no relations to further law statements: they yield stray low level hypotheses. The strong hypotheses occurring as initial assumptions of theories cannot be obtained by the techniques illustrated above. There are no known rules for inventing either high level concepts or the law statements that tie them up: unlike the finding of empirical generalizations, the creation of theoretical concepts and laws is not a rule-directed activity.

But before discussing the various kinds of law we should become acquainted with a number of specimens of law statement: to this end the next section is devoted.

Problems

6.2.1. Sociologists are fond of calling 'laws' statements like this: "The pressure on the group members to communicate with one another depends on the perceived discrepancy of opinion on an issue among the members of the group, and on the pressure of the group members to achieve uniformity of opinion". Is this a law statement or rather a hypothesis about the existence of a (so far unspecified) functional relation among three variables — i.e. a programmatic hypothesis? *Alternate Problem:* Collect some programmatic hypotheses occurring in journals of psychology and sociology.

6.2.2. Mention one pair of mutually irrelevant variables, another of favorably relevant, and a third of unfavorably relevant variables. *Alternate Problem:* Study the elucidation of the concepts of favorable and unfavorable relevance (i) with the help of the concept of function, and/or (ii) in terms of probability.

6.2.3. Study the analysis of correlations in search for laws. See M. Bunge, *The Myth of Simplicity* (Englewood Cliffs, N. J.: Prentice-Hall, 1963), Ch. 11, Sec. 5, and the bibliography cited therein. *Alternate Problem:* Most variables change in ways that are not obviously lawful. The law of variation, if it exists at all, can often be found by analyzing the variable into simple periodic motions of decreasing amplitude and increasing frequency and then testing for such a (harmonic) analysis. Examine this search for periodicities and find out whether they yield basic laws or rather their solutions.

6.2.4. Would irrelevance exist if every property were rigidly tied to every other property and consequently all laws constituted a single tight system — the block universe? What ontological hypotheses does the mere existence of fairly well corroborated laws each relating a few properties suggest?

6.2.5. Table 6.1 abbreviates ten singular propositions. State any of these propositions fully. Then contrast this set of propositions to the single statement of the corresponding law (SNELL's). For purposes of comparison write the latter simply '$L(i, r)$' and do not allow to be misled by the circumstance that SNELL's law is not usually written as an explicit function of the form "$r=f(i)$" but rather as an implicit function of the form "$f(i, r)=0$". The latter can easily be solved for r, namely thus: $r=\sin^{-1}(\sin i/n)$. *Alternate Problem:* Study the problem of goodness of fit of a polynomial to a set of data by means of the NEWTON-GREGORY interpolation formula.

6.2.6. Every pair of transparent substances (air-water, wine-quartz, olive oil-lucite, etc.) is characterized, from an optical viewpoint, by a given value of the refractive index. Are different laws at stake when n is assigned different values? *Alternate Problem:* Disclose the logical form of SNELL's law.

6.2.7. Discuss classification as a means for obtaining empirical generalizations about conjunctions or correlations of structures or functions, such as "Mammals are hot-blooded" or "The dental formula of the anthropoid apes and the Hominidae is 2.1.2.3/2.1.2.3". See the classic W. S. JEVONS, *The Principles of Science*, 2nd ed. (1877; N. York: Dover, 1958), Ch. XXX, esp. pp. 677 and 682.

6.2.8. Up to now those who have searched for laws of history have tried to draw empirical generalizations from available historical material. Is this method likely to yield anything else than isolated empirical generalizations? Would it not be possible to hypothesize models of evolving societies, with the help of social science?

6.2.9. E. HUSSERL, the founder of the phenomenological school, maintained that the essential laws are obtained with the method of "eidetic variation", whereby "free transformations" of "essential intuitions" are performed; the invariants of such transformations would be the essential laws. Would you be prepared to recognize essential laws from nonessential ones, and to establish a single essential law, with HUSSERL's method?

6.2.10. Could not the search for laws become a rule-directed activity? And if the proper rules were known, would it not be possible to entrust computers with the finding of laws out of data? Computers can, of course, be programmed to find the coefficients of polynomials given a set of data, but the problem is whether they would be able to find nontrivial functions and the basic equations which these functions solve. *Alternate Problem:* Suppose an association is suspected among certain variables. The first problem is to find out whether they are in fact correlated. If a high correlation is in fact obtained, the next problem is to discover whether it is genuine (systematic) or spurious (nonsensical). How do we proceed to solve this problem: do we take a larger sample or do we try to explain the trend in terms of mechanisms, i.e. of independently testable laws? What if we do not succeed in finding such a mechanism: shall we conclude that the correlation is spurious or shall be suspend judgment?

6.3. Kinds

There are as many kinds of scientific law as points of view or classification criteria we care to adopt. An enlightening viewpoint is the one that regards the qualitatively different levels — the so called integrative levels — into which reality may be analyzed: the physicochemical, the biological, the psychological, and the sociocultural levels (see Sec. 5.9). Every one of these levels may be characterized by variables and laws of its own, and the objective relations among the levels will be accounted for by inter-level laws. Let us, then, group the variables occurring in scientific research in the following way:

Physical variables φ_i, e.g., intensity of light.
Biological variables β_i, e.g., sex.
Psychological variables ψ_i, e.g., drive.
Sociological variables σ_i, e.g., division of labor.

The various constant, grounded and confirmed relations among such variables will be the set of known scientific laws. Those laws relating variables belonging to the same level (e.g., physical variable — physical variable relations) may be called *intralevel* laws. Those which relate variables belonging to different levels (e.g., sugar level in blood and fatigue) may be called *interlevel laws*.

A priori the following kinds of law (in respect of the level structure) might exist:

1. $\varphi_i = F(\varphi_j)$	*Physical & chemical laws*	*Intra-level laws*
2. $\beta_i = F(\beta_j)$	*Biological laws*	
3. $\psi_i = F(\psi_j)$	*Psychological laws*	
4. $\sigma_i = F(\sigma_j)$	*Sociological laws*	
5. $\beta_i = F(\varphi_j)$	*Biophysical & biochemical laws*	*Inter-level laws*
6. $\psi_i = F(\varphi_j)$	*Psychophysical & psychochemical laws*	
7. $\psi_i = F(\beta_j)$	*Psychobiological laws*	
8. $\sigma_i = F(\varphi_j)$	*Sociophysical laws*	
9. $\sigma_i = F(\beta_j)$	*Sociobiological laws*	
10. $\sigma_i = F(\psi_j)$	*Sociopsychological laws*	
11. $\psi_i = F(\varphi_j, \beta_k)$	*Psychobiophysical laws*	
12. $\sigma_i = F(\varphi_j, \beta_k)$	*Sociobiophysical laws*	
13. $\sigma_i = F(\varphi_j, \psi_k)$	*Sociopsychophysical laws*	
14. $\sigma_i = F(\beta_j, \psi_k)$	*Sociopsychobiological laws*	
15. $\sigma_i = F(\varphi_j, \beta_k, \psi_l)$	*Sociopsychobiophysical laws*	

These various possible relations are diagrammatically shown thus:

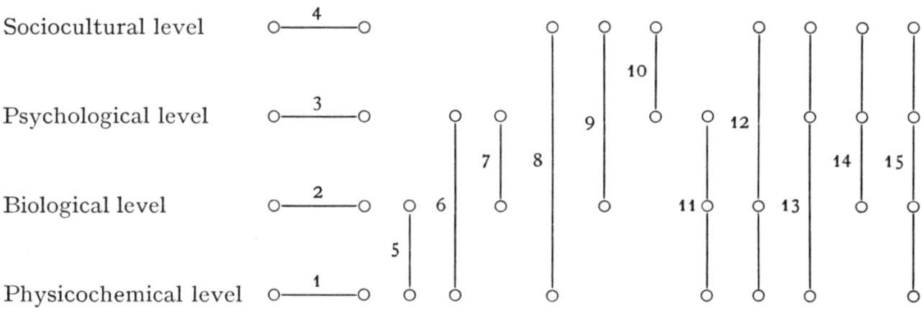

It might seem that the sets 12 through 15 of interlevel laws are empty, but this is not so. The 12th set is made up of the laws of social ecology. The 13th is exemplified by the laws of social psychology in which physical and social stimuli determine behavior variables. Instances of the 14th kind are those laws of social psychology in which biological variables (e.g., sex) and social variables (e.g., social status) determine behavioral variables. And instances of the 15th kind are those laws of social psychology covering the behavior of individuals subjected to the joint action of physical, biological, and social stimuli.

The laws of types 6, 8, 9, 12, and 13 skip over some intermediate levels. This would seem to violate the ontological principle not to jump over levels (see Sec. 5.9), a principle grounded on the study of the mechanisms connecting different levels. Thus, for instance, we know that a physical stimulus does not act directly on a mental state but must first set the organism in action since, after all, psychical phenomena are sets of special functions of the organism. Similarly, biological stimuli do not act directly on the social level: they are first felt by individuals; thus, food deprivation is felt as hunger. Hence, a link is missing in the case of biosociological laws, too. On the strength of the preceding considerations we might be tempted to ban all the law statements that omit variables belonging to intermediate levels. But this would be a mistaken strategy: we should keep such laws, only not as ultimates: we should demand their eventual analysis in terms of the missing variables. For example, we should request the analysis of a psychophysical law "$\psi = F(\varphi)$" in some such way as this: $\psi = G(\beta)$, $\beta = H(\varphi)$, whence the original law would become $\psi = G[H(\varphi)]$, i.e. the behavioral variables would be functions of functions of the physical variables rather than direct functions of them. In other words, the laws that skip over intermediate levels should be retained as global interlevel relations, and the detailed mechanisms of the relations among contiguous levels should be disclosed by finding out their corresponding laws. In short, the phenomenological approach should ultimately be supplemented by a deeper, representational approach (see Sec. 5.4).

Let us now mention a few scientific laws and show some of their traits.

Physical law: "The energy of an isolated system is constant." This statement is incomplete because it does not say in what respect the energy does not change; but through the context it is understood that the amount of energy is constant in time. There are various ways of accurately saying that a property, such as the total energy, remains constant in time (or with respect to some other variable). The most straightforward is to write '$\partial E/\partial t = 0$'. Another is to introduce two arbitrary time constants, t and t', and state that, for any t and t', if x is an isolated system during $t'-t$, then the total energy of x at t is the same as the total energy of x at t'. (In short: $t \neq t' \ \& \ I(x, t'-t) \rightarrow E(x, t) = E(x, t')$. Here, '$I$' is a two-place qualitative predicate standing for "isolated" and 'E' is a two-place quantitative predicate representing "energy". Actually a fixed reference frame is understood in this formula; if this presupposition is unearthed, an additional object variable must be included in I and E, which thereby become three-place predicates.)

Chemical law: "The water molecule consists of two hydrogen atoms and one oxygen atom." Notice that 'the' performs here the role of universal quantifier: we mean that each and every, hence all water molecules have the said composition. Consequently the expanded version is: "For every x, if x is a water molecule, then x is composed of two hydrogen atoms and one oxygen atom." We may regard the concept of composition as a *sui generis* functor and write it '$C(x)$', accordingly symbolizing the whole as: $(x)[W(x) \rightarrow C(x) = H_2O]$. If we recall that chemistry possesses over one million such composition laws, we are forced to admit that, of all sciences, it is the richest in law statements.If need be we may add that the composition of the water molecule is independent of space and time, i.e. that it is *spatiotemporally universal*, not only *referentially universal* (i.e., universally quantified with respect to the object variable). Spatiotemporal universality, a usual assumption regarding laws of nature, may be explicitly indicated by introducing the variable 'ϱ' for position in spacetime; each value of ϱ will be an ordered quadruple of numbers, one for time, three for the space coordinates. We may then write our chemical law as follows: $(\varrho)(x)[W(x, \varrho) \rightarrow C(x, \varrho) = H_2O]$, meaning "Everywhere and always, the composition of every bit of water is H_2O". We need not repeat the clause 'everywhere and always' whenever we state a law of nature if we assert once and for all the following metalaw or law of laws: "Laws are independent of location in spacetime." Or, stated in negative form: "Laws are neither dated nor placed." There is no harm in accepting this metanomological principle on condition that we realize how strong a metaphysical (ontological) hypothesis it is. But let us now continue with our stock of examples.

Geological law: "In the absence of foldings, the deeper geological strata are the older." Expanded form: "If x and y are different geological strata, and x and y are not folded, then: if x is deeper than y, then x is older than y." In symbols: $x \neq y \,\&\, S(x) \,\&\, S(y) \,\&\, -F(x) \,\&\, -F(y) \rightarrow [D(x, y) \rightarrow O(x, y)]$. Incidentally, geology is one of the sciences with fewer laws. It would be interesting to find out why this is so: are there few objective geological patterns, are physics and chemistry sufficient for most geological purposes, or is geology still in an underdeveloped stage?

Biological law: "Chromosomes are self-duplicating." Expanded form: "If x is a chromosome, then x duplicates itself." Symbolizing: $C(x) \rightarrow D(x)$.

Psychological law: "Inborn patterns of behavior are more stable than learned patterns." It is convenient to add the object variable here, an x ranging over all organisms. Otherwise the sentence might be understood as meaning the proposition that learned patterns in any organism are less stable than inborn patterns in the same (true) or in different organisms (false). The expanded form is then: "For every x, if x is an organism and y is an inborn behavior pattern of x and z is a learned behavior pattern of x, then y is more stable than z." Symbolically: $(x)(y)(z)[O(x) \,\&\, I(x, y) \,\&\, L(x, z) \rightarrow S(x, y, z)]$.

Sociological law: "Shepherd cultures are nomadic." Obvious symbolization: $(x)[S(x) \rightarrow N(x)]$. By the way, this passes for being *the* form of scientific laws.

Historical law: "The horde precedes the tribe and the tribe precedes the stratified society." Again, as in the case of the psychological law, the object variable is missing: what is meant is that, in the historical development of any given human group — call it x — the above-mentioned sequence pattern occurs. Possible symbolization: $G(x)\ \&\ t < t' < t'' \rightarrow H(x, t)\ \&\ T(x, t')\ \&\ S(x, t'')$.

We shall finally analyze an instructive case history: the story of ARCHIMEDES' *principle,* one of the earliest scientific laws. Leaving aside the uncertain tale concerning the crown and the bath tub, the problem faced by ARCHIMEDES was to account for the flotation of bodies. Available knowledge was insufficient to this end, although it did contain some loosely stated empirical generalizations that ARCHIMEDES must have used, such as "Solids displace liquids", "Bodies immersed in fluids lose weight", and "Buoyancy depends on the kind of liquid". These were vague and isolated generalizations from common experience. A merit of ARCHIMEDES' was to turn them into quantitative and mutually related laws. But to this end he had first to suspect the variables necessary and sufficient to account for flotation.

In his search for the relevant variables ARCHIMEDES may have been guided by the precept of atomistic philosophy, to select primary qualities as fundamental variables; and he may have eliminated various candidates, such as the viscosity and transparency of the fluid, or the form and composition of the floating solid, by means of a few empirical trials. At any rate ARCHIMEDES reduced the relevant variables to only three: hydrostatic pressure, buoyancy or upward thrust (loss of weight), and quantity of liquid displaced. Moreover, he further reduced these three variables to applications of a single concept: weight. We are now used to the search for quantitative variables, but in ARCHIMEDES' time the Platonic prejudice against the possibility of building a science of nature, and the Aristotelian qualitative and speculative physics, were dominant. ARCHIMEDES did not lay the foundation of hydrostatics and statics — the earliest chapters of theoretical physics — by merely applying some method: he had to invent the right approach himself.

The next problem was to "find" (i.e. to conceive) the law relating the three variables. (Actually several hypotheses, involving variables other than the above-mentioned ones, may have occurred to him and he may have discarded the irrelevant variables after empirically testing some of those working hypotheses. But no traces of the invention and discovery process are extant.) The first step may have been the assumption that pressure, buoyancy, and quantity of displaced liquid were all forces of a kind, expressible as weights. The original problem may then have been restated in this way: What is the weight that balances the loss of weight $W-W_a$ of a solid weighing W in a vacuum and W_a in the midst of the fluid? The question was, then, $(?X)(W-W_a=X)$. Clearly, X is the buoyancy, i.e. the upward force exerted by the fluid on the floating body and responsible for its flotation. That this force is a weight of some kind was presupposed by the question and follows from the principle of dimensional homogeneity (not stated by ARCHIMEDES). The next problem is to find out what is X the weight of.

That X is not a weight of the body seems clear, since the problem contains already the two relevant weights of the body, namely W and W_a. Nor could X be the weight of all the liquid since flotation is, within large bounds, independent of the amount of fluid. We may then surmise that X is related to the weight of the liquid displaced by the solid. This is not a hypothesis but rather a hypothesis schema or an infinite class of hypotheses, unless we specify the relation. Let us try the simplest of all conjectures, namely that X equals the weight of the displaced liquid. If it passes the test we shall keep it, otherwise we shall try a more complex assumption. Consequently we introduce in "$W-W_a=X$" the hypothesis "$X=W_f$", where 'W_f' stands for the weight of the displaced fluid. In this way we get: $W-W_a=W_f$. In words: "If a solid body is immersed in a fluid, it loses weight and its loss of weight equals the weight of the displaced fluid." As usual, the mathematical formula neglects the antecedent of this conditional.

This — ARCHIMEDES' principle — is a tentative solution to the problem "$(?X)(W-W_a=X)$". Before accepting that hypothesis as a law it must pass some tests. To test ARCHIMEDES' hypothesis we may proceed as follows. We first weigh a solid in a vacuum, i.e. we determine W. Next we immerse the body in a fluid and measure the new weight, W_a, of the body in it. Then we weigh the displaced fluid, obtaining a number W_f. Next we perform the subtraction $W-W_a$, a conceptual operation, and compare this number with W_f — a further mental operation. If the difference between the two numbers is less than the accepted experimental error, we conclude that ARCHIMEDES' principle has been confirmed for the chosen solid/fluid pair. The generalization of the principle, first to all the individual couples of a kind, then to all possible pairs, was presumably made after trying just a few pairs. A better test of the principle is its continued use as a means for obtaining specific gravities, because these values can be checked independently by the direct procedure of weighing and finding the volume of solids, and then dividing the weight by the volume.

Nowadays we slightly refine the principle adding the condition that the body be in equilibrium with the fluid. Also, we often replace "loss of weight" by "buoyancy" (both unobservable, though easily inferable properties). A modern version of the principle might read thus: "If a solid body is immersed in a fluid and is in equilibrium with it, then it is buoyed up with a force equal to the weight of the displaced fluid." In expanded form: "If x is a body and y is a fluid and x is immersed in y and x is in equilibrium with y, then the buoyancy force exerted on x by y equals the weight of the displaced fluid." In symbols: $B(x) \& F(y) \& I(x, y) \& E(x, y) \rightarrow F(x, y) = W_f$.

In physics books only the consequent of the conditional will be found. But an explicit statement of the conditions as *part* of the law statement has the advantage that it shows clearly what the *validity conditions* are. If these conditions are not met the conditional may by retained but it becomes *pointless*. This is the case, e.g., of fluids in motion: an upward current within the fluid will of course falsify both the antecedent and the consequent of the law, but not the whole conditional. Such cases are then not exceptions to the law but just cases outside the extension or domain of application of the law. But could there be exceptions to ARCHIMEDES' law? Might we not find a particular pair solid/fluid falsifying the law or at least requiring its transformation into an almost-all statement? Of course we might find exceptions, but to seek for them at this late date would be foolish: if they exist, let them pop up by accident. A constructive attitude has been taken instead, one which consists neither in looking for exceptions nor in piling up confirming evidences with the hope of increasing the degree of truth of the hypothesis by steadily enhancing its degree of confirmation. The constructive attitude has been to try to *systematize* or "theorify" the principle, i.e. to immerse it into a body of theory. And this has been achieved long ago: ARCHIMEDES' "principle" is now a theorem derived from fundamental laws of mechanics, which lend it a support it would lack had it remained as an isolated empirical or semiempirical conjecture. Let us rehearse an elementary derivation of the "principle": it will be instructive.

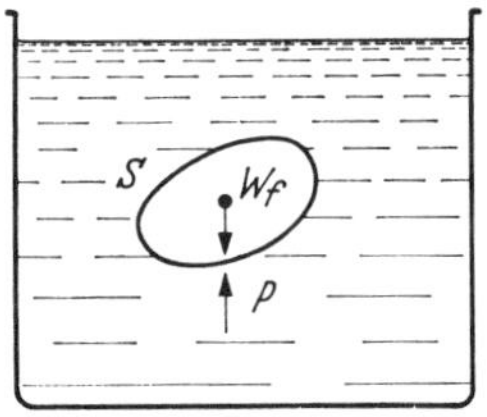

Fig. 6.6. A part of the liquid in equilibrium with the rest: the upward thrust P balances the weight W_f

Consider a homogeneous liquid at rest and imagine in the midst of it a region enclosed by the imaginary surface S: see Fig. 6.6. By hypothesis the liquid is at rest, and so are all its (macroscopic) parts, in particular S. Now, the fluid enclosed by S exerts a downward force, equal to its weight

W_f, on the fluid underneath. According to the principle of the equality of the action and the reaction (NEWTON's third principle), the force W_f is balanced by an upward push P originating in the liquid underneath S, i.e., $W_f = P$. Let us now substitute a solid body for the liquid enclosed by S; i.e. let us submerge a body of the same volume as S, and such that it is in equilibrium with the liquid. The liquid formerly enclosed in S will then be displaced and the solid will be buoyed up by the same push $P = W_f$ that was formerly exerted on S. On the other hand this push or buoyancy may be defined as the solid's loss of weight: $P =_{df} W - W_a$. Replacing this definition into the law "$W_f = P$" we get ARCHIMEDES' "principle".

What have we gained with this derivation of ARCHIMEDES' law? First, we now *understand* flotation as a particular case of balancing of forces. Second, we realize that ARCHIMEDES' law is not a mere empirical generalization which, like "Argentinians are fond of meat", might be false or become false without any readjustment in the mesh of laws: it has become firmly *entrenched* in mechanics. Third, as a consequence of the conversion of the law into a formula of mechanics, the law has gained *indirect support*: in addition to its direct supporters, namely the class of its instances, ARCHIMEDES' law has now as indirect supporters all the confirmations of the general principles of dynamics — to which in turn it lends its support. Every success of classical mechanics, like the correct account of an oscillatory motion or the accurate computation of the orbit of an artificial satellite, becomes an indirect supporter of ARCHIMEDES' law. And every failure of the same theory — e.g., for very small bodies — will throw doubt on the universality of ARCHIMEDES' principle and may even show in what domain it may be expected to be false.

Accordingly, nobody would receive a grant for the project of either collecting further direct confirmations of ARCHIMEDES' law (inductivist programme) or for exploring the universe in search of a particular solid and a particular liquid refuting the law (refutabilist programme). More fertile enterprises are these: to find the size of bodies (Brownian particles) for which the principle breaks down, to relax the condition of equilibrium and find a generalization to nonequilibrium conditions, or to investigate generalized theories of mechanics not involving the action-reaction principle (the law formula used in the derivation of ARCHIMEDES' principle). The theorification and the demarcation of the extension of a law are much more enlightening and fertile than purely empirical tests of it.

Problems

6.3.1. State a physical or a chemical law and and perform a logical and an epistemological analysis of it, i.e. show its logical form and investigate the degree of ostensiveness of the predicates occurring in it.

6.3.2. Do the same with a biological or a psychological law. *Alternate Problem:* Analyze an interlevel law.

6.3.3. Do the same with a sociological or a historical law. *Alternate Problem:* Classify the sciences into same-level and interlevel disciplines.

6.3.4. In his treatise *On Floating Bodies* ARCHIMEDES gave a derivation of his law but did not pose the problem of its test. Was it because (i) he did not deem it necessary for he regarded his axioms as self-evident?; or (ii) he thought (in common with so many 20th century thinkers) that mathematization guarantees factual truth?; or (iii) he did not consider it worth of a free man to mention that he had executed the empirical tests? See the collection *Greek Mathematics*, transl. I. THOMAS (London and Cambridge, Mass.: The Loeb Classical Library, 1941), pp. 249—251, and p. 31 for PLUTARCH's opinion concerning ARCHIMEDES' attitude toward utilitarian art. *Alternate Problem:* According to J. J. C. SMART, *Philosophy and Scientific Realism* (London: Routledge, 1963), Ch. III, there are no more biological laws than there are laws of engineering. Discuss this view.

6.3.5. Propose an example of systematization (theorification) of a law, i.e. of the conversion of an initially isolated hypothesis into an axiom or a theorem of a theory.

6.3.6. In the derivation of ARCHIMEDES' law it was assumed that the upward force P "felt" by the solid was the same as the one "felt" by the portion of fluid enclosed by S. In particular, it was not assumed that the push P would depend on any other property of the solid save its volume, which was also the volume of the displaced liquid. Would this assumption hold in a finalistic physics? If not, draw some moral concerning the relation of scientific research to general metaphysical (ontological) hypotheses.

6.3.7. Chemistry has probably more laws than physics, but the laws of chemistry are quite unrelated among one another at their own level: that is, it seems hardly possible to establish logical relations among them. It is the physical laws underlying the chemical laws that endow the latter with a sort of derivative systemicity: taken by themselves, as they were until not long ago, the laws of chemistry are quite unrelated. Comment on this situation and examine the opinion that every science deals with a particular net or system of laws. Also, speculate on the possibility that, if there are laws of history, then they may not be primarily but derivatively systematic, in the sense that the laws of sociology could give them a derivative systemicity.

6.3.8. Does the metanomological formula "Laws are independent of location in spacetime" mean that all laws are in force, so to say, in every corner of the universe even where there is nothing but space? And does it preclude the eventual extinction of some laws or the emergence of new laws?

6.3.9. Following a lead from the text, the *total support* $S(h)$ enjoyed by a hypothesis h might be defined as the numerosity or cardinality of the union of the sets of direct and indirect supporters of h, i.e. $S(h) =_{df}$ Card $[D(h) \cup I(h)]$. Examine this proposal. In particular, consider separately the four cases obtained by assuming that either set, D of direct supporters and I of indirect supporters, is finite or infinite. If $D(h)$ or $I(h)$ are infinite, are they sets of actual or rather of potential supporters? If the latter, how could they be determined?

6.3.10. Scientists are interested in demarcating the domain of validity and the inaccuracy of law statements rather than in measuring their degree of confirmation, which is the central concern of inductive logic. Does this divergence of interests point to a philosophical blindspot of scientists or to a lack of acquaintance of inductive logicians with the concern of research — or to something else?

6.4. Form and Content

The most obvious logical requirement we place on hypotheses in order to rank them as laws is *generality in some respect and to some extent.* (The requisite of well-formedness is already taken care of in the decision to regard a law as a scientific hypothesis). That is, we demand that at least one of the variables occurring in the law formula be prefixed by the operators 'for every', 'for almost every', or 'for most'; if the former is the case, i.e. if the law is a strict universal hypothesis, then we usually dispense with the explicit addition of the quantifier. If the law refers to an individual (as in the case of the geophysical laws, which refer to our planet), we require that the statement expresses the regular behavior of the individual as shown by a universal quantifier with respect to time; the quantifier may be unrestricted or bounded, and it may be manifest or tacit but it must be there: otherwise the proposition would be singular rather than general. If, on the other hand, the law formula does not refer to an individual but to a class, we may tolerate quasi-generality, as in "Most salts of the alkali metals have a high solubility in water", or "Most mammals have hairs". 'Most' and 'almost all' are not treated with respect by logicians, who lump them together with 'there is at least one', but in science their status is much higher than that of the existential operator: an almost-all-formula may be a law proper, and a most-formula may promise a universal law.

The important law of entropy increase is a typical almost-all-formula: "If a system is isolated, then in almost all cases it will go over to states of higher entropy." Countless theorems of statistical physics are prefixed by expressions like 'for almost all points' or 'for almost all trajectories'; and despite being quasi-universal they are regarded as perfectly respectable law formulas. In mathematics the equivalent expression 'with the exception of a zero-measure set' is frequently met in general theorems yet nobody would dream of denying generality to such statements although the set constituting the exception may be infinite. Strictly universal laws, i.e. law formulas with no exception and with an infinite scope, are very often stated — which does not prove that, as a matter of fact, they do hold with such unlimited generality. Macroscopic laws, which are actually averages or outcomes of microlaws, are certainly not without exception even though the rate of exceptions is not usually stated. At any rate quasi-universal law formulas are as worthy as strictly universal law statements, especially if (i) it is possible to account for the possible exceptions and (ii) they are not empirical generalizations but members of theories.

Among nonuniversal law statements those which are manifestly statistical are the most intriguing. The simplest among *statistical laws* are perhaps the percentage laws, such as "50 percent of the cars over five years old are useless in the U.S.A". This empirical generalization is not about every individual of a certain class: percentages (or, for that matter, relative frequencies) are not properties of individuals but nondistributive (nonhereditary) properties, i.e. collective properties that cannot be distributed among the collection's members. A brief analysis will make this clear and will raise important questions. The form of our statistical generalization is "The fraction f of A's are B's". Let 'Card (A)' and 'Card (B)' designate the number of members or cardinality of the sets A and B respectively. In our case Card (A) is the number of cars over five years old and Card (B) the number of useless cars, whether old or not. Then the set of cars that are both over five years old (A) and useless (B) is the intersection $A \cap B$. And the fraction of useless cars in the reference class (old cars) is accordingly Card $(A \cap B)$/Card (A). Consequently, our generalization can be written: Card $(A \cap B)/\text{Card}\,(A) = f$, where '$f$' designates a fraction between 0 and 1; in our example $f = 0.5$. Clearly, this statistical statement is not about single systems but about classes. Furthermore, it does not have the conditional form.

Individuals can be brought in if, instead of the empirical concept of percentage (or relative frequency), we employ the theoretical concept of probability. In fact, the statement about the fraction of useless cars in the class of old cars can be translated into the following nonequivalent *probability statement:* "The probability that an arbitrary member of A be in B equals p", where p is a (fixed) number close to the (fluctuating) frequency f. More precisely: If A is a nonempty set (i.e., $A \neq \emptyset$) and x is a member of A, then the probability that x is in $A \cap B$ equals p. Better: If A is a nonempty set, then the probability that x is in $A \cap B$, given that x is in A, equals p. In symbols: $A \neq \emptyset \rightarrow P(x \in A \cap B | x \in A) = p$. If a theory of car ageing were available, this theoretical statement would be contained in it and it would correspond to the given statistical generalization.

To call a probability statement a *translation* of the corresponding statement concerning percentages would be misleading, because the two formulas are not equivalent. In the first place, unlike the frequency statement its probability partner refers both to specified classes and to an unspecified individual. In the second place, the fact that the numerical variable p in the probability functor can be set equal to the observed fraction or percentage f does not mean that p is the same as f: (i) whereas p is a theoretical concept f is an empirical concept, and (ii) whereas the value of p is supposed to be fixed, the various empirically found values f are as many estimates of a single p value. (A percentage statement need not contain theoretical concepts although it may, whereas probability statements cannot help being at least half-theoretical even if they embody empirically found numbers.) In the third place, although f may be taken to be the same number as p, f does not mean the same as p: in the case of the statistical generalization it was a collective (nondistributive) property, whereas in the case of the probability statement it is a (potential) property of each and every member of the reference class A. Moreover, the introduction of an empirically found number like f into the probability statement must be justified by a methodological rule to the effect that the numerical value of a probability will be approximated by the corresponding (long run) frequency. In conclusion, we must distinguish *statistical statements* (concerning, e.g., relative frequencies, modes, or observed scatters) from *probability statements* (concerning, e.g., probabilities or parameters occurring in probability distributions). The former can be nontheoretical or rather semitheoretical statements whereas the latter are theoretical statements; the former refer to collective properties only, the latter to both individuals and classes. Statistical and probability statements may both be subsumed under the genus of *stochastic statements.*

Some die-hard classical determinists claim that stochastic statements do not deserve the name of law and are to be regarded, at their best, as temporary devices. This anachronistic view has no longer currency in physics, chemistry, and certain branches of biology (notably genetics), especially ever since these sciences found that all molar laws in their domains are stochastic laws deducible (at least in principle) from laws concerning single systems in conjunction with definite statistical hypotheses regarding, e.g., the compensation of random deviations. Yet the prejudice against stochastic laws still causes some harm in psychology and sociology, where it serves to attack the stochastic approach — without compensating for its loss by a scientific study of individuals. Now, it is impossible to account adequately for the behavior of an individual, be it an atom, a human subject, or a community, without taking into account the inner spontaneous fluctuations and the outer external disturbances, both of which have important random components. The way to master chance is to face it and to discover its laws rather than to deny its objective existence. Chance is an evil ghost only if regarded as lawless chaos or as an ultimate, not further analyzable, mode of being.

It will be noted that, in writing our probability law, and also some of the laws dealt with in the previous section, we left out the universal quantifier, i.e. we stated it as an *any*-statement rather than as an *all*-statement; in other words, we said something about an arbitrary member x of a set A rather than about every member of A. But, of course, we tacitly assumed that the law holds for every value of x: otherwise we would not have stated it. This intended meaning of the propositional function in question authorizes its universal generalization. In other words, applying to it the inference rule "What holds for any holds for all" we infer the universal statement: $(x)\,[A \neq \emptyset \rightarrow P(x \in A \cap B \mid x \in A) = p]$, which is a general conditional.

An any-statement like $P(x)$ is not equivalent to its universal generalization $(x)\,P(x)$ but equipollent to it in the sense that either is inferable from the other: that is, they are deduction-equivalent. Moreover, they are pragmatically equivalent as well: (i) to say that an arbitrary individual — i.e. any individual, or an individual picked out at random — has the property P is as effective as saying that every individual has the said property; (ii) in order to test the generalization "$(x)\,P(x)$" we pick arbitrary, i.e. unprivileged individuals. But while we are neither using nor testing "$(x)\,P(x)$" this formula is inequivalent to "$P(x)$". The differences between the two are formal and semantical and they deserve being pointed out because they are relevant to the logic of law statements. The syntactical difference between an any-statement and an all-statement is that the former is simple whereas the latter is complex. In the usual (extensional) interpretation, "$(x)\,P(x)$" is supposed to be the conjunction of singular statements obtained from "$P(x)$" upon assigning definite values to x. This expansion requires that x ranges over a denumerable universe, i.e. that the class $\{x \mid P(x)\}$ be countable. But this is a severe restriction that cannot be satisfied by formulas containing continuous variables. For example, the statement "Gravity is nearly constant at all points in this room" cannot be expanded as a conjunction of singular propositions each referring to a point in the said volume, because it constitutes a continuous set. This limitation of universal quantifiers to denumerable universes is, of course, pointless with respect to any-statements. Now, as regards manifest meaning '$P(x)$' is meaningless even if the value of P is specified: it has no fixed reference or, if preferred, its referent is the unspecified individual designated by 'x'. If a formula is meaningless then it is untestable, since in order to ascertain whether anything has in fact a given property or does not have it, that thing must either have the property or fail to have it. Let us apply these considerations to an elementary law statement.

GALILEI's law of falling bodies is usually written in the form

$$s(t) = \tfrac{1}{2} g t^2 + v_0 t + s_0 \qquad [6.3]$$

where '$s(t)$' designates the distance travelled by the body during the lapse t, 'g' stands for the acceleration of gravity, 'v_0' for the initial velocity and 's_0' for the initial position. We have just stated the semantical rules neccessary for endowing the mathematical formula [6.3] with a factual meaning; but this is insufficient because the formula is an any-statement. This is easily seen upon recalling that it refers to any freely falling body — i.e. by digging up the object variable. A fixed referent is pointed to by individualizing the body. This can be done either by naming it or by giving the values of its initial position s_0 and initial velocity v_0. In either way we rid [6.3] of the tacit object variable x but it still remains an open statement because it holds for any time. If we specialize the value of t we obtain a singular statement; we may also universally generalize [6.3] for all time, but with no obvious advantage: anyhow for purpose of inference we shall use the propositional function [6.3] because it can be treated as if it were a proposition proper. Consequently we respect the usual custom of writing universal statements as any-statements and supply the quantifiers when necessary for interpretation purposes.

Another question concerning the proper logical form of universal laws is this: why should universal laws be assigned the form of general conditionals? The reason for this is semantical rather than syntactical: namely, that the conditional form suggests *conditionality*, and conditionality is a characteristic of law statements as opposed to mere descriptive statements. When we assert "$p \rightarrow q$" we do not state that p is in fact the case but just that, *if* p obtains, then q is the case. Moreover, in the conditional the *necessary* and the *sufficient* conditions are clearly distinguished — almost pictorially — which is not the case of any of the equivalent forms employing symmetrical connectives, such as "$-p \vee q$" and "$-(p \,\&\, -q)$". Thus, the antecedent of "If it rains then the pavement becomes wet" is clearly the sufficient condition for the consequent. This is far from obvious in the equivalent forms "Either it does not rain or the pavement becomes wet" and "It is not the case that it rains and the pavement does not become wet". (Incidentally, to establish that a conditional holds is comparatively simple in formal science, where "$p \rightarrow q$" is true in a given system if q can be derived from p in the system. In factual science there is no such simple rule: we must be able to ascertain what are the *physically* sufficient conditions at stake, and logic is of little help to this end. In fact, we can regard "$p \rightarrow q$" as factually true if and only if we can show that p, which is logically sufficient for q, expresses in addition a physically sufficient situation for q to obtain — which can only be found out upon a theoretico-empirical investigation.) For that reason, then, the conditional is a convenient from to pour laws into.

But if we adopt conditionals we must take the so-called paradoxes of implication into the bargain: i.e. we must realize that they are not paradoxes at all. For example, "$p \rightarrow q$" is true for any false p; conditionals with a false antecedent are called *vacuously true*, and this is often regarded as paradoxical. Anyway the air of paradox should be thin in the sciences, where we never bother to state conditionals with antecedents known to be false. Take, for instance, the lawlike statement "If x and y are biological structures or functions, and x is more complex than y, then x is less subject to evolutionary changes than y". The antecedent is fulfillable and has an existential import rather than being an idle ghost-like assumption. In science "$p \rightarrow q$" is usually the formalization of "If p is *assumed*, then q holds". If the assumption p is false the conditional continues to hold but becomes *pointless*. We certainly do not require every antecedent to be true: we know that factual truths both strict and interesting are hardly attainable and we want to retain our freedom to hypothesize. Yet in science we do not stretch this freedom to the point of stating conditionals with manifestly false antecedents. Take, for instance, "Clairvoyants discover every secret", which may be symbolized "$(x)(y)[C(x) \,\&\, S(y) \rightarrow D(x, y)]$". This statement is vacuously true because, for all values of x, $C(x)$ is false: in fact, there are no clairvoyants. Yet we do not accept this truth, because it is pointless, just as the zoologist has no use for the truth "Centaurs are wise". Truths of this kind can be obtained by the million, by merely instructing a computer to match a given false or unfulfillable antecedent with arbitrary consequents, such as arithmetical statements.

In other words, we are not interested in accumulating pointless truths, truths concerning nonexistent entities or impossible conditions. When, in the course of research, we state a conditional we usually presuppose the possibility of its antecedent, i.e. we assume — tacitly or overtly — that the antecedent of a factual conditional could be physically realized. This is certainly not the case of statements the logical consequences of which we wish to explore: in such cases the maximum freedom is allowed. But it is certainly the case of law statements. Consider a law $L(x)$ referring to an unspecified object x of a certain kind; it will hold under certain conditions $C(x)$ rather than unconditionally. That is, we shall assume that if x satisfies the condition C, then x fulfils L as well:

$$C(x) \rightarrow L(x). \qquad [6.4]$$

To this statement we add, then, the assumption that the condition C is fulfillable: that is, we add the presupposition that it is possible that there exists at least one object x such that x satisfies the condition C. (We do not commit ourselves to the stronger assumption that there *is* such an object because further research might show that, as a matter of fact, C is unfulfillable.) That presupposition — like many another presupposition — will not take part in the deductive inferences in which the law statement takes part but it endows the latter with the existential import necessary to regard it as a law statement rather than as a pretence. Thus, in deducing consequences from a set of postulates referring to the mode of decay of some strange new particle, we do not employ the presupposition that this particle possibly exists, let alone that in fact it does exist; but when setting up the postulate set we do back it up by the existential hypothesis. Consequently, the introduction of the modal concept of possibility — which would be a nuisance as regards both inference and test — does not require the replacement of ordinary logic by some system of modal logic — never used in actual scientific inference.

If we were to adopt the strong version of the existential assumption we would complete [6.4] to

$$(\exists x)\, C(x) \dashv [C(y) \rightarrow L(y)] \qquad [6.5]$$

where the inverted turnstile '$\dashv$' is to be read 'is presupposed by' (see Sec. 5.1) and where two different letters have been used to designate the object variable in order to show clearly that the scope of the existential quantifier is limited to the first occurrence of the condition C. But since we must be prepared to surrender the clause $C(x)$, we had better adopt the weaker version of the existential assumption and consequently write

$$\diamond (\exists x)\, C(x) \dashv [C(y) \rightarrow L(y)] \qquad [6.6]$$

where '$\diamond$', read 'diamond', means "it is possible that".

A form like [6.6] may be called a *fulfillable* conditional. Even so it may turn out to be empty: the precondition may not be exactly fulfilled, but just approximately so. For example, the law of conservation of energy holds only for closed (insulated) systems, but only the whole universe is a perfectly closed system, every one of its parts leaking in some respect. In view of this the presupposition about the possible fulfilment of the antecendent of a law must be understood in a qualified way, namely thus: "Possibly there is at least one system such that it *approximately* fulfils the condition C". But then exactly the same could be said of the consequent L: that is, both the law statements and their preconditions are approximately fulfillable. But this metanomological statement concerns the truth of law statements and comes consequently after their test, which in turn is executed after their formulation. Therefore it does not affect our discussion of the form of law statements. In short, we may keep [6.6] as the standard form of universal law statements with the understanding that, strictly speaking, it refers to a more or less idealized model of the piece of reality it intends to refer to: it applies exactly to the model, more or less roughly to the model's real referent. We shall come back to this point in Sec. 6.5.

To pursue our study of the content of law statements let us have a specific example in mind — for example GALILEI's law [6.3]. We have seen that a falling body may be unambiguously characterized by a couple of numbers, namely its initial position and its initial velocity: this couple of numbers functions as the body's name. The possibility of specifying special circumstances like these is characteristic of low-level laws like GALILEI's but is lost in relation with higher level laws such as the general principle of dynamics which entails GALILEI's law. Specialized to the class of falling bodies, this principle states that their acceleration — which may be symbolized '$D^2 s$' — is a constant called g. (Unlike the parameters denoting the initial position and the initial velocity, g is not an individaul parameter but it specifies a class of motions, namely the set of uniformly accelerated motions. Free fall in a gravitational field is just a subclass of that class, and in this case g is called the acceleration of gravity.) In symbols, the high-level statement corresponding to GALILEI's law is

$$D^2 s(t) = g. \qquad [6.7]$$

('$D^2 s(t)$', also written '$d^2 s/dt^2$', is short for the second derivative of the distance with respect to time; [6.7] is then a differential equation. Most high-level law statements in the so-called exact sciences have the form of differential equations; but the strongest statements are integral equations.)

Any reference to *special circumstances*, such as initial position and initial velocity, is absent from [6.7]. Generally, high-level law statements do not contain references to specific individual characteristics or special circumstances. In order to bring such high-rank law statements into contact with data concerning individual features they must undergo a deep formal transformation, such as the one leading from [6.7] to [6.3], and they must be supplemented with some empirical information, such as, e.g., the particular values of the parameters g, v_0, and s_0, at the place where the measurements are conducted. Only a low level statement, such as [6.3], can absorb all the specific information necessary to confront it with special circumstances: high-level law statements remain aloof from experience. (In particular, differential equations must be integrated before thy can be tested. The process of integration introduces empirically determinable parameters such as v_0 and s_0.) This is one reason why high-level law formulas cannot be "abstracted" from empirical data but must be hypothesized. The law-data relation is one-way: from law-statements we may deduce particular statements into which empirical information can be fed; but no trick and no machine will turn a heap of data, however precise, numerous, and relevant, into a high-level statement. Only lowest-level statements, i.e. empirical generalizations, can be inferred from data — but then not unambiguously and they remain isolated until a stronger unifying principle is invented (see Sec. 6.3). Let us next explore further semantical aspects of the problem of laws.

Problems

6.4.1. Symbolize the qualitative law statement: "Any object attracts every other object." Notice, incidentally, that this is not generally true: certain objects, such as light waves, do not fall under the law of gravitation. Moreover, the relativistic version of the latter shows that the gravitational interaction is not external to the bodies concerned but depends on their state of stress — which in some cases leads to repulsion. This is just a reminder that every law formula has a limited domain of validity. Attempt to incorporate these qualifications in the antecedent of the above law.

6.4.2. Lucy learns at school that "Heat expands and cold contracts", whereupon she employs this statement to explain to Linus why days are longer in summer than in winter. Who is to be blamed and why?

6.4.3. In what sense or senses is LAVOISIER's law "In an isolated system the total mass of the reagents equals the total mass of the reaction products" universal?

6.4.4. BOYLE's law of ideal gases is usually written '$pV=$const". Unless the context is provided this is not a physical law but a mathematical formula — and this provided we presuppose that 'p' and 'V' are numerical variables. State the law in a more complete form, including a reference to its spatiotemporal universality — which, by the way, is a false assumption in this case.

6.4.5. Symbolize the following schemata of stochastic laws. (i) "The average B-value of the A's is b". (ii) "The spread [scatter] of the B-

values of A around the mean value of B is σ". (iii) "The distribution of B-values in the A's is D". Find illustrations of these forms.

6.4.6. Does the universal quantifier have a single ordinary language correlate? See G. J. WARNOCK, "Metaphysics in Logic", in A. FLEW, Ed., *Essays in Conceptual Analysis* (London: Macmillan, 1956), and Z. VENDLER, "Each and Every, Any and All", *Mind*, N. S., LXXI, 145 (1962). Finally, see the criticism of B. RUSSELL to the first essay, in "Logic and Ontology", reproduced in *My Philosophical Development* (London: George Allen and Unwin, 1959).

6.4.7. Examine the criticisms directed by the partisans of the gestalt psychology to the search for statistical laws and the design of statistical tests in the field of psychology. See, e.g., K. LEWIN, "The Conflict Between Aristotelian and Galileian Modes of Thought in Contemporary Psychology", *Journal of General Psychology*, **5**, 141 (1931) and J. G. TAYLOR, "Experimental Design: A Cloak for Intellectual Sterility", *Brit. J. Psych.* **49**, 106 (1958).

6.4.8. Propose a classification of scientific laws from a mathematical point of view, i.e. with regard to the mathematical concepts occurring essentially in the law statements. Take into account that at least the set concept will occur in any law statement, so that at bottom there are no nonmathematical laws.

6.4.9. Draw a parallel between a scientific law expressed overtly in mathematical form, and a song. Examine, in particular, whether in either case content determines form or vice versa.

6.4.10 The equations of the propagation of heat and electricity are mathematically the same. Does this entail that the laws of heat and electricity propagation are the same? Take into account that this is a general problem: the same differential equations recur or are expected to recur in all branches of physics. *Alternate Problem:* Examine the proposal to regard each symbol occurring in a quantitative law statement as a full magnitude rather than as the numerical variable(s) of a magnitude. This interpretation, sometimes called the "quantity calculus", was defended by E. A. GUGGENHEIM in "Units and Dimensions", *Phil. Mag.* **33**, 479 (1942).

6.5 Law Formula and Pattern

What, if anything, does the term 'law' refer to? A careful lexicographer might tell us that the term 'law' has no fixed usage but is used with a variety of senses: it is an ambiguous sign designating several concepts. The juristic concept of law will not concern us here: we are interested in the acceptations relevant to pure and applied science. In these fields 'law' covers the following concepts: (i) objective pattern; (ii) formula (proposition or propositional function) purporting to reproduce an objective pattern; (iii) formula referring to both an objective pattern and experience; (iv) metastatement referring to a law statement; and (v) rule based on a law statement. (See Fig. 6.7.) We shall avoid confusing these various meanings if we adopt the following conventions.

'*Law*' (or 'objective law', or 'nomic structure') designates an objective pattern of a class of facts (things, events, processes), i.e. a certain

Knowing	*Doing*
Metanomological formula (pattern of law statements)	*Grounded Rule* (technological prescription)
Law formula (knowledge of law)	*Nomopragmatic formula* (test & use of law formula)
Being	
Law (objective pattern)	

Fig. 6.7. Meanings of 'law' in science

constant relation or mesh of constant relations really obtaining in nature, whether we know it or not. A law, in this sense of nomic structure, is an extraconceptual object, like the flow of a river. But, unlike the flow of a river, its laws cannot be pointed to: they are imperceptible. In short, the concept of objective law is empirically meaningless; which shows that it is not trivial. We cannot exhibit a specimen of an objective law but we can utter a definite description such as "The law referred to by ARCHIMEDES' principle". Whether we acknowledge the existence of objective laws or not we need this concept, if only to argue against the philosophical hypothesis that there are laws underlying the law statements.

'*Law formula*' (or 'nomological statement') designates a proposition or a propositional function that is usually supposed to describe a law or a part of a law (nomic structure). A law formula is a conceptual object, namely, a scientific hypothesis satisfying certain requisites of generality, corroboration, and systemicity (see Sec. 6.6.) It goes without saying that the laws we find in scientific texts are mostly law statements.

'*Nomopragmatic formula*' designates a lawlike proposition or propositional function that refers at least partly to experience and, in particular, to scientific experience. Example: "If a body is left unsupported in the vicinity of the Earth it will be seen to fall down." This statement contains pragmatic terms: 'left' and 'seen'. Consequently it refers both to a class of objective facts and to our commerce with it. To say it with clumsy yet suggestive symbols: Nomopragmatic statement = Low-level law statement + Pragmatic terms.

'*Metanomological formula*' designates a law concerning the law formulas of a certain class, and expressing actual or desirable features of law formulas. Example: "The higher level law formulas are invariant with respect to the observer." Metanomological formulas are found in factual science and in metascience; they do not portray laws of nature or society but concern our scientific ideas about objective patterns.

'*Grounded rule*' designates a rule based on one or more law formulas and enabling us to achieve a predetermined goal (see Sec. 11.2). Example: "To prevent the rusting of iron keep it dry."

Objective laws, if granted an existence, are to be located in reality; law formulas, nomopragmatic formulas, and grounded rules, in pure and applied science; and metanomological formulas partly in science and partly in metascience. Let us now approach some philosophical problems concerning the former concepts; metanomological formulas will be discussed in Sec. 6.7, and grounded rules Vol.II, in Sec. 11.2.

The existence of objective laws seems to be tacitly accepted by most research workers, at least when they are engaged in research — but this, of course, is a matter for an empirical research into the philosophical beliefs of scientists and, whether our surmise is confirmed or not by such a research, it has no effect on the truth value of the hypothesis that scientific research presupposes that metaphysical hypothesis. The reality of objective patterns will be granted by anyone agreeing that the central goal of science is the *discovery* of objective patterns, a discovery which is judged to be attained at least in part when certain *inventions* — law statements — pass certain tests. The proposition "LAVOISIER discovered the law of conservation of mass" does not suggest that LAVOISIER made a search in some special law quarry and dug out, i.e. uncovered a ready-made thing called 'the law of conservation of mass'. What LAVOISIER did was to construct a conceptual object that had never existed before, namely the law statement that correctly reproduces the corresponding objective law. In short, law formulas are invented, but laws are discovered.

Law formulas may accordingly be characterized as *conceptual reconstructions of objective laws*. (This would be a definition proper if "objective law" where a primitive in a metascientific theory; but we have tacitly defined "objective law" as the referent of a law formula.) Such conceptual reconstructions are not mere images or reflections of objective laws but are genuine creations of the human mind — creations, to be sure, made with the help of preexisting conceptual material and aiming at truthfully reproducing objective patterns. In this respect a law formula does not differ from a singular empirical proposition such as "The sun is now overhead": this proposition is not discovered either. Facts are discovered sometimes; statements about facts, and a fortiori statements about the structure of facts, are made.

In a sense, the history of factual science is the story of the attempt to discover objective laws in nature and society. In each field of inquiry the results of this attempt constitute a time sequence of law formulas: $L_1, L_2, \ldots, L_n$. This movement is zigzagging but it does show a definite long-term trend of improvement: every one of the law formulas proposed to cover a given objective law may not constitute a better approximation

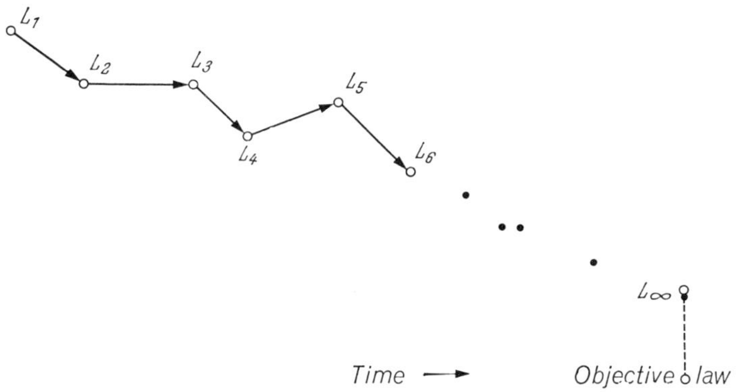

Fig. 6.8. A symbolic representation of the successive steps to the ideal limit of the completely true law statement, L_∞. Notice the temporary regression L_5, probably the effect of a wrong philosophy of science

than its predecessor but on the whole the sequence tends to an unknown and unattainable ideal limit of perfect adequacy to the objective pattern. (See Fig. 6.8.) It would be difficult to understand why such a painful process of successive (but not uniformly improving) approximations should proceed at all were it not on the (metaphysical) conviction that there are laws.

Every one of the successive approximations found in the search for laws has an *extension* or "domain of validity" of its own. Ex. 1: KEPLER's law of the refraction of light, "$i/r = n$", holds for small angles (see Sec. 6.2, Fig. 6.4). Ex. 2: GALILEO's law "The acceleration of gravity is constant" is a first-order approximation to a more complex law in which the acceleration of gravity depends upon the height and the Earth's radius (see Problem 6.5.2). Ex. 3: "In a large animal population, if matings are random the genotypic proportions do not vary from generation to generation (i.e., the population remains genetically stable)." This law (of HARDY-WEINBERG) becomes pointless if the conditions stipulated in the antecedent of the statement (large population and random mating) are not fulfilled; moreover, it is falsified by mutations and natural selection — i.e. strictly speaking it holds for no real population, except as a first approximation.

The above examples, which can be multiplied indefinitely, suggest the following moral: Every law formula has a limited extension or domain of validity — one beyond which it becomes definitely false. This is a sound metanomological formula: it warns againts the dogmatic belief in the unassailable truth of the last-found law formula. Every last-found formula is probably just a member of a sequence of hypotheses. The over-all trend of the sequence is one of relentless perfectioning, but a decrease in error does not eliminate error entirely. So far our metanomological formula concerning the domain of validity of law statements has only an empirical support; it will be given a theoretical justification in Ch. 8, on the basis of an examination of the way scientific systems are built — namely, by focusing on a handful of traits and discarding what are regarded as secondary variables.

A first-order approximation provides a basis for further explorations in search of higher-order approximations. Every discrepancy between a law formula and empirical findings, if interpreted in the light of some hypotheses, becomes a new source of information rather than being just unfavorable or negative evidence. Ex. 1: The deviations from rectilinear motion suggest the presence of forces. Ex. 2: The deviations from the law of the ideal gas are a hint regarding the size and interactions of molecules because that law assumes that molecules are point-like and do not interact. Ex. 3: If a large randomly mating population does not remain genetically stable, the HARDY-WEINBERG law will direct us to search for factors not taken into account by it, such as mutations and natural selection. In short, the search for laws is like a process of growth in which new stages of development result from the former stages and increase the capacity for coping with new problems.

The hypothesis that there are objective laws which we attempt to cover with our law statements solves a number of puzzles and, in turn, poses some difficult problems. One of them is whether every law formula corresponds to an objective law. The most likely answer is in the negative, and this for the following reasons: (i) in order to test a high-level law we must derive from it consequences close enough to experience, and this often requires the help of further high-level statements of a theory; (ii) we often succeed in compressing several law formulas in a single very strong axiom; moreover, most postulates of every quantitative theory can be thus compressed (namely, into a single variational principle). All we can hypothesize is that every system of law formulas, i.e. every *theory*, is a conceptual reconstruction of an unknown number of interrelated objective patterns.

The fact-law formula relation is far from simple. We cannot point to a given perceptible fact (a phenomenon) and say, for instance: 'Look at the facts covered by the law statement I just wrote on the blackboard'. This we cannot do because all known facts are highly complex: they are characterized by a large though unknown number of variables, whereas law formulas relate only a handful of variables (see Sec. 6.2). Law formulas, in fact, do not express "uniform relations among facts" (as the traditional formula has it) but invariant relations among *selected aspects* of facts, and these aspects are not usually on the surface.

Facts are so complex that, if we wish to find their laws, we must analyze them and abstract from most of their properties, focusing on a few at a time. Accordingly one and the same fact requires a number of

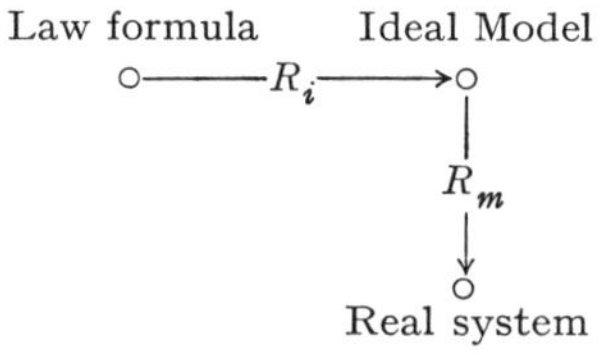

Fig. 6.9. The referents of a law formula: the immediate and the mediate

law formulas for its explanation. Moreover, it is likely that no single real fact will ever be exhaustively explained by a set of law formulas, however numerous. What a law formula can account for exhaustively is *an aspect of an ideal model* of a real system. Moreover, we may say that every law formula proper, in contrast with empirical generalizations, has two referents: an *immediate referent*, which is the schematic representation (the ideal model) of the real system, and the *mediate referent*, which is the real system itself. (See Fig. 6.9.) For example, the classical physics of solid bodies is consistent with at least three models of solid: the continuous substance, the system of unextended mass points and the system of extended atoms. The corresponding equations, which deal with solids as wholes, apply exactly to these idealized models, inexactly to real solids. We shall return to this question in Ch. 8.

Only low-level laws can — in conjunction with bits of information — describe selected aspects of perceptible facts (phenomena). Thus, the kinematical aspect of the motion of a bullet in air will be described, to a first approximation, by GALILEI's law of free fall — itself a deductive consequence of NEWTON's laws of motion; a better approximation will be obtained if the resistance of the air is represented in the ligh-level law statement and the latter is solved. Even so only the kinematical aspect will be covered by such laws, with neglect of the remaining aspects of the motion of the bullet, such as its heating, its loss of gases, the generation of sound waves, etc.

*The low-level laws describing perceptible phenomena contain parameters which, upon specification, allow for the *individualization* of the object concerned; in the case of the free fall these constants were the initial position and the initial velocity (see Sec. 6.4). Now, the values of many of the constants characterizing an individual object, though not *dependent* upon the observer, are *relative* to the conditions of observation. Thus, initial positions and velocities must be reckoned relative to a definite reference frame. Since there are infinitely many possible reference frames, there are also infinitely many different values for most of the constants entering in the low-level laws. A few such quantities, like the number of particles, the pressure, and the electric charge, are *invariant* under changes of reference frames, but most other quantities change alongside such transformations. In short, the low-level laws — by means of which phenomena are described — are relative to the reference system.

*The relativity of the low-level laws should not be interpreted in a subjectivistic sense: 'x is relative to the observer y's reference system' does not necessarily mean "x depends on the observer y". The infinitely many formulas mutually linked by changes in the reference system are all equivalent, at least in the small; in other words, all coordinate systems are locally equivalent, none is physically privileged, not even the one we happen to choose for convenience of observation or computation. Among the infinitely many possible reference systems two are often chosen in preference to all others: the object's proper frame — the one moving along it — and the laboratory frame, i.e. the one relative to which the observations and measurements are made. A physical object is best studied, to start with, in its proper frame: it is in relation to this unique frame that the values of properties such as mass and life-time are computed. But the empirical *test* of any such theoretical consideration requires relating it to the laboratory frame. The choice of the proper frame for theoretical purposes presupposes the ontological hypothesis of the (possible) autonomous existence of the object. And the choice of the laboratory frame for test purposes presupposes the epistemological hypothesis that statements about autonomously existing objects are testable only on condition that they are transformed into statements concerning object-laboratory relations.

*The equivalence of all reference systems is a postulate relativistic theories. When satisfied it has the following important result: The high-level laws hold in every reference frame (system of space and time coordinates). In other words, the high-level laws, unlike the low-level laws, are invariant with respect to changes in the choice of the reference system; in particular, they are independent of the observer. In short, whereas the low-level laws are *relative* to the reference frame, the high-level laws are *absolute*. The high-level laws are, therefore, the most faithful reflection of the objective laws.

*The preceding remarks can be reworded as follows. Let '$L=0$' symbolize a high-level law relating certain properties. L can usually be decomposed into two factors, A and S, in the following way: $L=AS$, where A is a certain operator that acts upon the solution S of the equation constituting the law statement. Relative to a given reference frame R we have, then, $L=AS=0$. Relative to a different reference frame, R^*, A will change into A^* and S into S^*, but in such a way that the change in A will be exactly compensated by the change in S, so that the high-level law L will be identical with its transformed L^*; that is, $L^*=A^*S^*=0$. Upon a change of frame of reference — in particular, upon a new choice of observation conditions — the structure of the high-level law (the way its elements are related to each other) does not change. But its solution will change and so will the *description* of phenomena, which is performed in terms of such noninvariant solutions. Incidentally, a natural mathematical tool for setting up invariant basic equations and invariant quantities — such as scalar products — is the tensor calculus.*

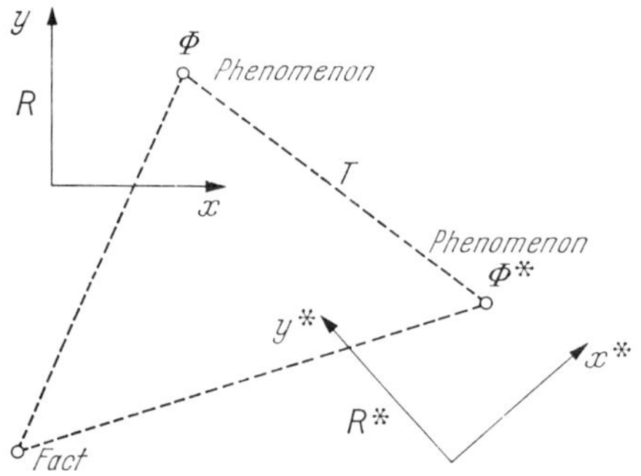

Fig. 6.10. A single fact f is viewed and described as phenomenon Φ relative to the frame R, and as a different phenomenon Φ^* relative to R^*. For the transformation T connecting Φ and Φ^* see Sec. 6.7

Suppose now a given high-level law L describes a certain aspect of a class of facts. Given a certain fact f in this class, a *potentially infinite class of phenomena* Φ will correspond to that single fact, since it can be viewed, in principle, from an infinity of viewpoints, i.e. it can be observed and described by operators tied to an infinity of reference frames (see Fig. 6.10). The relativity of low-level laws corresponds to the potential infinity of phenomena, whereas the absoluteness of the high-level laws corresponds to the uniqueness of the objective fact. This ruins phenomenalism.

Since science strives for objectivity, it must at the same time strive for high-level laws, i.e. for law formulas independent of appearances and circumstances. The introduction of nonobservational concepts and diaphenomenal law statements is then seen to be not only a device forced upon us by the unobservability of most of reality, but also a component of the search for objectivity. It is only when we use or test the law formulas that we must descend (deductively) from them in order to be able to specify the circumstances under which the use or test takes place.

As soon as individual data are introduced in a statement, experience comes in and we produce a *nomopragmatic formula*. Consider, for example, the low-level statement of the law of free fall, namely "$s(t) = \frac{1}{2}gt^2 + v_0 t + s_0$". In order to test it or to use it we must set up a certain reference frame involving a conventional time origin and a conventional position origin: it is with regard to these conventional zeros that we shall record the results of our measurements and of our predictions. Suppose a given measurement, made with 0.1 per cent error, yields the following figures: $g = (9.80 \pm 0.01)$ m/sec², $v_0 = (0 \pm 0.001)$m/sec, and $s_0 = (1 \pm 0.001)$ m. Introducing these data into the low-level law formula we can make the following singular prediction concerning the position the body will reach after 2 seconds: "$s(2) = (20.600 \pm 0.023)$ m". An ordinary language version of this result is something like this: "Two seconds after the body is released from position 1 relative to our reference system it reaches position 20.6 with an error of plus or minus 23 mm". This statement purports to be true of both an objective and an experienced fact.

In summary, just as we distinguish a factual proposition from the fact it refers to, we also distinguish the formulas called 'scientific laws' from their referents, which are patterns of reality in the case of nomological formulas, patterns of experienced reality (phenomena) in the case of nomopragmatic formulas, and patterns of law formulas in the case of metanomological formulas.

Problems

6.5.1. Discuss the van der Waals law for an ideal gas of interacting molecules of nonvanishing size. In what domain does it yield the simpler law statement known as BOYLE-MARIOTTE-CHARLES' law? In what domain is the finite volume of the molecules important? And in what domain do the intermolecular forces makes themselves "felt"? See J. M. H. LEVELT, *American Journal of Physics*, **28**, 192 (1960).

6.5.2. According to the elementary theory of gravitation, the acceleration of gravity g at a height h over the surface of a spherical body of mass M and radius R is

$$g = \frac{GM}{(R+h)^2} = \frac{GM}{R^2}\left(1 - 2\frac{h}{R} + 3\frac{h^2}{R^2} - \cdots\right),$$

where G is the universal constant of gravitation and '...' symbolizes the infinite but converging series of terms of the general form $(-1)^n (n+1)(h/R)^n$.

Point out the first-order and the second-order approximations and figure out whether the first-order approximation would be enough to deal with an artificial sattelite flying at a height $h=2R/10$ (approximately 1,200 km in the case of our planet). *Alternate Problem:* Perform a similar analysis of any other law known in various degrees of approximation, such as the law of swing of an ideal pendulum for arbitrary amplitudes.

6.5.3. In the absence of destructive forces and with unlimited energy and food supply, any collection of self-reproducing systems will grow by constant compound interest, i.e. according to MALTHUS' exponential law

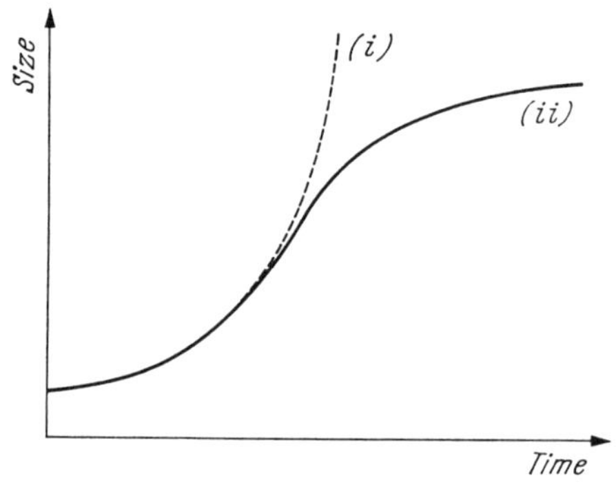

Fig. 6.11. Growth of systems of self-reproducing units. (i) The exponential growth law; (ii) the sigmoid growth law

[curve (i) in Fig. 6.11]. Real populations follow in most cases alternative growth laws; a common law is the one of decreasing rate of growth, or decreasing compound interest [curve (ii) of Fig. 6.11]. What can be inferred from the deviation of actual growth from the hypothesis of unlimited growth [curve (i)]?

6.5.4. Report on any of the following texts. A. N. WHITEHEAD, *Adventures of Ideas* (1933; N. York: Mentor Books, 1955), Ch. 7, especially pp. 115—123. A. SHIMONY, "Ontological Examination of Causation", *Review of Metaphysics*, 1, 52 (1947). M. BUNGE, *Metascientific Queries* (Springfield, Ill.: Charles C. Thomas, 1959), Ch. 4. *Alternate Problem:* Discuss the phenomenalist doctrine of Hume and Kant that laws refer to appearances (phenomana) and that "Natural science will never reveal to us the internal constitution of things" (KANT).

6.5.5. Perform a critical examination of the following doctrines concerning the nature of laws. (i) *Supernaturalism:* laws are norms superimposed on nature by a supernatural power (God, Logos, the Universal Spirit, etc.). (ii) *Nominalism:* the term 'law of nature' does not denote (E. BOUTROUX, G. K. CHESTERTON, P. W. BRIDGMAN). (iii) *Conventionalism:* laws are a priori schemata: handy and simple canvasses on which experience can be painted (I. KANT, H. POINCARÉ, P. DUHEM). (iv) *Empiricism:* scientific laws are mere mental schemata summarizing our actual and/or potential experience. (v) *Pragmatism*: laws are rules enabling us to act. (vi) *Naturalism:* laws are the ways of being and becoming: they just are; objective laws are approximately covered by law statements.

6.5.6. Comment on the following passage from G. ORWELL'S *1984* (N. York: Signet Books, 1950), p. 201, where the intellectual policeman expounds the philosophy of Big Brother: "We control matter because we control the mind. Reality is inside the skull. You will learn by degrees, WINSTON. There is nothing that we could not do. Invisibility, levitation — anything. I could float off this floor like a soap bubble if I wished to. I do not wish to, because the Party does not wish it. You must get rid of those nineteenth-century ideas about the laws of nature. We make the laws of nature".

6.5.7. Argue in favor or against of any of the following theses. (i) Laws shape the events. (ii) Events shape the laws. (iii) Laws are the shape of the events. *Alternate Problem:* Examine the following elucidations of the phrases 'it is physically necessary that' and 'it is physically possible that': (i) "It is physically possible that t" is equivalent to "t is deducible from a set of laws and data". (Difficulty: what about the causal laws?). (ii) "It is physically necessary that t" is equivalent to "t is deducible from a set of laws and data". (Difficulty: what about stochastic laws?)

6.5.8. Comment on the doctrine that scientific laws are generalizations of observations. For typical expositions of this view, consult the following. (i) C. S. PEIRCE, "The Laws of Nature and Hume's Argument Against Miracles", in P. P. WIENER, Ed., *Values in a Universe of Chance: Selected Writings of* C. S. PEIRCE (N. York: Doubleday Anchor Books, 1958): every scientific law "is a generalization of a collection of results of observations" (p. 289), and "of such a nature that from it can be drawn an endless series of prophecies, or predictions, respecting other observations not among those on which the law was based" (p. 290). (ii) H. REICHENBACH, *Modern Philosophy of Science* (London: Routledge & Kegan Paul, 1959): "a law is not a description of what is observed, but of what is observable" (p. 121). *Alternate Problem:* Locate the invariance requirements imposed on the basic laws in the traditional problem system of rest vs. change.

x	y
1	−8
−2	−2
5	9
3	1
−7	0

6.5.9. The above imaginary table displays a set of results of observations of two magnitudes, x and y, at successive instants of time. As can easily be checked, the average of each sequence is zero. This will warrant the following provisional generalizations: "Mean value of $x=0$", and "Mean value of $y=0$". From this we provisionally conclude that, although the two variables have in general different values at any given time — and, moreover, are not correlated, as can be verified — they "obey" the same *law of averages*. Note that in building our statistical generalizations we have compressed data. In general, when building averages we throw information away, and this information cannot be retrieved by analyzing the statistical generalization in question. Is this procedure consistent with the doctrine that scientific laws are faithful generalizations of observations? *Alternate Problem:* When one escapes jukebox noise by taking a supersonic jet, does he violate the basic laws of sound propagation?

6.5.10. Every statement is made up of conventional elements — symbols. In particular, law statements contain special signs such as mathematical symbols which, within bounds, can be chosen at will. Does this prove that the corresponding concepts are arbitrarily chosen? And does it prove that law statements are just handy conventions? If they were, what would be the point of testing them and trying to perfect them? *Alternate Problem:* Examine the widespread doctrine that, whereas the laws of classical physics were supposed to mirror an independently existing world, those of the relativity and quantum theories describe the world as seen by actual or possible observers, or even their own knowledge.

6.6. Requirements

"Sparrows are active", a general and true law of ordinary knowledge, is not rated as a scientific law because we do not understand why sparrows should be particularly active: the ethology of birds is not so far advanced as to have absorbed that commonsensical law in a mesh of scientific laws. Common sense knowledge contains a host of *empirical generalizations* of that kind and we rule much of our daily life in accordance with them. The following characteristics of commonsensical laws deserve being noticed: (i) they refer to daily life events; (ii) they presuppose no specialized knowledge; (iii) they have been subjected to no methodical tests; (iv) they are very often inductions, i.e. summaries of observed or inferred cases; and (v) they are isolated, i.e. nonsystematic.

Factual science, too, contains empirical generalizations. They differ from ordinary knowledge laws in the following respects: they go to some extent beyond daily life events, they are established with the help of specialized knowledge, and they are empirically tested. But, just as commonsensical laws, scientific empirical generalizations are *isolated* or nonsystematic, and most often generalizations of observed or inferred cases. Example 1: "Most intellectuals are liberal." Social psychology may eventually explain this generalization, which for the time being is accounted for in home-made terms such as "Intellectuals are liberal because they need freedom for their work", "Intellectuals are liberal because they tend to settle all disputes by reason", and so on. Such an explanation may be quite correct but it does not resort to systemic hypotheses and is therefore nonscientific. Example 2: "The mean distances of the various planets from the sun, when expressed in an adequate unit, fit the function '$d(n)=4+3.2^n$', where 'n' represents the order." This, BODE's "law", is a stray conjecture and was falsified by the finding of Neptune and Pluto. Notwithstanding, it yields an approximate value for the Neptune-sun distance — namely, 388 instead of the measured 300 — and ADAMS used that value in the computations that lead to the discovery of Neptune. There might be something in this partially true "law" — at present discredited — and it would be interesting to account for its hits as well as for its exceptions. Example 3: "Atomic nuclei with magic numbers are particularly stable." The "magic" numbers (of protons or neutrons) are 2, 8, 20, 50, 82, and 126; these numbers were found empirically — so far as nuclear research can be empirical. This finding stimulated and partly suggested the construction of the shell model of the nucleus, a task of which is to absorb the "magic" numbers regularity, i.e. to obtain it as a law deriving from higher-level statements.

Both commonsensical laws and scientific empirical generalizations are, then, isolated — very often inductions; but in science an effort is made to embed every item in a system. Why should we want to systematize or theorify empirical generalizations? First, because we want to *ground* them (see 5.5), and a way of doing this is to derive them from stronger assumptions belonging to some theory — i.e., to explain them. Second, because we want to *test* the empirical generalizations by checking whether they are consistent with the bulk of knowledge. Third, we do not want mere accidental generalities, short-term coincidences or chance conjunctions, like the sunspot-economic depression coincidences: we suppose that lawful events are in a sense *necessary*, i.e. that they could not have been otherwise under the same circumstances. Now, only a theory or, rather, a representational (nonphenomenological) theory can supply a mechanism showing that the events covered by our law statements are, or are not, systematically linked. (Systematization alone can provide the necessity denied by HUME to laws.) Fourth, we want to be able to make reliable *predictions* by means of our law statements, which is not possible unless they express laws. The examination of a couple of examples should make these points clear.

Example of a nonlaw statement: random time series. Marriage rates, the number of plane crashes, and many other variables are random statistical variables. In other words, if we plot their values over a number of months or years we obtain time series showing neither a long-term trend nor a systematic correlation among the successive items of the sequence: the variations are random, in the sense that successive values of the statistical variable are mutually independent or nearly so (see Fig. 6.12).

Of course we do find *some* pattern even in a time series, but it concerns the series *as a whole*. In fact, we will not be able to say that a given time

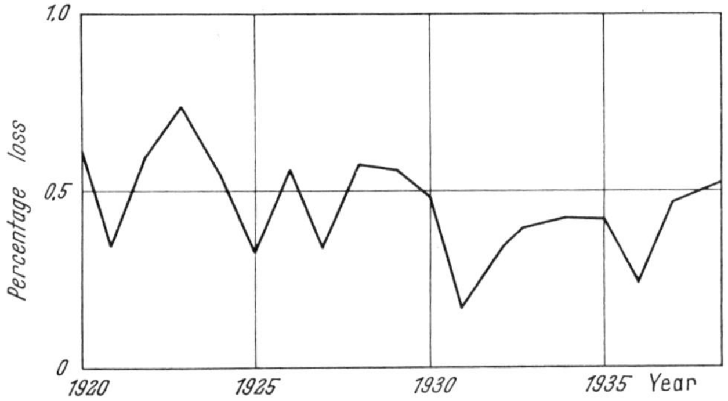

Fig. 6.12. British vessels sunk from 1920 to 1938: a random series. After G. U. YULE and M. G. KENDALL, *An Introduction to the Theory of Statistics* (1950), p. 614

series is random unless it shows certain *collective regularities*. *One of these regularities is the mean of the number p of peaks and troughs, which for a random series of n terms is $\bar{p}=\frac{2}{3}(n-2)$. Another collective regularity is that the mean variance of the number of peaks and troughs is $\overline{\sigma^2(p)}=(16n-29)/90$. If these regularities are not approximately satisfied we may suspect that the random series is not random and may look for a systematic trend in it: the application of the two formulas constitutes therefore a *test for randomness*. Which shows that over-all regularity is consistent with individual randomness and, moreover, that the latter gives rise to regularities on the whole. In other words randomness is a type of lawfulness rather than lawlessness.*

Every one of the individual events making up a random series may be lawful but, since no two individual events in such a chain are really linked, the series does not constitue a law. A time series is just a summary of the net effects of complex and independent processes in which a number of laws may be involved. In building a time series we select one class of events and inquire into its distribution in time instead of following up every single process; for example, instead of studying the lawful process of gestation of every individual we focus on the mutally independent births within a community. Such a selection is not arbitrary from the point of view of our interests but it is arbitrary as regards the course of natural events: as arbitrary as the grouping of stars into constellations. We should not be surprised on finding that an arbitrarily selected set of events fits no law: just as surprised as we would be if we found that the single events themselves (e.g., births) were lawless.

Yet not all time series are random: some express the net effect of the action of definite mechanisms, such as population growth, or the wearing out of machines: in such cases they will show definite *trends* and the theoretician will try to explain such trends by uncovering the mechanisms responsible for them (see Sec. 6.2). That is, nonrandom time series are summaries of data that can in principle be replaced by low-level theoretical laws: they are systematizable, whereas random time series are not. One wonders whether the failure of historians to find laws of history is not due to the fact that, upon focusing on observable gross events, they produce random time series. And one wonders whether they might not find laws of history if they hypothesized hidden mechanisms, just as the physicist and the biologist do. After all, the lawful character of every time series of events in which we happen to become interested would be as miraculous as the lawlesness of every single event of any such series.

Example of a law formula: "All dogs are born with a tail." Genetics is in a position to back up this commonsensical law and even to account for the occasional anomalies constituted by tail-less puppies. In fact, genetics can ground the generalization by explaining it and, by so doing, it slightly corrects the empirical generalization. Should a puppy be born without a tail we would explain this exception in terms of an improbable mutation — especially since we can induce such mutations in the laboratory. Therefore we will surrender universal generality in the benefit of truth and will restate the original, commonsensical law, as follows: "Almost all dogs are born with a tail." In general, both the empirical generalization and the eventual exceptions to it (possible according to theory even though they may not have been observed as yet) become subsumed under the theoretical law. And this law will show that, unless an accident (a mutation) happens, it is necessary for every puppy to have its tail. Upon systematization the commonsensical generalization "All dogs are born with a tail" suffers a slight change in scope ('all' has been replaced by 'almost all') and, above all, it changes its status. It is no longer a mere conjunction of singular propositions such as "Fido was born with a tail", "Snoopy was born with a tail", and so on, ranging over all the observed dogs. The fact that Fido was born with a tail is not independent from the fact that Snoopy, too, can wag its tail: the common possession of this property is assigned to a similar (dog-like) genotype, which in turn is accounted for by a common ancestry. In contrast with empirical generalizations, which express *constant conjunctions*, law statements express *necessary relations*.

Fig. 6.13 sketches the transformation of empirical generalizations into low-level theoretical laws — which contain theoretical concepts. For the sake of simplicity we have assumed an imaginary theory with a single

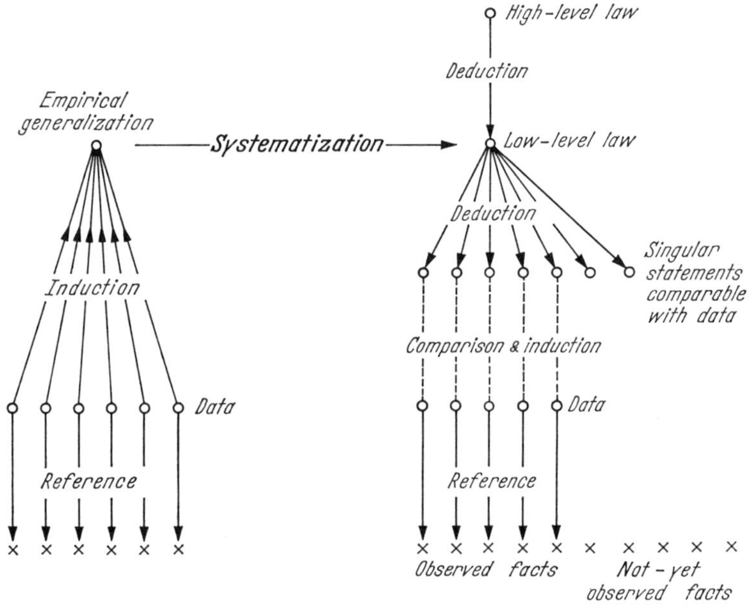

Fig. 6.13. (i) The empirical generalization as a summary of observation data. (ii) The transformation of the empirical generalization into a low-level law (a theorem of a theory)

high-level law statement. The hypotheses of the theory are: the high-level law, the low-level law representing the empirical generalization, and all the singular statements derivable from the low-level law and comparable with the evidence. Notice, firstly, that the theory yields singular statements, such as predictions, referring to possible facts that have not yet been observed: such statements clearly jump over the evidence at hand. Secondly, in the case of the empirical generalization two relations enter: *reference* (linking observed facts and evidence) and *induction* (linking evidence with generalization). In the case of the system, on the other hand, we have three relations: *reference* (facts-evidence), *nondeductive inference* (comparison and induction, relating evidence to prediction), and *deduction* (of singular statements from low-level law, and of the latter from the higher-level law).

Once a given field of knowledge has been theorified the only important distinction among law statements is the one they hold in the theory's logical hierarchy, and this place is determined by the relation of deducibility. Take, for instance, the law of the ideal gas, "$pV=nRT$", which relates three variables (p, V, and T), one parameter (n) and one universal constant (R). This law is deducible from two sets of high-level propositions: the laws of analytical mechanics, and certain statistical hypotheses regarding the randomness (independence) of the individual molecules' paths. In turn, the law is the vertex of a small tree each branch of which is a special law. These special laws were originally introduced as empirical

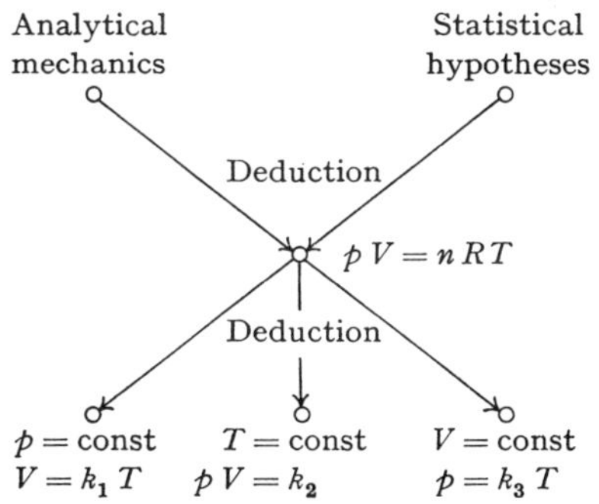

Fig. 6.14. The general law of the ideal gas subsumes three special laws, characterized each by one parameter

generalizations covering the observed behavior of real gases. (See Fig. 6.14.) Another example: with the advent of the atomic theories it became possible to explain the behavior of the various materials, i.e. to derive the special laws characterizing the behavior of the various substances: in this way the previously unexplained parameters occurring in the empirical generalizations are accounted for. In short, theorification reduces the vast array of kinds of law statements to just two logically interesting classes: highest-level laws (axioms or postulates) and lower-level laws (theorems). A finer distinction among the latter, into intermediate-level and lowest-level laws, will depend on the complexity of the theory in which they occur.

Granting that empirical generalizations are not long-term desiderata of research but rather raw materials posing the problem of theory construction, how are we to obtain the theoretical laws? We saw before (6.12) that the search for high-level hypotheses is not a rule-directed activity. Yet in view of certain widely held opinions it may be worth while to mention two procedures by which theoretical laws are *not* secured. One obviously inadequate procedure is *induction:* given a set of empirical generalizations of a certain kind we may occasionally set up a set of higher-level generalizations subsuming them all, but this will provide no theoretical law for the simple reason that theoretical laws contain non-observational concepts introduced by the theory and unnecessary for the statement of the empirical generalizations. Whereas empirical inductions sum up and generalize what is perceived, theoretical laws refer to what is unperceived. Moreover, quite apart from the fact that nonobservational concepts are unobtainable from experience, it is impossible to deduce hypotheses from empirical data: from a datum (or set of data) e we validly infer a unlimited number of conditionals $h \rightarrow e$, not however the hypotheses h themselves, let alone the best of them. Should the reader have the slightest doubt about the impotence of induction to yield high-level laws, he should attempt to apply it to a specific instance; for example, he should try to derive the high-level laws in Fig. 6.14 from the corresponding low-level laws.

Another useless procedure for obtaining high-level laws is mathematization of the empirical generalization by, say, the interpolation techniques (see Sec. 6.2). This is a good procedure for *packaging* and *generalizing* empirical information, but it neither produces the high-level constructs necessary for having a theory nor the necessity that is supposed to be expressed by law statements. In fact any time series, like the random variation of the total sound volume at Piccadilly Circus, may be plotted as a function of time even though there is little or no connection among the individual sounds of the sequence. Such a mathematical function cannot be absorbed by a theory and cannot be used for prediction. Needless to say, computers face the same limitation: their sole advantage is that they can handle huge amounts of data, but laws are not found by processing data — it is rather the other way around: significant data are sought in the light of laws. We conclude that neither mathematization nor induction lead by themselves to the establishment of theoretical laws.

We may now state explicitly the requirements we have tacitly imposed on a hypothesis for calling it a scientific law. One is *genuine generality* in some respect (i.e. with regard to some variable) and to some extent (i.e., between 'most' and 'all'). (The qualifier 'genuine' is intended to exclude pseudouniversal statements like "$(x)\,[x = c \rightarrow P(x)]$".) A second requisite is *empirical corroboration* to a degree regarded as satisfactory at the time the promotion takes place. This condition presupposes factual reference but not empirical meaning. Needless to say, what was regarded as satisfactory empirical confirmation at one time may turn out to be deficient and consequently the promotion may be annuled: there is no tenure for law statements. The third and last requisite will be *systemicity* or belonging to some scientific system, whether full-fledged or developing. Genuine generality, which characterizes the wider class of *lawlike* statements, and empirical confirmation, which is required for assigning a measure of truth, are insufficient for upgrading commonsensical laws and empirical generalizations, but systemicity will prevent such an unjustified promotion. Solid generalizations, like "All ravens are black", and

"Man's gestation period is about nine months (whenever it is not less)" will not be accorded the status of law statements until some tested theory does not back them.

The above is condensed in the following *Definition:* A scientific hypothesis (a grounded and testable formula) is a *law* statement if and only if (i) it is *general* in some respect and to some extent; (ii) it has been empirically *confirmed* in some domain in a satisfactory way, and (iii) it belongs to some scientific *system.*

We have arrived at the door of scientific theory; but before knocking at it we should deal with metalaws and with lawfulness.

Problems

6.6.1. Law statements are usually supposed to be true or at least demonstrable as true. See H. REICHENBACH, *Elements of Symbolic Logic* (N. York: Macmillan, 1947), p. 368. Should truth without qualification be required from synthetic (nonformal) statements? For the approximate and consequently provisional character of physical laws, see P. DUHEM, *The Aim and Structure of Physical Theory* (1914; New York: Atheneum, 1962), pp. 165ff.

6.6.2. Illustrate the process whereby an empirical generalization was subsequently converted into a low-level theoretical law. Employ a good history of science.

6.6.3. Mention a couple of empirical generalizations regarding personality or society and try to fit them into a theory known to you. *Alternate Problem:* All seven U.S.A. Presidents elected at 20-year intervals in zero-numbered years from President HARRISON (1840) to President KENNEDY (1960) died in office. Is this a law statement? Has it any predictive power?

6.6.4. Report on the predominance of empirical generalizations over laws in contemporary social science. See R. K. MERTON, *Social Theory and Social Structure,* 2nd ed. (Glencoe, Ill.: The Free Press, 1957), pp. 95—100.

6.6.5. Exemplify the following classification of law statements. (i) Relations among directly observable variables. (ii) Relations among directly observable and directly unobservable variables. (iii) Relations among variables that are not directly observable. See H. FEIGL, "Existential Hypotheses", *Philosophy of Science,* **17**, 35 (1950).

6.6.6. Should we set up as a desideratum the subsumption of every empirical generalization under scientific theories or is it possible for theories to show that some of our empirical generalizations (and even some of our singular propositions advanced on the strength of experience) are in fact false?

6.6.7. "The quantity of electric charge in an isolated system is constant in time" and "The quantity of electricity in a system does not depend on the frame of reference" are among the most solid laws. They are not only empirically verified but derive from higher-level theoretical laws. They might be corrected in the future, but for the time being there is no indication of this. Rather on the contrary: those law statements are instrumental in making important discoveries (of, e.g., new "fundamental" particles). Yet in order to wind up a certain cosmological theory (the steady-state theory) it has been assumed that electric charged is created out of nothing. What status would you assign to this conjecture?

6.6.8. Universal synthetic (i.e. nonanalytic) statements may warrant counterfactual inference or they may not have such a power. (See 5.3.) In the former case, from "All P's are Q's" we can infer "If c, which is not a P, were a P, then c would be a Q". In the latter case such an inference is impossible. It has been held that counterfactual power distinguishes law statements from nonlaw statements. See R. M. Chisholm, "Law Statements and Counterfactual Inference", *Analysis*, **15**, 97 (1955). Examine whether a statistical law of the kind discussed in Sec. 6.4 (the one about useless cars) complies with this requisite. Try whether counterfactual power might not be employed to characterize generalizations that are not true by chance — in case we know that they cannot be true by chance.

6.6.9. Discuss the requirements for lawfulness laid down by S. Körner, "On Laws of Nature", *Mind*, N. S., LXII, 216 (1953). *Alternate Problem:* Point out the differences between a statistical generalization derived from empirical data (such as, e.g., a statement of the hypnotizability-intelligence inverse correlation) and a stochastic law (such as, e.g., Maxwell's law of velocity distribution). And find out why theoretical statistical laws are usually not mentioned in philosophical discussions of statistical statements.

6.6.10. Whenever we state a relation among two or more variables referring to properties of a class of systems we presuppose that the said relation holds *ceteris paribus* (other things being equal). Question 1: What does this mean in terms of variables? Question 2: Is it true — as is usually held — that the *ceteris paribus* condition is a limitation characterizing the social sciences and from which the physical sciences are free? *Alternate Problem:* The requirements we now place on law statements have originated only recently. Recall that in 1676 R. HOOKE formulated his law "Stress is proportional to strain" in the form: *ceiiinosssttuu*, an anagram of *Ut tensio, sic vis*. Are those requirements like ly to stay with us forever?

6.7. *Laws of Laws

Consider one of the various possible formulations of the covariance principle: "The basic physical law formulas are [or should be] invariant with respect to (general continuous) coordinate transformations". This principle may be interpreted as holding that, since the choice of representation (e.g., of coordinate system) is subjective, it ought not to influence the statement of the highest-level law statements, however much it will determine the form of the solutions to the former equations. Notice, first, that the above principle is general: it refers to every conceivable basic law formula: it is, then, lawlike. Second, the principle is in fact satisfied by a number of important high-level formulas; and, whenever a basic formula is found which does not comply with the principle, efforts are made to modify the given formula so that it will fit the principle — which witnesses to the normative function it fulfils. Third, the principle is far from being a stray proposition: it is entrenched in all relativistic theories, where it plays the role of superpostulate. According to the definition of law formula given in the last section, the principle of covariance is then a full-fledged scientific law. Yet it refers to no event or process in reality: in fact, it refers to law formulas and states an actual or a desirable characteristic of them. It is not, therefore, on a par with the formulas it refers to: logically, hence linguistically, the covariance principle belongs to a level higher than the level of its referents. It is, in short, a *metanomological formula* (see Sec. 6.5).

Quite often it is not clear, in the scientific literature, whether a given law formula is an object proposition or a metaproposition, i.e. whether it refers to facts or to a further proposition. At other times the same idea may be expressed both as an object sentence and as a metasentence. For example, the relativity principle in classical mechanics may be stated either as "All inertial systems are equivalent" (an object proposition) or as "NEWTON's laws of motion hold in all inertial systems" (a metaproposition). Moreover, it is sometimes possible to dispense with a law formula in the postulate basis of a theory on condition that it be reintroduced as a superpostulate, i.e. as a metanomological formula belonging to the corresponding metatheory. Thus, e.g., NEWTON's action-and-reaction principle can be regarded as a postulate proper, but it can also be dispensed with if the following metanomological formula is adopted at least tacitly: "Every force law is to be such that the force exerted by one particle on a second particle equals, with changed sign, the force that the latter exerts on the former". The saving of a postulate by this procedure is, of course, illusory; but it is a historical fact that this course has been advocated.

Advanced scientific theories abound with metanomological statements, i.e. formulas that, while complying with all the requisites of lawfulness, do not reproduce real patterns at the conceptual level but rather *describe or prescribe basic traits of law formulas.* Unfortunately their special logical status is never shown, whence serious muddles often occur. A recent case was the shrinking of the field of validity of the law of conservation of parity, or mirror invariance. The failure to clearly indicate *what* does not remain invariant under a reflection (a change of the coordinates x_i into $-x_i$) results in obscurity as to *where* are asymmetries to be found: whether in facts, in laws, or in both. Apparently 'nonconservation of parity' refers to certain law formulas, and the asymmetry has testable consequences that can be compared with certain facts; in other words, 'nonconservation of parity' is in this case an ambiguous phrase, since it refers both to certain laws and to certain sets of facts. Strangely enough, the ambiguity is never pointed out.

Two genera of metanomological statement can be distinguished: descriptive and prescriptive. (i) *Descriptive* metanomological formulas are statements about actual properties, performance (domain of validity), or usefulness of object law statements. Example: "The laws of history are statistical." (ii) *Prescriptive* metanomological formulas are statements about *desirable* logical, epistemological, or methodological properties of law formulas. Example: "The transition probabilities among different states of a system ought to be the same in all reference frames (or ought to be independent of both observation conditions and representation)." Prescriptive metanomological formulas do not state what, in fact, are the characteristics of law formulas, but rather what they ought to be: they are programmatic, they are not written after the law formulas have been found and consequent upon their examination, but before they are sought: they guide research by limiting the set of candidates to law statement to those formulas complying with certain requisites. Yet clearly, if for some reason a characteristic of a set of law formulas has been found desirable for all formulas in a field of science, then the descriptive statement will be reworded as a prescriptive statement. In other words, prescriptive metanomological formulas, though programmatic, are not a priori.

Conspicuous metanomological formulas are those which state the invariance of a set of laws under certain transformations to which the independent variables are subjected. Take the form

$$L = A\,S = 0, \qquad [6.9]$$

which symbolizes in a compact form most quantitative law formulas (see Sec. 6.5). Both the operator A and the operand S (the solution) will, in general, depend on a number of variables. We shall lump all the independent variables occurring in [6.9] in the single symbol 'v', whence

$$L(v) = A(v)\, S(v) = 0. \qquad [6.10]$$

The symbol '$A(v)$' has no independent meaning if [6.10] is in fact a high-level law statement. On the other hand facts and, in particular, experientiable facts (phenomena) may be described with the help of the solutions $S(v)$. Now, the description of one and the same fact can be made in infinitely many ways, depending on the "point of view"; each mode of observation and description will be characterized by a particular reference frame and a particular set of scales on which the variables v are recknoned. These infinitely many possible descriptions are related to one another via a certain transformation; in other words, there will be a transformation leading from one set of descriptions to another (see Fig. 6.10). Calling $T(v)$ the operator performing the passage among the various descriptions and $S^*(v)$ a transformed description, we may write

$$S^*(v) = T(v)\, S(v). \qquad [6.10]$$

(This is an any-statement: by specifying the value of v we obtain particular transformations and descriptions.) The group of transformations $T(v)$ is infinite but not arbitrary: we want to multiply the number of possible descriptions of facts but we also want to *keep the law formulas invariant under changes in mode of description.* Otherwise we could not claim that our law formulas are adequate descriptions of objective laws: they would be tied to the observer, i.e. to phenomena rather than to objective facts.

In other words, we either *find* that our basic (high-level) law formulas comply with the condition of invariance under certain groups of transformations (Galilei, Lorentz, Hamilton, Gauge, etc.), or we *impose* upon them the invariance condition. In the former case we have a descriptive, in the latter a prescriptive metanomological formula concerning our law statement. And in either case we will have restricted the group of transformations to those leaving the given law formula invariant, and we shall write

$$TL \equiv L^* = L = 0. \qquad [6.11]$$

(We may take a further step by imposing upon T the condition that it has an inverse T^{-1} defined implicitly by $T^{-1}\,T = \mathrm{I}$ (identity). Then $TL \equiv TAS = TAT^{-1}TS = A^*S^* = 0$, where $A^* = TAT^{-1}$ is the transformed operator.)

Such *law-invariant transformations* occur in mechanics (Galilei, Lorentz and canonical transformations), electromagnetic theory (gauge and Lorentz transformations), and quantum theory (unitary transformations).

They are essential to physical theory but they need be assigned no meaning beyond this, which is primarily epistemological: a given set of high-level law statements will explain not only one class of phenomena but an infinite class of classes of phenomena. In other words, one and the same set of phenomena, referred to in a mediate way by a given law formula, is describable in infinitely many ways, one for every possible "point of view"; and all these different descriptions will be equivalent among one another as long as they are related by law-preserving transformations. In short, what is essential for scientific theorizing is not the phenomenon and its descriptions but the underlying fact and its explanation.

On what grounds do we accept or reject metanomological formulas? Plainly, descriptive metanomological formulas will be accepted when true and rejected when false. From prescriptive metanomological formulas, on the other hand, fertility and a vaguely felt philosophical soundness will be required, since programs can be neither true nor false. Now, descriptive metanomological formulas can be of two kinds: *analytic* and *synthetic*. The statement that a certain statement has a given formal property (e.g., that it remains unchanged upon the exchange of t and $-t$) can be checked with paper and pencil only: hence all statements of invariance, or symmetry properties of law formulas, are analytic. Analytic metanomological formulas will, then, be validated just as mathematical theorems. And, if the underlying law formulas are found to be inconsistent with empirical data, the corresponding analytic metanomological formula may become *pointless* (in case it is restricted to stating something about the specific law formula that has been found to fail), or it may be retained — in case its scope is a whole class of law formulas. In no case will experience with facts be competent to judge it and, in particular, to refute it: only "experience" with law formulas is relevant to this case.

Metanomological formulas of the synthetic species are, by definition, sensitive to experience and indeed in two ways. In the first place, if certain metanomological principles are not satisfied, law formulas with squarely false consequences may be obtained; conversely, a proposed metanomological formula will be rejected if it prohibits true law formulas. A second test is success in restricting the number of conceivable hypotheses and thus narrowing down the set of candidates to the title of law formula. For example, if NEWTON's action-and-reaction axiom is stated as a metanomological formula, it will lead to admitting as possible, among others, any force law of the form "$F(x, y)=f(x-y)$, where 'x' and 'y' designate the positions of two mass points and 'f' stands for an odd function of their mutual distance; for, upon exchanging x and y, we get $F(y, x)=f(y-x)=f[-(x-y)]=-f(x-y)$, in accordance with the principle. On the other hand, the principle will prohibit force laws such as $F(x, y) = f(x - y)$, where f is an even function of the mutual distance; it will also eliminate candidates such as $F(x, y) = k\,x\,y$ and $F(x, y) = k\,x/y$, where k is a constant. And we would reject, or rather restrict the domain of validity of the principle if we found sufficiently true law formulas that do not satisfy it. This is, in fact, what happened with electrodynamics, where law formulas occur that force us to restrict the action-and-reaction principle to instantaneously acting forces.

Metanomological formulas of the prescriptive kind cannot be true or false but rather fruitful, barren, noxious, or amusing. Example of an amusing metanomological formula (actually advanced by philosophers who regard themselves as law givers rather than as law students): "No predicate should occur in a law formula which cannot occur in an empirical evidence (observational proposition)". Example of a noxious formula: "No law formula inconsistent with X-ism (fill in the name of any dogmatic philosophy) shall be accepted". Example of a barren formula: "All law formulas ought to be written in either Greek or Gothic characters". Example of a fruitful formula: "All law formulas containing mediating variables that do not denote properties ought to be eventually derived from law formulas containing exclusively, or predominantly, variables denoting properties (i.e. hypothetical constructs)." Metanomological formulas, whether fruitful or not, are clearly seen to belong to the strategy of theory construction and to shade over into the philosophy of science.

Some metanomological formulas of the prescriptive genus lay down the possible forms of law formulas; this is the case of the rival formulas "The fundamental equations of physics must be integral equations" and "The fundamental equations of physics must be differential equations". Other formulas concern the nature of the variables related by the law formulas; e.g., "Any kind of variable may enter a law formula as long as rules of interpretation are appended to the variables so that the formula has testable consequences". The validity of any such proposals consists essentially in their fruitfulness. Their evaluation is accordingly a very delicate matter. Metanomological formulas of the wrong kind may misguide or even block whole research fields. Witness the requirement that law formulas ought to be mere generalizations of observation reports: this prejudice is still holding in check theoretical developments in biology and in the sciences of man. Such dangers can only be avoided by regarding metanomological formulas as *tentative guides* to be corrected as soon as they mislead or shrink the scope of research. Be that as it may, metanomological formulas act as constraints on the possible law formulas: they are not enough to obtain law statements but they function as heuristic clues — usually of a negative kind.

Let us finally mention that the partition of metanomological formulas into descriptive and prescriptive can be shifted. Thus, e.g., instead of saying 'Law formulas *ought to* have the property P' we may say '*Proper* [or correct, or well-conceived] law formulas *have* the property P'. The word 'proper' is here both descriptive and persuasive, like 'good' in 'Good kids go early to bed'. What is important from a pragmatic point of view is the function actually performed by a sentence rather than its linguistic form. Thus, in the case of metanomological formulas some have been so successful that they are regarded as paradigms to be imitated; their mere statement has then prescriptive force no matter what their wording is — much in the same way as an advertisement of the form "x smokes y". However, the pragmatic equivalence between some norms and certain persuasive sentences should not make us forget that norms are not tested in the same way as propositions are: tests of fertility (or conduciveness to given goals) are different from tests for truth.

To conclude. Law formulas are not allowed to crop up haphazardly but are subjected to higher level laws. These — the metanomological formulas — are included partly in scientific theory, partly in metascience. Let us leave this almost virgin soil and face the problem of the lawfulness of reality.

Problems

6.7.1. Examine the following propositions: decide whether they are law statements, metalaw statements, or metanomological statements. (i) "The laws of economics cannot be deduced from psychological laws alone". (ii) "The laws of nature are such that it is impossible to build a perpetual motion machine".

6.7.2. Examine the invariance conditions imposed upon psychophysical laws by R. D. LUCE, "On the Possible Psychophysical Laws", *Psychological Review*, **66**, 81 (1959).

6.7.3. In any law statement in which the divergence of a vector occurs it is possible to add the curl of an arbitrary vector to the given vector because the divergence of the curl of a vector vanishes. What kind of a statement is this? And why is such a possibility not exploited unless there is evidence that the additional vector represents a physical property: for the sake of simplicity of for the sake of testability?

6.7.4. Characterize the following sentence, which is usually regarded as a law of nature: 'In nature only symmetrical or antisymmetrical states are realized' — meaning "The wave functions occurring in the wave equations that account for the known particles and fields are either symmetrical or antisymmetrical". *Alternate Problem:* Examine the methodological status of the metanomological formula stated in Sec. 5.5: "Every law formula has a limited domain of validity (or, rather, a limited extension)". Is it both confirmable and refutable or just confirmable?

6.7.5. Characterize the following statements. (i) H. JEFFREYS, *Scientific Inference*, 2nd ed. (Cambridge: University Press, 1957), p. 36: "The set of all possible forms of scientific laws is finite and enumerable, and their initial probabilities form the terms of a convergent series of sum 1. We shall call this principle the *simplicity postulate.*" (ii) H. J. BHABHA, "On the Postulational Basis of the Theory of Elementary Particles", *Reviews of Modern Physics*, **21**, 451 (1949), p. 453: ... "the equations of motion shall only contain universal constants besides the wave function ψ, its complex conjugate and their derivatives".

6.7.6. Examine the statement "Laws are independent of location in spacetime". What evidence do we have in support of it? Do we know for certain that it holds or do we assume this? Decide whether it is fertile and whether it could be refuted, and if so how.

6.7.7. Examine the device of "eliminating" a postulate by rewording it as a metanomological statement and pushing it to the body of the theory's presuppositions.

6.7.8. Reword the current discussions on intervening variables and hypothetical constructs in psychology as a conflict between sets of metanomological statements of the prescriptive kind. *Alternate Problem:* Elucidate the concept of conflicting prescriptions or norms.

6.7.9. Consider the following statement by W. HEISENBERG, cofounder of the quantum theory, in *Daedalus*, **87**, 95 (1958): "The laws of nature which we formulate mathematically in quantum theory deal no longer with the particles themselves but with our knowledge of the elementary particles". Must we accordingly say that the laws stated by the quantum theory are not physical but epistemological laws? Or shall we say instead that they are metanomological statements? Or, finally, shall we say that some are object law statements and other metanomological statements? In either case: how shall they be tested? Hint: do not let yourself discourage. *Alternate Problem:* The laws characterizing ideal materials, such as rigid bodies, must hold in all reference frames, if their behavior is not to depend on the mode of description. Study to which extent and how this is achieved. See C. TRUESDELL and R. TOUPIN, "The Classical Field Theories", Ch. G, in S. FLÜGGE, Ed., *Encyclopedia of Physics* (Berlin: Springer, 1960), Vol. III/1.

6.7.10. Discuss the nature of the following theorem: "Relativistically invariant field equations are invariant under the combined charge, time, and parity inversions" (Lüders-Pauli combined parity theorem). See M. BUNGE, *The Myth of Simplicity* (Englewood Cliffs, N. J.: Prentice-Hall, 1963), Ch. 12. *Alternate Problem:* Study the problem of the pragmatic equivalence (inequivalence) of logically or semantically inequivalent (equivalent) formulas.

6.8. The Rule of Law

How would a geologist react if told that — as the Latin fable has it — the mountain gave birth to a mouse? He would, of course, laugh at it because the alleged fact fits no law of nature, nay is incompatible with the known laws of nature: being a scientist he is supposed to abide by the principle that *Every event fits a set of laws*. Stated negatively, as a scientist the geologist is not supposed to believe in miracles, i.e. in "violations" of objective laws. Only man-made "laws" (the rules of social life) can be violated; law statements are not violated but refuted.

In nature, and partly in culture as well, not everything that is logically possible is physically possible and therefore bound to occur in the long run. The pervasive and constant restrictions on logical possibilities are precisely the objective laws. To say that anything might happen, or that there are no limits to nature's whims, is to grant the existence of lawless events. This possibility is not encouraged by science: even socially unlawful behavior is accounted for by laws.

On the other hand every scientist knows, or suspects, that law formulas, unlike objective patterns, may be either irrelevant to certain facts or falsified by fresh evidence or by fresh theoretical argument. In both cases — irrelevance and falsification — certain facts fall outside the set of laws concerned — not however outside every law. Thus, e.g., the number of planets in a solar system, and their distances to the sun, are accidental relative to their laws of motion. But they will fall under the laws of an adequate theory of the origin of solar systems, i.e. they will cease to be accidental in this alternative context. Likewise, what is at stake in the "violation" of a law is a shortcoming of a hypothesis, a corrigible assumption going beyond the evidence at hand. In other words, scientists are prepared to meet exceptions to law formulas, not however to objective laws.

What, then, is the normal attitude of a scientist faced with an infrequent event? His first attempt will be to fit it into a known, though perhaps off-the-track law formula. Failing this he will try to devise a more comprehensive pattern wherein the exception may fit, thereby ceasing to be an exception. Should this attempt fail he would not despair of the lawfulness of reality but rather of his own wits. A couple of examples will make this clear.

Suppose a botanist saw a dwarf specimen of sequoia. If he were reasonably certain, on the strength of certain age tests, that the tree is both old and underdeveloped, he would try to explain this unexpected and therefore striking fact — i.e. he would try to "legalize" it. To this end he would try to discover the properties "responsible" for dwarfness, such as the lack of central roots or an abnormal hormone concentration; in turn he might try an explanation of either in terms of environment or of mutations. At any rate the botanist would try to explain the exception to the empirical generalization "Old sequoias are gigantic" by finding

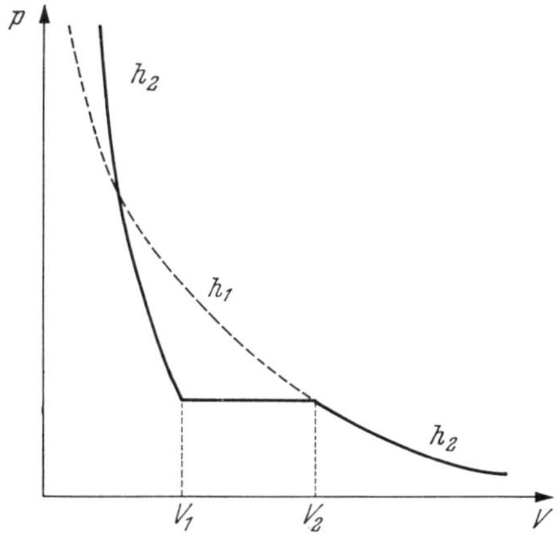

Fig. 6.15. Isothermal compression of a real gas. The first observations were consistent with h_1. Exceptions (deviations from h_1) were then found. The new hypothesis, h_2, accounts for these exceptions, which are in turn interpreted

a more exact and comprehensive pattern making room for occasional dwarf sequoias.

Another example: an empirical generalization h_1 is first established and subsequently exceptions to it in a certain domain are found. If the experimenter is reasonably certain that the new observations are more exact than the former, he will replace h_1 by a new hypothesis h_2 embodying the exceptions to h_1. Moreover, he will try to *explain* why facts fit h_2 rather than h_1: that is, he will try to subsume the new empirical generalization under a theoretical law accounting for the mechanism of the process or will try to explain the deviation from h_1 as due to the interference by another law. The pattern is always this: the exception to a given law formula is enthroned as an instance or a consequence of a more comprehensive and exact formula. Thus, e.g., after Boyle proposed his law of the ideal gas it was found that in a certain volume range changes of volume are not accompanied by changes of internal pressure. (Had Boyle's observations been very accurate he would not have discovered his law.) Accordingly, Boyle's hypothesis h_1 was replaced by a more exact hypothesis, h_2 (see Fig. 6.15). But physicists did not remain content with replacing one law by another: they explained the failure of h_1 in the interval $[V_1, V_2]$ with the help of the atomic theory. At V_2 the system ceases to be a pure gas: drops of liquid begin to form, as a consequence of which fewer molecules collide against the wall; and at V_1 no gas remains, and the steeply ascending part of the curve corresponds to the small compression of liquids.

Exceptions, if confirmed, are not to be dissimulated since they stimulate the search for new law formulas. (Scientific progress, like moral progress, is possible only on the basis of recognizing imperfections.) This manner of dealing with exceptions involves the acknowledgment of the principle that exceptions are not absolute, that every exception is such in relation to a set of law formulas. But this presupposes the ontological principle of lawfulness: *All facts are lawful* (see Sec. 5.9).

Loosely speaking, this means that every fact is an instance of a law. But since only general propositions have instances, and objective laws are not propositions, facts cannot exemplify laws. We might say, instead, that every statement of fact is a substitution instance of a statement of law. But this would be doubly false: first, because we daily utter contrary to law statements; second, because only lowest-level logical consequences of the high-level law formulas can be compared with empirical evidence. A better wording seems to be this: *Any fact fits a set of laws* or, if preferred, *Any fact can or could eventually be explained by a set of law formulas* (and a set of empirical data).

There is a troubling ambiguity about the term 'fact': shall it cover both simple and complex facts, shall it be applicable to single facts and to set of facts as well? Accordingly, shall we assume that every fact, however complex, is lawful and that any set of fact we please will satisfy some set of laws? Clearly, we cannot put limitations on the complexity of facts, because every event requires the interplay of at least two objects and every real object has a number of properties. Moreover, we cannot presume to know them all, so that we must assume that real complexity is at least as large as the complexity our theories allow us to discern. Hence we must assume that lawfulness, if it applies at all, must apply to every single fact, no matter how complex it be. It is not so with *arbitrary sets of facts* — for instance, with the sets of events that command the interest of a historian. It would be absurd to assume that any given sequence of events that happens to catch our imagination "obeys", *qua* sequence, a given law or a set of laws. We may well assume that every member of the sequence is lawful, but the sequence as a whole need not be lawful: it will be lawful only as long as it constitutes a *single process*, i.e. a chain (whether strong or stochastic) of events such that every one of them produces its successor or exerts an influence upon it.

In the elucidation of the principle of lawfulness we have smuggled in the concept of *set of laws*. This is because no known fact can be accounted for by a single law formula: every known fact "obeys" a number of law formulas. It is only by conventionally selecting one class of aspects, e.g. the mechanical one, that we may delude ourselves into believing that one, or even a handful, of law formulas will exhaustively account for a given real fact. The concepts of mechanical fact, electrical fact, and so on, are as many abstractions with no exact real counterpart. Every fact has a number of aspects, one of which may be definitely predominant over the others. For example, a low-energy collision between two bodies will in general be an almost exclusively mechanical fact, but even so there will be subsidiary nonmechanical facts in this; e.g., the generation of some heat, or the disturbance of electrical equilibrium in the collision zones. Accordingly a number of laws will subsidiarily operate in a collision in addition to mechanical laws. A second reason for insisting that facts fit sets of laws rather than single laws is, that the rigorous explanation of even a single aspect of facts usually requires one or more theories, i.e. systems of law formulas.

But why should we accept the principle of lawfulness? Might not there be absolutely lawless facts, i.e. facts fitting no set of laws? And could not we, men of the space age, change the laws of nature to some extent? That there might be spontaneous, indeterminate swervings from law lines has often been maintained from Epicurus on, but not argued cogently. It has even been held that microphysics has abandoned the principle of lawfulness, that the fluctuations characteristic of the quantum theory are entirely chaotic. But this is mistaken: it is only from the *laws* of the quantum theory that we conclude that quantum events show a characteristic statistical fluctuation — which, incidentally, is computable. To establish that there are lawless events takes proving that such events fall outside *every possible* set of law formulas. And who dares attempt this?

No scientist, if presented with an anomalous fact, could be justified in lazily concluding that the given fact is absolutely lawless, because no scientist can marshall all the available and discoverable laws. And only a complete knowledge of laws could establish the lawlessness of a fact, because 'lawless' means precisely "fitting no set of laws". The mutually contradictory hypotheses: "All facts are lawful" and "Not all facts are lawful" (i.e., "Some facts are lawless") are not equally likely in the light of available evidence. "All facts are lawful" has so far been confirmed by science and, moreover, it may be regarded as confirmable only but not as refutable. On the other hand, "Not all facts are lawful" will never be established if, as is likely, the physical universe is infinite in at least one respect (temporally or spatially).

There is an apparent exception to the principle of lawfulness: the set of theorems in statistical mechanics beginning with the phrase 'Almost always' or some logical equivalent of it, such as 'For almost all points' or 'For almost every trajectory'. Any statement of this kind allows for a countable infinity of exceptions to the law stated after the disturbing prefix. What about these possible exceptions: are they realized in nature? For instance: are spontaneous decreases of entropy occurring in some isolated system? In case some were: would they be recognized as exceptions or rather, owing to their improbable and unsystematic repetition, interpreted as experimental errors? And in case they were recognized as anomalies: would they be absolutely lawless or would they "obey" alternative laws — i.e. would it be possible to subsume them under strictly universal laws? These questions are not answered by present-day physics. They are not answered mainly because they are not being asked. At any rate, the existence of laws of the almost-all type does not refute the principle of lawfulness, which does not state that every event fits the known laws, but rather that no fact falls outside some pattern, whether known or unknown. And an unexplained exception is a challenge to the discovery of new laws rather than a disconfirmation of the lawfulness principle.

Even supposing that nature does not "violate" its own laws, it might be asked whether man could change them. According to an ancient belief man can in fact "violate" some laws of nature (e.g., by acting *contra natura*) — but then Nature takes revenge on the violator. What is really meant in this case is that social conventions regarded as "natural", because they are old, can be violated but at the price of punishment. What about laws proper? The current cultivated opinion is that laws proper cannot be changed by man but that man can temporarily check or suspend the operation of a few laws. For example, we can prevent the normal growth of certain populations by controlling their food supply or their reproduction rate or even by suppressing them. But, of course, in this case we use, as it were, some laws to counteract others.

It may be argued, though, that in addition to checking and suspending the operation of certain laws man can elicit the emergence of *new laws* by just producing radically new events or things. In fact, if something new is produced this novelty will be either (i) in *number* (e.g., a new car in a standard series), or (ii) in *arrangement* (e.g., a new car model), or (iii) in *quality* (e.g., a new elementary particle, or a new polymer, or a new hybrid plant not previously found in nature). Now radical novelty, i.e. the novelty that has presumably emerged for the first time in the history of the universe, is characterized either by new properties or by new relations among preëxisting properties. If the properties are new the principle of lawfulness will direct us to hypothesize that they will be related in an invariant way, i.e. that they will enter laws — new laws of course. If the properties had previously been exemplified, then the relations among them will constitute a new law. In either case genuine novelty in things or events accompanies genuine novelty in laws. In short, the emergence of qualitative novelty overlaps with the emergence of new laws. This is only too clear in the case of objects under man's direct control, namely, social relations. Whenever man has created a new form of society he has at the same time created new laws of society, though retaining some of the older laws, probably those which are most directly tied to basic biological and psychological laws. But the emergence of new laws entails no violation of the lawfulness principle. It only poses the tricky problem of the laws of emergence of laws (see Sec. 9.5).

What ground is there to accept the principle of lawfulness: might it not be just the carrot that keeps science going? In addition to its heuristic value, the following reasons can be adduced in support of it. First, it has been confirmed, whereas the assumptions concerning lawless facts have all been disproved in due course. Second, the search for law — the marrow of scientific research — presupposes not only that there are laws (weak principle) but more particularly the strong principle that nothing occurs outside laws. The dogmatic assertion that there must be certain fields — e.g., Mind — that are inherently lawless is scientifically and philosophically crippling because it precludes from the start the search for laws in the given field. In this way, in fact, the presumptive lawless field is excluded from science. Because there is no science proper without law formulas. The mere pursuit of science therefore presupposes the principle of lawfulness (see Sec. 5.9).

What is the status of the principle of lawfulness? It has been maintained that it is an analytic proposition, i.e. a logically true formula. This would indeed be the case if "law" and "fact" were defined in terms of each other. For instance, if we defined "fact" as "that which fits a set of laws", the principle "Every fact is lawful" would become a tautology. We can take this decision but we must abstain from taking it because the principle of lawfulness, as usually understood, is far from being hollow and there is no advantage in emptying it. Another objection to the principle might be that, although it is confirmable, it cannot be refuted. In fact, if a seeming exception to the principle were to appear we would immediately shield it by saying that, given time, a truer law formula subsuming the apparent anomaly will be found. And no science lover could sensibly reject this protective and programmatic hypothesis, because to kill the principle of lawfulness would be a worse crime than killing the golden egg hen: it is not just a piece of knowledge but a motor of knowledge. The only way out, then, is to acknowledge irrefutable hypotheses — among them the lawfulness principle — as long as they further research instead of blocking it. (The need for irrefutable but confirmable and grounded hypotheses in science is argued in 5.8.)

Supposing we accept the principle of lawfulness we must ask what kind of a principle it is. It is not a scientific hypothesis since it refers to no particular class of facts and is not fully testable by experience. The principle is not metascientific either as it does not refer to science although it is relevant to it. (At first sight it might count as a metanomological principle since it refers to laws. But this identification would be wrong, for clearly the term 'lawful' occurring in the sentences expressing our principle means "fitting objective laws" rather than "agreeing with law statements".) The principle of lawfulness, being neither scientific nor logical nor epistemological, must be ontological — with the proviso that the term 'ontology' is to be stripped of its traditional meaning of "science of being as such", independent of factual science, and be understood instead as a discipline dealing in a scientific spirit with wide categories having a factual reference, such as law, time, and organization, and with law formulas that are not restricted to special fields, like "Nothing will grow forever". In short, then, the principle of lawfulness is an ontological principle presupposed and confirmed by scientific research.

The extent to which the principle of lawfulness drives research can be gauged on realizing that it is the basis for a methodological principle that, in one way or other, has always inspired research. This is the following

Rule. Search for laws without allowing yourself to be stopped either by failure (discovery of exceptions) or by success (finding of as yet exceptionless law formulas).

This completes our account of laws. We can now approach the subject of systems of law formulas, i.e. scientific theories.

Problems

6.8.1. Allergy and anaphylaxis were initially regarded as exceptions to the laws of immunity and are now regarded as a kind of immunity reaction. Account for this development in terms of lawfulness.

6.8.2. Accidental or occasional parthenogenesis ("virgin birth") has been regarded as either impossible or miraculous and in either case outside the order of law. What is its present status? Hint: gather and discuss information regarding the embryology of lizards. *Alternate Problem:* The genetic code can suffer spontaneous changes (mutations) as a result of "proton errors". Are such "errors" random departures from law?

6.8.3. Comment on MONTESQUIEU's phrase "Laws are the necessary relations that derive from the nature of things" (opening sentence of

L'esprit des lois). *Alternate Problem:* Discuss Chap. X, "On Miracles", of D. HUME's *An Enquiry Concerning Human Understanding* (many editions).

6.8.4. Examine the following fragment from B. RUSSELL's "On Scientific Method in Philosophy" (1914), repr. in *Mysticism and Logic* (London: Penguin, 1953), p. 99: ... "what is surprising in physics is not the existence of general laws, but their extreme simplicity. It is not the uniformity of Nature that should surprise us, for, by sufficient analytic ingenuity, any conceivable course of Nature might be shown to exhibit uniformity. What should surprise us is the fact that the uniformity is simple enough for us to be able to discover it." What kind of law formulas did RUSSELL have in mind: empirical generalizations of the kind of empirical curves, or theoretical laws? And what kind of simplicity did he have in mind: syntactical, semantical, epistemological, or pragmatic? For an analysis of simplicity, see M. BUNGE, *The Myth of Symplicity* (Englewood Cliffs, N. J.: Prentice-Hall, 1963), esp. Chs. 4 and 5. *Alternate Problem:* Work out the differences beween trend line and law. Take into account that, whereas laws are usually assumed to be eternal (which is controvertible), no trend line seems to continue indefinitely: the longer a trend line has gone on the more likely it is to stop or change (G. G. SIMPSON).

6.8.5. Examine A. TOYNBEE's criticism of antinomianism, i.e. the disbelief in laws of history. Cf. his *A Study of History*, in the *Abridgment* of Volumes VII—X by D. C. SOMERWELL (New York and London: Oxford University Press, 1957), pp. 265 ff. *Alternate Problem:* What do physicists mean by 'anomalous': fitting no law or disconfirming an otherwise well supported law?

6.8.6. Describe some traits of an imaginary world without laws. In particular discuss whether we might act deliberately and effectively and whether we might learn from experience in such a world. *Alternate Problem:* Discuss the conjecture that our unusual experiences, as well as the anomalies found in the course of scientific work, are cases of breakdown of laws of nature and, in particular, result from the interference with "psychical phenomena" such as psychokinesis.

6.8.7. Is every set of facts lawful? In particular, is every set of successive events of a kind lawful? Recall the case of the random time series (see 6.6). Try to set up a criterion of lawfulness for sets of facts and relate this with the problem of the laws of history. *Alternate Problem:* Could experience disprove the principle of lawfulness? See W. WHEWELL, *Philosophy of the Inductive Sciences*, 2nd ed. (London: Parker, 1847), I, p. 253.

6.8.8. H. REICHENBACH, in his *Elements of Symbolic Logic* (N. York: Macmillan, 1947), p. 393, proposed the following definitions

Df. 1. p is physically necessary $=_{df}$ 'p' is a nomological statement
Df. 2. p is physically impossible $=_{df}$ '$-p$' is a nomological statement
Df. 3. p is physically possible $=_{df}$ neither 'p' nor '$-p$' are nomological statements.

Discuss the following questions. (i) Probability law formulas would, according to Df. 1, cover necessary events alone — unless we capriciously stipulated that 'nomological statement' does not designate a probability law formula. Since either result seems undesirable, what are we to do with Df. 1 and its consequence, Df. 2? (ii) Df. 3 seems to equate "physically possible" with "lawless", which would consecrate miracles as physically possible. Is there some way out? (iii) Propose your own definition of "physically possible". *Alternate Problem:* Science studies a number of kinds of random events. Does 'randomness' mean "according to no law"? See J. VENN, *The Logic of Chance,* 3rd ed. (1888; N. York: Chelsea Publishing Co., 1962), Ch. V, and M. BUNGE, *The Myth of Simplicity* (Englewood Cliffs, N. J.: Prentice-Hall, 1963), Ch. 11.

6.8.9. Assuming that the stock of objective laws might change, two modes of change are conceivable: (i) new laws emerge, and (ii) preëxisting laws change. Question 1: What could the change of a law consist of? Question 2: Would such a change be discontinuous or continuous? Question 3: Would the emergence of new laws and the change of preëxisting laws be lawful? If so, would there be laws of change of laws? *Alternate Problem:* Many things are said to be arbitrary: for instance, names are arbitrary in the sense that they need not match any property of the named objects. Are these cases of lawlessness?

6.8.10 Discuss the following fragment from H. WEYL's *Symmetry* (Princeton: Princeton University Press, 1952), p. 26: "If nature were all lawfulness then every phenomenon would share the full symmetry of the universal laws of nature as formulated in the theory of relativity. The mere fact that this is not so proves that contingency is an essential feature of the world". Question 1: Is WEYL right in presupposing that theoretical laws describe phenomena and, indeed, that every phenomenon is exhaustively describable in terms of a single law formula or even of a single theory? Question 2: Are symmetries in the high-level laws necessarily retained by their low-level consequences, by means of which phenomena are described? Consider, e.g., the temporal symmetry of the laws of mechanics, i.e. their invariance under the exchange of t and $-t$. Question 3: Is lawfulness to be identified with the mathematical symmetry exhibited by the fundamental laws of relativity theories (when written in a form adequate to produce the mathematical symmetry among the time and space coordinates)? *Alternate Problem:* Discuss the role of laws in knowing the essence of actuals and in foretelling possibles.

Bibliography

ASHBY, W. ROSS: An introduction to cybernetics, ch. 7. London: Chapman & Hall 1956.

BRIDGMAN, P. W.: Dimensional analysis. New Haven, Conn.: Yale University Press. 1922.

BUNGE, M.: Metascientific queries, ch. 4. Springfield (Ill.): Ch. C. Thomas 1959.

— Causality, 2nd. ed., ch. 10. New York: Meridian Books 1963.

— The myth of simplicity, chs. 9—12. Englewood Cliffs, N. J.: Prentice-Hall 1963.

EXNER, F.: Vorlesungen über die Physikalischen Grundlagen der Naturwissenschaften, 2nd ed., part IV. Leipzig u. Wien: Franz Deuticke 1922.

HOLT, R. R.: Individuality and generalization in the psychology of personality. J. Person. **30**, 377 (1962).

JEVONS, W. STANLEY: The principles of science, 2nd ed. (1877), chs. xxi, xxii, xxix and xxxi. New York: Dover 1958.

KANT, I.: Prolegomena to any future metaphysics (1783). Indianapolis: Bobbs-Merrill 1950.

KNEALE, W.: Probability and induction, secs. 16—20. Oxford: Clarendon Press 1949.

MEHLBERG, H.: The reach of science, part II, ch. 2. Toronto: University Toronto Press 1958.

MILL, J. STUART: A system of logic, 8th ed. (1875), Bk III, ch. iv. London: Longmans, Green & Co. 1952.

POPPER, K. R.: The poverty of historicism, 2nd ed., secs. 20 and 26—28. London: Routledge & Kegan Paul 1960.

— The open society and its enemies, 4th ed., I, ch. 5. London: Routledge & Kegan Paul 1962.

SCHRÖDINGER, E.: Science: theory and man (1935), ch. vi. New York: Dover 1957.

YULE, G. U., and M. G. KENDALL: An introduction to the theory of statistics, 14th ed., chs. 26 and 27. New York: Hafner 1950.

Chapter 7

Theory: Statics

Syntheses are beyond infant science just as they are absent from infant thought. Scientific research, just as infant inquiry, starts with questioning but, unlike infant questioning, it culminates with the construction of closely knit systems of ideas, i.e. theories. A peculiarity of 20th century science is that the most important scientific activity — the deepest and most fertile — is centered around theories rather than around stray questions, data, classifications, or stray conjectures. Problems are posed and data are gathered in the light of theories and with the hope of conceiving new hypotheses that may in turn be expanded or synthesized into theories; observations, measurements and experiments are executed not only to collect information and generate hypotheses but also to test theories and find their domain of truth; and action itself, to the extent to which it is conscious action, relies more and more on theories — for better or for worse. In short, an emphasis on system — on empirically testable theory, of course — rather than on raw experience is what characterizes contemporary science.

Scientific theory can be studied either as an activity or as a finished even though not final product of that activity. But it is hopeless to try to understand the dynamics of theory construction before knowing what theories are. Whence the unhistorical approach chosen here: we shall start by examining the peculiar formal and semantical traits of theories and shall then, in the next Chapter, cast a glance at theory construction and theory reconstruction. The problems of theory application and validation will concern us in Volume II.

7.1. The Nervous System of Science

The childhood of every science is characterized by its concentration on the search for singular data, classifications, relevant variables, and isolated hypotheses establishing relationships among these variables, as well as accounting for those data. As long as a science remains in this semiempirical stage it lacks logical unity: a formula in one department is a self-contained idea that cannot be logically related to formulas in other departments of a science. Consequently the test of any one of them may not affect the others. In short, while in the semiempirical — pretheoretical — stage, the ideas of a science are neither mutually enriched nor mutually controlled.

As research develops, relations among the previously isolated hypotheses are discovered or invented and entirely new, stronger hypotheses are introduced which not only include the old hypotheses but yield unexpected generalizations: as a result one or more *systems of hypotheses* are constituted. These systems are syntheses encompassing what is known, what is suspected, and what can be predicted concerning a given subject matter. Such syntheses, characterized by the relation of deducibility holding among some of its formulas, are called hypothetico-deductive systems, models, or simply theories. Outstanding examples of scientific theories are NEWTON's mechanics, DARWIN's theory of evolution, and HULL's theory of behavior.

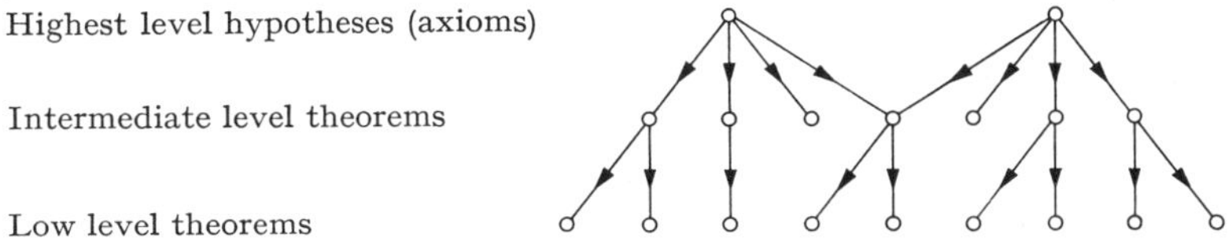

Fig. 7.1. Graphic representation of an imaginary axiomatic theory with two initial premises (axioms). The arrows symbolize the relation of logical consequence or entailment $\vdash$. When two arrows lead to a node, the deduction of a formula from two higher level formulas taken jointly is represented

In ordinary language and in ordinary metascience 'hypothesis', 'law' and 'theory' are often exchanged; and sometimes laws and theories are taken to be the manhood of hypotheses. In advanced science and in contemporary metascience the three terms are usually distinguished: 'law' or 'law formula' designates a hypothesis of a certain kind — namely, non-singular, non-isolated, referring to a pattern, and corroborated; and 'theory' designates a system of hypotheses, among which law formulas are conspicuous — so much so that the core of a theory is a system of law formulas. In order to minimize confusions we will provisionally adopt the following characterization: A set of scientific hypotheses is a scientific theory if and only if it refers to a given factual subject matter and every member of the set is either an initial assumption (axiom, subsidiary assumption, or datum) or a logical consequence of one or more initial assumptions. This characterization will be refined in the suceeding sections. It suggests tree diagrams such as the one on Fig. 7.1, which represents an imaginary theory and illustrates the subclass of theories known as axiomatic theories, which contain no assumptions other than the axioms. Everything in these theories hangs from the initial assumptions or highest level hypotheses. If the name 'basic assumption' is preferred to 'highest level hypothesis', the tree may be turned upside down; in this case a form of the tree diagram familiar from algebra is obtained.

A heap of uncoördinated hypotheses, though better than no hypothesis at all, may be likened to a huge mass of protoplasm without a nervous system. It is inefficient, unenlightening and, moreover, it does not account for the actual interrelations that obtain among some of the real patterns (objective laws). The progress of science, therefore, does not consist in an unchecked heaping of isolated generalizations, let alone of single data. The progress of science involves always, among other things, an increase in *systemicity* or coordination. Systemicity has, among others, the following virtues: (i) a factual proposition can acquire full meaning only within a context and by virtue of its logical relations with other items of this context, whereas an isolated proposition — to the extent to which it can exist at all — is hardly meaningful; (ii) by being absorbed in a theory, a hypothesis is supported (or disclaimed) by a wider field of facts: namely, by the whole field covered by the theory, whereas an isolated hypothesis has just the support of its cases, if any. In short, systematization renders the meaning of hypotheses more precise and enhances their testability. In addition it explains most of the hypotheses by subsuming them under stronger assumptions (axioms and intermediate level theorems).

The relative bulk and adequacy of theoretical work measures then the degree of advancement of a science, much as the relative bulk and efficiency of the nervous system is an index of biological progress. This is why psychology and sociology, despite their huge store of empirical data and low-level generalizations, are regarded as being still in an underdeveloped stage: because they do not abound with theories wide and deep enough to account for the available empirical material. Yet in these as well as in other underdeveloped departments of inquiry theorizing is frequently regarded as a luxury and data gathering — i.e., description — as the only decent occupation, to the point that theory (speculation) is opposed to research (data hunting). This paleoscientific attitude, encouraged by a primitive kind of empiricist philosophy, is largely responsible for the backwardness of the sciences of man. In fact, this view ignores that data are meaningful and can be significant only in a theoretical context, and that the haphazard accumulation of data, or even of information-packaging generalizations, when unaccompanied by a theoretical processing capable of accounting for the former and of guiding research, is largely a waste of time. One cannot know whether a datum is significant until one is able to interpret it, and data interpretation requires theories. Furthermore, only theories will suggest the search for information not given by sense perception: just imagine whether the search for the genetic code would have been possible without a genetic theory. Besides, it is instructive to reflect on the finding that the octopus obtains nearly the same raw information from the environment as we do, and that dogs receive even more sensory information of certain kinds, despite which they have evolved no science.

Since the data-gathering-and-packaging view of science ignores the aims of theorizing, it will be convenient to state such aims explicitly. The basic desiderata of scientific theory construction are the following. (i) To *systematize knowledge* by establishing logical relations among previously disconnected items; in particular, to explain empirical generalizations by deriving them from higher-level hypotheses. (ii) To *explain facts* by means of systems of hypotheses entailing the propositions that express the facts concerned. (iii) To *increase knowledge* by deriving new propositions (e.g., predictions) from the premises in conjunction with relevant information. (iv) To *enhance the testability* of the hypotheses, by subjecting each of them to the control of the other hypotheses of the system.

The above list of basic aims of scientific theorizing may be used as a criterion for telling scientific theories from *pseudotheories*. No set of conjectures which fails to constitute a hypothetico-deductive system proper, does not provide explanation and prediction, and is not testable, counts as a scientific factual theory. This is the least that must be required from a scientific theory, yet many doctrines that pass for theories — such as Gestalt psychology, psychoanalysis, and parapsychology — do not meet all these requirements.

A few scientific theories comply not only with the basic desiderata i—iv but also with the following, additional goals. (v) To *guide research* either (*a*) by posing or reformulating fruitful problems, or (*b*) by suggesting the gathering of new data which would be unthinkable without the theory, or (*c*) by suggesting entire new lines of investigation. (vi) To *offer a map of a chunk of reality*, i.e. a representation or model (usually symbolic rather than iconic) of real objects and not just a summary of actual data and a device for producing new data (predictions).

The theories that satisfy all six desiderata i—vi are usually regarded as *great* scientific theories. Among them, the greatest are those which generate an entirely *new mode of thinking:* these are the giants of scientific knowledge. An example of a grand scientific theory, that revolutionized the mode of thinking in a number of departments, was DARWIN's theory of evolution: it posed the problems of the origin of individual differences and of the mechanisms of formation of new species and their geographic distribution; it accustomed biologists, and later on sociologists, psychologists, linguists, and others, to think in terms of origin, fitness, and evolution. Little wonder, then, that great theories are regarded as the summit of science.

When should theorizing begin? This question presupposes that theorizing should begin at some time. Apparently this presupposition is refuted by the existence of pretheoretical disciplines like geography and pre-evolutionary biological systematics, which are strictly descriptive and taxonomic. Yet, since they do not state and test hypotheses, they have no occasion to employ the scientific method, and consequently they are not sciences but nonscientific disciplines, however exact they may be. They provide and even systematize data that can be used by science, yet they are not sciences themselves but at most protosciences. Thus, geography provides geology and sociology with information, and pre-evolutionary systematics provides biology with information. Such data become problems that must be solved by building theories. No theory, no science.

Our original question, 'When should theorizing begin ?', has then only one sensible, though sibylline, answer: 'Theorizing should begin as soon as possible'. Speculative minds would begin theorizing about, say, life on other planets, *before* any information had been gathered about the physical conditions on those bodies. Down-to-earth minds, on the other hand, would defer theorizing on that problem until *after* a bulky mass of actual observations of extraterrestrial organisms were secured, and would brand as nonscientific any attempt to interpret indirect, astronomical data, as evidences in favor or against conjectures concerning extraterrestrial life. Theorizing on insufficient information may lead to irrelevant or immature theories, just as gathering data in the light of no theory leads to irrelevant or at least nonunderstood information. Either may be wasteful, yet without some preliminary speculation nobody would even conceive the idea of gathering data of a certain kind — for example, data relevant to the hypothesis that there are organisms outside our planet.

A motivation for theorizing is the systematization of a body of data gathered in the light of some stray conjectures. What is the bulk of data necessary to begin theorizing ? When is it neither too early nor too late to begin ? Nobody can tell. It all depends on the novelty of the field and on the existence of theoretically-bent scientists prepared to take the risk of advancing theories that may not account for the data or that may succumb at the first onslaught from fresh information gathered in order to test the theories: this takes moral courage, particularly in an era dominated by the criterion of success, which is best secured by not attacking big problems. Two things, though, seem certain: namely, that premature theorizing is likely to be wrong — but not sterile — and that a long deferred beginning of theorizing is worse than any number of failures, because (i) it encourages the blind accumulation of information that may turn out to be mostly useless, and (ii) a large bulk of information may render the beginning of theorizing next to impossible.

Too many unpatterned data are a nuisance because, whenever a large body of information is available, no simple model is likely to be constructed — and simple models is all we can afford in a beginning. (Computers can "handle" — i.e. store, sort out, group and even package — any amount of data but will not discern any patterns in them save those initially fed into the machines by the operators, such as, e.g., polynomials. Only brains trained in an environment with a cultural tradition can discern or, rather, hypothesize patterns.) In the beginning only simple models are constructed which, if scientific, can eventually be improved on, i.e. complicated in order to better fit the data and to get a deeper insight. From the beginning to the end only *models* or theoretical schematizations are built — i.e. conceptual systems that attempt to *represent* some interrelated aspects of real systems. (A model, in this sense of the term, need not be visualizable or picturable: it may be symbolic or abstract in an epistemological sense. Visualizable models are a subclass of models.)

Whereas the experimenter is concerned with investigating actual systems — e.g., real fluids — in all relevant detail, and with ascertaining the extent to which available theories account for such concrete objects, the theorist does not handle real systems at all, but creates idealizations of them which bear only some resemblance with actual systems. Indeed, every scientific theory refers to just a few aspects of the real system it schematizes — e.g., to the mechanical aspects of it. Furthermore, for the account of any such aspect only a few variables are introduced, the remaining variables being unknown or neglected. Finally, only a few relations among such relevant variables are introduced to account for the objective patterns of the structure and behavior of the real referent.

In short, scientific theories deal with ideal models that are supposed to represent, in a more or less symbolic way and to some approximation, certain aspects of real systems — never all of them. For example, a theory of economic transactions may idealize people as behaving in a purely rational way and may disregard every non-economic feature of people. This is why, although factual theories are often called *models*, it is truer to say that factual theories involve models and that these models, rather than the theories themselves, are supposed to represent the referents of the theories. A theory as a whole *refers* to a system or, rather, to a class of systems, and the model involved in the theory *represents* the system: see Fig. 7.2. *Rival* theories will have the same referent but will depict it differently; and their corresponding models may but need not be different. Thus, a collection of point particles may serve as a model either in a classical or in a relativistic theory.

No theory portrays literally a real thing, event, or process. First, because most aspects are unknown or irrelevant, or relevant but secondary (or rather thought to be secondary). Second, because no theory is a portrait proper: scientific theories are constructions with materials (concepts, hypotheses, and logical relations) that are essentially different from their referents and most of them nonpictorial. (A photo, too, is essentially different from the object it portrays, but on the other hand it is made of perceptible material.) For one thing logic and mathematics — the cement that binds scientific concepts — are symbolic. And so are many of the substantive concepts of scientific theories, such as "electric charge" and "drive". Of course most such substantive concepts have a referent; nonetheless they are symbolic much as a written word, in contrast with a hieroglyph, is a symbol or noniconic sign that may stand for (refer to) a real object. In short, factual theories are *sketchy* and *symbolic* rather than complete and iconic reconstructions of real systems.

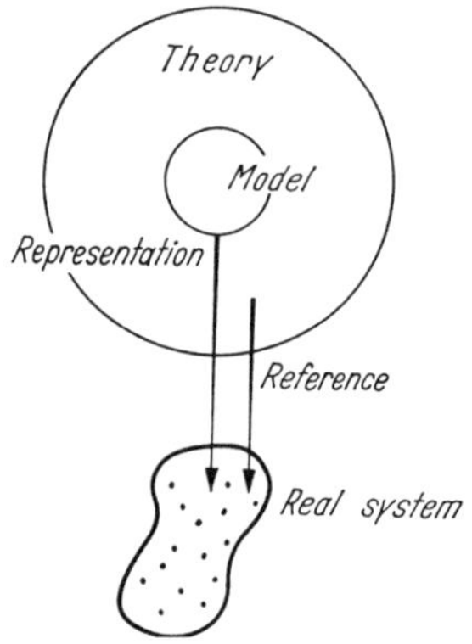

Fig. 7.2. Factual theories are not models but involve models. A model is an idealized representation of a class of real objects

It is often asked whether "theoretical entities" exist objectively. Of course they do not. What may exist is the individual thing, event, or property referred to by the theory containing a given theoretical predicate. Moreover, not all of the components of a factual theory may individually have a real referent. Think of a particle moving in a force field: these two things — particle and field — which by their interaction constitute a single system, may exist separately up to a point (i.e. as long as the field is not the particle's own field). But what about the concepts that make up the theoretical model of this system, such as "instantaneous particle position" and "field strength at a given point"? Of course they have no separate existence: they are concepts — factually meaningful concepts to be sure, but concepts. Moreover, it is doubtful whether we can assign them individual material counterparts: rather, the *whole* theoretical picture may have a counterpart in the real world: namely, the *whole* physical system (particle-in-a-field). The various parts, properties and relations of such a real system have no separate existence. No doubt a speck of dust zigzagging in a room occupies a position at each instant and the electromagnetic field in a TV tube has a given intensity at each point. But eliminate the speck of dust and the particle position will go; turn off the field in the tube and its intensity will go. There are no independent properties just as there are no things without properties. To ask whether position and force exist is like asking whether the relation "more" exists or whether the thought function exists: it is to indulge in the sin of reifying predicates. What may exist are moving particles, thinking individuals, and so on. It is we who, in order to explain the behavior of concrete objects, build conceptual models which, however accurate, cannot claim that every one of their ingredients has a separate real counterpart. In short, the correspondence between theoretical models and their referents is not a point-wise but a global, *system-to-system* correspondence.

The complexity of the theory-reality relation, the fact that theories are not just data summaries or snaphots of things, suggest that the earlier the theoretical work begins in a field the easier its growth will be. In the beginning of a science, when few facts and few generalizations are known, and in addition the standards of rigor are tolerant, it is comparatively easy to set up theories, particularly with the help of the symbolic techniques of mathematics. As more facts become known, partly through the intervention of the still crude theories, the latter can be improved; in turn, the improved theories will be instrumental in gathering fresh significant information, which would otherwise not be sought. If, on the other hand, theory construction is indefinitely delayed either because the need for it is not felt or because of fear of failure, a woolly set of beliefs and a mess of blind procedures takes its place since, anyhow, we cannot dispense with ideas. Accordingly the heap of data may become so imposing and disorderly that only an army of geniuses could guess where to begin. This is what has happened with the haphazard accumulation, along millenia, of imprecise and superficial — yet very often "inside" — information concerning human behavior. Let us then try theories from an early stage on: anyhow they will be superseded by more complex and truer theories.

The early theories in any field occupied by data-gatherers are apt to be derided by them because they are much too artificial: too sketchy and too inaccurate. This kind of facile criticism will hardly push theoretical work ahead and it is based on a misconception of the nature of theories. No one has ever seen a solid body or a fluid as depicted by rational mechanics: real materials behave quite differently from their theoretical models; yet these are indispensable even for designing experiments concerning solids and fluids. The right answer to the criticism of artificiality is not to start with real things and summarize information concerning them but to introduce further theoretical predicates and further hypotheses relating them with the aim of enriching the conceptual model, and to design experiments which will mimic ideal theoretical conditions as far as possible. No scientist should reject a theory because it does not faithfully represent its object in its entirety. Scientific theories are all (i) *partial*, in the sense that they deal with only some aspects of their referents, and (ii) *approximate*, in the sense that they are not free from error.

The perfect scientific theory (complete and entirely accurate) does not exist and will never exist. The first member of this conjunction is easy to establish with the help of empirical information regarding the performance of available theories: none of them accounts for all the traits of their referents and to any desired degree of accuracy — and if some theory seems to be perfect just wait long enough. The most that can be said of any given theory is that it satisfactorily covers certain aspects of a given class of objects to within an error smaller that the experimental error now tolerated. The second member of the above thesis — "No perfect theory will ever be built" — might be regarded as irrefutable for, if perfection were claimed for any theory, we might argue that the appearance of perfection is a delusion stemming from our present limitations, and that future experience will sooner or later disclose inaccuracies in it. This argument, however, smacks of dogmatism. We need not resort to an a posteriori argument, still less to one depending on future experience: the imperfection, hence the eventual failure of any scientific theory, can be established a priori. The argument runs as follows.

Every scientific theory is built, from the start, as an idealization of real systems or situations. That is, the very building of a scientific theory involves *simplifications* both in the selection of relevant variables and in the hypothesizing of relations (e.g., law statements) among them. Such simplifications are made whether or not we realize that they amount to errors — not mistakes but just discrepancies with actual fact. Moreover, this is not a mere descriptive statement concerning actual habits of theory construction: it is a rule of theory construction that as many simplifications as needed are to be made at the start, relaxing them gradually and only according as they are shown to constitute too brutal amputations. Such simplifications are, of course, deliberate departures from truth.

The idealizations inherent in the initial assumptions will be propagated to their testable consequences; moreover, auxiliary simplifying assumptions known to be false (e.g., that the earth is flat, or that it is perfectly spherical) must often be added to the axioms proper in order to derive theorems comparable with empirical data. When such theorems are subjected to empirical tests a discrepancy with the outcomes of such tests will sooner or later emerge — the finer the empirical techniques the sooner. And such a discrepancy will force the theoretician to gradually complicate the theoretical picture or even to make an entirely fresh start. But the improved and more complex theory will no less refer immediately to an ideal model than the previous theory. This cannot be helped, because theories are conceptual systems rather than bundles of experiences.

Since there can be no perfect theory in factual science, let us try to build better and better theories, i.e. let us attempt to contribute to theoretical progress — which would be impossible if perfection were attainable. Now, a condition for being able to ascertain whether a given theory constitutes an improvement over some other theory is to have some canons or standards of theory construction and some canons of theory evaluation. The requisites of a good scientific theory will be investigated in the following sections; the criteria of theory weighing must be left to Vol. II, Ch. 15.

Problems

7.1.1. Summarize a scientific theory expounded in some science textbook and point out (i) its referent or object, (ii) its basic asumptions (axioms), and (iii) its domain of validity.

7.1.2. A *doctrine* is a set of beliefs, taught as true, and which is not supposed to undergo substantial change in the light of new experience. Moreover, etymologically 'doctrine' means what is taught rather than what is to be worked out and subjected to test. (i) What traits are shared by doctrines and scientific theories and what distinguishes them? (ii) Doctrines are often endlessly argued over: does the same hold for theories? *Alternate Problem:* What is the specific difference between scientific theorizing and other types of search for unity amidst diversity?

7.1.3. Why did the alchemists fail and why did the nuclear physicists succeed in transmuting "base" metals into gold? Was it because the former had not made enough observations or because they did not care to build a testable theory of the transmutation mechanism? Consult some history of science. Hint: Try hard, because most histories of science presuppose that ideas play little if any role in science.

7.1.4. How and when did the following modes of thinking originate? (i) The consideration of the whole universe as a purely physical and self-sustaining system. (ii) The consideration of organisms as physical systems. (iii) The consideration of mental processes as brain processes. (iv) The search for economic determinants of social patterns. (v) Thinking in statistical rather than in individual terms. (vi) Thinking in terms of sets. (vii) The search for quantitative variables.

7.1.5. Consider (i) a set of data, such as those occurring in a telephone directory; (ii) a set of propositions of any level, related by symmetrical logical connectives such as 'and' and 'or'; (iii) the Gestalt hypothesis that learning proceeds by flashes of insight largely independent of previous experience; (iv) a Constitution. Which, if any, count as theories, and why? *Alternate Problem:* Examine the view that a theory just tidies up a mass of data.

7.1.6. Elucidate the concepts of point of view, approach, and conception in relation with the concept of theory. *Alternate Problem:* Which comes first: theory or data? Caution: examine the question itself, as it might have an unwarranted presupposition.

7.1.7. Point out cases of antitheoretical bias in some discipline. Hint: peruse recent volume of professional journals and uncover some of the roots of that attitude. See, e.g., B. F. SKINNER, "Are Theories of Learning Necessary?", *Psychological Review*, **57**, 193 (1950), where psychological theorizing, even in physiological psychology, is criticized as such and where an "acceptable scientific program" is characterized by data gathering (finding of relevant manipulable variables and, in some cases, establishing their relations, i.e. empirical generalizations). For the opposite view, that doing something without a theory is not a scientific experiment but "mere busy work", see K. M. DALLENBACH, "The Place of Theory in Science", *Psychological Review*, **60**, 33 (1953). *Alternate Problem:* Nowadays physicists are usually identified as either experimentalists or theorists. Why is a similar partition unusual in the other sciences?

7.1.8. T. PARSONS proposed a typology of human actions, an "action frame of reference" which he described as "a logically articulated conceptual scheme". No measurable variables and law formulas proper occur in this doctrine; moreover, the latter does not purport to deal with real facts but is presented as a "logical framework in which we describe and think about the phenomena of action". Can this be regarded as a sociological theory? See T. PARSONS and E. A. SHILS, *Towards a General Theory of Action* (Cambridge, Mass.: Harvard University Press, 1951). For a resumé, see T. PARSONS, "The Social System: A General Theory of Action", in R. R. GRINKER, Ed., *Toward a Unified Theory of Human Behavior* (New York: Basic Books, 1956). For criticism see M. BLACK, Ed., *The Social Theories of Talcott Parsons: A Critical Examination* (Englewood Cliffs, N. J.: Prentice-Hall, 1961) and R. RUDNER, *Philosophy of Social Science* (Englewood Cliffs, N. J.: Prentice-Hall, 1966).

7.1.9. Decision theory, game theory and other modern theories of human behavior are often described as *normative* or prescriptive, rather than *descriptive*, because they do not represent actual behavior but an ideally correct behavior in certain areas — much as logic, which sets standards for deductive inference but does not account for actual reasoning, a problem left to the psychology of thinking. Is the inference from '*x* describes an ideal model' to '*x* is normative' correct? *Alternate Problem:* To what extent can factual theories be said to be descriptive?

7.1.10. An entity is, by definition, a real existent. And anything theoretical is, by definition, unreal. Are we then justified in speaking about *theoretical entities* and in subsequently asking whether they exist? If not, coin a better name. *Alternate Problem:* The view that scientific theory is just an a posteriori rationalization of empirical data was popularized by F. BACON in his *Novum Organum* of 1620, that is, before modern science produced scientific theories proper. How come that this philosophy of pretheoretical science is still so often regarded as essentially correct?

7.2. Conceptual Unity

A scientific theory is a system of hypotheses that is supposed to give a partial and approximate account of a bit of reality. At this point we shall emphasize that a theory is a *system*, i.e. a unified body, and not just a set of formulas: see Fig. 7.3. Now, the systemicity or unity of scientific theories is both formal (syntactical) and material (semantical).

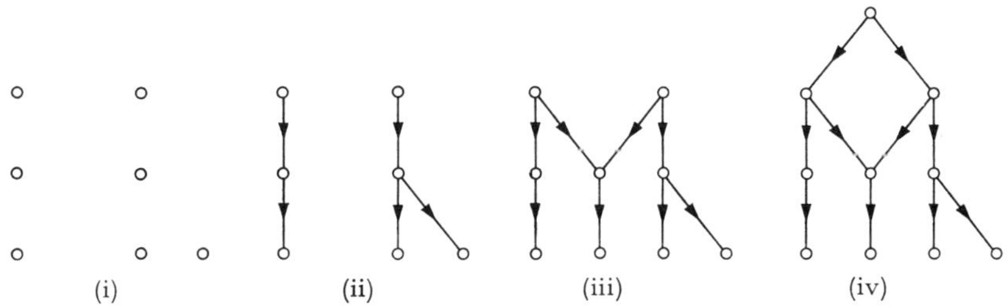

Fig. 7.3. Different degrees of systemicity of a set of formulas: (i) a set of disconnected formulas (no system); (ii) two mutually unrelated systems; (iii) and (iv) systems

The *formal unity* of a scientific theory consists in the existence of logical relations among the formulas of the theory, such that no formula remains isolated. In short, the formal unity of a theory consists in being a hypothetico-deductive system, i.e. a body the parts of which are either initial hypotheses or logical consequences thereof. On the other hand, what may be called the *material unity* of a scientific theory requires a common reference of its parts, which reflects itself in the conspicuous recurrence of certain key concepts. These concepts are distributed among the formulas of the theory so that no specific concept of the theory remains isolated from the others. Thus, e.g., in a theory of electrons all the variables refer to electrons and to the things (e.g., external fields) that may influence the behavior of electrons — not to bridges or to human expectations; and the hypotheses of the theory are just relations among such variables. The reference to *external* objects — or rather to objects that are taken to be real *ex hypothesi* — renders the objectivity of theories possible even though it is insufficient for this purpose. And the unity of reference, i.e. the reference to a *single set of objects*, gives factual theories their conceptual unity or, as we may also say, their *semantical consistency*. In other words, every nonlogical theory is about some nonempty set which, far from being an arbitrary collection, is characterized by certain mutually related properties. No such condition of singleness and existence is imposed on logical theories.

A consideration of semantical consistency should precede the treatment of syntactical consistency. Indeed, before we attempt to establish deductive relations among some of the formulas of a set we must find out whether such relations are at all possible, and such a possibility depends on certain semantical properties. First and foremost on a common reference to a given *universe of discourse* or *reference set*. This should be clear from the following counterexample. Consider the formulas.

For every x in U, Px — i.e., $(x)(x \in U \rightarrow Px)$, or briefly $(x)_U\, Px$.

For some x in V, Qx — i.e., $(\exists x)(x \in V \,\&\, Qx)$, or briefly $(\exists x)_V\, Qx$.

These formulas state certain properties of the members of two different sets, U and V, but they say nothing about the relations among them. For all we know U might be the set of unicorns and V the set of volitions. Unless we add a third formula stating, say, that V is a part of U or conversely, we will be unable to stipulate that the above are formulas of one and the same theory, and this simply because we will fail to combine them in order to deduce some consequence from them.

A *common universe of discourse* or reference set is therefore necessary for establishing logical relationships among the members of a set of formulas, whether in mathematics or in physics. (The fact that in the case of abstract theories, such as Boolean algebra and group theory, and semi-abstract theories such as set theory, the nature of the elements of the reference or basic set is not specified, does not entail that the reference set itself is not fixed in advance. And when the universe of discourse is not indicated, it is understood that it is an arbitrary set — as in the case of the theory of identity, which holds for every conceivable set.)

A second factor of semantical consistency is the condition that the predicates of the theory match in meaning, i.e. that they *belong to the same semantical family*. Or, as we shall also say, the predicates of the theory must be *semantically homogeneous*. This condition automatically excludes from scientific theories statements such as "Magnets are lovely", which is syntactically self-consistent but semantically inconsistent: it mixes heterogeneous predicates. The requirement of semantical homogeneity applies also, of course, to sets of formulas every one of which is both syntactically and semantically consistent. For example, "Every Olympus reflects relations proper to the human group which begets it", and "There are infinitely many primes", are each syntactically and semantically consistent but, taken together, they form a set which lacks semantical homogeneity and cannot therefore be regarded as part of the same theory.

The requirement of semantical homogeneity is easier to exemplify than to state. Accordingly, there will be border cases in which it will not be easy to decide whether semantical heterogeneity is just a result of lack of insight into a deeper lying unity: after all, interdisciplinary theories, such as those of biochemistry and social psychology, are born from bringing together concepts initially belonging to separate semantical families. As a consequence, the requirement of semantical homogeneity should be regarded as a rough and fallible guide. The ultimate decision concerning the semantical homogeneity or heterogeneity of a set of predicates will rest on the existence or nonexistence of verisimilar theories in which the concepts are related, rather than on semantical considerations. Be that as it may, the requirement of semantical homogeneity reflects and consecrates the multiplicity and separateness of theories, and this in turn seems somehow to correspond to the comparative independence among the various aspects of the world. If reality were either a solid block (radical *monism*) or a heap of altogether heterogeneous items (radical *pluralism*) it would not be possible to group properties into disconnected or at least loosely related families.

A third condition for securing semantical consistency is that the predicates of the theory be all those and only those which occur in the theory's initial assumptions and in the definitions. This will be called the condition of *semantical closure*. It is designed to hinder the smuggling in of predicates foreign to the field covered by the theory, and which are sometimes introduced at the level of the theorems. Semantical closure precludes the double interpretation of one and the same symbol occurring within a given theory — as happened with the artificial distinction between inert and gravitational mass. (See Problem 7.2.10.) We shall see later on (Sec. 5) that part of the confusion concerning the foundations of the quantum theory may be traced to a violation of the condition of semantical closure.

Were it not for the requirement of semantical closure we might be tempted to try the following move. Suppose we build a theory, subject it to empirical tests, and find it wanting. We might attempt to save the theory by adding an *ad hoc* assumption containing, say, a predicate designating a property that had not been taken into account to begin with but which nicely explains away the unfavorable result. This can be done either by adding the new hypothesis to the postulate set or by changing the inadequate theorem t into the weaker but empirically confirmed statement $t \vee u$, where u is the *ad hoc* hypothesis. (Such a transformation is logically legitimate: t entails $t \vee u$, where u is an arbitrary statement.) The addition of the *ad hoc* hypothesis to the axioms of the theory can be rejected if the new assumption has no independent testable consequences (see Sec. 5.6). That is, the trick would produce a different theory which could be dismissed because one of its assumptions serves no purpose other than protecting the other axioms. But the second manoeuvre, consisting in using the rule of addition to replace the false theorem t by the trivially true theorem $t \vee u$ — where 'u' may stand for "London is

Axiom 1	C_1 —— C_2	Axiom 1′	C_1 —— C_2
Axiom 2	C_3 —— C_4	Axiom 2′	C_2 —— C_3
	(i)		(ii)

Fig. 7.4. (i) Axioms 1 and 2 do not jointly constitute a theory because the four primitive concepts, C_1 to C_4, are not distributed among them. (ii) Axioms 1′ and 2′ do satisfy the condition of conceptual connectedness: one of the three primitives, namely C_2, is shared by the two axioms

the capital of Great Britain" — cannot be avoided by resorting to testability. The principle of semantical closure, on the other hand, prevents such a manoeuvre. But it does not prevent the addition of a statement formed with predicates germane to the theory yet not deriving from the theory's initial assumptions. Only formal unity (systemicity) can avoid this.

The condition of semantic closure does not prohibit the *exportation* of concepts introduced by a given theory: the closure, like a semipermeable membrane, works only one way, to prevent the entry of new concepts once the theory has been formulated, in order to preclude an endless readjustment of the theory either to the data or to some philosophical tenet. The diffusion of basic concepts throughout science — a process enhancing the integration of scientific knowledge — is not hindered by the principle.

A fourth factor of semantical consistency is the requirement that the key concepts of the theory "hang together": this we shall call the condition of *conceptual connectedness*. In other words, the primitive (or undefined) concepts of the theory must somehow be distributed among the axioms, so that the latter may dovetail. (See Fig. 7.4.) More precisely, we shall say that an axiom set is *conceptually connected* if and only if no primitive concept occurs in a single axiom of the set. No primitive dovetailing (conceptual connectedness), no axiom dovetailing; no axiom dovetailing, no deducibility relation; no deducibility, no system at all.

The property of conceptual connectedness is related to but not identical with the condition of a common reference set (first requirement). In fact, by unearthing the object variables of a set of formulas we might show that they do have one object variable in common, e.g., that they all contain the term 'body'. But some of the formulas might refer to mechanical properties of bodies, others to the cost of their production, and so on. A common reference class is necessary but not sufficient to ensure conceptual connectedness. And in turn conceptual connectedness does not warrant a common universe of discourse. Thus, "Cats are independent" and "The more independent a person the less adjusted it is" share the predicate 'is independent', but the reference set of the first proposition is the class of cats, whereas the second statement refers to persons.

It is desirable, but not mandatory, that the axioms of a theory be independent at the propositional level, in the sense that they be not interdeducible (see Sec. 6). But if the axioms were mutually independent at the conceptual level as well, i.e. if no two axioms of the "theory" shared some primitives, no deduction would be possible. Conceptual connectedness is, then, necessary for derivability, which is of course essential for having a theory. Conceptual connectedness is desirable for the set of all scientific theories in a given field — e.g., psychology — since it renders their mutual enrichment, control, and support possible. Moreover, the conceptual connectedness of the whole of factual science is desirable even though it cannot be required of currently available theories. Chemistry became a science when it established contact with physics by importing the basic physical concepts of mass and atom, and psychology became a science when it borrowed concepts from physiology. But the desideratum of conceptual connectedness should not be exaggerated to the point of excluding from science all specific concepts not exportable to other sciences, or of using only generic concepts such as "object", "determine", and "change".

In sum we have, then, four factors of conceptual unity or semantical consistency, which are jointly necessary (though not sufficient) for having hypothetico-deductive systems, or theories for short. These factors are: (i) a single universe of discourse or reference set, which is often indicated by the name of the theory (e.g., Rigid Body Dynamics) although it may be disregarded in the listing of the primitive concepts; (ii) semantical homogeneity (of the predicates), or belonging to the same semantical family — a condition which is typically violated by crackpot theories; (iii) semantical closure or the prohibition to introduce opportunistically new predicates — unless it is clearly stated that a new theory is being propounded; (iv) conceptual connectedness, or the equitable distribution of predicates among the formulas.

It might be asked whether the semantical consistency or conceptual unity of a theory or of a group of theories is just a logical property or whether it somehow reflects an objectively existing unity, e.g. a physical interconnection. The following example will show that semantical consistency or conceptual unity need not reflect physical unity. Let x and y be the numerical variables of quantitative concepts representing each a physical property of a class of systems; suppose further that both depend on a third variable, t, which will be regarded as the independent variable and may be thought of as time. Consider first the pair of equations

$$x=f(t), \quad y=g(t).$$

This pair satisfies all four conditions of semantical consistency: it refers to a given class of systems, the three variables are semantically homogeneous (they are all physical variables), no new variable has been allowed to descend upon them as a *deus ex machina*, and finally one of them, namely t, is shared by both formulas. Yet the preceding set of formulas may not represent a concrete system (e.g., an atom or an economy), because x and y do not determine one another: in fact, the values taken on by x are independent of those taken by y and conversely: x and y run parallel in time. We conclude that conceptual unity need not reflect physical systemicity — i.e. it is not a sufficient condition of physical togetherness. On the other hand, a system of equations such as

$$ax+by=f(t), \quad cx+dy=g(t)$$

does represent a physical system if the variables involved represent physical properties of systems of a kind, because in this case x, y and t determine one another. (This example shows, by the way, that although form does not determine content it must match it.)

We saw that conceptual unity is necessary for securing formal unity and deducibility; but it is insufficient. In order to be able to deduce something from the initial premises these must be *precise and rich*. From imprecise premises nothing can be concluded unambiguously: impreciseness is the semantical homologue of contradiction, in that both lead to an unwanted proliferation of consequences. And from poor axioms little or nothing can be concluded.

Thus, no geometry can be built on assumptions such as "Space is unfathomable", "Space is a form of existence", or "Space is the sensorium of the Deity": they are much too imprecise. A modest formula for the distance between two arbitrary points will, on the other hand, be a good candidate for a geometrical axiom both because of its preciseness and its richness. Nor do negative statements qualify as axioms although they may occur as theorems, and that for a similar reason: a negative proposition is much too indefinite and consequently enjoys a weak testability. An affirmative proposition, on the other hand, represents a definite commitment and, if factually meaningful, it directs us to search in the world of facts for the referent that will support or undermine it. Likewise, a proposition such as "The universe looks everywhere and always much the same" is useless as an axiom of cosmology not only because it is rather vague but also because it is poor: in fact, it is little more than a (hasty) generalization of observation records. The axioms of factual science are not empirical generalizations but high level law statements which, far from summarizing data, can explain both data and generalizations of observation reports.

In order to secure precision it is sufficient, though not strictly necessary, to employ quantitative concepts such as length and population density. And in order to secure rich axioms we tend to maximize (i) the *range of the variables*, i.e. the domain embraced by the logical quantifiers; (ii) the *comprehensiveness* of the concepts, and (iii) the *degree of abstraction* of the concepts. The effectiveness of the first device is obvious: the larger the range of a formula the more instances are subsumed under it. Now, of two formulas with the same degree of generality — two universal statements, say — the richest will be the one containing the more general concepts. (For concept generality see Sec. 2.3.) Thus, e.g., "Men are self-reproducing", "Mammals are self-reproducing", "Vertebrates are self-reproducing" and "Organisms are self-reproducing" are all universal propositions but of an increasing richness because the extension of "organism" includes that of "vertebrate", the latter that of "mammal" and this in turn that of "man". At the same time these concepts are ordered in respect of epistemological complexity or degree of abstraction: interpreting '$<$' as "less abstract than" we may write

$$\text{"man"} < \text{"mammal"} < \text{"vertebrate"} < \text{"organism"}.$$

That abstraction or estrangement from immediate experience is a source of richness for the premises of a theory and should therefore be coveted is understandable on two counts. First, if we wish to build rich theories we must choose basic concepts holding *many relations* among one another, so that they can occur in many formulas. And abstract and comprehensive concepts, such as "organism", fulfil this condition better than concrete and singular terms such as 'Snoopy'. Second, if we want to explain experience we must *rise above it* by analyzing it in nonexperiential terms. We may derive hypotheses concerning colored bodies from hypotheses regarding colorless molecules but not vice versa. Accordingly a phenomenal predicate such as 'red' will not occur as a primitive concept in chemistry, which needs on the other hand hypothetical constructs such as "atom" among its primitive concepts. The closer a concept is to experience the less apt it will be to enter the axioms of a theory. This is one reason why theories are not "abstracted" from observation. Another reason is that science, and in particular physics, teaches us that experiential facts, such as observable events, are macroscopic outcomes of myriads of microscopic (hence transempirical) events. Accordingly, the important distinction between primary (physical, objective) properties and secondary (phenomenal, subject-centered) properties is not arbitrary, and the explanation of secondary properties in terms of primary properties is rooted in the structure of the world.

The (epistemologically) abstract character of the basic or primitive concepts of scientific theories is a reason that low-level formulas, such as empirical generalizations, are usually "discovered" earlier than the axioms from which they may be made to derive. It is, indeed, a historical fact — or, rather, a generalization from remarks concerning intellectual history — that, by and large, theorems come in time before axioms, even though axioms are logically prior to theorems. The historical pathway has almost always been the following: (i) establishing low-level generalizations (future theorems); (ii) generalizing the foregoing (i.e., generalization of particular theorems); (iii) "discovering" logical relations among known theorems; (iv) "discovering" that some of them might serve as axioms, or inventing higher-level premises out of which the available body of knowledge can be derived; (v) logical systematization of the body, i.e. axiomatization (full or partial) and, eventually, formalization. The last two stages, axiomatization and formalization (see Sec. 8.3), are rarely reached in factual science, partly because of the pressure of fresh substantive problems, partly because the importance of axiomatization and formalization is underrated, perhaps because it is not realized that clear thinking about known facts and old ideas is not inferior to muddled thinking about the newest discovery.

That historical trend, from the particular to the general and from the concrete and empirical to the transempirical, is exactly the reverse of the logical order. A first consequence: specialists in the logical analysis of the final outcomes of research usually do not have a feeling for the actual making of science and, conversely, historians of science are seldom aware of the logical and epistemological peculiarities of scientific theories. A second consequence: the axioms, contrary to the Aristotelian tradition, are not more but far less self-evident than most of the theorems they entail. 'Axiomatic' is then misused as a synonym for 'self-evident'. Whoever feels the slightest doubt about this should shake it off by taking a look at the initial assumptions of any of the advanced theories of the day. A third consequence: in the process of scientific theory construction, the groping in the search for axioms is chiefly, though not solely, controlled by the available lower-level hypotheses, which the axioms must entail at least approximately. Only abstract mathematical theories are frequently built without preëxisting weaker formulas; but even in this case the known formulas act either as motivations or as controls of the stronger ones. But goals do not supply means: the lower-level hypotheses the theory must yield cannot unambiguously suggest the axioms. These have to be invented; they are not free creations, because they are controlled by the lower-level hypotheses, but they are creations all right.

In short, the primitive concepts of a scientific theory should be as precise and rich as possible, and the primitive formulas (axioms) of it should be semantically consistent, i.e. they should constitute a unified conceptual whole.

We can now study the logical form or structure of theories.

Problems

7.2.1. Examine the rule that the axioms of a science be self-evident. Is this requirement consistent with the program of explaining the experientiable in terms of what is not experientiable yet is supposed to occur objectively?

7.2.2. If the axioms need not be self-evident, can they be arbitrary? In other words, is the choice of axioms determined by nothing (undetermined) save possibly taste?

7.2.3. Examine the characterization of axioms as initial truths. Is this characterization correct with regard to (i) formal science and (ii) factual science?

7.2.4. Examine the widespread opinion that the primitives of a theory should be well-understood concepts, if possible taken over from ordinary knowledge, as would be the case with the concepts of space and time. *Alternate Problem:* Examine the commandment that the primitives of a theory shall be observational concepts and, in particular, "operationally defined" concepts. (For operational "definitions" see Sec. 3.6.)

7.2.5. Is it desirable or even possible for any given scientific theory to be "an island unto itself"? And going to the other extreme: is a universal, all-purpose scientific theory of reality desirable or even possible?

7.2.6. If phenomena (observable facts) were all there is, i.e. if the world were not largely hidden to the senses, would we still employ transempirical concepts? In other words, are theoretical constructs necessary only because there are unperceivable (e.g., atomic) entities, or for some other reason as well?

7.2.7. Elucidate further the concept of conceptual connectedness of a theory sketched in the text. *Alternate Problem:* How is conceptual unity related to the rule of parsimony in introducing *ad hoc* hypotheses? Recall Sec. 5.6.

7.2.8. Take any two time-dependent variables such as sound intensity and crop yield. Time can be eliminated between the two relations, giving a direct relation between the two variables. Will this represent a system? *Alternate Problem:* State the theses of monism and pluralism in terms of families of semantically homogeneous predicates. For monism and pluralism see, e.g., W. JAMES, *Essays in Radical Empiricism* and *A Pluralistic Universe* (1909; New York: Longmans, Green, 1958).

7.2.9. Elucidate the concept of richness of a proposition and relate it to the concept of logical strength dealt with in 5.7. Take into account that, whereas strength is a syntactical (formal) property, richness is both syntactical and semantical.

7.2.10. Newtonian mechanics and gravitation theory employ a single mass concept; but mass occurs in some terms as a coefficient of acceleration and in others as a factor of the gravitational force. For example, the equation of motion of a body moving in a homogeneous gravitational field of strength g is: $ma=mg$, whereas if the field is electrical the law is: $ma=eE$. In the former case we eliminate m and simplify to: $a=g$. This, which is typical of the gravitational field, is not possible in the second case. This circumstance suggested to E. MACH the idea that there are actually two kinds of mass: *inertial mass* (occurring as a coefficient of acceleration) and *gravitational mass* (occurring as a factor of the gravitational force). The distinction, however, occurs neither in the classical nor in the relativistic theory of gravitation but it is demanded by operationalism and is therefore conspicuous in text books. In fact, there are essentially two kinds of procedures for *measuring* mass values: one which does not employ gravitation (e.g., the spring scale) and another which does (e.g., the weights scale). Is the distinction between two kinds of mass in keeping with the condition of semantical closure? *Alternate Problem:* Elucidate the concept of *nontrivial theory* in terms of the kinds of object (universe of discourse) and problems a theory can handle. Take into account that in principle any field of facts — e.g., cow milking and soccer — can be made the object of a theory.

7.3. Deducibility

In order to study the deducibility relation it is convenient to watch how it works in axiomatic theories because these are the best organized. An *axiomatic theory* may be pictured as a web hanging from its initial assumptions (see Fig. 7.1). These are a handful of comparatively rich and precise formulas (propositions and/or propositional functions), called axioms or postulates, which satisfy the condition of conceptual unity (see 7.2). Below the axioms stand all the remaining hypotheses of the theory, which are called its theorems even if the theory has a factual content, since the term 'theorem', just as the term 'axiom', indicates a certain logical status independent of content.

In a completely axiomatized theory all the theorems can be derived from the initial assumptions alone by purely formal (logical or mathematical) means, i.e. by application of rules of deductive inference. In other words, given the axioms of the theory and the rules of inference presupposed by the theory (i.e., the underlying systems of logic and mathematics), all the theorems remain uniquely determined even if none of them has actually been derived. Logical predetermination is the only effective predetermination and it is peculiar to axiomatic theories.

In formal science an *axiom* or postulate is an unproved assumption the function of which is to help prove other formulas of the theory. In factual science, too, an axiom is unproved and helps in proving other statements, but its introduction is justified to the extent to which these other statements (the theorems) are somehow validated by experience. The most ambitious task a theoretician can set himself is to invent an axiom set to cover exhaustively a given field of knowledge. (Whether this aim is fully attainable is another story.) To this end, no amont of observational data will help him: empirical information is precisely what he wishes to account for and will therefore function as a stimulus and as a control of theory construction but not as its source or even as its guide (see 7.2). All the theorist can do in order to invent a nice theory is to exploit to the full whatever genetic and cultural legacy he has been supplied with. (More on this in Sec. 8.1.)

Once the axioms of the theory have been laid down the theorist can either (i) derive new theorems from them, or (ii) establish connections with other fields of inquiry, or (iii) try to modify some of the axioms with a view to obtain a more perspicuous or a more economical or a richer system, or (iv) try to specify their factual and/or empirical meaning, if any. Conversely, given a set of more or less disconnected partial results of research (such as, e.g., the various generalizations concerning nuclear structure known before the shell model was introduced), the task of the theoretician will be to introduce (invent) an axiom basis from which they can be derived, i.e., converted into theorems. Once such a basis is invented it may be expected that new, previously unknown propositions, will be derivable. (If the axioms existed by themselves in a Platonic realm of ideas, or if any given formula suggested by itself the postulate set from which it can be derived, we might speak of *discovering* axioms. But, since axioms are not found ready-made and are never uniquely determined by the preexisting lower-level formulas, why should we hesitate to speak about *inventing* or *creating* axioms?)

In every theory a group of axioms or highest-level formulas can be discerned, but not every theory is axiomatic, i.e. self-contained. For one thing, no factual theory is fully axiomatic and, as we shall see in a while, axiomatization does not exhaust the content of a factual theory. The reason of this impossibility is that factual theories must be open to experience. More precisely, factual theories cannot dispense with singular propositions (data) that are neither among the axioms nor are logically derivable from them. It would not do to place such empirical information among the axioms of the theory: first, because they entail nothing of interest; second, because data constitute an open set and, as it were, a living set since it grows and is kept young by the replacement of obsolete bits of information by newer, more accurate ones. On the other hand, if an axiom is changed or added a new theory is produced. Much the same holds for the auxiliary assumptions (e.g., the simplifying hypotheses) that are joined to the axioms opportunistically, whenever the theory is applied to a particular case. Since data and auxiliary assumptions are indispensable for the derivation of singular statements, and since these are what can be compared with experience, we must admit them into our theories; but, since data and auxiliary assumptions are neither axioms nor theorems, they cannot belong to the axiomatic core of theories. In other words, factual theories cannot be fully axiomatized.

The premises of a factual theory are of either of the following kinds: (i) *initial assumptions* or axioms, and (ii) *subsidiary premises* such as special hypotheses, lemmas, and data. Hypotheses which are proved as theorems in alternative theories, and which are borrowed from them, are to be called *lemmas*; they occur frequently in the deductions of factual science. Thus, a computation in physiology may require the introduction of formulas from analysis and from mechanics. And special hypotheses, as well as particular informations, are of course indispensable to explain and predict with the help of a theory. For example, if we wish to predict what will happen to a particular individual under given circumstances, we need certain specific information regarding both the individual and the circumstances.

Subsidiary premises (special hypotheses, lemmas, data) are not and could not be entered in the top list of assumptions of a theory but are adjoined as they are needed for inference. There is no limit to the number of subsidiary premises in a nonaxiomatic theory: it is permissible to borrow them from any field whatever as long as they comply with the requisites of semantic closure and homogeneity (see Sec. 2), i.e. on condition that they introduce no foreign elements and that they expedite inferences. Aside from this, the greatest freedom for the introduction of relevant premises prevails in the work with hypothetico-deductive systems — as long as they are not fully axiomatized, i.e. self-contained or closed.

This freedom to invent auxiliary premises and to use empirical data ends as soon as we espouse a fixed set of initial assumptions in order to build a *fully axiomatized* system, i.e. a hypothetico-deductive system the sole hypotheses of which are given initially. In fact, since an axiomatic system consists only of a set of axioms and its logical consequences, it admits no subsidiary premises (special hypotheses, lemmas, data) as are usually adjoined along the way when a factual theory is employed to solve a specific problem. It follows that only a part — the general body or *core* — of scientific factual theories is axiomatizable. Equivalently: factual theories are essentially but not exhaustively axiomatizable. It also follows that, as soon as an axiomatic theory is applied to a specific problem, such as computing a prediction, its axiomatic shell must be broken so as to let further premises pour in. Strictly axiomatic theories in factual science are important for foundational research but it is impossible to work with them alone precisely because they are closed systems. In practice *semi-axiomatic* theories, i.e. theories with an axiomatized *core* and allowing the introduction of subsidiary premises, are preferable. In summary:

- *Formulas of Axiomatic Theories*
  - Axioms & Definitions
  - Theorems

- *Formulas of Semiaxiomatic Theories*
  - Premises
    - Axioms & Definitions
    - Subsidiary premises
  - Theorems

Now that the advantages and limitations of axiomatization have been shown we can take a closer look at the relation of logical consequence. When a formula t is derivable from a set of assumptions A by means of the rule(s) R, we write: $A \vdash_R t$. (The sign '$\vdash_R$' does not designate an operator which, applied to the assumptions A, yields the logical consequence or theorem t: for, given A, there are usually infinitely many t's such that $A \vdash_R t$.) The relation $\vdash_R$ is specified by enumerating the rules of inference admitted by the theory; in other words, to specify the relation of entailment $\vdash_R$ amounts to mentioning the logic and the mathematics underlying the theory. Since usually *ordinary* (two-valued) logic and the mathematical theories respecting it are presupposed by factual theories, we may drop the explicit mention of the rule(s), simplifying the former formula to "$A \vdash t$", which is read 'A entails t' or 'A logically implies t' — not to be confused with 'A implies t', i.e. '$A \rightarrow t$'. Entailment, the glue that keeps the formulas of a theory together, is stronger than implication: *If $A \vdash t$, then $A \rightarrow t$ but not conversely*. (Generalized to a set $\{S, A\}$: If $\{S, A\} \vdash t$, then $S \vdash A \rightarrow t$.) A common strategy for proving conditional statements is therefore trying to prove that the antecedent entails the consequent.

Clearly, the assertion of A — or just its hypothesizing for the sake of argument — commits us (logically) to accept the consequence(s) t of A: if A entails t then it is inconsistent to assert both A and $-t$. In other words, regardless of the truth value of A (which may be unknown), the conjunction $A \& -t$ is logically false. Consequently its negate, i.e. $-(A \& -t)$, is logically true. Now, it is easy to see (by assigning truth values to A and t) that $-(A \& -t)$ is the same as $A \rightarrow t$. Hence, to say that A entails t is the same as saying that $A \rightarrow t$ is logically true or, also, that $A \rightarrow t$ is an *analytic conditional* — not just a contingent conditional, i.e. one that may or may not be true depending on the correspondence of A and t with fact.

*Three other statements of the same idea (the so-called *deduction theorem*) are the following. (i) A necessary and sufficient condition for a formula A to entail t is that the conditional "$A \rightarrow t$" belong to the set L of logical formulas or logical identities. Briefly: $A \vdash t$ iff $(A \rightarrow t \in \mathsf{L})$. (ii) Extension to a *set* of premises: A necessary and sufficient condition for a set of assumptions $A = \{A_1, A_2, \ldots, A_n\}$ to have the logical consequence t is that A contains a subset $A' = \{A_1, A_2, \ldots, A_k\}$, with $k \leqq n$, such that $[(A_1 \& A_2 \& \ldots \& A_k) \rightarrow t] \in \mathsf{L}$. (iii) The same rewritten with the help of the exportation rule: A necessary and sufficient condition for a set of assumptions $A = \{A_1, A_2, \ldots, A_n\}$ to have the logical consequence t is that A contains a subset $A' = \{A_1, A_2, \ldots, A_k\}$, with $k \leqq n$, such that $[A_1 \rightarrow (A_2 \rightarrow (\ldots \rightarrow (A_k \rightarrow t)))] \in \mathsf{L}$.*

From a purely formal point of view *proving* and *deducing* are one and the same thing: if we deduce t from A with the help of the rule(s) R we prove t in the system based on A and R. (Notice the relativity of proof to system.) But if A happens to be the base of a factual theory, the proof of t on the strength of A and R will not prove that t is *true* to fact. All it shows is that, *if* A is (or were) true and R is (or were) adopted, then t necessarily follows (or would necessarily follow). The validity of the *argument* leading from A to t via R will have been established, not the truth of t itself. This must be kept in mind in relation to the mathematical theories of factual science, in order to avoid taking a mathematical proof of a factual statement for its empirical validation.

Up to now we have classed the formulas of an axiomatic theory into initial assumptions (axioms) and their logical consequences (theorems); and we have tacitly lumped definitions together with axioms because they too are posited, not derived. Yet not all of the axioms of a theory are equally important and not all of its theorems are on the same footing. In most theories a small subset of initial asumptions, quite often a single axiom, can be regarded as essential or *central*. Accordingly, if the peripheral axioms are changed while the central axiom(s) are kept, we do not tend to speak of different theories — which they certainly are — but rather of different *formulations* or *versions* of one and the same theory. Thus, for example, the central axiom of the elementary probability theory is: "If A and B are disjoint, then the probability of the union of A and B equals the sum of the probabilities of A and B" (see Sec. 7.5). The remaining axioms might be changed without essentially affecting the theory, since their chief function is to fix the range, which is entirely conventional. Similarly, the central axiom of Newtonian mechanics is the formula "Force = Mass × Acceleration".

As to the theorems of a theory, their relative importance depends on many factors. But there is a dichotomy that applies universally because it is of a logical nature: namely, the partition of logical consequences into general theorems, or *theorems* for short, and *instances* thereof. An instance of the theorem "$s(t)=(\frac{1}{2})\,gt^2$" is: "$s(10)=(\frac{1}{2})\cdot 981\cdot(10)^2$". A theorem has an unlimited number of instances. These instances, rather than the theorem, can refer to singular facts. (Partly for this reason the instances of theorems are occasionally called *facts*. But this is a misnomer in every philosophy save objective idealism, because it obliterates the difference between an idea and its referent.) The theorems of a factual theory discharge an inferential function in addition to bridging the gap between the high-brow axioms and singular propositions close to experience. In fact, even in an axiomatized theory the initial assumptions A are insufficient for the derivation of all their consequences. The usual procedure is this. First the immediate consequences or corollaries of the axioms A are derived; for instance, from the axiom "$p \to p \vee q$" the corollary "$p \to p \vee p$" is immediately obtained. Then some important or strong theorem t_1 is derived, which is henceforth used in conjunction with A to derive further consequences, and so on. In other words, every new theorem can be added to the base A, which thus enriched enables the derivation of further consequences. For example, once t_1 is obtained we might be able to prove $t_1 \to t_2$ on the strength of A and t_1; finally, by *modus ponens* we detach t_2.

Let us now focus our attention on theories as *wholes*, i.e. as sets ordered by the relation of logical consequence. Every member of a scientific theory is a (quantified or unquantified) propositional function of some kind or other; hence every scientific theory is a structured *set of formulas* that, quite apart from their content, come under the sway of the branch of logic known as the *predicate calculus* (with identity), or PC= for short. The phrase 'with identity' means that the PC is extended to include the rather trivial but important theory of identity, which specifies the meaning of '='. Any set F of formulas of the predicate calculus, paired to the deduction relation $\vdash$, constitutes a *theory* T. In symbols,

$$\mathsf{T}=\langle F, \vdash\rangle. \qquad [7.1]$$

Deduction within the set F produces no new formula, i.e. no formula outside F. In other words, the set F of formulas of a theory is *closed under deduction*. (Similarly, the set of nonnegative integers is closed under addition — not however under subtraction, which may produce negative integers.) From a logical point of view, then, a theory, whether scientific or not, can be characterized by the following *Definition*: A *theory* is a set of formulas of the predicate calculus (with identity) closed under deduction in this calculus. (The restriction is usually added that the predicate calculus be of first order, i.e. that it deal with properties of individuals only. We do not place this restriction because in science we need higher-order functions, such as "is a dynamical variable", and quantification over predicates, as in "for all forces".)

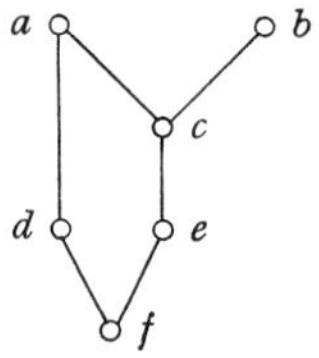

Fig. 7.5. A filter with basis $B=\{a, b\}$. The element c is the greatest lower bound of a and b, i.e. $c=a\cap b$, and f the greatest lower of bound d and e, but also the greatest lower bound of d and f, and of e and f

*An equivalent definition is this. T is a *theory*, or deductive system, if and only if

*D*1 If p and $p\rightarrow q$ are in T, then q is in T for every p and every q.

*D*2 If p and q are in T, then $p \,\&\, q$ is in T for every p and every q.

Such a set is an example of a *filter*, a very general algebraic structure. (See Fig. 7.5.) In general, given a partially ordered set P, — i.e. a system constituted by an abstract set and the abstract relation $\leqq$ in it — if F is a nonempty subset of P, then F is a *filter* if and only if

*F*1 If x is in F and x precedes y, then y is in F for every x and every y.

*F*2 The greatest lower bound of any two members of F is in F.

*F*3 The empty set is not included in F.

If F is interpreted as a set of formulas (e.g., propositional functions), $\leqq$ as implication, and the greatest lower bound as conjunction, it can be proved that every nonempty filter is a deductive system and conversely. The study of the structure of theories (deductive systems) is thereby taken up by abstract algebra.*

The above definitions apply to any theory, whether axiomatized or not. Let us now distinguish, within the set F of formulas of a theory T, the subset A of its initial assumptions or axioms, and let us rule that no premises other than those constituting the base A shall be introduced: that is, let us consider the special case of *axiomatic* theories. The axiom set A, or *base* of the theory, is countable and usually, but not always, finite. (If the base of a theory is a finite set, i.e. if the assumptions are finite in number, the theory is called *finitely axiomatizable*.) Call $Cn(A)$ the infinite set of all the logical consequences or theorems and instances thereof derivable from A. Since any formula entails itself (i.e., the relation $\vdash$ is reflexive), $Cn(A)$ contains A. In other words, $Cn(A)$ is the set of initial assumptions plus all their consequences: in sum, the set of axioms and theorems. In short, $Cn(A)$ is the set F of all the formulas of the theory based on A:

$$F = Cn(A). \qquad [7.2]$$

(In turn, $F = Cn(F)$, whence an axiomatic theory contains all its own consequences, i.e. it is deductively closed.) We may accordingly adopt the following *Definition*: An *axiomatic theory* with base A is the set of all the logical concequences of A.

The simplest case occurs when nothing is assumed, i.e. when the base of the theory is the empty set $\emptyset$. If we ask 'What follows from nothing?' the counterintuitive yet right answer is that (ordinary) logic follows from nothing. That is, there are no assumptions from which logic follows; or, also, nothing (logically) precedes logic; or, again, logic is self-contained. (The autonomy of logic is thereby proved; see Sec. 5.9.) Briefly,

$$Cn(\emptyset) = \mathsf{L}, \quad \text{or} \quad \emptyset \vdash \mathsf{L}. \qquad [7.3]$$

The dual of this theorem is: Everything follows from logically false assumptions. That is, the set U of all formulas follows from any unit set $\{A\}$ such that the negate of A is logically true:

$$\text{If} \quad -A \in \mathsf{L}, \quad \text{then} \quad Cn(A) = U. \qquad [7.4]$$

A first moral: Beware of formulas than can explain everything, both in the given field and elsewhere. A second moral: Do not strengthen your assumptions at all cost. (See Sec. 5.7 for limitations on logical strength.)

What, apart from logical identities (tautologies), can be derived from a unit set $A = \{e\}$ constituted by a simple (atomic) factual proposition, such as an empirical report? Obviously, e itself; in addition, the useless conditional $h \rightarrow e$, where h is an arbitrary statement, i.e. one not determined by e. That is, isolated bits of information are deductively sterile. Let us next take a set of N empirical evidences $\{e_i\}$ as our base. Nothing but the assumptions themselves, harmless conjunctions and disjunctions thereof, and indeterminate conditionals $h_i \rightarrow e_i$, will be derivable. We may continue to pile up simple low-level propositions to no avail: a set of logically isolated informations, however accurate and numerous, is no more of a theory than a bunch of gossips. Logical consequences — hence proofs — can begin to flow from a set of assumptions if these transcend data. Then even a single initial assumption — as long as it is rich (see Sec. 2) — may entail infinitely many logical consequences. In short, *data generate no theories*.

The simplest possible pregnant base A consists of a conditional and its antecedent or the negate of its consequent. In these cases, the immediate logical consequences are

$$A_1 = \{p, p \rightarrow q\} \vdash q \qquad (\textit{Modus ponens}), \qquad [7.5]$$

$$A_2 = \{-q, p \rightarrow q\} \vdash -p \qquad (\textit{Modus tollens}). \qquad [7.6]$$

*These immediate consequences by no means exhaust all the deductions from the given axioms. In order to obtain the *whole class* $Cn(A)$ of consequences of A, i.e. the entire set F of formulas of a theory with base A, we may apply the following *Rule*: Conjoin the assumptions and expand it into a conjunction of disjunctions (i.e., put it in conjunctive normal form). Every conjunct and every conjunction of conjuncts will be a logical consequence of the given axioms. In the case of [7.5] we have the chain

$$(p \,\&\, p \rightarrow q) \leftrightarrow [p \,\&\, (-p \vee q)]$$

$$[p \,\&\, (-p \vee q)] \leftrightarrow [p \vee (q \,\&\, -q)] \,\&\, (-p \vee q)$$

$$[p \vee (q \,\&\, -q)] \,\&\, (-p \vee q) \leftrightarrow (p \vee q) \,\&\, (p \vee -q) \,\&\, (-p \vee q),$$

whence

$$F_1 = Cn(A_1) = \{p \vee q, p \vee -q, -p \vee q, (p \vee q) \,\&\, (p \vee -q), (p \vee q) \,\&\, (-p \vee q), (p \vee -q) \,\&\, (-p \vee q), (p \vee q) \,\&\, (p \vee -q) \,\&\, (-p \vee q)\}.$$

Since

$$(-p \vee q) \leftrightarrow (p \rightarrow q),$$

$$(p \vee q) \,\&\, (p \vee -q) \leftrightarrow p \vee (q \,\&\, -q) \leftrightarrow p$$

$$(p \vee q) \,\&\, (-p \vee q) \leftrightarrow (p \,\&\, -p) \vee q \leftrightarrow q$$

$$(p \vee -q) \,\&\, (-p \vee q) \leftrightarrow (p \leftrightarrow q)$$

$$(p \vee q) \,\&\, (p \vee -q) \,\&\, (-p \vee q) \leftrightarrow [(p \vee q) \,\&\, (p \vee -q)] \,\&\, [(p \vee q) \,\&\, (-p \vee q)] \leftrightarrow p \,\&\, q,$$

we finally have the (minimal) family initiated by A_1:

$$F_1 = Cn(A_1) = \{p, p \rightarrow q;\, q, p \,\&\, q, p \vee q, p \vee -q, p \leftrightarrow q\} \qquad [7.7]$$

as the sum total of formulas of the microtheory with base $A_1 = \{p, p \rightarrow q\}$. Similarly for [7.6]:

$$(-q \,\&\, p \rightarrow q) \leftrightarrow [(-q \vee p) \,\&\, (-q \vee -p) \,\&\, (-p \vee q)]$$

whence

$$F_2 = Cn(A_2) = \{p \vee -q, -p \vee q, -p \vee -q, (p \vee -q) \& (-p \vee q), (p \vee -q) \& (-p \vee -q), (-p \vee q) \& (-p \vee -q), (p \vee -q) \& (-p \vee q) \& (-p \vee -q)\}.$$

Recalling well-known equivalences we finally get

$$F_2 = Cn(A_2) = \{-q, p \to q; -p, -p \& -q, p \vee -q, -p \vee -q, p \leftrightarrow q\}. \quad [7.8]$$

*Actually F_1 and F_2 above are the *smallest* sets of distinct logical consequences of the two-member bases A_1 and A_2: by repeated application of the rule of addition ("p entails $p \vee q$"), arbitrary propositions r, s, t, ... can be brought into the systems, which can then expand into infinite sets. Such a growth will have to be checked by the principles of semantic homogeneity and closure (Sec. 2), which will screen out those newcomers which are not germane to the original base. This rule of endogamy will effectively check the size of the sets of formulas, but the occurrence of quantitative concepts in any of the theory's initial assumptions will suffice to yield infinitely many logical consequences since some theorems will have infinitely many instances.

*Let us now state two simple theorems of the *calculus of axiomatizable systems* (A. Tarski). Let A_1 and A_2 be two axiom bases and $F_1 = Cn(A_1)$, $F_2 = Cn(A_2)$ the corresponding set of formulas, i.e. the corresponding deductive systems. Then the following theorems can be proved.

Theorem 1. $F_1 \subseteqq F_2$, i.e. $Cn(A_1) \subseteqq Cn(A_2)$, if and only if $A_2 \to A_1$. [7.9]
In algebraic terms: A filter F_2 with basis A_2 is *finer* than a filter F_1 with basis A_1 if and only if every member of A_1 follows from an element of A_2. In words: A necessary and sufficient condition for a theory to include another theory is that the axiom base of the former implies the axiom base of the latter. A_2 is then said to be an *extension* of A_1, or equivalently F_1 a *subtheory* of F_2. For example, the mechanics of a single mass point is included in the mechanics of a system of interacting mass points, the base of which differs from the former's base essentially by the adjunction of the axiom of the equality of the action and the reaction.

Theorem 2. $F_1 = F_2$, i.e. $Cn(A_1) = Cn(A_2)$, if and only if $A_1 \leftrightarrow A_2$. [7.10]
In general: Two filter bases are *equivalent* if and only if they generate the same filter. In words: A necessary and sufficient condition for two theories to be equivalent is that their axiom bases be equivalent. This equivalence, let it be noted, is formal and not semantical; i.e., A_1 and A_2 may not have the same interpretation. For example, wave mechanics

and matrix mechanics are mathematically equivalent but they are built with primitives that are partly different; consequently they are not semantically equivalent: their physical meaning is not the same.*

Given the base A of a theory, the set $F = Cn(A)$ of the logical consequences of A remains fully determined. But 'determined' does not mean "given": the conclusions $Cn(A)$ are not *separately* contained in the premises A but must be *detached* from them with the help of rules of inference (e.g., computation algorithms) and *ad hoc* tricks, such as the addition of a convenient premise which is by no means dictated by the premises but rather by the goal (the theorem to be proved). That is, a theory T is not *given* by its base A but must be *constructed* step by step on the basis of A and with the help of whatever may legitimately expedite the inference process. Viewed from a historical point of view, then, a theory is a growing set of formulas: at any given time only a finite part of the total set F is known; the complement of this *actual* set, i.e. the set of *potential* (derivable but as yet not deduced) formulas, is determined by the base A but not given by it.

When studying the logical structure of theories we take the whole set F of formulas, i.e. the union of the actual and the potential formulas of a theory. When the known and the unknown, the old and the new are thus lumped together, deductive inference seems to add nothing new to the initial assumptions: in this view, perfectly legitimate but partial, novelty is excluded *ab initio* by identifying "determined by A" with "given by A". Things look different when deduction is treated as a *process* rather than as a logical relation, i.e. when the epistemological and the historical viewpoints are substituted for the logical viewpoint. In this case the novelty introduced by nontrivial deduction is as patent as in the case where hydrogen and oxygen, the constituents of water, are detached from it through electrolysis. And the novelty is further increased when the logical consequences are evaluated, no longer on the strength of the premises and the inference rules, but in the light of empirical data: metastatements concerning the empirical truth value of the conclusions are then added, which are usually more or less different from the metastatements concerning the true value of the same formulas on the strength of the premises. In short, the empirical test of our premises (a bit of original knowledge) supplies further original knowledge. Needless to say, the logical view of deduction as a timeless relation (entailment) is different from but compatible with the epistemological view of deduction as a method for the growth of knowledge.

Problems

7.3.1 Is there any difference between an axiom and a premise? Recall that, by definition, a premise is any formula that is introduced in a logical derivation, whether it was listed among the initial assumptions or not; definitions may occur among the premises of an argument.

7.3.2. In the partition of formulas of a theory, used in the text, definitions were lumped together with axioms. Is this justified? See Sec. 3.4.

7.3.3. Is it possible to build a theory on a single axiom? Do not think of axioms as unanalyzed propositional formulas but as formulas of the predicate calculus with identity. *Alternate Problem:* Are SCHWANN's cell theory and RUSSELL's theory of types theories proper, i.e. hypothetico-deductive systems?

7.3.4. Find out whether the relation of entailment, $\vdash$, has all the properties of the relation $\leqq$ as stipulated by the axioms of the theory P of *partial order*, expounded on pp. 414—415. Take into consideration that the symbol '$\leqq$' can be read 'precedes' but can be interpreted in a number of ways. *Alternate Problem:* Study the relations among two or more theories in a given field, e.g., algebra.

7.3.5. What operational meaning, if any, can be attached to the sentence: 'A set of assumptions logically determines all the consequences that follow from it with the help of the acknowledged rules of inference even if no single consequence has so far been in fact deduced'? See M. BUNGE, *Intuition and Science* (Englewood Cliffs, N. J.: Prentice-Hall, 1962), pp. 42—49. *Alternate Problem:* Any set of empirically testable formulas can be derived from infinitely many different sets of assumptions. Does this indeterminacy render the basic assumptions untestable and therefore arbitrary?

7.3.6. Build a miniature nondeductive "system" (set of formulas) and a miniature deductive system. *Alternate Problem:* The problem-solver is rarely premise-conscious and interested in the structure of theories: he seizes upon any proposition that can suit his purpose. Can the theory critic and the theory builder afford to proceed in the same way?

7.3.7. Is it possible to recognize hypotheses of various levels (i.e., axioms, intermediate-level theorems and low-level theorems) in psychoanalysis, so as to establish deductive relations among them? In other words, is psychoanalysis a theory proper? See, on the psychoanalytic side, E. FRENKEL-BRUNSWIK, "Confirmation of Psychoanalytic Theories", in P. G. FRANK, Ed., *The Validation of Scientific Theories* (Boston: Beacon Press, 1956); on the critical side, E. NAGEL, "Methodological

Issues in Psychoanalytic Theory", in S. Hook, Ed., *Psychoanalysis, Scientific Method, and Philosophy* (New York: New York University Press, 1959). *Alternate Problem:* What is the snag in [7.3]?

7.3.8. Discuss and illustrate the process of derivation of scientific theories from stronger theories. *Alternate Problem:* Discuss examples of the relation of subtheory to theory.

7.3.9. Since the concept of entailment is relative to the set R of rules of inference accepted in advance — e.g., built in the logical theory underlying the given theory — by changing R to a different set, R', we may also change the set of consequences $Cn(A)$ to a different set, $Cn'(A)$. Thus, for instance, one and the same set of axioms A will lead to alternative sets of theorems according as two-valued or three-valued logic is adopted, and consequently two different theories will result. Discuss (i) whether the confrontation of the testable consequences of such alternative theories with empirical data will be able to force a decision among them, taking into account that the statements in observational language, too, will have to be interpreted in terms of the non-ordinary system of logic; (ii) whether there would be any advantage in adopting non-ordinary systems of logic; (iii) whether such an adoption in one given field (e.g., in quantum mechanics) should not force us to change the logic underlying every other field of science or at least in adjacent fields, in order to preserve the possibility of building bridges between the various theories — which is after all the main justification for wishing to retain the logical unity of science.

7.3.10. Report on the present state of the algebraic and the logical theories of theories (calculi of systems). See A. Tarski, *Logic, Semantics, Metamathematics*, transl. J. H. Woodger (Oxford: Clarendon Press, 1956), Ch. XII. A. Tarski, A. Mostowski and R. M. Robinson, *Undecidable Theories* (Amsterdam: North-Holland, 1953). P. C. Rosenbloom, *The Elements of Mathematical Logic* (New York: Dover, 1950), Ch. II, Sec. 3. E. W. Beth, *The Foundations of Mathematics* (Amsterdam: North-Holland, 1959), Ch. 19. H. B. Curry, *Foundations of Mathematical Logic* (New York: McGraw-Hill, 1963), Ch. 2 and the bibliography cited therein. *Alternate Problem:* Given a set A of specific assumptions and presuppositions, and an arbitrary set B of formulas, the set of consequences $F = Cn(A \cup B)$ may be regarded as a *syntactically* legitimate family generated by A. Clearly, there is no end to the number of families of this sort that a given axiom basis can generate. See whether the requirements of conceptual unity (Sec. 7.2) select a single *semantically* legitimate subfamily of F.

7.4. Abstract Theory and Interpretation

A set of symbols will be called *semantically abstract* if and only if none of them are interpreted or meaningful, *semiabstract* if some but not all of the symbols are interpreted, and *interpreted* if they are all interpreted. The semantical, not the epistemological concept of abstraction is here involved: we are not merely dealing with symbols the referents of which are far from ordinary experience — e.g. 'osmotic pressure'. Semantically abstract symbols are nondenoting signs, such as the unspecified operation symbol '$\circ$' in abstract algebra, which has no fixed meaning and can therefore be interpreted a posteriori in a number of ways. Semantically abstract symbols are also epistemologically abstract, but the converse is not true. Thus, the highest-level concepts of theoretical physics are epistemologically abstract but they are not meaningless: they have at least a specific mathematical interpretation (usually as numerical variables) and most of them mean physical properties. For example, even if the potentials of field theories are assigned no physical meaning, i.e. even when they are not made to refer to objective physical properties but are regarded just as mathematical auxiliaries, they are regarded as real functions, so that they are attached a mathematical meaning.

An *abstract theory* is a deductive system of schemata (propositional functions) in which only uninterpreted (abstract) symbols occur, such as the reference class U of nondescript elements $x, y, \ldots$ and the binary operation $\circ$ in an algebra $\mathsf{A} = \langle U, \circ \rangle$. The initial assumptions or axioms of an abstract theory may not contain free individual variables; universal and existential formulas will qualify as members of such a theory as long as the nature of the individual variables and the meaning of the predicates are left unspecified. Thus, e.g., a formula such as

For every x and every y, if x and y are in U, then $x \circ y = y \circ x$

will do as a propositional function or schema on condition that U and $\circ$ are not interpreted. Accordingly the unproved initial assumptions of an abstract theory are neither true nor false — let alone self-evident. They are just formal conditions to which the primitive symbols (U and $\circ$ in the above example) are subjected. In other words, the primitive (undefined) symbols of an abstract theory are not interpreted but abstract: they are non-denoting symbols, hence it might be said that they are not concepts at all but just symbols. Consequently, abstract theories are meaningless.

In presenting an abstract theory we usually mention how its primitives may be read. Thus, when expounding an algebraic theory involving the otherwise unspecified operation symbols '$\oplus$' and '$\otimes$', we may remark that, in certain interpretations of the theory, the former will mean the arithmetical addition functor and the latter the arithmetical product functor. But these will be regarded as extrasystematic or extra-theoretical remarks or comments with exclusively psychological or didactic aims, such as facilitating understanding and arousing interest.

Abstract theories are characteristic of modern logic and mathematics. But they also occur in the initial stages of the logical reconstruction (formalization) of factual theories, such as mechanics or genetics. There are at least two reasons for valuing abstract theories. One reason is that every abstract theory contains in seed an unlimited number of interpreted theories or models: abstract theories are *generic* by being uncommitted. In other words, a single abstract theory may underlie a number of specific (interpreted) theories; and once this is discovered formal derivations can be made once and for all for the entire set of interpreted theories having the same skeleton. This has not only the advantage of economy but also the virtue of rigor: in fact, if a derivation is purely formal, if no intuitive elements occur in it, then the risk of error is greatly diminished; on the other hand, purely formal derivations are apt to be more difficult to perform — which is a practical drawback of theirs. A second reason for valuing abstract theories is of a metascientific character, namely, that the *structure of a scientific theory*, whether formal or not, is precisely an abstract theory: all else is interpretation and application. Therefore the logic of scientific theories is, strictly speaking, the study of the underlying abstract theories; moreover, since theories constitute the nucleus of contemporary science, the structure of science is primarily a subject for logic and mathematics including the respective metadisciplines. (Recall the theorem stated in Sec. 3, according to which every deductive system is, as regards its structure, a filter.)

In this section and in the next we shall sketch the core of two elementary abstract theories that can be handled as tools in the analysis of science: namely, the theory of partial order and probability theory. They are not only *formal*, in the sense of being logically independent of experience, but they are also semantically abstract, in that they are uninterpreted, hence susceptible to an unlimited number of interpretations.

In its bare essentials the theory of *partial order* is a set of formulas that relate two sets of primitive symbols: a nonempty set U of nondescript elements $x, y, z, \ldots$, and a binary relation symbol, '$\leqq$', which can be read 'precedes'. We may therefore symbolize the theory P of partial order as the relational system: $\mathsf{P}=\langle U, \leqq\rangle$. The axioms of the theory, i.e. the conditions to which the primitive symbols are subjected, are the following:

$P1$ For every x in U, $x\leqq x$ (*Reflexivity*), [7.11]

$P2$ For every x and y in U, $x\leqq y \,\&\, y\leqq x \rightarrow x=y$ (*Antisymmetry*), [7.12]

$P3$ For every x, y and z in U, $x\leqq y \,\&\, y\leqq z \rightarrow x\leqq z$ (*Transitivity*). [7.13]

Look at some partially ordered sets:

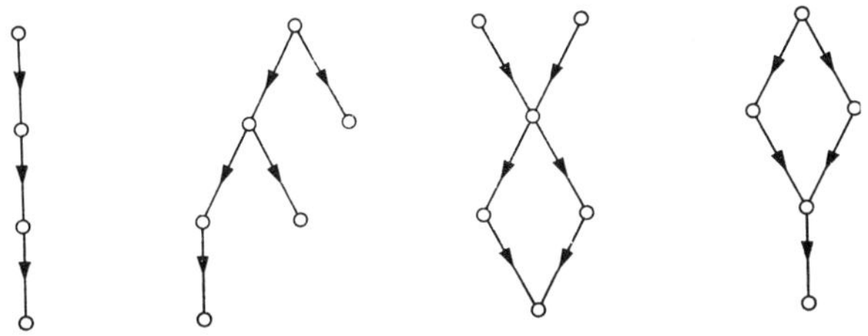

Fig. 7.6. Partially ordered sets. Every member of the set is represented by a small circle; every line represents the relation $\leqq$. The first diagram on the left is a chain, i.e. it represents a set in which the relation $<$ holds. *The first and fourth sets are more than just partially ordered sets: they are lattices, because for every pair of elements there exist both a greatest lower bound and a least upper bound*

We shall simplify the symbolization by adopting the following notational *Convention:* '$(x)_U$' stands for 'For every x in U'. With this abbreviation, the base of the abstract theory of ordered sets is the following list of nonsignificant symbols.

Primitives (predicate base)

1. *Logical primitives:* all those of elementary logic and set theory.

2. *Extralogical (specific) primitives:* the abstract set U and the binary relation $\leqq$.

Axioms (initial assumptions)

$P1$ $(x)_U(x \leqq x)$ (Reflexivity), [7.11′]

$P2$ $(x)_U(y)_U[x \leqq y \;\&\; y \leqq x \rightarrow x = y]$ (Antisymmetry), [7.12′]

$P3$ $(x)_U(y)_U(z)_U[x \leqq y \;\&\; y \leqq z \rightarrow x \leqq z]$ (Transitivity). [7.13′]

Definition 1: A set U is said to be *partially ordered* if and only if the above axioms are satisfied. (*Total* ordering is obtained by adding the condition that the relation $\leqq$ holds among *any* two elements of the set: $(x)_U(y)_U(x \leqq y \vee y \leqq x)$. In this case the order is *linear*, i.e. the set is a chain.)

From the above schema a few theorems can be derived, such as, e.g.,

$$T1 \quad (x)_U(y)_U(z)_U[x \leqq z \leqq y \;\&\; x = y \rightarrow x = y = z].$$

The ulterior development of the theory requires the introduction of new concepts. Since in an axiomatic theory no new concepts can be introduced which are not listed among the primitives, the construction of concepts must be done via definitions in terms of the primitives. For example, we may introduce the stronger relation $<$ of strict precedence (generating a *strict* partial ordering) by means of the following

Definition 2:

$$(x)_U (y)_U [x < y =_{df} x \leqq y \ \& \ x \neq y]. \qquad [7.14]$$

This enables us to state, e.g.,

$T2 \quad (x)_U - (x < x)$

and

$T3 \quad (x)_U (y)_U (z)_U [x < y \ \& \ y < z \rightarrow x < z].$

Further consequences of interest are obtained by introducing the concepts of immediate successor, first and last element, greatest lower bound, least upper bound, and so on. In this way — by widening the predicate base — new, richer abstract theories are built, such as lattice theory and filter theory. The theory of partial order is contained in all of them — or, as one usually says, all those theories are *based* on the theory of partial order.

Now, the abstract theory sketched above can be interpreted in an unlimited number of ways. In other words, there is an unlimited number of relations of precedence of different nature but with the same logical structure. Or, again, infinitely many interpretations of the primitives of the theory of partial order *satisfy* its axioms; every one of them is called a *model* or *representation* of the abstract theory. Every model of our theory will be characterized by a pair of *interpretation assumptions:* one for the primitive U, another for the primitive $\leqq$. An interpretation assumption for a symbol s is a statement of the form "$I(s) = m$", where 'm' stands for a meaningful array of signs. A few models of our abstract theory P are generated by adjoining to it the following interpretation assumptions.

1. *Set-theoretical interpretation:* set inclusion
   $I(U) =$ set of subsets
   $I(\leqq) = \subseteqq$ [improper inclusion].
2. *Logical interpretation 1:* implication
   $I(U) =$ set of statements
   $I(\leqq) = \rightarrow$ [implication].
3. *Logical interpretation 2:* deduction
   $I(U) =$ set of statements
   $I(\leqq) = \vdash$ [entailment].
4. *Arithmetical interpretation:* divisibility
   $I(U) =$ set of integers
   $I(\leqq) =$ divides [e.g., '$3 \leqq 9$' is read '3 divides 9'].
5. *Geometrical interpretation:* alignment
   $I(U) =$ points on a line
   $I(\leqq) =$ to the left of or coincident with.

6. *Physical interpretation:* thermal order
   $I(U)$ = set of bodies
   $I(\leqq)$ = less hot than or as hot as.

7. *Political interpretation:* power
   $I(U)$ = set of human organizations
   $I(\leqq)$ = wields power over.

8. *Value-theoretical model:* valuation
   $I(U)$ = all conceivable objects
   $I(\leqq)$ = less valuable than or as valuable as.

Let us now give a general characterization of the concept of *interpretation* of an abstract theory. Let U be the universe of discourse of a theory T, and $P_1, P_2, \ldots, P_n$ the primitive predicates of the theory, every one of which applies to members of the reference set U. Let further $I(U)$ be an interpretation of U, and $I(P_1), I(P_2), \ldots, I(P_n)$ interpretations of the primitives — not arbitrary but acceptable interpretations. For an interpretation of the primitives to be *acceptable* in factual science it must satisfy the following conditions. First, the interpretation should not be *ad hoc*, i.e. invented for, say, proving the consistency of the given abstract theory but otherwise isolated from the rest of science: scientific theories are not exercises in semantics. (*Ad hoc* interpretations are on the other hand metatheoretically useful: see Sec. 7.6.) Second, the interpreted symbols must satisfy the theory's axioms: i.e., the formal relations satisfied by the primitives must hold for their interpretations as well. Thus, if A and B are unspecified ("undefined") sets and the abstract theory stipulates that $A \subseteq B$, then $I(A) \subseteq I(B)$ must hold as well. In the case of factual theories we shall not demand that the interpreted axioms be completely true, but just verisimilar. Every set $\{I(U), I(P_1), \ldots, I(P_n)\}$ satisfying the preceding conditions is an *interpretation* of the abstract theory concerned: it is *a* (not *the*) *code* for deciphering the formalism.

A true interpretation of an abstract theory will be called a *conceptual model* of it just in case the primitive symbols are made to correspond to concepts existing in some theoretical context but having no real reference. In other words, a conceptual model is an interpretation of the given abstract theory in terms of concepts belonging to a nonabstract theory. For example, the theory of strict partial order, which belongs to abstract algebra, can be interpreted, in arithmetic, as the theory of the relation "less than" among numbers. All the interpretations 1—5 above yield conceptual models of the same syntactical system; and the mathematical formalism of a physical theory is a conceptual or ideal model of the underlying abstract theory. Model theory, a branch of metamathematics born in the 1950's, is concerned with conceptual models.

Consider now a theory which is either abstract (uninterpreted) or a conceptual model of an abstract theory — in short, a theory the primitives of which have either no reference at all or no extraconceptual reference. Suppose the primitives of such a theory are assigned a factual meaning and/or an empirical meaning (see Sec. 3.5). In other words, assume a set of interpretation assumptions is added to the given theory, by virtue of which every primitive becomes a nonformal predicate — e.g., a concept standing for some physical property or other. Such a set of interpretation assumptions can be an assorted mixture of rules of designation, interpretation postulates, and referitions (recall Sec. 3.6). The primitives will not thereby be *defined* — let alone operationally. They will be *interpreted* if they were abstract to begin with; and they will acquire an *additional*, nonformal yet perhaps nonempirical meaning if they belong to a conceptual model, as when the numerical value m in mechanics is interpreted as the value of the mass of a system.

When such a nonformal interpretation is conferred upon the primitives of a formalism we shall say that a *factual model* of the latter is produced. Thus, the interpretations 6—8 above yield three different factual models of the abstract theory of partial order. Again, every physical theory may be regarded as a factual interpretation of its mathematical formalism. Thus, the various physical interpretations of the mathematical skeleton or formalism of quantum mechanics are as many factual models of it. Too often the interpretation is not formulated explicitly: it is taken for granted that the reader realizes what the *intended interpretation* of the formalism is, i.e. how its symbols should ultimately be "read" or what is the interesting application (model) the abstract theorist has in mind. But this intention may be far from obvious or much too ambiguous, in which case it may be doubted whether the theory has a content at all or at least a single intended interpretation. The critically minded physicist will then complain that the theory is physically meaningless or that it has no physical content — and he may uncritically demand that a set of "operational definitions" of the primitives be appended to it — which is logically impossible (see Sec. 3.6).

Furthermore, there is room to doubt that the sole addition of interpretation assumptions will, save in trivial cases, give a formalism an unambiguous meaning. What the interpretation assumptions effect is to outline possible *models* of a formalism. But, since extralogical concepts are usually vague (recall Sec. 3.1), a factual model is apt to be surrounded by some fog. In other words, interpretation rules may suggest rather than exhaustively fix the meanings of extralogical primitives: they just sketch the semantical profile of a theory. Even a semantical system, such as MAXWELL's electromagnetic theory, may not be interpretable in an exact and unique way. Meanings are not born full-fledged but appear to grow and evolve. In particular, the interpretation of a scientific theory matures in the course of the deductive development, application, and critical examination of the theory. This applies to the semantical senses of 'meaning' (see Secs. 2.3 and 3.6) and, a fortiori, to its psychological sense — namely, "understanding". The existence of meaning indeterminacy — a sign of a theory's immaturity — accounts for some of the controversies over the interpretations of young theories and goes to show that foundational research and controversy, far from being a luxury, should be encouraged in the interest of the maturation of science. But enough of propaganda: let us resume our discussion of the kinds of model.

In addition to the aforementioned pure cases of the conceptual and the factual models we find *mixed models* or semi-interpreted theories in factual science. These are theories some predicates of which represent real (or presumably real) properties whereas the remaining properties are given no factual or empirical interpretation. These variables which refer to no extratheoretical things, properties or relations, are given a mathematical (hence a conceptual) interpretation. Many high-level factual theories contain such variables performing a purely mathematical job. Thus the partition function is not a physical property in classical thermodynamics but rather as an intervening variable with a purely syntactical function; the same holds for the potentials in electromagnetic theory and for the wave functions in the usual interpretation of quantum mechanics: none of these primitives are assigned direct physical meanings. These theories are, accordingly, mixed models of the underlying abstract theories. But it would be mistaken to regard them as syntactical systems, i.e. as abstract theories: their universes of discourse are sets of physical systems, and all their primitives have at least a mathematical meaning.

A small but rapidly increasing set of factual theories are *mixed models* characterized by the definite predominance of conceptually interpreted variables over factually interpreted primitives. Among such semi-interpreted systems we recognize: (i) information theory, which can be applied to open systems of any kind, from radars to organisms; (ii) the fundamental theory of servomechanisms, which is the same for electrical and for mechanical automatic control devices; (iii) network theory, which can be applied to electric circuits and mechanical vibrating systems alike. All three theories can in turn be subsumed under a general black box theory in which the most general relation among an input function $I(t)$ and an output function $O(t)$ is hypothesized and worked out. In such a theory t is the only factually interpreted primitive: in fact, it stands for time. The input I and the output O are noncommittal symbols interpreted only as numerical variables (conceptual interpretation), but with no fixed factual meaning. The incomplete interpretation that characterizes mixed models is a source of their generality — as well as of their low testability. And the fact that they can be interpreted (applied) in so many different ways shows that law formulas belonging to different fields may be formally identical.

Thus far we have dealt with *theoretical models*. All such models are mental creations, no matter how closely they may represent real objects. A real system, in turn, may be regarded as a *material model* of a theory. Thus, an electric switching system may be regarded as a material model or physical analog of the propositional calculus and may accordingly be used as a deduction aid.

The variety of senses of the word 'model' can be bewildering unless we care to specify what sense we have in mind. Table 7.1 summarizes the chief senses we have discussed.

Table 7.1. *Some Senses of 'Model'*

- *Model*
  - *Theoretical*
    - *Ideal model* involved in a theory
      - *Iconic* (e.g., the lock-and-key model of enzymes)
      - *Symbolic* (e.g., the democratic society in political science)
    - *Interpretation* of an abstract theory
      - *Conceptual* (e.g., an arithmetical interpretation of group theory)
      - *Factual* (e.g., a physical interpretation of Euclidean geometry)
      - *Mixed* (e.g., general automata theory)
  - *Material* (e.g., hydraulic model of a country's economy)

Before we go any further it will be convenient to become more familiar with abstract theories and their interpretations.

Problems

7.4.1. Write down the axioms of the theory of linear order (strict partial order). Show that this theory is contained in the theory of partial order sketched in the text. Propose an arithmetical, a geometrical, a physical and a biological interpretation. *Alternate Problem:* Is set theory abstract or semiabstract? Compare it with the theory of partial order.

7.4.2. Examine the general theory of groups with the following foundation.

Primitives: A nonempty set U, a binary operation $\circ$, a unary operation $^{-1}$ (which applied to an arbitrary element x of U yields its inverse, x^{-1}), and an individual constant e (the neutral element).

Axioms:

$$G1 \quad (x)_U (y)_U (z)_U [x \circ (y \circ z) = (x \circ y) \circ z], \qquad [7.15]$$

$$G2 \quad (x)_U (x \circ e = x), \qquad [7.16]$$

$$G3 \quad (x)_U (x \circ x^{-1} = e). \qquad [7.17]$$

Try to give an alternative formulation, e.g. by replacing axioms $G2$ and $G3$ by statements concerning the existence of the neutral element e and of the inverse elements x^{-1}. Set up a definition of the group concept; to this end recall Sec. 3.3 on the so-called postulational definitions. Build conceptual (e.g., arithmetical, geometrical and set-theoretical) models of group theory as well as factual interpretations of it. Hints: find out whether the addition and multiplication of integers satisfy the axioms of group theory; see whether the translations of a point on a plane form a group; and study the set of rotations of a mass point around a fixed center through a variable angle. *Alternate Problem:* Discuss the relevance of group theory to epistemology, in particular to the idea of invariance under changes of observation conditions. See Sec. 6.5 and J. Ullmo, "The Agreement Between Mathematics and Physical Phenomena", in *The Critical Approach*, Ed. by M. Bunge in honor of Karl Popper (New York: The Free Press, 1961).

7.4.3. Characterize and exemplify the concepts of formal theory (as distinct from factual) and of abstract theory (in contrast to interpreted theory). *Alternate Problem:* Report on either of the following texts. (i) E. V. Huntington, "The Fundamental Propositions of Algebra", repr. in J. W. A. Young, Ed., *Monographs on Topics of Modern Mathematics* (1911; New York: Dover, 1955), Secs. II and III. (ii) E. V. Huntington, "The Method of Postulates", *Philosophy of Science*, **4**, 482 (1937). (iii) C. I. Lewis and C. H. Langford, *Symbolic Logic*, 2nd ed. (New York: Dover, 1959), Ch. XI. (iv) W. and M. Kneale, *The Development of Logic* (Oxford: Clarendon Press, 1962), Ch. VI, Sec. 4.

7.4.4. One and the same input-output relation can be materialized in infinitely many ways. For example, any of an infinity of amplifiers can amplify a given signal in a preassigned way. What does this suggest — aside from sayings such as "All paths lead to Rome"? *Alternate Problem:* Set up the interpretation rules needed to establish a one-one correspondence between the propositional calculus and the qualitative theory of circuits. Hint: two switches in series correspond to binary conjunction, and two switches in parallel to binary disjunction.

7.4.5. Find out whether preference (i.e., the relation "is preferred to" or, if preferred, the relation "is more valuable than") satisfies the axioms of strict partial order. *Alternate Problem:* Examine the material model of some scientific theory, such as the Moniac (hydraulic model of KEYNES' theory of national economics).

7.4.6. Take G. PEANO's axiomatics of elementary arithmetic and perform the following tasks. (i) Build the corresponding abstract theory (i.e., disinterpret) and formulate it with the symbols of the predicate calculus. (ii) Find two models of the abstract theory. (iii) Find out whether PEANO's axioms unambiguously characterize the concept of natural number, i.e. whether they supply a (contextual) *definition* of the set of positive integers. For purposes of reference PEANO's postulates are given in plain words:

N 1 Zero is a natural number. [7.18]

N 2 The successor of any natural number is a natural number. [7.19]

N 3 No two different natural numbers have the same successor. [7.20]

N 4 Zero is the successor of no natural number. [7.21]

N 5 If a property is such that (i) it applies to zero and (ii) if it is true of an arbitrary number then it is true of its successor, then the property belongs to every natural number. [7.22]

Notice that the axiom set functions as a stipulation for the following specific (extralogical) primitives: "natural number" (a unary predicate), "zero" (an individual constant), and "the successor of" (which can be analyzed either as a unary predicate or as a binary relation). "Zero" may be replaced by "one". *Alternate Problem:* Are the axioms of set theory physically interpretable? In particular, could there be physical models of the axioms of infinity and choice? Distinguish "physically possible" from "feasible".

7.4.7. EUCLID was convinced that his *Elements* disclosed the geometry of physical space. Did he, as a matter of fact, build a physical theory or else an abstract theory, or a conceptual model of an abstract theory that was unearthed much later? *Alternate Problem:* The partisans of the phenomenological approach held, around 1900, that a good physical theory is a set of differential equations with no underlying hypothesis and such that the equations cover the empirical data. Differential equations were even opposed to atomistic hypotheses. Was such an opposition correct?

7.4.8. Discuss the role of the disclosure of abstract theories in the unification of science at the syntactical level. Keep in mind the case of theories that study *dual concepts*, such as the theories of the relations $\leqq$ and $\geqq$, $\subsetneqq$ and $\supsetneqq$, or the sum and the subtraction of segments. Every such theory is characterized by some duality law or other. For example, in the theory of ordered sets we have the following *duality principle* (actually a metaaxiom): Every theorem concerning an ordered pair $\langle U, \leqq \rangle$ holds for its dual $\langle U, \geqq \rangle$ if the symbols '$\leqq$' and '$\geqq$' are exchanged in the theorem. Find out further illustrations in projective geometry and in logic.

7.4.9. According to an influential view, the primitives of a factual theory are not interpreted: only the so-called observation terms of the theory are interpreted, whereas the "theoretical terms" acquire a meaning indirectly, through their relations with the "observation terms". What kind of meaning is involved in this view: factual or empirical? Is it true that all theories contain observational concepts? See R. CARNAP, "Testability and Meaning", *Philosophy of Science*, **3**, 419 (1936) and **4**, 1 (1937), and *Foundations of Logic and Mathematics* (Chicago: University of Chicago Press, 1939), Secs. 23 and 24. For criticisms see (i) P. K. FEYERABEND, "Carnaps Theorie der Interpretation theoretischer Systeme", *Theoria*, XXI, 55 (1955) and "On the Interpretation of Scientific Theories", *Atti del XII Congresso Internazionale di Filosofia* (Firenze: Sansoni, 1960), V and (ii) H. PUTNAM, "What Theories are Not", in E. NAGEL, P. SUPPES and A. TARSKI, Eds., *Logic, Methodology and Philosophy of Science* (Stanford: Stanford University Press, 1962).

7.4.10. To what extent can the formalism and the interpretation of a theory be changed independently from one another? *Alternate Problem:* Report on the present state of the theory of models (either conceptual or factual). See the collective volume cited in Problem 7.4.9 and H. FREUDENTHAL, Ed., *The Concept and Role of the Model in Mathematics and Natural and Social Sciences* (Dordrecht: Reidel, 1961), and A. ROBINSON, *Introduction to Model Theory and to the Metamathematics of Algebra* (Amsterdam: North-Holland, 1963).

7.5. *Probability: Calculus, Models, Misinterpretations

Our second example of an abstract theory will be the calculus of probability. This theory is of paramount importance in factual science and in metascience: being abstract, it can be interpreted and misinterpreted in an unlimited number of ways.

Elementary probability theory presupposes elementary logic and set theory, and elementary arithmetic. One of the two primitives of the theory is, as usual, a nonempty reference set U, the members of which we shall call a, b, c, ... The theory, then, will predicate something about these nondescript elements and the collections they can constitute. If $U=\{a, b\}$, the subsets of U are $\{a\}$, $\{b\}$, and U itself. (We count only the nonempty subsets of U.) Calling $S(U)$ the set of all the nonempty subsets of U, in the particular case of the two-member set we have $S(U)=\{\{a\}, \{b\}, \{a, b\}\}$. If U happens to have three elements instead of two, $S(U)$ will consist of $7=2^3-1$ nonempty subsets. The set $S(U)=\{A, B, C, \ldots\}$ of all the nonempty subsets of a basic set U is called the power set of U. In general, the cardinality or numerosity of such a collection of subsets of a basic set U with N elements will be 2^N-1. It may help to interpret either the elements of U or its subsets as possible

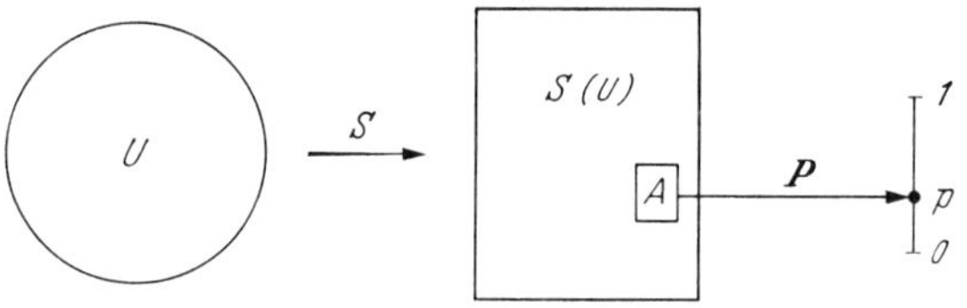

Fig. 7.7. The mapping of the subsets of U into the basic real numbers interval

events of a kind. For example, we may think of A as the event $A=\langle$Head, Tail$\rangle$ in two successive flips of a coin. But, since our aim is to build an abstract theory, any such interpretation must be handled as a *heuristic aid:* one or more definite contents will be poured into the empty shell once it has been constructed.

The other specific or extralogical primitive of the theory is a set function P "defined" (given) on $S(U)$. (We may think of $P(A)$ as the weight of A with respect to S, where A is part of a body S. But, again, this provisional interpretation is just a heuristic aid.) The set function or probability measure P establishes a correspondence between the power set $S(U)$ and the unit interval $[0, 1]$ (see Fig. 7.7). In other words, the subsets A of U are represented by numbers p in the $[0, 1]$ interval. The number p corresponding to the subset A is the numerical value of the function P at A, i.e. $P(A)=p$. The *structure* of the probability concept is, then, that of a function with one object variable running over the nonempty subsets of the basic set, and one numerical variable running over the $[0, 1]$ interval, which is the image of U under P.

Yet the above are all extrasystematic asides. The base of the abstract elementary probability theory boils down to the following structure (A. N. Kolmogoroff).

Primitives: U, a nonempty set; P, "defined" on $S(U)$ and taking real values.

Axioms

$P1$ $(A)_{S(U)}[P(A) \geqq 0]$, i.e. $P(A)=p$ is a nonnegative number. [7.23]

$P2$ $P(U)=1$, i.e. the probability of the basic set equals one. [7.24]

$P3$ $(A)_{S(U)}(B)_{S(U)}[A \cap B = \emptyset \rightarrow P(A \cup B) = P(A) + P(B)]$, i.e. the probability of the measure of the union of two sets with no common elements equals the sum of their separate measures. [7.25]

The sentences accompanying the preceding formulas are added for didactic purposes: they do not belong to the axiomatic theory. The symbol '$(A)_{S(U)}$' is to be read 'for every A in the collection of non-empty subsets of U'. (Incidentally, the occurrence of class variables such as A, which run over the set $S(U)$, shows that the elementary probability theory is nonelementary in a logical sense.)

We shall now derive a few important theorems that are useful in the theory of nondeductive inference. As is customary, the quantifiers will be dispensed with: since all our premises are universal, we can freely pass from *all* to *any* and finally back to *all*. The derivations will be *formal*, in the sense that they make no use of specific interpretations in terms of coins, urns, balls, or other heuristic devices.

Thm. 1. The probability of the complement equals the complement of the probability:

$$P(\bar{A}) = 1 - P(A). \qquad [7.26]$$

Proof. From $P2$ and the (implicit) definition of complementary set, namely $\bar{A} \cup A = U$, we get $P(\bar{A} \cup A) = 1$. Applying this to $P3$ we obtain $P(\bar{A} \cup A) = P(\bar{A}) + P(A) = 1$, which can be rewritten in the form [7.26].

Thm. 2. The numerical value of a probability lies between 0 and 1:

$$0 \leqq P(A) \leqq 1. \qquad [7.27]$$

Proof. The first inequality follows from $P1$, which can be rewritten as $0 \leqq P(A)$. To prove the second inequality we examine Theorem 1. By $P1$, $P(\bar{A}) \geqq 0$, whence the second-hand member of Theorem 1, i.e. $1 - P(A)$, is nonnegative as well. In other words, $P(A) \leqq 1$. Q.E.D.

Thm. 3. The probability measure of the union of any two sets equals the sum of their separate measures minus the measure of the overlap [which would otherwise be counted twice]:

$$P(A \cup B) = P(A) + P(B) - P(A \cap B). \qquad [7.28]$$

Proof. We decompose B into two disjoint sets, namely thus: the set of the elements which are in both B and A, and the set of the elements which are in B but not in A: $B = (A \cap B) \cup (\bar{A} \cap B)$. (See Fig. 7.8). Finally we introduce this partition into the arguments of $P(B)$ and $P(A\cup B)$ and apply $P3$:

$$P(B)=P[(A\cap B)\cup(\bar{A}\cap B)]=P(A\cap B)+P(\bar{A}\cap B),$$

$$\begin{aligned}P(A\cup B)&=P[A\cup[(A\cap B)\cup(\bar{A}\cap B)]]=P[[A\cup(A\cap B)]\cup(\bar{A}\cap B)]\\&=P[A\cup(\bar{A}\cap B)]\\&=P(A)+P(\bar{A}\cap B).\end{aligned}$$

Use has been made of the equality "$A\cup(A\cap B)=A$", which follows from set theory and is represented in Fig. 7.8. Subtracting the last two

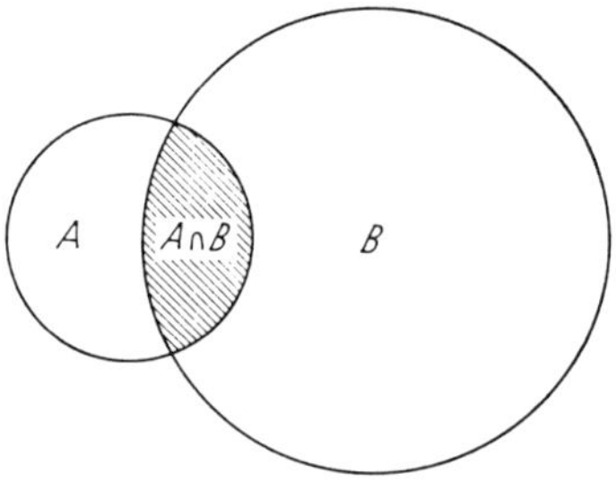

Fig. 7.8. Decomposition of B into two disjoint sets in order to be able to apply $P3$. This figure is an extrasystematic pedagogic crutch: look at it askance

equations we get rid of $P(\bar{A}\cap B)$; rearranging the extant terms we get Theorem 3.

Notice, first, that although in the derivation we have used only axioms $P1-P3$ and elementary set-theoretical and arithmetical relations, the latter were not suggested by the axiom base: they were extra premises extracted from theories presupposed by the probability calculus. The choice of those extra premises was hinted at by the desired results and was controlled by no rule save this: "Try any premise that might be relevant to the proof, as long as the premise belongs to any of the theories presupposed by the theory concerned". In short, the theorems do not "flow" from the axioms and the definitions by mechanically applying rules of inference: deduction is opportunistic rather than principled, and the rules of inference are not so much rules that *direct* inferences as rules that *validate* them: they permit and prohibit but do not tell how to proceed. Secondly, note that Theorem 3 is more general than axiom $P3$, which holds in the special case when $A\cap B=\emptyset$. The reason that a theorem may be stronger than an axiom of the same theory is, again, that behind axioms $P1-P3$ lie elementary set theory and arithmetic. Situations of this kind are apt to arise in factual theories formulated with the help of mathematical theories: comparatively modest axioms may give rise to strong theorems in a theory backed by a rich mathematical structure. To take advantage of this is to apply the cuckoo technique of using ready-made conceptual nests (see Sec. 8.2).

We now develop a further portion of probability theory by introducing the concept of conditional probability. This we do via the following

Df. 1 (*conditional probability*)

$$\text{If} \quad P(B) \neq 0, \quad \text{then} \quad P(A \mid B) =_{df} \frac{P(A \cap B)}{P(B)}. \qquad [7.29]$$

'$P(A \mid B)$' is read 'the (conditional) probability of A given B'. Next we state

Thm. 4. If $P(A) \neq 0$ and $P(B) \neq 0$, then

$$P(A \cap B) = P(A \mid B) \cdot P(B) = P(B \mid A) \cdot P(A). \qquad [7.30]$$

Proof. The first equality is an immediate consequence (corollary) of Df. 1. The second equality is obtained from Df. 1 by exchanging A and B. In fact, if in Df. 1 we substitute B for A and recall that $A \cap B = B \cap A$, we get

$$\text{If} \quad P(A) \neq 0, \quad \text{then} \quad P(B \mid A) =_{df} \frac{P(A \cap B)}{P(A)}, \qquad [7.31]$$

which gives the wanted equality.

Thm. 5 (*Bayes*). If $P(A) \neq 0$ and $P(B) \neq 0$, then

$$P(A \mid B) = \frac{P(A) \cdot P(B \mid A)}{P(B)}. \qquad [7.32]$$

This theorem follows immediately from Theorem 4 by just keeping out the first-hand member. BAYES' theorem is central in the theory of plausible inference, in which A and B are interpreted as sets of propositions.

Df. 2 (*irrelevance*). B is irrelevant to A if and only if $P(A \mid B) = P(A)$. Upon trading A by B the following dual proposition is obtained:

Df. 2′ (*irrelevance*). A is irrelevant to B if and only if $P(B \mid A) = P(B)$.

Df. 3 (*independence*). A is independent of B if and only if either B is irrelevant to A or A is irrelevant to B. Briefly,

$$A \text{ is independent of } B =_{df} [P(A \mid B) = P(A) \vee P(B \mid A) = P(B)]. \qquad [7.33]$$

Thm. 6. If A is independent of B, then B is independent of A. [That is, the relation of independence is symmetrical.]

Proof. Substitute A for B in Df. 3 and recall the commutativity of the disjunction.

Thm. 7. If A is independent of B, then

$$P(A \cap B) = P(A) \cdot P(B). \qquad [7.34]$$

Proof. Apply Df. 3 to Theorem 4.

Thm. 8. If A is independent of B, then

$$P(A \cup B) = P(A) + P(B) - P(A) \cdot P(B). \qquad [7.35]$$

Proof. Substitute Theorem 7 in Theorem 3.

Heretofore we have dealt with the *abstract* probability theory. In most elementary textbooks on probability one or more of the following interpretations of this concept, leading to the corresponding models, are merrily mixed.

1. *Semantic interpretation*

$I(U)$ = set of all the propositions of a given kind

$I[P(A)]$ = likelihood that the proposition(s) A be true.

2. *Statistical interpretation*

$I(U)$ = set of empirical data of a given kind

$I[P(A)]$ = relative frequency of the data (or datum) A in the universe of data U.

3. *Physical interpretation* 1

$I(U)$ = set of random events of a given kind

$I[P(A)]$ = probability of the event(s) A.

4. *Physical interpretation* 2

$I(U)$ = set of events of a given kind

$I[P(A)]$ = propensity or disposition of the event(s) A to occur.

5. *Psychological interpretation*

$I(U)$ = set of judgments of a given kind (about, e.g., the possible outcomes of an action)

$I[P(A)]$ = credibility (or degree of confidence, or certainty) of the judgment(s) A.

6. *Ontological interpretation*

$I(U)$ = set of possibilities (possible facts) of a kind

$I[P(A)]$ = weight of the possibility of A.

The semantic and the statistical interpretations yield conceptual models of the probability calculus; the remaining interpretations listed above, when added to the calculus, supply factual models thereof (see Sec. 7.4). The ontological interpretation would seem to be a generalization of the disposition interpretation; with this interpretation, the probability theory becomes the quantitative theory of possibility, to be contrasted to the qualitative theories of possibility, viz., the various systems of modal logic. The physical and the ontological interpretations regard probability as an objective property of things — or rather of sets or sequences of facts —, whereas the psychological interpretation, by rendering probability the complement of uncertainty, considers it as a subjective attribute. Subjective probabilities, so often confused with objective probabilities, are conceptually and numerically different from them: the probability of a given fact is seldom known with complete accuracy and therefore it is not identical with the degree of rational credibility in the judgment expressing the said fact; if there were no such differences between objective and subjective probabilities nobody would play games of chance for money.

Alternative models of the abstract probability theory are then possible. Strictly speaking, then, the question 'What is probability?' is unambiguous only if no interpretation is involved or if a definite interpretation is indicated. In the former case a correct answer could be by way of the following *axiomatic definition*: A set function P on a given basic set U is a *probability* (measure) if and only if it satisfies the postulates $P1-P3$ above. Should it be objected that this answer sheds no light on the *meaning* of the term 'probability', an adequate rejoinder might be this. Right — but abstract symbols are meaningless, and when definite meanings are attached to them it is seen that several different contents appear. Consequently, instead of asking 'What is *the* meaning of 'P'?', one should ask 'What meanings have, as a matter of fact, correctly been assigned to 'P' up to now?'

The restriction to *correct* interpretations suffices to discard certain popular misinterpretations, such as the probability theory of truth. This theory, encouraged by a number of ordinary languages — notably German, in which 'probability' is synonymous with 'verisimilitude' — equates the degree of truth of a statement with the probability — that the statement be *true*. Since the truth concept is made an ingredient of the probability concept, it is circular to identify probability with degree of truth. (This becomes apparent when the idea is symbolized: $V(s) =_{df} P[V(s) = 1]$. The notion of truth, which is supposed to be defined, reappears in the definiens.) The concept of probability can be used to elucidate a number of other concepts (see Sec. 3.2) but it is not a universal primitive. In particular, it cannot be used to elucidate the notion of (partial) truth, because this concept is logically prior to the concept of probability statement: we must have an independent concept of truth if we want to know what the truth value of a given probability statement is.

The distinction between the probability theory and its various possible interpretations dissolves the old, yet still alive, controversy between the upholders of the subjectivist interpretation of probability as a degree of belief and the partisans of the empiricist interpretation of probability as a relative frequency. Subjectivists and empiricists are bound to talk through one another since the two are defending alternative possible interpretations of the abstract probability theory. Both contenders are wrong in claiming that their own is the right interpretation, and so are those, more tolerant, who grant that there are two possible models of probability (or two probability concepts). The recognition of the possibility of any number of interpretations of the abstract probability calculus dissolves a barren disputation and, in addition, it stimulates the search for further interpretations of the same empty structure; on top of this it reassures us in our confidence in the unity of the mathematical theory of probability — which unity was never shattered by the controversy.

But a *possible* interpretation need not be *true* to fact, hence acceptable. We have seen before (Sec. 7.4) that a factual interpretation of an abstract axiom set must satisfy the very same axioms of the latter — at least approximately. Now, the frequency interpretation of probability has been satisfactorily confirmed: frequencies do, in fact, combine like probabilities — allowing for typical fluctuations. (Moreover, the probability calculus was initially built as an idealization of generalizations concerning frequencies.) On the other hand we do not know yet whether subjectively estimated uncertainties do in fact satisfy the axioms of the probability calculus, as it has always been taken for granted. Until further research, of a psychological nature, has been conducted the probability measure of certainty or belief remains uncertain. In short, the psychological interpretation of probability yields a possible model of the probability calculus; whether it is in fact true remains to be established by psychology. Hence, of the two unjustified contenders, the subjectivist and the frequentist, the former is the least justified.

Let us now return to the general problem of the components of an interpreted theory (semantical system). Disregarding the presuppositions of the theory, the basis of it consists of (i) a list of specific *primitives*, (ii) a list of *interpretation* (or correspondence) assumptions that confer a meaning to some of the primitives (partially interpreted theory) or to all of them (fully interpreted theory), and (iii) a list of *axioms* (unproved assumptions), i.e. of basic formulas interconnecting the primitives. Just as an abstract theory is a set of assumptions partially ordered by the relation $\vdash$ of deducibility, an interpreted theory is a set of assumptions partially ordered by $\vdash$ and with a content given by a set I_n of interpretation assumptions. In brief,

Theory	*Structure*	$\mathsf{S} = \langle A, \vdash \rangle$	*Syntactical system or abstract theory*
	Model	$\mathsf{M}_n(S) = \langle A, I_n, \vdash \rangle$	*Semantical system or interpreted theory.*

There are as many theoretical models $\mathsf{M}_n(S)$ of a given abstract structure S as sets I_n of rules of interpretation (or correspondence) of the primitive symbols. Thus, one and the same set of equations may recur in the most varied fields; i.e., one and the same structure may be endowed with a variety of contents — whence all scientists ought to learn roughly the same mathematics.

Moreover, in a given theory one and the same symbol may be attached *two different factual interpretations*, one objective and the other operational. Thus, the atomic physicist must distinguish between a symbol referring to a system unperturbed by measurement (e.g., an atom in a stationary state) and a symbol referring to a system deeply perturbed by measurement (e.g., the same atom bombarded by X-rays). In the former case the symbol will refer to a property of a free object, in the latter the same symbol will refer to a property of "the same" object when coupled to a measurement apparatus — hence to a different object. In general, we shall say that a symbol is assigned an *objective interpretation* if a correspondence is established between the symbol and a property of a real system independently of test conditions. And a symbol will be said to acquire an *operational interpretation* if a correspondence is stipulated between the symbol and results of an actual or possible operation designed to observe or measure "the same" property of "the same" real system. The two interpretations confer factual meanings upon the symbol and the two may be assigned to it as long as they are kept separate; and none of them counts as a definition (see Ch. 3).

The above distinction becomes mandatory in the cases of atomic and subatomic objects and of organisms: in either case the systems are apt to be perturbed by observation or measurement operations. Thus, a microscopic particle will be somewhat deviated from its spontaneous course by the light beam aiming at tracking it, and a patient's blood pressure may increase at the mere thought that it will be measured. In short, an object interacting with our means of observation may be *objectively* different from the free object. The difference can in principle be accounted for by theory if the said interaction is incorporated into it: in this way our image of the free object can be checked empirically even though we have no empirical access to free objects.

The distinction between the objective and the operational interpretation of a theoretical symbol makes no sense in the context of operationalism, for which it is meaningless to speak of free objects in the sense of being independent of the operator. But this should not be a deterrent since operationalism originates in confusing reference with evidence and measurement with definition (see Secs. 3.5 to 3.7). An unclear distinction between objective and operational meaning can lead to confusion and even to semantical inconsistencies. In keeping with Sec. 7.2 and with the above, we may say that a theory is *semantically inconsistent* if shifts of meaning occur in it — as when a symbol is initially assigned an objective meaning and all of a sudden it is interpreted operationally. In other words, a theory is semantically inconsistent if it contains statements involving concepts that are neither among the theory's declared primitives nor are defined in terms of the latter but are smuggled into the theory.

The usual interpretations of the quantum theory exhibit semantical inconsistency. Indeed, one usually starts by considering an autonomously existing object, such as a hydrogen atom in the sun's interior, and ends up by reinterpreting the consequences of such a starting point in terms of interactions of the given object (e.g., the distant atom) with a measuring device, or even with the operator's mind. No such interactions and ensuing perturbations had been assumed to begin with, as is clear from an examination of the initial assumptions, which contain solely variables referring to the object itself and none specifying the composition, state and mode of action of the measurement apparatus. Yet the latter and sometimes even the observer's conciousness and perceptions are smuggled into the interpretation of the logical consequences of the initial formulas. No wonder that such shifts of meaning should encourage a wild proliferation of opinions concerning the quantum theory, from naive realism to spiritualism. If a unitary interpretation of the quantum theory in terms of laboratory operations (preparation, observation and measurement) is adopted, then it must be acknowledged that such a theory will not apply to autonomously existing objects, such as atoms that are not being subjected to any such operations. But no such consistent interpretation has so far been presented. (It would have to include the consideration of measurement set-ups from the start and it would have to dispense altogether with terms which, like 'stationary state', can be assigned an objective but not an operational interpretation.) What is more, nobody seems prepared to renounce the application of the quantum theory to nonlaboratory fields such as astrophysics, which after all is nothing but the physics of celestial objects. Wanted: A semantically consistent and thoroughly physical interpretation of quantum mechanics that could in principle be applied to either autonomous objects or — *mutatis mutandis* — to objects under observation, and such that the latter are regarded as just further physical systems interacting with macroobjects in a purely objective way, without involving the observer's mind.

Let us hasten to remark that the wish for a semantically consistent interpretation of the quantum theory has nothing to do with either the desire to eliminate the stochastic features of the theory or with the craving for iconic (pictorial or mechanical) models of it. We have learned to live with chance: we have discovered some of its laws and have therefore come to realize that chance is not chaos. It certainly would be interesting to be able to analyze the stochastic laws of quantum mechanics in terms of nonprobabilistic laws referring to deeper-seated events, but we must not take such a possibility for granted and, above all, even if it were realized we should not regard it as the analysis ending all analyses. As to the wish to visualize, we must remember that we have no sense experience of most things and that the whole point of the atomic theories is to explain the perceptible by the nonperceptible. Moreover, even if a pictorial interpretation can be given of some features of a theory, it need not be iconic or photographic: it may be symbolic or nonfigurative. A visualizable interpretation of a set of symbols may be just a psychologically (heuristic) helpful device, and all we should expect from it is that it be consistent and useful.

*Yet if the theory's referents are spatiotemporal systems we should not be surprised upon finding that a more or less hazy spatial picture of it *can* be given: at least we should not ban the *search* for such a picture just because conventionalist philosophers claim that a scientific theory is nothing but a symbolic predicting device. Moreover, if a search for such spacetime (yet symbolic) pictures were to fail, we should not conclude that they are impossible in principle. Thus, the failure of either the particle picture or the wave picture to adequately represent objects on the atomic level does not prove the failure of alternative pictures and consequently does not doom the search for them. Research programs can fail but need not be sterile, since we learn from failures; on the other hand the dogmatic prohibition to engage in certain research programmes — for example, because they do not match certain philosophical tenets is worse than any number of failures because it curtails the freedom of research, whithout which science withers to death.

To conclude. Every abstract theory, or structure, can be interpreted in a number of ways, either partially or fully. Conversely, if a theory is divested of the interpretation assumptions that confer a content upon it, a purely syntactic schema remains. The simplest interpretation an abstract theory can be given is to regard the theory's objects (the members of the universe of discourse U) as natural numbers; every syntactically consistent structure allows for such an interpretation, which yields what is miscalled a *denumerable model*. A factual model can be built instead of or above such a conceptual model if the appropriate interpretation clues (referitions and semantical hypotheses) are added. But not every interpretation is acceptable. In order for an interpretation to be adequate, hence acceptable, it must not be *ad hoc* (artificial), it must satisfy (if only approximately) the theory's initial assumptions, and it must satisfy the condition of semantic closure, which is necessary to avoid semantical inconsistencies (double-talk). Only under such conditions is a (partially) interpreted theory entitled to be subject to test; and, of course, only favorably tested theories are candidates to (partial) truth.

Problems

7.5.1. Build a mathematical (probability) model of the sequence of physical events that happen when three coins are flipped simultaneously a number of times. Construct the reference class consisting of events such as $\langle H, T, H\rangle$, and compute the probabilities of the following subsets: (i) the subset in which at least one head occurs; (ii) the subset in which at least two tails occur; (iii) the union of (i) and (ii); (iv) the intersection of (i) and (ii). Finally, mention some of the real aspects you have neglected in constructing the model. See, e.g., J. G. Kemeny, J. L. Snell and G. L. Thompson, *Introduction to Finite Mathematics* (Englewood Cliffs, N. J.: Prentice-Hall, 1956), Ch. IV. *Alternate Problem:* Propose geometrical models of probability.

7.5.2. Suppose you take a normal coin and toss it. The probability of "heads" in any single random flipping is $\frac{1}{2}$. Is this probability (i) a physical property of the coin-tossing procedure, (ii) a subjective estimate of the event, or (iii) both? And is any particular outcome ("head" or "tail") relevant to the constant probability value $\frac{1}{2}$? That is, when a "head" comes up does this result force us to change the probability value from $\frac{1}{2}$ to 1? If not, why? *Alternate Problem:* Are probability statements any less certain than nonprobability statements? I.e., does probability involve uncertainty?

7.5.3. Determine whether the following is an instance of (i) induction, (ii) statistical control of hypotheses, or (iii) using probability theory as a matrix for hypotheses conception. Background factual knowledge: an urn contains 7 balls. Problem: guess the most likely composition of the urn. Observation allowed: random extraction with replacement after each drawing. Suppose for definiteness that three successive extractions yield 2 black and 1 colored balls. Since there are 7 balls altogether, of which at least two are black and one colored, the possible hypotheses are

h_1 = "Urn contains 2 black and 5 colored balls"
h_2 = "Urn contains 3 black and 4 colored balls"
h_3 = "Urn contains 4 black and 3 colored balls"
h_4 = "Urn contains 5 black and 2 colored balls"
h_5 = "Urn contains 6 black and 1 colored balls".

7.5.4. There is a tendency to regard every percentage as the empirical counterpart of a probability. Is in fact every percentage interpretable as a measured value of a probability, or is it necessary for such an interpretation to satisfy certain conditions? If the latter is the case, which are the conditions?

7.5.5. Does the use of the phrase 'expectation value' (for weighted average) in physics commit us to the subjectivist interpretation of probability?

7.5.6. Discuss the following theses concerning probability theory and statistical theory (mathematical statistics). (i) Statistical theory is a branch of probability theory, which is in turn a branch of measure theory. (ii) Probability theory is identical with statistical theory, which in turn is a natural science (the mathematics of mass phenomena). (iii) Probability theory is a mathematical theory and statistical theory is an application of probability theory to scientific problems requiring stochastic models, via the interpretation of 'probability' as relative frequency in a random sequence.

7.5.7. Does the frequency *interpretation* of probability commit us to the *definition* of probability as the limit of a frequency? Is such a definition mathematically tenable? See M. BUNGE, "A Critique of the Frequentist Theory of Probability", *Congresso Internacional de Filosofía* (São Paulo: Instituto Brasileiro de Filosofía, 1956), III, 787. *Alternate Problem:* Examine the thesis that there are two concepts of probability: the logical and the statistical ones. See R. CARNAP, *Logical Foundations of Probability* (Chicago: University of Chicago Press, 1950).

7.5.8. It is often taken for granted that probability theory is a kind of logic (the "logic of chance") and, indeed, a more general logic than the ordinary two-valued logic: it is claimed, in fact, that it is a multiple valued logic involving an infinity of truth values between straight falsity ($p=0$) and straight truth ($p=1$), and that its arguments or deductions are probable rather than conclusive. Examine this opinion. In particular, determine whether the probability calculus does not use ordinary (two-valued) logic and whether it constitutes an adequate elucidation of the concept of degree of truth. See M. BUNGE, *The Myth of Simplicity* (Englewood Cliffs, N. J.: Prentice-Hall, 1963), pp. 129 to 131.

7.5.9. Suppose you bring a certain microbe into the classroom. Then the probability that the microbe be in the classroom equals unity. Now ask someone to search for the microbe. Will the probability of that person *finding* the microbe equal the probability of the microbe *being present* in the classroom? Take this as an exercise in epistemology with an eye on the next problem. By the way, the problem can be stated and solved rigorously within search theory, an operations research theory. See, e.g., J. DE GUENIN, "Optimum Distribution of Effort", *Operations Research*, **9**, 1 (1961).

7.5.10. In microphysics the position of a particle is represented by a random variable, i.e. a variable the values of which are assigned probabilities. In the physical literature one finds phrases like these: (i) 'The probability that the electron *lies* within the interval $[x, x+dx]$ equals $\varrho(x)\,dx$', and (ii) 'The probability that the electron *be found* in the interval $[x, x+dx]$ when its position is *measured* equals $\varrho(x)\,dx$'. Are these two phrases equivalent? Can the two probability be conceptually and numerically identical? Do the two problems — that of computing the probability of actual presence and the probabilities of finding — belong to the same problem system? Hint: Recall the difference between the objective and the operational interpretations of a symbol, sketched in the text, and remember Problem 7.5.9. *Alternate Problem:* Elucidate the semantic concept of *partial satisfaction*, which has been employed in the text with regard to factual models. Hint: start by considering the case of propositional functions with an arithmetical interpretation. For example, $x^2=4$ is *exactly* satisfied by $x=\pm 2$, and it is *approximately* satisfied by $x=\pm 1.99$.

7.6. Formal Desiderata

We close this chapter by reviewing the formal requirements a perfect theory should meet: formal consistency (both internal and external), primitive-independence, and axiom-independence. We shall expound and evaluate these desiderata.

Formal inner consistency is the foremost desideratum. A set of formulas is said to be *formally consistent* if and only if it contains no contradictions. Equivalently: a set of formulas is formally inconsistent if and only if it contains both a formula and its contradictory, and consequently the conjunction of the two, i.e. a self-contradiction. For example, the set of formulas: $x+y=z$, $x-y=z$, $y>0$, is inconsistent because subtracting the second from the first we obtain $y=0$, which contradicts the third. Consistency can be achieved in this case by dropping the third formula. Usually the attainment of consistency requires more effort; some people never get it.

Why do we cherish formal consistency? Because inconsistency, i.e. contradiction, generates an unending sequence of formulas, both true and false: in fact, by Theorem [7.4] a logical falsehood entails anything. If a contradiction is allowed to remain in a set of formulas, then (i) the distinction between truth by derivation and falsity by derivation becomes blurred, and (ii) any empirical report whatsoever can be made to count as favorable evidence either for one hypothesis of the set or for another. Since we wish to avoid such easy manoeuvres as permitted by inconsistency, we require formal consistency as an *absolute* logical condition for every scientific theory. A further rationale is to be able to find at least one model for every one of our theories: in fact, model theory teaches that a theory admits of a model if and only if it is formally consistent.

In addition to internal consistency we require, or at least strive for, *external consistency*, i.e. for the consistency of every factual theory with every other (noncompeting) theory in the same field or in adjacent fields. We want external consistency for the following reasons. First, the explanation and the prediction of any real fact require the cooperation of a number of theories, roughly one for every aspect of the fact; just think of the number of theories involved in the prediction of the orbit of an artificial satellite. If such theories were not mutually consistent anything could be derived from the conjunction of their initial assumptions and accordingly the "explanations" would be much too accomodating and the "predictions" would have no test value, since for every forecast we could compute its negate. Second, external consistency is handled as a truth test. Thus, a chemical theory inconsistent with physics, or a sociological theory inconsistent with biology, should be rejected without waiting for empirical tests. In short, the *over-all consistency* of science is desirable: it is necessary for the systematization of real facts, which are complex, and it renders the mutual control of its various parts possible.

How can we recognize inconsistency? In simple cases the seal of consistency is just the truth of the separate assumptions of the system. But what about an abstract (uninterpreted) system or a set of high-level interpreted hypotheses that cannot be immediately recognized as true — as is the case of scientific theories? The axiom base may be so complex that a contradiction may hide in it; we may derive a number of mutually consistent and even true theorems before producing a pair of mutually contradictory propositions. The *definition* of consistency is of no help in this case, which is the case of almost every scientific theory. In addition to the definition we need a *criterion* of consistency and, if possible, a decision procedure, i.e., a rule that can be applied "mechanically" to yield an unambiguous decision, that is, a yes or no answer, to every question of the form "Is T consistent?".

A simple consistency criterion for sets of propositions is BERNAYS' criterion, which involves an arithmetical interpretation of the propositional calculus. BERNAYS' *decision criterion* works as follows. First, write down all the axioms of the given theory with the sole help of negation and disjunction. Second, assign alternatively the values 0 and 1 to each of the component propositions, with the conventions "$-0=1$", "$-1=0$", and "$0.1-1.0=0$". Third, treat every disjunction as an arithmetical product and evaluate accordingly the total value of each axiom. (For example, the value of "$p \vee -p$" is $1.(-1)=1.0=0$.) Fourth, declare the system to be consistent if and only if all its axioms are assigned the same number, inconsistent otherwise. Take, for example, the set of formulas: $A=\{p, -q, p \to q\}$. The third assumption is clearly inconsistent with the joint assertion of the first two formulas since to assert $p \to q$ amounts to asserting that it is false that p is true and q is false; moreover, from the first and the third we derive (by *modus ponens*) q, which is the contradictory of the second premise. BERNAYS' rule gives

$A1$	p	1	0
$A2$	$-q$	1	0
$A3$	$-p \vee q$	$0.0=0$	$1.1=1$,

i.e. inconsistent.

Notice that the subtheory with base $A1$ and $A2$ is consistent, so that these axioms may be salvaged if they are worth while; if, on the other hand, $A3$ is better confirmed that either $A1$ or $A2$, we shall retain the former. That is, the requirement of consistency does not force us to give up an entire theory: it may be worth while to try to *eliminate the inconsistencies* in a theory containing assumptions that have rendered distinguished service. This requires spotting and removing the assumptions responsible for the inconsistency. A partially consistent but also partially true and fruitful theory is more valuable than a completely consistent but pointless or false or even just shallow theory. This is not to say that consistency can be dispensed with but that it is a desideratum which, as every other desideratum, may not be initially satisfied (see Sec. 8.4).

The decision procedure expounded above can only be applied if a truth value can be assigned to each and every axiom. This condition is not met if either the theory's axioms are propositional functions or they have not been tested for truth. In such cases one assigns the primitives of the theory an *ad hoc* interpretation and tests for the truth of the ensuing model. If the interpreted axioms are somehow recognized as true the abstract theory itself is pronounced consistent. When the primitives can be assigned arithmetical meanings the task is easy — once the adequate interpretation rules have been set up. In short, an abstract theory is pronounced consistent if and only if it is *satisfiable*, i.e. if there is at least one true interpretation of its primitives. The foundation of this rule is intuitive: the formulas of a set can be true all at once only if they do not contradict one another. In other words, no inconsistent set of formulas has a model; again, a sufficient condition for a theory to be consistent is that it has a model. Reason: by definition of "model" every formula in the original set must hold in its models, a condition that cannot be satisfied by a contradiction.

As an example let us test whether the axioms for group theory in Problem 7.4.2 are mutually consistent. We rewrite them here for the sake of convenience.

$$G1 \quad (x)_U (y)_U (z)_U [x \circ (y \circ z) = (x \circ y) \circ z],$$
$$G2 \quad (x)_U (x \circ e = x),$$
$$G3 \quad (x)_U (x \circ x^{-1} = e).$$

The individual variables x, y, z run over the set U; the constant e, the unary operator $^{-1}$ and the binary operator $\circ$ are unspecified primitives. Nothing is said about these symbols save how they combine with one another; consequently $G1$—$G3$ are propositional functions. Let us build two models of this abstract theory.

Model 1 (arithmetical addition)	*Model* 2 (arithmetical multiplication)
$I(U)=$ set of integers	$I(U)=$ set of integers
$I(e)=0$	$I(e)=1$
$I(\circ)=+$	$I(\circ)=\cdot$
$I(x^{-1})=-x$	$I(x^{-1})=1/x$.

With these interpretations, $G1$ becomes the associative law for the sum of integers (Model 1) or for the product of integers (Model 2); $G2$ becomes a partial characterization of the number zero (Model 1) or of unity (Model 2); and $G3$ becomes an implicit definition of subtraction (Model 1) or of division (Model 2). This completes the proof of the consistency of the axiom set $G1$—$G3$.

A third formal property axiomatic theories are desired to possess is *primitive-independence:* i.e., the theory's primitives should be mutually independent. A set of concepts is said to be *independent in a given theory* if and only if none of them are interdefinable, even though some or even all of them may be defined in alternative theories. A *test* of primitive-independence is the following (A. Padoa). Suppose every primitive of a theory has been assigned a meaning. Reinterpret the primitive under test and check whether the axioms are still satisfied after this partial reinterpretation; if they are, the reinterpreted symbol is independent of the other primitives of the theory. This procedure is successively applied to every one of the primitives of the theory until their independence is proved or disproved. The ground of this rule is the following: upon a partial reinterpretation the meaning of the primitive under test is changed arbitrarily from the outside, i.e. its meaning is not specified (let alone defined) by the remaining basic concepts of the theory, which are therefore independent of it. Example: Let U and $\circ$ be the basic concepts of a theory, where U is a two-member set and $\circ$ a binary operation. A possible model is determined by the interpretation rules: $I_1(U)=\{0, 1\}$, $I_1(\circ)=+$. Now we reinterpret $\circ$ but not U: $I_2(U)=I_1(U)$, $I_2(\circ)=\times$, which makes as much sense as the foregoing model. Conclusion: U and $\circ$ are, in fact, mutually independent ideas.

*A more thorough independence test of basic concepts is the following (adapted from J. C. C. McKinsey). Suppose T is a theory with the following candidates to primitive concepts: a set U of subsets S_i, a set of relations R_i, and a set of operations O_i. Any one of the subsets, say S_1, will be said to be *dependent* on the remaining primitives of the theory if and only if it can be proved in T that

$$Df.\ 1 \qquad x \text{ is in } S_1 \leftrightarrow F(x; S_2, \ldots; R_1, \ldots; O_1, \ldots). \qquad [7.36]$$

Similarly, R_1 will depend on the other candidates iff T contains a theorem such as (supposing for definiteness that R_1 is a binary relation)

$$Df.\ 2 \qquad R_1(x, y) \leftrightarrow G(x, y; S_1, \ldots; R_2, \ldots; O_1, \ldots). \qquad [7.37]$$

And O_1 will be dependent iff T contains a theorem such as (supposing O_1 is a binary operation)

$$Df.\ 3 \qquad x\, O_1 y = H(x, y; S_1, \ldots; R_1, \ldots; O_2, \ldots) \qquad [7.38]$$

where F, G and H are single-valued functions. A primitive will be said *independent* iff it is not dependent.

*The above set of definitions is clearly inadequate as an independence *criterion*, since the existence of the theorems involved might not be proved within our lifetime. McKinsey's *test*, based on the above definitions, works as follows. First, interpret all the candidates to primitive status, i.e., find a model $M_1(\mathsf{T})$ of the theory T. Call $I_1(S_i)$, $I_1(R_i)$ and $I_1(O_i)$ the interpretations of the S_i, the R_i and the O_i respectively. Next find a second model of the theory, such that only one of the primitives, say S_1, is reinterpreted while the remaining candidates ($i \neq 1$) keep their original meaning; call $I_2(S_1)$ this reinterpretation of S_1. If such a partial reinterpretation of T is found, i.e. if the meaning of 'S_1' can be altered without changing the truth value, then S_1 does not depend on the other primitives, i.e. it is independent as suspected. But if a reinterpretation of S_1 is not possible without making adjustments in the other basic concepts, S_1 depends on them. Similarly for the remaining candidates.

*Justification of the test: if S_1 were dependent on the other primitives, then by definition [7.36]

$$x \text{ is in } S_1 \leftrightarrow F(x; S_2, \ldots; R_1, \ldots; O_1, \ldots).$$

Since the formal relations among the theory's basic concepts must remain invariant upon reinterpretation, we shall have

$$x \text{ is in } I_1(S_1) \leftrightarrow F(x; I_1(S_2), \ldots; I_1(R_1), \ldots; I_1(O_1), \ldots)$$

and, after the partial reinterpretation (of 'S_1' alone),

$$x \text{ is in } I_2(S_1) \leftrightarrow F(x; I_1(S_2), \ldots; I_1(R_1), \ldots; I_1(O_1), \ldots)$$

whence

$$x \text{ is in } I_1(S_1) \leftrightarrow x \text{ is in } I_2(S_1).$$

But this equivalence is false since, by hypothesis, the two interpretations of S_1 are different, i.e. the set $I_1(S_1)$ does not contain all the elements of the set $I_2(S_1)$. Consequently, the dependence assumption is false — i.e., S_1 does not depend on the other primitives. Similarly for the relation and the operation concepts.*

Tests for primitive-independence might seem to be a luxury but, as a matter of fact, they are both theoretically and pragmatically valuable. The former because the test establishes what exactly are the necessary and sufficient basic concepts of a theory and directs primarily the attention of the foundational worker to such symbols rather than to the derivative ones. The practical utility of discussing primitive-independence consists in that it discourages attempts to dispense with certain basic concepts which look suspect to some philosophical school, and it clarifies discussions on the meaning (e.g., the physical interpretation) of theoretical symbols. For example, a metalogical discussion of the status of "mass" and "force" in classical mechanics would have saved us from the misguided attempts to define them explicitly, hence to eliminate them as basic concepts. In fact PADOA's technique, available since 1899, shows that they are independent primitives in Newtonian mechanics. To see this recall that m and f are related by the Newton-Euler law "$f = ma$". If we assign f the (numerical) interpretation 0, we get $0 = ma$. This equation can be satisfied in infinitely many ways, by assigning arbitrary values to m. Since an infinity of (numerical) interpretations of m are consistent with the interpretation $I(f) = 0$, we conclude that neither mass can be defined in terms of force nor conversely. MACH's interpretation of classical mechanics is, then, logically wrong, and so are the hundreds of textbooks which follow it as it if were the last word.

A fourth formal property perfect theories should have is *axiom-independence*. An axiom set is independent if and only if its members are not interdeducible, i.e. if none of them is derivable from the others (even though it may turn out to be a theorem in a different theory). Also: A statement s is independent of a set A of assumptions iff $\{A, -s\}$ is consistent. Axiom-independence is the analogue, for basic statements, of primitive-independence, which can be true of basic concepts. The usual test of axiom-independence closely parallels the test for primitive-independence; it consists of the following rules. Assign an interpretation to each primitive in the theory in such a way that all the axioms except one of them, say A_i, are satisfied; in other words, find a model in which A_i does not fit. If such an interpretation is found, A_i cannot be a consequence of the remaining assumptions; for, if it were, it would be automatically satisfied as a consequence of the satisfaction of the remaining axioms. In short, if all axioms except A_i are satisfied by a given interpretation of the primitives, then A_i is independent. In this way every member of the axiom system is tested until complete independence is proved (or disproved). An equivalent procedure is this: Form a new system constituted by $-A_i$ and the remaining axioms; if the ensuing system is consistent, we have shown that the presence of A_i does not disturb the remaining assumptions, i.e. that A_i is independent of its companions. For, if A_i followed from them, a contradiction should appear upon conjoining them with $-A_i$. This is a syntactical test, whereas the former was a semantical test.

It is desirable for the axiom base of a theory to be independent although it is not mandatory. A first reason is scientific in character: if the basic assumptions of a theory are mutually independent, we can generate an unlimited number of alternative systems by successively replacing any of its axioms by different assumptions referring to the same objects. In particular we can replace any axiom by its negate. This was how the earliest non-Euclidean geometries were in fact generated: namely, by replacing the Euclidean parallels axiom by alternative postulates. The value of axiom-independence in the construction and remodeling of theories becomes especially apparent in the face of data which definitely refute one or more axioms but are neutral with regard to the others (precisely because they are independent). In such cases honest, not *ad hoc* adjustments of the theory become possible. A second reason for valuing axiom-independence is metatheoretical: it is a principle of good ordering to set all the initial assumptions apart from their logical consequences, if only to be able to check the derivations at any time. A third reason is pragmatic: by reducing a system to independent components we avoid stating the same idea twice, once as an axiom and then again as a theorem: independence is time-saving. And, needless to say, axiom-independence and primitive-independence, which are formal properties, do not in the least impair the conceptual unity which every theory is supposed to have (see Sec. 7.2).

Formal consistency, both internal and external, primitive-independence, and axiom-independence, are all we can expect from a good theory as far as its form is concerned. When a theory satisfies these conditions and, in addition, it contains an explicit and exhaustive enumeration of its primitives and its rules (both syntactical and semantical), it is said to be *formalized*. A formalized theory may or may not have a factual content and should not be taken for a *formal* system — in the philosophical sense of 'formal' (see Sec. 1.4). Formal theories, whether formalized or not, lack an objective referent; and among them abstract theories, whether formalized or not, have no definite content or meaning (recall Sec. 7.4). Formalization bears on organization, not on content; and formalized theories, whether in logic or in physics, are nothing but ideally organized theories. Like most ideals, formalization does not command our interest because it can be found everywhere but precisely because it is rarely met with. We shall return to this topic in Sec. 8.3.

Further formal desiderata have been proposed from time to time, among them economy and completeness, but there are good reasons for not acknowledging them. *Formal economy* in a theory's foundations means paucity of primitive concepts and syntactical simplicity of the basic relations among them, i.e. of the theory's axioms. Even if there were adequate measures of formal simplicity, and however desirable it is for asthetic reasons and for metatheoretical purposes (such as consistency tests), formal economy should not guide theory construction. One reason for this is that paucity of extralogical predicates, i.e. economy of the basis of primitive concepts, can always be attained by just disregarding concepts, i.e. by ignoring traits of the theory's objects. Similary, by neglecting complexities in the relations among such aspects the basic assumptions can be simplified as much as desired. But then truth will suffer. Certainly, redundant primitives are to be avoided as well as unnecessary complications in their mutual initial relations (axioms). But such complications are taken care of by the methodological rule: "Do not hold arbitrary (ungrounded and untestable) opinions". In factual science, simplicity is valuable as long as it does not conflict with optimal foundation, testability, and corroboration. Moreover, economy at the base does not coincide with either straightforwardness or perspicuity, whence it does not contribute to ease of interpretation.

Let us finally deal with completeness. A theory is said to be *deductively complete* if and only if no new formula can be introduced in the theory which cannot be derived from its axioms, eventually with the help of definitions. Equivalently: a theory is deductively complete if and only if, for every pair of mutually contradictory formulas, either one or the other is derivable in the theory. If a new formula which is not so derivable is added to the theory, a contradiction will arise if the theory is complete, but no catastrophe will happen if the theory is incomplete. In other words, a theory is (formally) complete if it cannot be enlarged without contradiction; it is incomplete if it can be enlarged in that way: if it can grow while remaining consistent. A test for formal completeness is therefore the following: When a premise is added to the original set of assumptions, the latter is complete if either the enriched set is inconsistent or if the new formula turns out to be derivable from the old set. Caution: let us not mistake formal completeness for semantical completeness or exhaustion of the field concerned. As we shall see in a minute, semantical completeness or coverage requires formal incompleteness.

With the exception of certain elementary formal theories, such as the propositional calculus and the first-order predicate calculus, all consistent and sufficiently rich theories — such as elementary arithmetic — *must* be incomplete, i.e. they can grow without contradiction: this is the substance of one of K. GÖDEL's incompleteness theorems. Since consistency is an absolute desideratum, and since we want the richest possible theories, we must relinquish formal completeness as a desideratum for logical and mathematical theories. A fortiori we must abandon it in relation with factual science, since the latter presupposes formal science. But there is a difference: whereas in formal science we part reluctantly with completeness, regarding it as a lost paradise, in factual science we should regard formal completeness as definitely undesirable. In fact, the ability to absorb new premises germane to the basic assumptions — such as auxiliary hypotheses and empirical data — is a desideratum of factual theories because we want to apply them to particular cases, and every such application requires the introduction of fresh information and fresh assumptions. A formally complete theory cannot, by definition, grow by accepting formulas won outside the theory. Now, recalling the characterization of axiomatic theories (see Sec. 7.3), we realize that formal completeness is equivalent to *axiomatizability*. We confirm accordingly our previous conclusion: full axiomatization is undesirable in factual science. Stated in constructive terms: Axiomatization (formal completeness) is desirable for the *core* of every factual theory only, i.e. for the set of its general assumptions. Formal incompleteness (partial axiomatizability) is overcompensated for by growth and application ability, i.e. by openness to fresh experience (see Sec. 7.3).

Now, if a theory is incomplete — as every consistent and reasonably rich factual theory will be — then it will be unable to cope with every problem germane to it. In other words, incompleteness is accompanied by some extent of *unsolvability* (relative to the given theory). In fact, suppose a given problem Π is well-stated and well-conceived in the framework provided by a theory T. The problem Π will be said to be *solvable in* T if and only if T contains formulas from which a solution $S(\Pi)$ to Π can be derived. Otherwise Π will be unsolvable in T, although it may turn out to be solvable in an alternative theory. In short, Π is solvable in T if and only if $\mathsf{T} \vdash S(\Pi)$. Now, if T is formally and semantically complete it will allow for a derivation of the solution. If, on the other hand, T is incomplete in some sense — as befits a well-developed factual theory — then $S(\Pi)$ may not be derivable in T. In other words, not *every* problem is solvable in a given factual theory even if it can be posed in its context. The remedy will consist in building a richer theory, in which the given problem is solvable; but the new theory will be incomplete as well and will leave out still other problems. In view of the preceding results, it is surprising that any given factual theory — such as the present quantum theories — should be advertised as complete, hence in no need of supplementation with other theories. The effect of such an unwarranted belief is to ban as meaningless all those problems that cannot be solved within the available theories. On the other hand, the acknowledgment of the essential incompleteness of every theory that includes arithmetic invites handling presently unsolvable problems by building new, stronger, theories.

This completes our sketchy treatment of the syntax and the semantics of theories, a quickly expanding field of the science of science. Let us now turn from the analysis of the inners of finished theories to the process of theory construction and the relation of theory to fact.

Problems

7.6.1. Someone found the (unlikely) inductive generalization: "Every Monday the weather is fine". In order to explain this low-level hypothesis he stated the following axioms:

A 1 The weather changes every day from fine to bad or conversely.

A 2 Every Sunday the weather is bad.

Show that a contradiction hides in these axioms. *Alternate Problem:* Find a domain of individuals where the axiom system: $\{(x)(\exists y)Rxy, -Rxx\}$ is satisfied and another where it is not satisfied when R is interpreted as $<$.

7.6.2. If a consistent system of logically true propositions (tautologies) is given an arithmetical interpretation, then all the formulas in the set get the value 0, no matter what values (0 or 1) the propositional constituents of every formula take. Conversely, if all the axioms of a system get the value 0, then it is a consistent system of tautologies. Show that this is the case with the axioms of the *propositional calculus* proposed by HILBERT and ACKERMANN, i.e.

$$PC1 \quad (p \vee p) \rightarrow p, \qquad [7.39]$$

$$PC2 \quad p \rightarrow (p \vee q), \qquad [7.40]$$

$$PC3 \quad (p \vee q) \rightarrow (q \vee p), \qquad [7.41]$$

$$PC4 \quad (p \rightarrow q) \rightarrow [(r \vee p) \rightarrow (r \vee q)]. \qquad [7.42]$$

7.6.3. Consider the axiom set

$$A1 \quad p \to q,$$
$$A2 \quad q \to r,$$
$$A3 \quad r.$$

Is it consistent? Is it independent? Is it complete? *Alternate Problem:* How is a system of linear algebraic equations tested for compatibility (consistency)?

7.6.4. What did SPINOZA mean when he wrote that his system of ethics was handled in the manner of geometry *(more geometric tractata)*? That he had employed geometry, in the sense that the axioms and the derivations were mathematical? Or that his system was organized in imitation of Euclidean geometry, the sole axiomatic system known at the time (in the 1660's)? *Alternate Problem:* Symbolize SPINOZA's axiom set and find whether it is consistent, independent and complete.

7.6.5. Change any of the postulates [7.23] to [7.25] of the elementary probability theory (Sec. 7.5) without making adjustments in the other two. For instance, change the first postulate, requiring that the probability measure be a nonpositive number, or change the second postulate into "$P(U)=0$". Is the new system consistent? And is every such change bound to produce inconsistencies?

7.6.6. A. TARSKI, in *Logic, Semantics, Metamathematics* (Oxford: Clarendon Press, 1956), p. 296, has pointed out the parallel between two sets of metatheoretical concepts:

Primitive	Axiom
Mutual independence of primitives	Mutual independence of axioms
Completeness of a concept set	Completeness of an axiom set
Definability	Deducibility
Definition	Proof

Comment on this parallel and decide whether the concept of referition (see Sec. 3.6) has a similar analogue. *Alternate Problem:* How is a theory enriched: by expanding it or by shrinking its primitive base and/or its axiom base?

7.6.7. Study A. TARSKI's work on primitive completeness, op. cit., pp. 308ff., and discuss its relevance to theory construction. *Alternate Problem:* Every consistent theory of a certain kind can be completed without changing its predicate base (LINDENBAUM's theorem). What is the scope of this theorem? And does it hold for factual theories presupposing mathematics?

7.6.8. The less the number of axioms of a theory the less the danger of inconsistency. If a theory's base is shrunk to a single self-consistent statement (e.g., a variational principle) the risk of inconsistency becomes nil. Describe this procedure and find out how far it is used in contemporary science. And decide whether such a technique yields the simplest possible theory. *Alternate Problem:* Study the role of contradiction in science as a motivation for new theoretical work aiming at avoiding inconsistency.

7.6.9. Sometimes, when a theory does not yield the correct results in some field, an *ad hoc* hypothesis is added to it which will produce the desired result while leading to incorrect results in other fields. For instance, in solid state theory a fictitious spin variable is added to the energy operator, which yields the correct energy levels but the wrong spin value. Examine this kind of procedure in the light of the requirement of logical consistency. *Alternate Problem:* Does every definition work as a criterion? What properties should definitions have in order to serve as criteria?

7.6.10. Program a computer to perform either of the following tasks in relation with a given axiomatic theory: proof of primitive-independence, of axiom-independence, and of consistency. *Alternate Problem:* the properties of consistency, primitive-independence, and axiom-independence, are syntactical and so are the definitions of the corresponding concepts. But the *recognition* of those properties, i.e. their test criteria, are semantical since they depend on the concept of model (or of satisfaction, or of truth). Does this make syntax dependent on semantics?

Bibliography

Beth, E. W.: The foundations of mathematics. Amsterdam: North-Holland 1959.

— Formal methods. Dordrecht: D. Reidel 1962.

Braithwaite, R. B.: Scientific explanation, chs. I and II. Cambridge: Cambridge University Press 1953.

Bunge, M.: Foundations of physics, Chs. 1 and 5. Berlin-Heidelberg-New York: Springer-Verlag 1967.

Carnap, R.: Foundations of logic and mathematics. Chicago: University of Chicago Press 1939.

— Introduction to symbolic logic and its applications, part two. New York: Dover 1958.

Copi, I.: Symbolic logic 2nd ed., ch. 6. New York: Macmillan 1965.

Curry, H. B.: Foundations of mathematical logic, ch. 2. New York: McGraw-Hill Book Co. 1963.

FREUDENTHAL, H. (ed.): The concept and the role of the model in mathematics and natural and social sciences. Dordrecht: D. Reidel 1961.

HENKIN, L., P. SUPPES, and A. TARSKI (eds.): The axiomatic method. Amsterdam: North-Holland 1959.

HILBERT, D., and P. BERNAYS: Grundlagen der Mathematik, vol. I. Berlin: Springer 1934.

LEWIS, C. I., and C. H. LANGFORD: Symbolic logic, 2nd ed., chs. XI and XII. New York: Dover 1959.

MARTIN, R.: Truth and denotation, ch. I. Chicago: University of Chicago Press 1958.

McKINSEY, J. C. C.: On the independence of undefined ideas. Bull. Amer. math. Soc. **41**, 291 (1935).

MOSTOWSKI, A.: Thirty years of foundational studies, being fasc. XVII of acta philosophica fennica (1965).

NAGEL, E.: The structure of science, ch. 5. New York and Burlingame: Harcourt, Brace & World 1961.

— P. SUPPES, and A. TARSKI (eds.): Logic, methodology and philosophy of science. Stanford: Stanford University Press 1962.

PADOA, A.: Essai d'une théorie algebrique des nombres entiers, précedé d'une introduction logique à une théorie deductive quelconque, secs. 8—18. Bibliothèque du Congrès Internat. de Philosophie **3**, 309 (1900).

POPPER, K. R.: The propensity interpretation of probability. Brit. J. phil. Sci. **10**, 25 (1959).

— Creative and non-creative definitions in the calculus of probability. Synthèse **15**, 167 (1963).

RASIOWA, H., and R. SIKORSKI: The mathematics of metamathematics, ch. V. Warszawa: Państwowe Wydawnictwo Naukowe 1963.

ROSENBLOOM, P.: The elements of mathematical logic, ch. ii, sec. 3. New York: Dover 1950.

STOLL, R. R.: Sets, logic and axiomatic theories. San Francisco and London: W. H. Freeman and Co. 1961.

— Set theory and logic. San Francisco and London: W. H. Freeman and Co. 1961.

SUPPES, P.: Introduction to logic, ch. 12. Princeton: D. van Nostrand Co. 1957.

TARSKI, A.: Logic, semantics, metamathematics, chs. III, V, XII, XVI. Oxford: Clarendon Press 1956.

WOODGER, J. H.: The technique of theory construction. Chicago: University of Chicago Press 1939.

— Biology and language. Cambridge: Cambridge University Press 1952.

Chapter 8

Theory: Dynamics

Since the goal of theorizing is the building of mature and verisimilar theories, and since what distinguishes a theory from alternative fragments of scientific knowledge are certain logical and semantical properties, such as deducibility and conceptual unity, we had to investigate the logic and the semantics of theories before we could cast a glance at the actual process of theory construction and at the way theories account for their referents. The form/content dichotomy we have adopted is methodologically convenient but it is psychologically and historically artificial, since in factual science one conceives interpreted theories from the start, employing to this end the conceptual tools that seem most adequate for the representation of matters of fact. Abstract theories, or purely conceptual models thereof, occur only, if at all, in the logical reconstruction (formalization) of existing factual theories — and, of course, in formal science. It is equally artificial to start considering full-fledged theories, as we have done, rather than theory embryos: if we wish to know how science is made we cannot limit ourselves to a philosophy of science textbooks — but we cannot ignore the finished product either since it is the aim of theoretical research. Let us then correct the static picture a purely logical approach is apt to give, by taking a glimpse at the building of theories and their relation to the world. The relation of theory to scientific experience will concern us in greater detail in Volume II, Part IV.

8.1. Theory Construction

Theories can be built, remodeled, logically reconstructed, applied, demolished, or forgotten. The construction of a scientific theory is always the building of a more or less refined and consistent system of statements that unifies, enlarges and deepens ideas which, in the pretheoretical stage, had been more or less intuitive, blurred, sketchy and disconnected. In the beginnings of a science such a pretheoretical stage is just ordinary knowledge; in established disciplines a pretheoretical stage can occur in individual fields only: there is always some theory or other nearby to help in the construction, either by lending materials or by suggesting approaches. At any rate, theory construction does not proceed in a vacuum but in a preëxistent matrix. When the matrix is poor the first theorizing efforts can be very hard; this is the case of fields characterized by the blind accumulation of information and by its inevitable companion, conceptual fuzziness.

The conceptual items employed in the construction of a theory depend not only on the problem system the theory is meant to solve but also on the available knowledge and techniques. This is why different investigators may, on quite different considerations, arrive independently at the same theory. In addition to such a social property we must reckon with the theorist's peculiar background, pets, phobias, and style of thinking: otherwise we will be at a loss to explain the complementary phenomenon of the different theories that are often advanced to meet the challenge of one and the same problem system.

There are no established techniques for theory construction: the theorist is entitled to use any conceptual means whatsoever as long as the obviously wrong ones do not occur in the finished product or can be eliminated from it. Such means as are employed in theory construction but may not show up in its final presentation are called *heuristic* ideas. The caloric hypothesis was a heuristic aid in the building of thermodynamics, and so was the homunculus fiction in early genetics. In addition to hypotheses and theories, thought experiments — i.e. experiments that are planned but not executed — play often a heuristic role. (Oddly enough, thought experiments are sometimes believed to prove theories — something not even real experiments can.) Finally, certain particularly suggestive examples may take part in shaping a theory (heuristic cases, as distinct from illustrations). In any case, theory construction is not usually done in a neat, logic-conscious way and it results in what may be called a *natural* or intuitive theory, a rather untidy draft that ulterior developments, applications and discussions may gradually organize and clarify — or destroy.

At the draft stage the chief requirement the theorist will pose is that the theory covers in a unified way a fair portion of the field in view, i.e. the set of hitherto disconnected, or loosely related, data and low-level generalizations at hand. To account for them, to systematize them will be his "boundary condition", even though his ultimate aim may be the more ambitious one of accounting, in addition, for as yet undiscovered facts and generalizations. Metaphorically speaking, a scientific theory is like a mollusc with a soft core — the set of high-level hypotheses — sorrounded by a somewhat harder but porous shell — the set of low-level or empirical generalizations (see Fig. 8.1). Through the pores of the periferal part, information about the extrenal world pours in, which enables the central core to produce new particular items (predictions and retrodictions) that are projected to the external world. Sometimes the excreta are more valuable than the food.

In the early stages of theory building the scientist may not worry too much about conditions of a logical kind, with the sole exception of (i) semantical consistency (homogeneity and connectedness of the basic concepts) and (ii) formal consistency, as recognized in a natural or intuitive way. (For semantical consistency, see Sec. 7.2; for formal consistency, see Sec. 7.6.) The theory may eventually be subjected to logical reconstruction or formalization, i.e. rebuilt in such a way that its presuppositions, primitives, axioms, and rules be all as fully and clearly stated as found possible at a given moment. This process of theory refinement is most often never started: most scientists are more interested in finding new consequences (theorems) or in checking them against

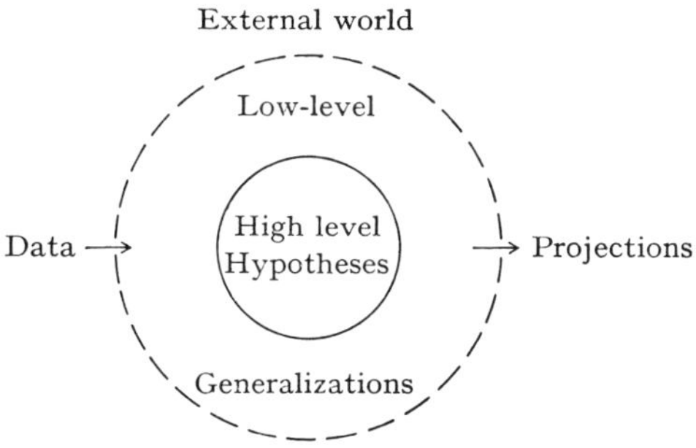

Fig. 8.1. Scientific theory as an organism feeding on data concerning the external world, into which it projects back new information (predictions and retrodictions)

data than in a neat logical organization: the latter is usually taken up by elementary textbook authors, who rarely master the necessary logical and philosophical tools and often look upon the theories accepted during the previous decade as final results rather than as objects of research.

If the process of theory refinement is started at all it may never be completed, if only because (i) theory formalization requires logical and metascientific techniques that are in the process of being invented or perfected, and (ii) the interpretation and understanding of any given theory is largely conditioned by the state of adjacent theories, which by changing modify that interpretation. Examples: recent work in classical mechanics and thermodynamics. At any rate, a whole gamut may be recognized between the draft stage and a careful (yet not final) logical reconstruction. As regards logical organization we should then recognize degrees within the following rough partition:

Knowledge
- *Systematic scientific knowledge*
  - *Formalized theory*
  - *Natural theory*
- *Presystematic scientific knowledge (protoscience)*
- *Common knowledge*

Whatever the stage of formalization it may have reached, a theory may be remodeled or readjusted in an attempt to improve its adaptation to facts. Should this move fail it may be demolished either in the process of criticism or in the endeavor to reduce it to sets of lowest-level formulas that, being close to experience, would gratify the immature yearning for certainty. Theory construction is so far a more or less spontaneous process; on the other hand, theory formalization, criticism, and demolition involve the deliberate use of definite logical and methodological principles and techniques — not to mention philosophical views. We shall meet some of them at the end of this section and in Sec. 8.3.

Natural or unformalized theories can be built in a number of ways, each way yielding a kind of theory and none of them being rule-directed to the point of constituting a method each. At one end of this array of approaches and theories we find the mere organization of existing empirical generalizations stated in verbal form, with the characteristic vagueness, ambiguity and shallowness of ordinary language. Most theories in the behavioral sciences are still of this fuzzy kind, although a vigorous movement of logical cleansing and mathematization is under way. At the other end we find theories based on strong high-level, non-empirical hypotheses stated mathematically and accordingly benefiting from the neatness and deductive power characteristic of mathematics (see Sec. 8.2). Most physical theories and an increasing number of biological and behavioral theories are of this kind.

Between verbal theories and mathematical "models" there is a wide range of intermediate degrees of formalization. Verbal theories often contain mathematical *concepts*, such as those of set inclusion, tree, numerical variable, ratio, and proportionality — only, by definition of "verbal theory" such systems will not include mathematically stated *relations* among such concepts. Consequently, their statements (e.g., law statements) will be rather loose, difficult to pin down, to work out, and to test. (The occurrence of *quantitative* concepts in a theory is neither necessary nor sufficient for it to be mathematical, let alone scientific. A theory with nonquantitative concepts, such as those of set and order relation, can be mathematical: see Sec. 7.3; on the other hand, a theory with quantitative concepts, such as "cost" and "price", will still be verbal if it does not link these concepts mathematically, e.g., through functions and equations.) In the behavioral sciences a number of quantitative concepts (e.g., "number of reinforcements") and pseudo-quantitative concepts (e.g., "psychical energy") are handled, but comparatively few mathematical relations among them have been suggested and corroborated. When some of a theory's initial assumptions are stated in a mathematical form we may speak of a semimathematical theory; examples of this kind of theory abound in classical economics.

Consider the following nonquantitative hypothesis containing quantitative concepts: "The greater the population density the more numerous the social contacts (or the larger the contact frequency)". An infinite family of possible mathematizations of this hypothesis is obtained by introducing an unspecified monotonously increasing function f relating the number c of interpersonal contacts (in a period of time) and the population density d: $c=f(d)$. But as long as the precise form of the function f is left unspecified, the foregoing relation is just a matrix — a propositional function. Methodologically speaking, it is a programmatic hypothesis inviting research on a definite problem: that of finding the precise form of the relation f (see Fig. 8.2). In the behavioral sciences many *programmatic schemas*, i.e. theory schemata containing

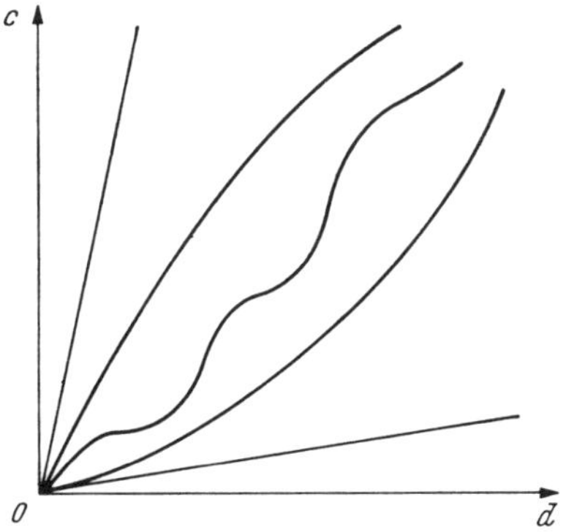

Fig. 8.2. A few among infinitely many definite relations fitting the schema "The greater the population density the more numerous the social contacts". The pretence of continuity has been added for the sake of simplicity

indeterminate assumptions of that kind, or even less precise ones (like "c depends on d"), still pass for theories. Programmatic schemas certainly have to be built in every preliminary stage of exploration and even afterwards, when the attempt is made to unify sets of theories. But they do not exhaust theory construction if only because they contain barely meaningful and, consequently, barely testable statements. Only definite and meaningful relations can be tested (see Sec. 5.6).

Granting that a fully meaningful and testable theory will contain only precise, nonelusive relations among concepts, what kind of concepts should it contain? Clearly, *theoretical concepts*, i.e. concepts introduced by the theory or at least considerably refined by it — subject to the obvious condition of minimal intensional and extensional vagueness (see Sec. 3.7). Some theoretical concepts preëxist theory in a rough or intuitive state and are only refined upon theorification: thus "distance" and "heredity". Were it not for such links with ordinary knowledge, theorizing could never begin; moreover, a theory is sometimes built or rebuilt in order primarily to clarify certain concepts. But other concepts, such as "surface tension", "enzyme", and "habit strength", are born with theories: they have weak links, if any, with ordinary knowledge. Those which, like "force field", "gene", and "theory", have no antecedents in ordinary knowledge, characterize the most original, profound, and fertile theories. Similarly with hypotheses: in underdeveloped theories the initial assumptions are little more than refinements and extensions of empirical generalizations. Mature theories, on the other hand, are characterized by nonobservational hypotheses (see Sec. 5.4); and the ripest among them, by representational hypotheses going beyond input-output relationships, as will be seen in Sec. 8.5.

The hypotheses of scientific theories, which cannot help containing theoretical concepts, are sometimes regarded with suspicion on the assumption that, by departing from experience, they could be, at their best, extremely uncertain extrapolations, and at their worst mere fairy-tales. According to such a view there is a single and simple rule for building theories: namely, to start from elementary experiences, i.e. from particulars, generalizing them cautiously (induction). The whole history of science has unwittingly been counterfeited under the influence of this immature philosophy of protoscience, in such a way that observation replaced reasoning, induction took the place of invention, and patience that of talent; and problems, the sparks of both experimenting and theorizing, had no place and no homologue in this picture. Every great revolution in ideas, such as evolution or the quantum, came then as a shock since it clearly was not "deduced" from observation, and was accordingly rejected in the name of methodology or again adulterated to conform to the preconceived (hence nonempirical) pattern. This idealess history of science, unconcerned with problems and hypotheses — the core of scientific research — is slowly being replaced by a truer account backed up by a post-Baconian philosophy of science. But we need not resort to the new, still largely unwritten history of science in order to refute the data systematization view of scientific theory: an analysis of the process of theory construction will suffice.

There are several reasons why a theory cannot be a summary of data, let alone a synthesis of elementary (atomic) experiences. One reason is purely logical: by definition a theory is a set of formulas partially ordered by the deducibility relation; and no such relation holds among particular statements referring to a fact each: deduction currents can only be set up among formulas on different potential levels, so to say: we have to rise above experience and build a transempirical multi-level system if we are to deduce formulas about experience (see Sec. 7.3). A second reason is that there *are* no isolable, self-contained, simple experiential items (atoms). Indeed, what we experience is always a whole and, as such, it is of no use to science: scientifically relevant experience is not pure experience — the uncontaminated experience of our ancestors living one million years ago — but conceptualized or interpreted experience: an experience mixed with anticipations and ideas and rooted in a conceptual background. Analysis, a conceptual operation, enables us to decompose the given experiential whole into a system of more or less distinguishable units and to disclose their mutual relations. But the analysis of experience does not stop at the decomposition of experiential wholes into experiential units or percepts; the conceptual processing of experience leads always to some idea or other: a problem, a conjecture, a whole cluster of hypotheses. If we further elaborate the initial idea with the tools taken from our conceptual background, it may develop into a system of interrelated but clearly distinguishable hypotheses — a theory. A third reason is that, in the processing of experience and in the invention of ideas, most particulars are *discarded* and the rest are *disfigured* rather than carefully collected and packaged. Percepts, which anyhow are products of analysis rather than raw experiences, are mostly discarded in the process of selecting relevant items. And those that are picked out become transmuted into ideas, which are in turn anything but faithful reproductions of the given. A posteriori we discriminate and sort out the ideas and come to realize that some of their component units — concepts — have no experiential counterpart, this being why they have a chance of participating in the explanation of experience.

A theory, in sum, is an original creation rather than an arrangement of elements each carefully anchored to experience (e.g., operationally "defined"). Particulars are not the building materials but rather, once sifted, an occasion for theorizing and a test of theories: they pose problems and they constitute as many conditions that the theory must satisfy to some extent. If the theory is true, particulars will be found as logical consequences of the theory's general assumptions in conjunction with further particulars. The singular case has a number of functions: it will pose a problem, it may suggest a conjecture, it may refute it, and it will illustrate a theory: but it cannot generate it.

The pathway is not from data to theory but data to problem, from problem to hypothesis, from hypotheses to theory; and back from theory and evidence to a projection that can be checked by another piece of evidence — with the help of further theories (see Fig. 8.3). This is a picture of theory construction compatible with logic: since nothing can be validly concluded from a set of evidence-statements alone (see Sec. 7.3), something must be created beyond experience if we are to explain it. Data are to be found at the two ends of the process: at the beginning and at the end, as triggers and testers of theorizing, never as building blocks of theories.

The point of departure of a theory is usually the problem of synthesizing and explaining a set of more or less disconnected generalizations — either empirical or not — that may be called the theory's *founder hypotheses.* The unification of a set of founding hypotheses requires (i) their *refinement* — e.g., their casting in more sophisticated theoretical terms allowing for more precision, or (ii) the introduction of radically *new* and stronger hypotheses establishing logical links among the founder hypotheses and permitting their deduction. NEWTON's invention of rational mechanics required both operations. The founder hypotheses he chose were a set of disconnected laws of motion, chiefly GALILEI's laws of falling bodies and of ballistics, KEPLER's laws of planetary motion

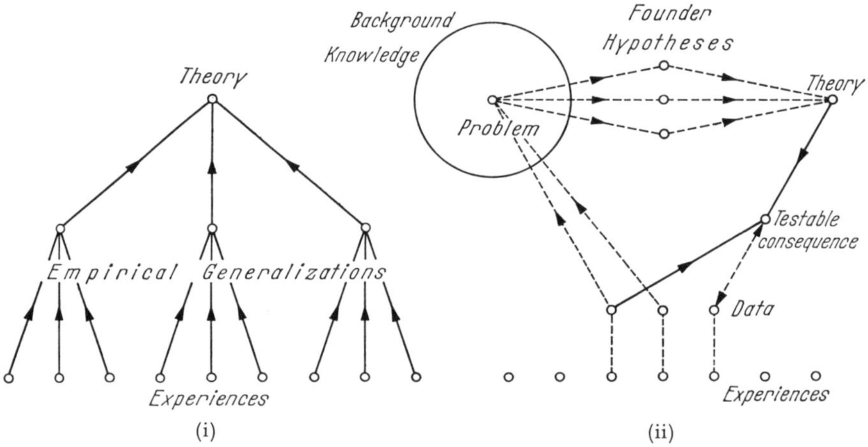

Fig. 8.3. (i) The naive view of theory construction: experience as the source of theory. (ii) A more realistic (but still simplified) view: theory is prompted and tested by data anchored to experience

and HUYGENS' laws of small oscillations. NEWTON did not just *generalize* these laws by induction, as the ascension model of theory construction would have it: he did not perform an inductive generalization encompassing the founder hypotheses as instances, if only because there are no such higher order kinematical laws subsuming the kinematical laws of GALILEI, KEPLER and HUYGENS. In order to synthesize and derive them NEWTON invented a dynamical theory: he introduced a system of entirely fresh hypotheses which did not describe observable particle trajectories, as the kinematical laws did, but referred to imperceptible forces and masses and to their kinematical manifestations (accelerations). The kinematical laws — the old ones as well as any number of new ones — were deduced from the dynamical laws with the help of a new calculus — as the available deduction technique was too weak. Much in the same way, N. RASHEVSKY has in our days attempted to derive the integral laws of human behavior from the differential laws regarding the neural mechanisms that transmit and process the observable stimuli.

The extent to which this goal has been achieved is debatable, but this is independent of the programme's value.

In the process of theorification the founder hypotheses may suffer modifications. In the case of mechanics, they were found to be, for the most part, approximate solutions to the "exact" (yet not completely true, as we now know) highest-level laws. Theory can not only fuse together and explain the early stray hypotheses but can correct them and, by so doing, it can improve on the description of experience. In short, theory assimilates, enriches and corrects experience. And this it can do only because it introduces newly begotten transempirical concepts. The inexact concepts of common knowledge, however useful they are for the prescientific description of ordinary experience, are insufficient to frame the high-brow hypotheses of science, not only because they are much too vague but also because they correspond to superficial traits of reality. What things really are cannot be perceived but must be hypothesized: we must think out more or less idealized models made with theoretical constructs, and then test them.

Take, for instance, decision theory — or, more exactly, the part of it known as *Bernoullian utility theory* — the aim of which is to account for and rationalize choice. The only *observation concept* of this theory is that of simple alternative; it may be elucidated as an ordered pair $\langle x, y\rangle$ interpreted as a couple of mutually exclusive outcomes of an action (e.g., a game). The theory introduces two *directly inferable* (but strictly speaking unobservable) relations: preference and indifference. (From acts of choice, conjoined with the assumption of constant preference, we infer preference ratings: if a subject in fact chooses x rather than y, we infer that he prefers x to y, and so forth. A purely behavioristic approach would have to dispense with these unobservables. The scientific attitude is to accept them and to put them under the custody of a testable theory, i.e. to link them with more or less reliable objective indicators.) Finally, the basic *constructs* or highest-level theoretical concepts of the theory are two: one is the utility function u, whose value at x measures the strength of the subject's preferences for the outcome x; the other construct is the probability p of this outcome x. (This probability will be objective in some theories, subjective in others.) The axioms of the theory may be taken to be the following two. (There are, of course, alternative choices of axioms.)

*U*1. For every pair $\langle x, y\rangle$ of outcomes, the subject prefers x to y, or is indifferent between x and y, if and only if: $u(x) \geqq u(y)$. [Keeping the inequality sign only, this might pass for a definition of "preference", an unobservable, in terms of "utility", another unobservable.]

*U*2. For every pair $\langle x, y\rangle$ of outcomes, and every value of the probability p of the outcome x, the utility of the alternative $\langle x, y\rangle$

equals the weighted average of the individual utilities, i.e. $p \cdot u(x) + (1-p) \cdot u(y)$. [Note that $1-p$ is the probability of the outcome not $-x$, i.e. of y.]

These axioms relate, then, the six specific primitives of the theory, which can be arranged in order of epistemological sophistication:

Constructs (highest level theoretical concepts)	Utility function u of the organism concerned
	Probability $P(x) = p$ of the outcome x
Intermediate level theoretical concepts	Preference $>$
	Indifference $=$
Low level theoretical concepts	Alternative pair $\langle x, y \rangle$
	Outcomes $x, y, \ldots$.

The concepts of outcome and alternative pair have been included among the theoretical concepts because, although they were not introduced by decision theory, they are certainly elucidated by it. Strictly speaking, no theory contains concepts other than theoretical concepts.

The preceding example exhibits clearly three conspicuous and intertwined traits of theorizing: *simplification*, *invention*, and *generalization*. Simplification bears first of all on the empirical material and it results in the selection of a few variables that, for some reason or other, are assumed to be the essential ones, and a few key relations among them. The less variables we handle the more refined or complex hypotheses tying them up we may afford. Conversely, the larger the number of variables the more schematic, hence less exact will have to be the assumptions concerning their interrelationship. This is one reason why technological theories cannot afford to contain as complex relations as the underlying scientific theories: precisely because, having to deal with more realistic models of things, they must include a larger number of variables. Anyway, when we simplify the given empirical information we discard particulars rather than systematize them all.

But if we just eliminated or ironed out details we would build nothing: we would do no more than simplifying the original description, The key variables and the key relations among them are not given and they do not emerge from simplification: they are newly introduced or at least refined. Moreover, some of them are far from being suggested by the empirical knowledge at hand: they are invented, as are all constructs. (Invention is compatible with simplification if the two bear on different objects: the empirical material is simplified thanks to the enrichment of the body of knowledge via construct invention.) Invention and guesswork culminate in a model of the system being studied. This model, rather than the intended real referent (which may turn out to be unreal), is the proper object of the theory. As the model involves a substantial deliberate simplification of empirical knowledge, as well as original constructs not found in experience, it cannot be expected to be perfect. That is, we should know *a priori*, from an analysis of the very process of theory construction, that every factual theory is at best approximately true, just because it involves too many simplifications and some inventions that are bound to be inadequate to some extent because they cannot be fully controlled either by experience or by logic.

Owing to the elimination of particulars — to the focusing on essentials — and to the introduction of rich transempirical concepts, the theory can be much more general than any set of particular descriptions and empirical generalizations that prompted its construction. Thus, a single theory of electromagnetic waves, centered in the model of the electromagnetic wave — characterized in turn by three variables only: amplitude, wavelength and phase — covers the whole spectrum from the far infrared to gamma rays, and thereby unifies fields of research that were unrelated before the theory was introduced, i.e. while those research areas kept much too close to experience. A simplification of the given — or rather the sought —, married to the invention of constructs, is necessary to build any theory stronger than the set of its founder hypotheses.

Is it possible to stipulate a set of rules sufficient for constructing theories? Many philosophers have tried it ever since the Renaissance but none have succeeded, although books have been written on the unborn logic of scientific discovery and on the equally unborn technique of theory construction. Is this failure circumstantial or can we hope to build a theory of theory construction in the future, and even to state it in such a way that we could program computers which, by mechanically applying the golden rules, could segregate scientific theories? There is no such hope, even though much is being written about the imminence of such a revolution. The reason for this is that invention, the kernel of theory construction, is not a mechanizable procedure. Invention is certainly a subject of study — a sorely neglected one, partly because scientists are not supposed to invent anything but just to look hard at hard facts and to carefully note what they perceive. Moreover, scientific invention is not free like poetic invention but is subject to logical, methodological, and even philosophical strictures: it is controlled by rules but not propelled by rules. In short, scientific theory construction is not a rule-directed activity although it is rule-controlled. The study of scientific invention, a subject interesting by itself, does not promise to yield a set of rules both necessary and sufficient for inventing new theories, and this for the following reasons.

First, if you have got an original mind then you need no theory construction rules; if, on the other hand, you feel lost without such rules, then you have got no brains enough to do original work. Second, the very idea of a set of prescriptions for acquiring originality is self-contradictory: if you invent a genuinely new idea it is because it was born with the help of no rule: it was not inferred from preëxistent ideas in accordance with some recipe — even though it was not born in an intellectual vacuum either and is most likely "subject" to some set of as yet unknown psychological laws. Third, the proposal of inventing a set of rules for theory invention leads to an infinite regress: in effect, what about the rules themselves: if we rule that there must be rules for producing them we are stuck, otherwise we are forced to grant originality somewhere along the line.

In short, we have no theory construction techniques and, if the above reasoning is cogent, we shall never have them. This is why we have automata theories but no robot design contemplates the manufacture of a theorizing robot. The only thing we know about theory construction is that it is not achieved by manipulating data, with or without the assistance of computers, but by *inventing an ideal schematization* of the object of the theory and then *gradually complicating it* — i.e. introducing further theoretical concepts and more complex relations among them, as required and as allowed by our imagination. But this is as much of a recipe as the advice "Get yourself a powerful imagination and subject to the control of reason and experience".

There are, to be sure, a few more or less definite advices and desiderata that can be given to help theory construction. Among the former, the following may be mentioned: "Do not begin to theorize until you have a bunch of clearly stated problems and a handful of empirical generalizations concerning the field of your choice", "Do not postpone theorizing to a point where the crowd of undigested data — many of them possibly irrelevant — will confuse you altogether", "Do not choose down-to-earth or observational concepts as basic (primitive) units, but strong transempirical ones", "Shun inscrutability", "Keep a reasonable compatibility with well-corroborated theories", "Do not choose deductively sterile assumptions (vague and/or singular formulas), but the strongest ones compatible with facts", "Do not choose either disjunctive or modal propositions as postulates: commit yourself" — and so on. But these are all negative recommendations: they discourage unpromising paths but they do not point to promising ones. In addition to such negative injunctions we can think of a set of desiderata, such as internal consistency, maximum predictability, and explanatory depth (see Sec. 15.7). But neither negative advice nor positive desiderata constitute rules for the effective construction of scientific theories, however effective they can be in the *evaluation* of theoretical programmes and of finished theories. Recommendations and desiderata will be helpful but they cannot replace the conception of a new theory: to this end the proper organs are required rather than rules which, anywhow, are logically impossible.

Should anyone allege that he has actually produced a scientific theory with the sole help of data and a set of rules of discovery and/or invention, we would have to conclude either that he does not quite know what 'theory' means, or that he has cheated himself. This negative conclusion regarding the "logic of scientific discovery" and the "technique of theory construction" need not discourage but should on the contrary stimulate inquiries into the *actual* mechanics of scientific conception and into the desiderata or ideals scientific theories should satisfy — as well as into the reasons why such constructs should comply with such and such conditions rather than with others. Such a study requires a multidisciplinary approach: it calls for the cooperation of logicians, philosophers, psychologists, sociologists, and historians of science conversant with real scientific theories. The results of such an inquiry might help scientists to understand what they are doing and might thereby help them improve their own work, or at least it might help them avoid fruitless strategies. But there is no more a technique for inventing scientific constructs than for begetting geniuses.

There are, on the other hand, two precise techniques for spoiling a theory: one for cutting out its experiential roots, the other for demolishing it altogether. In order to sap the empirical support of a theory it is sufficient to eliminate all the lower-level theoretical concepts of the theory (e.g., "trajectory"), leaving only the higher-level ones (e.g., "mass"): in this way the theory becomes untestable. Think of a psychological theory without either behavioral or physiological concepts and playing the uncontrollable game of "soul", "consciousness", "complex", "motivation", and "repression": it is sheer speculation. Usually speculative theories do not have an empirical motivation to begin with: such a "basis" is avoided altogether or sought for a posteriori, in the form of illustrations or favorable cases. Hence theory sapping may never have been applied; yet it is conceivable as a move to save some theory from empirical refutation.

On the other hand, the demolition of theories has seriously been proposed by conventionalists, empiricists, pragmatists, and other philosophers holding the views (i) that ordinary experience is safer than scientific theory and therefore preferable to it, and (ii) that the sole goal of theorizing ought to be the production of economic information packages and guides to action. *A slightly modified version of the best available technique (W. Craig's) for theory demolition runs as follows.

Take a theory proper, i.e. a hypothetico-deductive system T that cannot fail to contain theoretical concepts. Derive and collect all the lowest-level theorems, i.e. those which can be confronted with empirical information. (Pretend that no auxiliary theories are needed to this end.) Next replace in them every theoretical term by the corresponding empirical expression. For example, if one of the theorems is about a light *ray* (nonempirical) falling on a *plane* (nonempirical) at a definite *angle* α (nonempirical), translate it into a statement concerning a narrow light *beam* falling on an *interface* at an angle *interval* $[\alpha-\varepsilon, \alpha+\varepsilon]$, where ε is the experimental error. That is, take every "net result" t of the theory

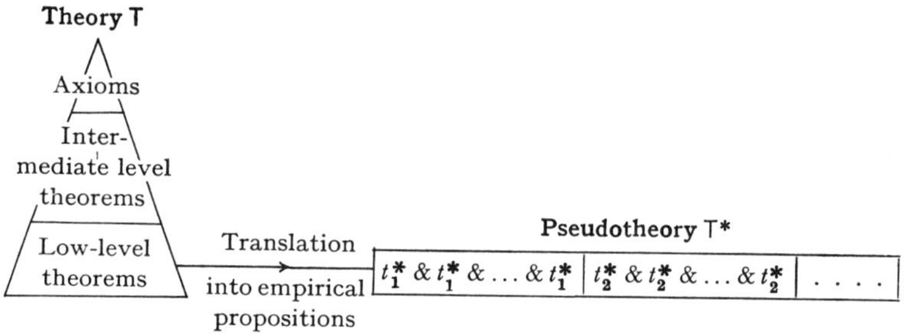

Fig. 8.4. The theory demolition technique. The case of a theory without continuous variables has been illustrated

and translate it into a statement t^* couched in the experimenter's language, just as you would proceed if, instead of rebuilding the theory, you were to test it. (CRAIG's original procedure assumes the CARNAP-BRAITHWAITE-HEMPEL doctrine that (i) every theory contains purely observational predicates alongside the theoretical ones, and (ii) by sheer deduction within a theory one can obtain purely observational statements t^*, i.e. theorems in which only observational predicates occur. These assumptions have not been made here because (i) is unrealistic and (ii) is false: the very universe of discourse of a theory is a more or less idealized model of the real thing, and even the most modest theorem refers to a member of the universe or basic set of the theory.) Now handle these observation statements t^* as the building blocks of a mock theory T* free from allegedely unsafe concepts ("auxiliary expressions"), hence acceptable to anti-theoretically biased philosophers: see Fig. 8.4. More precisely, take as axioms of the new "theory" T* the conjunctions $t^* \& t^* \& \ldots \& t^*$ of the theorems of T translated into an empirical language. If the original theory T is rich enough there will be infinitely many theorems with one empirical translation each. Moreover, if T contains at least one continuous magnitude, such as length, the set $\{t\}$ of theorems will be nondenumerably infinite; but the set $\{t^*\}$ of its possible translations into empirical terms will be denumerably infinite because measurement yields only fractionary numbers, which are denumerable (see Sec. 13.2). Anyhow, the expurgated "theory" T* will contain infinitely many axioms. This infinite (and pointless) formal complexity will overcompensate its epistemological poverty.*

*The procedure outlined above is not a technique for *building* theories of a certain kind — namely, theories satisfying narrow empiricist or conventionalist demands. Firstly, the expurgated "theory" T* is not a theory at all since it does not allow for new deductions: not being a set of formulas partially ordered by the deducibility relation, T* fails to satisfy the definitions of "theory" accepted in logic and in science (see Sec. 7.3). Secondly, the translation T* of T contains no theoretical concepts and, consequently, it does not apply to the ideal model or schematization which constitutes the marrow of theories (see Sec. 7.1). Accordingly, T and T* will not refer to the same thing; for example, if T refers to electrons T* may happen to refer to their visible tracks in a cloud. This difference in reference — consequently in meaning — is, of course, characteristic of every theory about empirically inaccessible (yet supposedly real) objects, from electrons to nations. Thirdly, the pseudo-theory T* cannot be built directly out of data, hence it fails to satisfy the requirement of inductivism: in fact, T* presupposes T, which is a multilevel system that has to be torn down before its debris can be arranged as a ground floor "theory" T*. This theory demolition technique, then, far from proving that theoretical concepts and nonobservational hypotheses and theories are dispensable, actually fires a *coup de grâce* at every attempt to trivialize science by shaving off its transempirical concepts and hypotheses.*

Let us now turn to a constructive task, namely, to a study of the use of mathematical ideas as tools for theory construction.

Problems

8.1.1. Discuss some speculative theory advanced in the belief that it is scientific and find out what distinguishes scientific from non-scientific speculation. Recall Secs. 5.5 and 5.6. *Alternate Problem:* Determine whether the sets of formulas proposed by G. Sommerhoff in *Analytical Biology* (Oxford: University Press, 1950), pp. 88ff. (directive correlation) or by G. Karlsson in *Social Mechanisms* (Glencoe, Ill.: The Free Press, 1958), pp. 134ff. (social interaction) constitute theories or rather programmatic schemas.

8.1.2. Discuss the account of the successive operations involved in theory construction according to P. Duhem, *The Aim and Structure of*

Physical Theory (1914; New York: Atheneum, 1962), Part I, Ch. 2. *Alternate Problem:* Discuss the contrast between the zigzagging course of actual theory construction and its smooth final presentation — i.e. between the natural and the formalized or semiformalized theory.

8.1.3. Evaluate the relative importance of discovery and invention. Hint: Begin by characterizing and exemplifying the concepts of discovery and invention and draw a distinction between experimental and theoretical science. *Alternate Problem:* What the theoretician is "given" is a set of candidates to founder hypotheses. His task is to "discover" higher level (logically and/or epistemologically stronger) hypotheses from which the founder hypotheses can be deduced. Are there any rules for inferring back (retroducing) such initial assumptions? In other words: is there a logic of theory invention, or can we at least expect to invent such a logic? Recall the remark by A. EINSTEIN, cit. by R. S. SHANKLAND, *Nature,* **171**, 101 (1953): "There is, of course, no logical way leading to the establishment of a theory but only groping constructive attempts controlled by careful consideration of factual knowledge". Is there a logical justification for EINSTEIN's "of course"?

8.1.4. Granting that neither data nor logic lead to theory, what about general law statements (e.g., conservation of energy) and metanomological statements (e.g., independence of law statements vis à vis of the observer)? Do they play a role in theory construction? If so in what way: by steering or by preventing derailments? *Alternate Problem:* Discuss the negative recommendations for theory construction advanced in the text. Concerning the avoidance of negative statements as axioms, recall Sec. 7.2. With regard to modal statements, recall that "Possibly p" can be conjoined with "Possibly not-p" without contradiction, so that one and the same datum may support both of them at once.

8.1.5. Examine the view that every scientific theory is a simple summary of empirical findings. Is it true? Is it an efficient guide to theory construction? Or is it rather a theory contraceptive? *Alternate Problem:* Nine years before J. C. MAXWELL published his monumental theory of the electromagnetic field he stipulated the conditions any such theory should satisfy, hence the aim of such a theoretical project. They were: the new theory should (i) yield the known laws in their domain of validity, (ii) extend them beyond this domain, and (iii) relate them. See, e.g., his paper "On FARADAY's Lines of Force" (1856), in *Scientific Papers* (Cambridge: University Press, 1890), I, pp. 155—156: "No electric theory can now be put forth, unless it shews the connexion not only between electricity at rest and current electricity, but between the attraction and inductive effects of electricity in both states. Such a theory must accurately satisfy those laws, the mathematical form of which is known, and must afford the means of calculating the effects in the limiting cases where the known formulas are inapplicable". His own theory did fulfil these desiderata stipulated in advance. Relate this case to the views that all theorizing is just a systematization of observation reports and is done with no previous plan and no conditions imposed beforehand on the outcome.

8.1.6. What is more convenient for the advancement of science: building limited but testable theories, or grandiose but vague and nearly untestable schemas in the tradition of ARISTOTLE's physics and psychology? *Alternate Problem:* Examine the complaint that contemporary psychology and sociology, by choosing to build theories piecemeal-wise, and indeed theories involving highly idealized models of matters of fact, constitute a step backwards relative to 19th century psychosophy and sociosophy.

8.1.7. Take a scientific theory, individualize its specific primitives, and group them into low-level ("observation") concepts, intermediate-level ("directly inferable") concepts, and high-level concepts (constructs). *Alternate Problem:* E. MACH, B. RUSSELL and A. N. WHITEHEAD at one time, P. W. BRIDGMAN, R. CARNAP at one time, S. S. STEVENS, N. GOODMAN and several other eminent thinkers have forcefully proposed the programme of replacing theoretical concepts and formulas by observation terms and formulas or by logical constructions out of sense-data. Evaluate this programme in the light of (i) the history of British empiricism, (ii) the history of logical positivism as culminating with the Vienna Circle, (iii) the history of contemporary physics, chemistry, and biology, and (iv) CRAIG's theorem concerning the alleged eliminability of theoretical concepts.

8.1.8. Pick up a scientific theory, choose a couple of low-level theorems of it and see whether they can be regarded as observation statements proper or whether they contain theoretical concepts. In the language of the text: Take a couple of theorems, t_1 and t_2, of a scientific theory (an existing one, not one of those philosophers are fond of inventing *ad hoc* for illustrating their views), and determine whether these statements can be directly confronted with empirical data or are in need of being translated into quasiobservation statements t_1^* and t_2^* respectively. *Alternate Problem:* Examine the popular doctrine according to which if deduction from the high-level assumptions of a theory proceeds far enough it will automatically end up in observation statements — i.e. that deduction, if carried far enough, will take us from theory to experience. See R. Carnap, *Foundations of Logic and Mathematics, International Encyclopedia of Unified Science* I/3 (Chicago: University of Chicago Press, 1939) and R. B. Braithwaite, "Models in the Empirical Sciences", in E. Nagel, P. Suppes and A. Tarski, Eds., *Logic, Methodology and Philosophy of Science* (Stanford: Stanford University Press, 1962).

8.1.9. Examine either of the following papers. (i) F. P. Ramsey, "Theories", in *The Foundations of Mathematics* (London: Routledge & Kegan Paul, 1931); (ii) W. Craig, "Replacement of Auxiliary Expressions", *Philosophical Review*, **65**, 38 (1956). Determine, to begin with, whether either of these theories of theory applies to existing scientific theories or whether they assume that any scientific theory contains formulas in which no theoretical ("auxiliary") concept occurs. (Recall Problems 8.1.8.) Since the present writer knows of no such theory, Craig's technique was expounded in the text with the addition that every theorem be translated into a semiempirical language — which is not the theory's language. Discuss the shift of meaning effected when passing from T to T^*. Consult H. Putnam's "Craig's Theorem", *Journal of Philosophy*, **62**, 251 (1965). *Alternate Problem:* According to Craig's work, the "replacement" of theoretical concepts is paid for by an infinite formal complexity. Discuss the bearing of this result on simplicism. In particular, determine whether simplicism is compatible with phenomenalism, as both are represented, e.g., in N. Goodman's *The Structure of Appearance* (Cambridge, Mass.: Harvard University Press, 1951). See M. Bunge, *The Myth of Simplicity* (Englewood Cliffs, N. J.: Prentice-Hall, 1963), Chs. 4 and 5.

8.1.10. Discuss the following theses concerning the philosopher's task in relation with scientific theory. (i) "It is the duty of philosophers to keep guard at the gate of scientific theory in order to prevent the entry of elements far removed from experience". (ii) "Philosophers should not meddle with scientific theories: they had better learn some and ape them". (iii) "Philosophers have a right to critically examine every scientific theory, and even to propose new avenues of approach to theory construction, provided they are familliar with the subject of their inquiry". *Alternate Problem:* Why do so many theoreticians favor pretheoretical or even antitheoretical philosophies of science?

8.2. Mathematization

Mathematicians are employed nowadays by almost every department of science. In the underdeveloped sciences they are often summoned when the observations are over, in order to process the data. The assumption is that data processing is a numerical manipulation that will compress and organize bits of significant information into laws; some go as far as to think that the mathematician will be able to churn the data until a full-fledged theory comes out. The assumption is mistaken: the mathematician should not be called in *articulo mortis*, when all the data-gathering has been done without the benefit of clear ideas, and he should not be expected to distil a theory out of data, because theories are not emanations of data. If the mathematician performs his statistical duty he may or may not be useful: this will depend on the scientist who planned the observations, on the ideas the data are supposed to support or undermine, and on the significance of the data. Should the mathematician succeed in producing a theory to account for a bunch of data, he would be doing the work that the scientist himself is supposed to do, i.e. he would invent a set of hypotheses somehow entailing the data. In this process mathematics will have provided a number of ready-made forms into which substantive ideas will have been poured. The moral is clear: the scientist himself should learn mathematics and he should consult the mathematician only with regard to specific mathematical problems, which he must be able to pose in a purely formal way. Mathematics, in short, is most valuable in science as a tool for theory construction, not as a substitute for theory.

Given a class of facts — or rather a set of data supposed to report on the facts — a number of theories can be built to represent them. The most accurate, though not necessarily the truest of them will be mathematical theories — or, rather, substantive theories involving some mathematics. A mathematical theory of factual science is a theory the initial assumptions of which are expressed mathematically — are mathematical formulas as far as their form is concerned — so that their consequences can be derived by the largely standardized mathematical procedures of deduction (computation). A first, rough dichotomy of mathematical theories is into nonnumerical and numerical, the latter being those containing numerical variables. The key concepts of nonnumerical mathematical theories are algebraic (e.g., ordering relation) or set-theoretical (e.g., class inclusion); they are the simplest, yet they have only recently been developed: we still have to learn how to think in terms of these basic theories. Numerical mathematical theories, by far more complex and widespread in science, make mostly use of classical algebra and analysis. They may in turn be divided — with an eye on the ontological problem of determinism — into functional and stochastic. The assumptions of the functional theories are formulas relating numerical variables representing each some property of the ideal model handled by the theory; for example, in one theory the organism may be schematized as a function mapping a set of inputs onto a set of outputs. In stochastic (or probabilistic) theories the formulas are functional as well but at least one of the variables is a random (chance, stochastic) variable, i.e. a variable which assumes every one of its values with a certain probability. In short, we have the following fan:

Mathematical theories — *Nonnumerical*
Mathematical theories — *Numerical* — *Functional*
Mathematical theories — *Numerical* — *Stochastic*

In the sciences of man, theories stated in mathematical terms are often called *mathematical models*, perhaps to avoid the ambiguity of the term 'mathematical theory', which may be taken to designate a part of mathematics or a factual theory using mathematical formulas. Another rationale of the preference for 'model' over 'theory' may be the wish to warn the reader that a rough schema or idealized sketch rather than a faithful portrait of a bit of reality is being constructed; but this warning should be unnecessary for anyone familiar with scientific theories. A third rationale may be the wish to distinguish one's own unrealistic but sober speculation from the wordy and muddled "systems" of the traditional kind, which still pass for theories in the sciences of man. A fourth may be the shame some scientists educated in the empiricist tradition feel for theories: scientists are supposed to deal with "hard facts" rather than with "mere theories", but everyone has a right to relax playing with "models". Physicists, who began to learn the language of mathematics along with mathematicians, hardly speak of mathematical *models* or even of *mathematical* theories in their field: they take for granted that the most important physical ideas must be expressed mathematically. Accordingly they just speak of theoretical physics — to distinguish it from experimental physics — and keep the term 'mathematical physics' to designate the investigation of purely mathematical problems arising in theoretical physics and having an instrumental interest for physicists but hardly any value for pure mathematicians except as possible problem sources. The same process of maturation will probably occur in what are now called 'mathematical biology', 'mathematical psychology', and 'mathematical sociology': they will eventually be known as theoretical biology, theoretical psychology, and theoretical sociology respectively. This should happen when the present revolutionary process of mathematization in these sciences becomes so well advanced that no theoretician in any of these fields need apologize for inventing sketchy models and for stating the properties of such models in a mathematical language.

A mathematically trained scientist will spontaneously tend to frame his theories in mathematical terms, if only because he will find this clearer and easier than the corresponding verbal formulation — which will be impossible in many cases. Having once tasted accuracy, deductive power and formal elegance, his palate refuses raw food and he tends to equate *premathematical* with *pretheoretical*. Mathematization, in fact, does not so much depend on the subject matter or object as on (i) the state of the discipline, (ii) the scientist's realization that theories cannot be built unless details are forgotten and schematic models are invented, and (iii) the scientist's mathematical background. It is futile to attempt the mathematization of a discipline unless there is a minimum body of scientific theory and conceptual clarity: imagine trying to mathematize chemistry before LAVOISIER. Once a minimum degree of maturity is attained mathematization will make a further clarification and growth possible; in the very "translation" of verbal statements into mathematical formulas the concepts will be sharpened. Consequently, failure to mathematize a field may be an index of either the field's muddled state or of the theorist's limitations. A social scientist with a classical background and outlook may love vague pomposity and abuse the mathematical approach to theory construction, whereas a sociologist with a mathematical training will naturally tend to think in mathematical terms. And this alone will help him conceive an ideal model of his object — i.e., a set of definite relations among definite variables.

For example, when confronted with the problem of communication within animal or human groups, the mathematically trained social scientist may propose a model centered on the hypothesis that contacts among individuals are random (as assumed in A. RAPOPORT'S random choice model). The traditionally-minded behavioral scientist may frown upon this gross oversimplification: do not we know that we never choose other people accidentally?; where do love and hatred come into?; and are not randomness and choice mutually incompatible? The modern oriented scientist will probably reply, first, that our meeting and choosing (in general, communicating with) other people *is*, demonstrably, largely random; second, that a random choice model has the definite advantage of showing to what extent our actual communication behavior is *not* random, since deviations of the model's predictions from empirical evidence will suggest deliberate choice rather than accidental encounter; third, that such discrepancies may show the way to more adequate theories of communication within groups: to more complex models allowing for nonrandom behavior; fourth, that a random model is comparatively simple and is therefore bound to be tried before any more sophisticated model is built.

A start in mathematization or mathematical modeling, however unrealistic, is better than either a prolix but unenlightening description or a grandiose verbal sketch. Those who reproach mathematical behavioral scientists for being unrealistic exhibit (i) a lack of the historical perspective necessary for passing judgment over a young enterprise, (ii) a lack of acquaintance with the chief aim of theorizing, which is not the most detailed covering or summarizing of experience but an understanding of it, and (iii) a lack of acquaintance with the nature of factual theory, the kernel of which is a model, i.e. a more or less schematic conceptual representation of a complex real system (see Sec. 7.1).

There is no sin in simplicity as long as it is not taken for an absolute aim. Simplicity, a trait of young theories, cannot be a goal of theorizing because simplification amounts to neglecting actual traits, so that the simplest models are apt to be the less realistic (see Sec. 8.1). But simplification is a *means* for making theoretical work possible; once a theory has

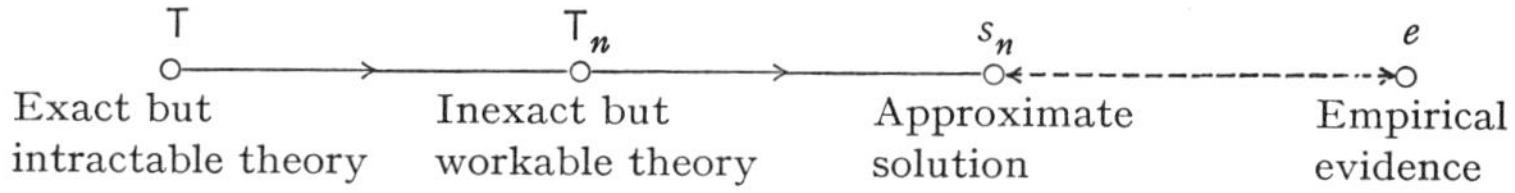

Fig. 8.5. The empirical test of a theory with highly complex (but also believed to be highly accurate) initial assumptions requires the building of a simpler, intermediate theory, with effectively solvable problems

shown to be approximately true it may be complicated, i.e. enriched in an attempt to improve its adequacy. Thus, gas theory began picturing a gas as a collection of free point-particles; the next improvement was to replace the point-particles by elastic spheres and to introduce weak forces among them; then the inner structure of the molecules began to be taken into account, and once this is begun there is no end in view.

The simplifications inherent in theorizing are particularly apparent in the case of the mathematical theories of factual science. Simplification will occur here, first of all, in the selection of the basic variables and their mutual relations — i.e. in the very building of the model; even so, the model may be much too complex for testable consequences to be drawn from the assumptions. In this case more or less brutal simplifications have to be introduced, most of which consist in linearizing or even neglecting entire terms in some equations. *For example, the axioms of fluid dynamics and of gravitation theory are nonlinear; and even the exact solutions to linear equations may be too complex for purposes of interpretation, application, or test, so that the experimentalist and the technologist will usually add simplifications to those made by the theoretician. Accordingly in such cases the comparison of the given theory T with the experimental evidence e is not made directly via the deduction of testable consequences (e.g., predictions) of T, but via an *intermediate theory* T_n constituting an n-th order approximation to T and yielding solutions s_n that are exact in T_n but only approximate in T (see Fig. 8.5). In short, even when a theory has been built by sacrificing heaps of details, further simplifications may be needed in order to work it out unless new, more powerful mathematical computation techniques are invented.*

Generalization, too, is particularly conspicuous in the case of mathematical models. This is easily understood if it is recalled that generalization can be achieved (i) by neglecting or abstracting from certain aspects and/or (ii) by disinterpretation or emptying of factual content. The first avenue to generalization is a reward for ignoring actual complexity and is, as we have seen, a conspicuous trait of mathematical modeling. The second generalization technique — abstraction — can be applied as soon as a mathematical "model" has been built, by simply ignoring the interpretation rules that give it a specific content. For example, it is possible to set up a general contagion theory covering all kinds of processes of the spreading of something: epidemics, habits, rumors, beliefs, and what not. The basic assumptions of the theory are :(i) Every "infected" individual becomes a contagion agent, and (ii) There will be the more "infected" individuals the more candidates (i.e., "uninfected" individuals) there remain. A single axiom can join these two ideas, namely: The rate of change of the fraction of "infected" individuals is proportional to both the fraction of already "infected" individuals and to the fraction of still "uninfected" individuals. (The function of the quotation marks is here to emphasize that no specific process is being referred to.)

*Save for the lack of precise symbols the last but one statement is essentially mathematical because it involves definite variables and a definite relation between them which can easily be translated into mathematical terms. Call $C(t)$ the fraction of individuals "infected" at time t and $DC(t)$ the rate of increase of this fraction also at time t. ($DC(t)$ is not an independent concept but is definable in terms of C and t via the limit concept; in fact, $DC(t)$ is the derivative of $C(t)$ with respect to t. Were it not for the infinitesimal calculus we would have to introduce contagion rate as a separate concept, i.e. as another primitive.) With these symbols our axiom is symbolized in the following equation

$$DC(t) = kC(1-C), \quad 0 \leqq C \leqq 1. \qquad [8.1]$$

(The second formula, fixing the range of C, is actually a separate axiom.) In writing the above formulas we have tacitly assumed the following interpretation rules:

$$I(t) = \text{time}, \quad I(C) = \text{contagion fraction}.$$

The constant k has, for the time being, only a mathematical interpretation: it is a proportionality coefficient. In order to be able to apply the calculus we have made an *auxiliary assumption* typical of the mathematization procedure: we have, in fact, made the pretence that $C(t)$ and

$DC(t)$, which are actually discrete variables, can be approximated by continuous variables without serious distortion (see Fig. 8.6). This simplification is made in open violation of observational data but it does not amount to a mere discarding of particulars: it consists in the introduction of a theoretical concept — namely $C(t)$ — by means of which a model is built which is reasonably adequate for a large community, since in this case the step-wise variation of $DC(t)$ is approximated by a continuous variation.*

*In the contagion equation [8.1] we can read some traits of the contagion process that were far from obvious in the ordinary language formulation of the basic assumption. First, while $C=0$, $DC=0$; i.e., in order for the contagion process to start there must be at least one carrier. Second, the maximum contagion rate, namely $k/4$, is reached when half of the population has been struck. Third, from this point onwards the contagion rate decreases, reaching 0 when the whole population is "sick" (see Fig. 8.7). The mere mathematical statement of the basic assumption, in conjunction with the elementary technique of graphical representation, has been a source of new information. If we wish to make full use of the deductive power of mathematics, though, we must solve for $C(t)$. And finally we must plot this solution and confront the theoretical curve with empirical data: this is a routine treatment applied to every mathematical model. The result is

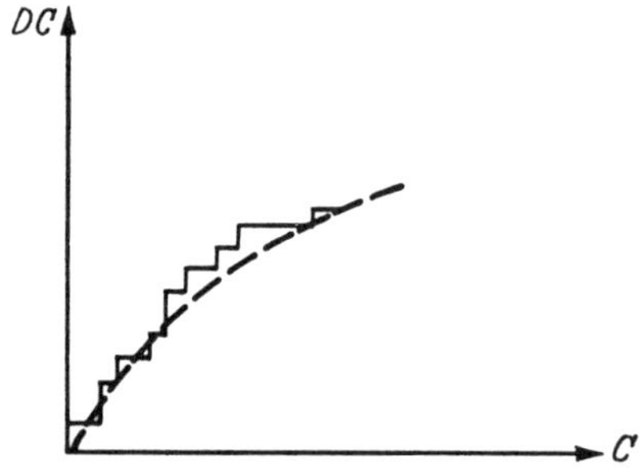

Fig. 8.6. The actual contagion process in a population is stepwise (full line); its mathematical representation is continuous (dotted line)

$$C(t)=1/(1+A\,e^{-kt}) \qquad [8.2]$$

where the value of the integration constant A can be determined from the initial condition $C(0)$. The graph representing the function $C(t)$ is the logistic curve, which describes a number of social and even historical "contagion" processes, such as the territorial expansion of empires (see Fig. 8.7). If the solution stands the empirical test we pronounce the basic assumption [8.1] confirmed — until further notice.*

One should not stop at confirmation, i.e. at the lack of unfavorable evidence, but should try to furnish *positive evidence of a deeper kind.*

In the case of contagion theory this is achieved if a plausible contagion *mechanism* is found. In other words, even though a phenomenological contagion theory stands the test of experience, we try to support (and understand) it by building an underlying *representational* theory. A reasonable hypothesis for such a deeper theory is that contagion is by contact. Now, 'contact' is a vague term: we have to refine it if we wish to theorify it. The least committal way of doing this is to introduce the quantitative concept of probability of contact, p, between an "infected" and an "uninfected" individual. An even deeper theory will specify the mode of contact and the kind of "germ" transmitted during it; but the introduction of the new primitive, p, will suffice to build a

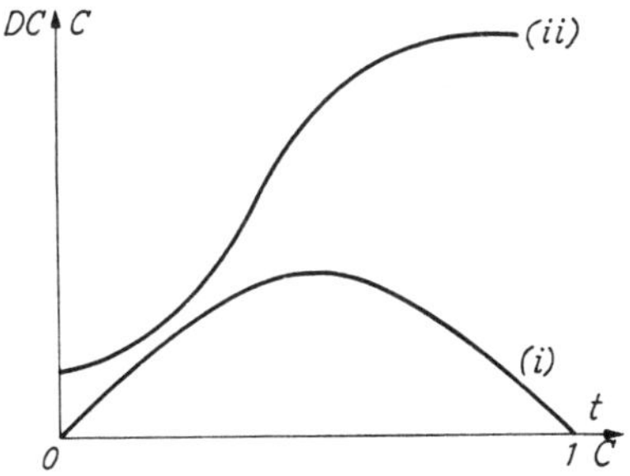

Fig. 8.7. (i) The contagion curve that visualizes equation [8.1]. (ii) The logistic curve that represents function [8.2]

deeper theory, pointing to a level beneath the level of the net effect of the contacts among individuals. In fact, the commonsensical hypothesis "Contagion is by contact" can now be spelt out as follows: (i) The rate of increase of the "infected" fraction is proportional to the probability of contacts between "infected" and "uninfected" individuals; (ii) The probability of contact is proportional to the fraction of both "infected" and "uninfected" individuals.

*The base of the new theory is

$$C1 \quad DC(t)=k_1 p, \qquad [8.3]$$

$$C2 \quad p=k_2 C(1-C), \quad 0<k_2\leqq 1, \quad 0\leqq C\leqq 1 \qquad [8.4]$$

with the rules of interpretation included in the previous theory plus

$I(p)$ = probability of contact between an "infected" and an "uninfected" individual.*

The previous theory had two primitives — t and C — and one main axiom. The new theory has three primitives — t, C and p — and two axioms. The mathematical theory behind the former was analysis; the second theory presupposes, in addition, probability theory and is therefore formally richer. Since the new — stochastic — theory has more formulas, it is susceptible to more tests. Thus, the new variable may perhaps be independently controlled, and the two basic assumptions of the theory may perhaps be separately tested. By the same token — i.e. for having more consequences that could be upset by observation — the new theory is far less certain than the previous one. The less we talk the wiser we may hope to look. But the second theory is to be preferred to the first precisely because it is more fully testable and because it provides a deeper insight into the contagion process.

The departure from the early, phenomenological approach has led to a deeper analysis. But it cannot be the final analysis: we have hypothesized that contagion is by contact but we have failed to specify the stuff (or else the information) that is being transmitted during the contact, and we have given no details concerning the contact process. In particular, we have treated the contact probability as an ultimate, whereas a deeper theory would regard it as a function of, say, the spatial configuration and mobility of the agents as well as of their immunity and of the "stuff's" virulence. In short, probability would be regarded as a physical property deriving from more fundamental properties: as a sort of net effect of a number of factors. As soon as any such specifications are made, i.e. as soon as new hypotheses are introduced concerning the contagion mechanism, a deeper and more realistic theory can be obtained. But a price must be paid for this: the wide coverage of the initial phenomenological theory will be lost because now the specific *differences* rather than the similarities among contagion processes will be stressed: in some cases we will have information transference, in others energy transmission, in still others germs transmission, and so on. We will know more about less. (For the phenomenological/representational alternative see Sec. 8.5; for the coverage/depth complementarity see Sec. 9.6).

We are now in a position to assess the advantages of mathematical theories over verbal ones. Let us list them.

1. *Help in theory construction.* Mathematics provides ready-made symbols that, properly interpreted, can be used by any science; just recall the concepts of set intersection and of probability. The laying of factual assumptions in an adequate mathematical nest and letting mathematics do the hatching is the more valuable the poorer the substantive assumptions are: bulky theories can be generated by handling modest initial assumptions with powerful mathematical tools. The advantage of this cuckoo technique should not blind us to the substantial poverty of the initial ideas — as exemplified by much of contemporary psychology and physics. On the other hand, as soon as the number of variables increases and the relations among them become more complex, the use of mathematics becomes mandatory: it is merely

impossible to handle a complex model without its assistance. This is one reason why the basic equations of theoretical physics cannot be popularized, although brave attempts are sometimes made to couch them in ordinary language. Just think of one of MAXWELL's equations, namely $\nabla \times H = \frac{1}{c}\frac{\partial D}{\partial t} + 4\pi j$. If mathematization were a kind of translation — just an accurate symbolic rewording — it should be possible to render the preceding equation in plain words — but it is not.

2. *Accuracy*. Take a proposition like "The growth of knowledge depends on the bulk of available knowledge", which is not more imprecise than the typical conjectures of most of behavioral science. It may be taken to mean either that the available knowledge causes or at least makes possible its own growth or that both variables, knowledge and the rate of growth of knowledge, depend on a third unmentioned variable. Even assuming that the first interpretation is intended, the form of the stated dependence is left unspecified: the verbal statement is consistent with infinitely many mathematizations, all of which fall under the schema "$DK = f(K)$", where 'DK' designates the rate of growth of the body of knowledge K, and f the dependence alluded to (see Problem 8.2.8). Generalizing: an infinity of accurate theories may hide behind a single verbal theory. In less optimistic terms: verbal theories are not fully interpreted theories but, rather, partially interpreted systems; only mathematical "models" can be or become fully semantical systems. (For the difference between abstract theory and model, see Sec. 7.4.) Consequently, the nonmathematical theories abounding in the sciences of man are, strictly speaking, indefinite schemata: every one of them is, potentially, an infinite set of theories — each of them accurate but unborn. Accordingly, the semantical indeterminacy of premathematical theories is infinite: they can at best serve as preliminary guides for the construction of definite theories. But let us repeat a word of caution: preciseness is not to be equated with quantitative accuracy. Every quantitative statement is exact but the converse is not true, as shown by any statement concerning class inclusion or precedence. Moral: nonquantitative mathematical theories serve, as regard preciseness, the same purpose as the quantitative ones.

3. *Deductive power*. In ordinary language it is as difficult to make elaborate deductions — or even to check them — as to state complex relationships; and it is equally difficult to ascertain whether a given conclusion follows from the explicit assumptions alone, with no hidden extra premise. Only the most obvious implications can be recognized intuitively; the nonobvious ones may be the most important and they require special deduction techniques. Just think that NEWTON had to invent the differential calculus in order to prove his conjecture that his

equations of motion, in conjunction with his law of gravity, did entail the Keplerian orbits.

4. *Testability.* Preciseness and deductive power make up testability: the more logical consequences a theory has and the more accurate they are, the better it lends itself to empirical tests. Verbal theories, on the other hand, can be short of irrefutable. A statement like "y depends on x" is confirmed whenever y is not independent of x, whence a variety of empirical findings is consistent with it. But such a cheap confirmation is hardly valuable except as a preliminary tip, as a suggestion to look for a precise relation of dependence between x and y.

5. *Metatheoretical advantages.* Inconsistencies and lack of independence of the primitives and the initial assumptions can best be brought to light through a precise formulation in mathematical terms.

6. *Comparison with rival theories.* The region of disagreement of alternative theories can best be spotted, and the comparison of the respective virtues and faults can best be made if they are mathematized. An additional bonus is that emotional overtones and extrascientific considerations in such a comparison can be minimized, though perhaps not eliminated.

After praising the virtues of mathematization we should warn against a couple of mistaken beliefs held by uncritical partisans of mathematization, from PYTHAGORAS to EDDINGTON. The first tenet is that mathematization coincides with formulation in terms of quantitative mathematics such as classical algebra and analysis. Surely, most people are more used to classical mathematics, which is predominantly quantitative, and therefore find it easier to state hypotheses in quantitative terms; moreover, quantitative hypotheses are the most precise of all. But the application of quantitative mathematics requires an advanced level of empirical research, and the pace of research could be hurried if we tried to handle the weaker tools of relational mathematics while things are not mature for a more precise approach.

A second tenet is that the mathematical treatment of a subject incorporates it into mathematics. For this reason rational mechanics — the earliest fully mathematized field of physics — was regarded as a branch of mathematics until recently. The positive results of this confusion are in sight: some of the most powerful mathematical brains applied themselves to mechanics and pushed the subject ahead. The harm done was less but nonvanishing: a large number of purely academic problems were attacked, and the science was regarded as immune to empirical refutation, to the point that criticisms of Newtonian mechanics were sometimes taken as insults to mathematicians. That the view is mistaken should be obvious to anyone recalling that, unlike theories of pure mathematics, factual theories contain interpretation assumptions establishing a correspondence between symbols and concepts that are supposed to represent nonmathematical objects — e.g., physical properties. The view that "number and figure" generate truth by themselves is partly based on the widespread confusion between *truth* and *validity*. If a theory's assumptions are factually true then its theorems are factually true as well; the validity of the deduction ensures such a conservation of truth. But, of course, we cannot know whether the axioms are true: all we can hope for is that some of their consequences will be found reasonably consistent with the empirical evidence. Moreover, the converse does not hold: one and the same theorem can be entailed by infinitely many alternative axiom bases: if $\{A\}\vdash t$ then also $\{A, B\}\vdash t$, where B is an arbitrary formula. Mathematization ensures the validity of the derivations and their easy checking. But the axioms, even if impeccably formulated, might be false; moreover, strictly speaking they are false since they refer to a highly idealized model.

The belief that mathematics warrants truth is so widespread that it may be worthwhile to exhibit a mathematically correct but factually empty theory: the following *axiomatic theory of phantoms*.

Specific primitives

U, a set of phantoms; E, the phantom energy; d, the ectoplasm density; t, the phantom age; and N, the number of wickednesses performed by a phantom in unit time.

Axioms

A 1. For every x in U, the energy of x is directly proportional to the density of the ectoplasm of x and inversely proportional to the age of x:

$$E=k_1\, d/t, \qquad t>0, \qquad k_1>0.$$

A 2. For every x in U, the ectoplasm density of x is a linear function of the number of wickednesses x performs per unit time:

$$d=k_2\, N+d_0, \qquad k_2>0, \qquad d_0>0.$$

A 3. For every x in U the average wickedness number per unit time performed by x is constant:

$$N_{\mathrm{AV}}=k_3, \qquad k_3>0.$$

The most interesting theorems follow immediately.

T 1. By *A* 1 and *A* 2,

$$E=k_1\,(k_2 N+d_0)/t.$$

T 2. By *A* 3, after a period T the number of wickednesses increases to

$$N(T)=N_{\mathrm{AV}}\, T=k_3\, T.$$

$T3$. From $T1$ and $T2$ we obtain the phantom energy after a period T:

$$E(T) = k_1 (k_2 k_3 T + d_0)/T = k_1 k_2 k_3 + k_1 d_0/T.$$

$T4$. By $T3$, for T approaching infinity the energy tends to

$$E_\infty = k_1 k_2 k_3 = \text{const.}$$

That is, phantoms keep alive by wickedness. This agrees with the evidence. On the other hand $T4$ is counterintuitive in view of $A1$. In other words, $A1$ itself is counterintuitive — as every high-level assumption should be.

From the above it should be clear that mathematization is not coincident with scientific method. There is no scientific method without empirical test, and mathematics enhances testability but not actual test, since it is alien to empirical procedures. By itself, then, mathematics is not sufficient to constitute science; but a point is reached in every discipline beyond which no important progress is achieved unless mathematical tools are employed in the building and working out of theories. In other words, although mathematics does not warrant scientificity, the scientific method does involve an increasing use of mathematical concepts and theories: mathematics is necessary for the maturation of science.

Still, mathematization is not the final stage to which a theory can be brought. What crowns theorization is complete formalization, to which we now turn.

Problems

8.2.1. Peruse the volumes of the *Bulletin of Mathematical Biophysics*, *Journal of Theoretical Biology*, *Journal of Mathematical Psychology*, or *Operations Research*, and report briefly on any of the mathematical theories exposed in them.

8.2.2. Summarize and comment on any of the theories expounded in the following works. (i) D'A. W. THOMPSON, *On Growth and Form*, 2nd. ed. (Cambridge: Cambridge University Press, 1942), especially on magnitude and on the theory of transformations. (ii) R. MCNAUGHTON, "A Metrical Concept of Happiness", *Philosophy and Phenomenological Research*, **14**, 172 (1953). (iii) N. RASHEVSKY, *Mathematical Biophysics*, rev. ed. (New York: Dover, 1960), 2 vols., and *Mathematical Biology of Social Behavior*, rev. ed. (Chicago: University of Chicago Press, 1959). (iv) H. SOLOMON, Ed., *Mathematical Thinking in the Measurement of Behavior* (Glencoe, Ill.: Free Press, 1960). (v) L. B. LEOPOLD, "Rivers", *American Scientist*, **50**, 511 (1962). (vi) "Symposium on Mathematical

Theories of Biological Phenomena", *Annals of the New York Academy of Sciences*, **96**, 895 (1962). (vii) J. G. KEMENY and J. L. SNELL, *Mathematical Models in the Social Sciences* (Boston: Ginn, 1962). (viii) R. D. LUCE, R. BUSH and E. GALANTER, *Handbook of Mathematical Psychology* (New York: Wiley, 1963ff.), 3 vols, and *Readings in Mathematical Psychology* (New York: Wiley, 1963ff.), 2 vols. (viii) J. S. COLEMAN, *Introduction to Mathematical Sociology* (New York: Free Press, 1964).

8.2.3. Many people oppose the construction of mathematical "models" in the sciences of man. Why? Examine, among others, the following alternative (but mutually compatible) explanations. (i) Because they think everything human is much too complex to be mathematized: "You can neither compress the richness of man in a formula nor subject his free will to law". (ii) Because they are ignorant of mathematics, hence afraid of it. (iii) Because the building of mathematical "models" in the sciences of man (*a*) exhibits by contrast the theoretical emptiness of the traditional schools and (*b*) endangers the various traditional ideologies by accustoming people to think in accurate terms. *Alternate Problem:* Outline and exemplify the possible tasks of mathematical logic and algebra in biology. Take a look at J. H. WOODGER's work.

8.2.4. Examine KANT's argument against the possibility of turning psychology into a science: all science proper is quantitative; psychology cannot be quantified; hence psychology cannot become scientific. *Alternate Problem:* It has been claimed that a mathematical "field theory of parapsychology" would establish this undertaking as a science. Evaluate this claim. See G. D. WASSERMANN, "An Outline of a Field Theory of Organismic Form and Behaviour", in G. E. W. WOLSTENHOLME and E. C. P. MILLAR, Eds., *Ciba Foundation Symposium on Extrasensory Perception* (Boston: Little, Brown and Co., 1956).

8.2.5. A common argument against the use of mathematics in building theories about man is that many of them have failed. Does this establish the futility of the approach itself or rather the limited success of the attempts made so far? In general: do failures in the realization of scientific programmes establish the latter's inadequacy? If not, how else can we assess them? *Alternate Problem:* It is usually held or implied that the proper mathematics of psychology and sociology is "finite" mathematics. Is this true forever?

8.2.6. Mathematize the following miniature sociological theory found in H. ZETTERBERG's *On Theory and Verification in Sociology* (1954), repr. in P. LAZARSFELD and M. ROSENBERG, *The Language of Social Science* (Glencoe, Ill.: The Free Press, 1955).

A 1. The greater the division of labor, the greater the solidarity.

A 2. The greater the solidarity, the more the uniformity.

A 3. The greater the number of members of the group, the greater the division of labor in it.

Try the following mathematization of the imprecise relation statement "the greater x, the greater y": $y = kx$, with k a constant. Derive some theorems. Finally compare your results with those ZETTERBERG obtained without using mathematics.

8.2.7. The hypothesis that the rate of growth of the body of knowledge K is proportional to the size of knowledge itself can be written thus:

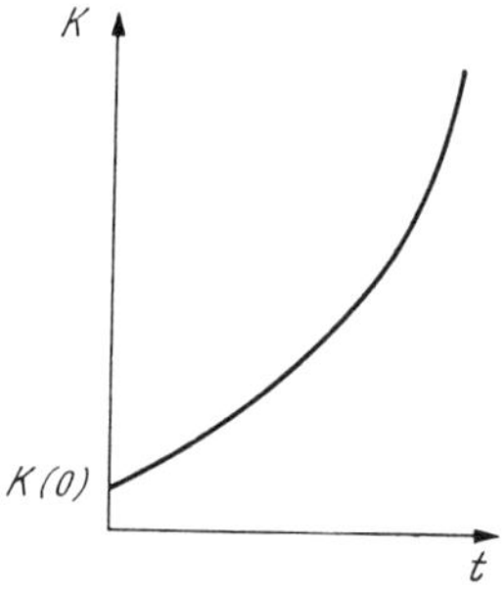

Fig. 8.8. The exponential growth of scientific knowledge as gauged by the bulk of scientific literature

$DK = aK$, where a is a parameter characterizing the kind of knowledge. Recalling that $DK =_{df} dK/dt$, the equation for the growth of knowledge can be rewritten in the form: $dK = aK\,dt$, which can be read: the increase dK of knowledge during the time interval dt is proportional to the size K of the available knowledge and to the time interval. With the help of the integral calculus we deduce: $K(t) = K(0)\,e^{at}$, where $K(0)$ is the body of knowledge available at the time $t = 0$. The graph of the preceding function, shown in Fig. 8.8, is the actual growth curve of the number of learned journals during the last centrury or so. It is also the so-called law of accelerated cultural growth (the relative acceleration being a^2), which anthropologists have found by plotting data. Queries: (i) to what discipline do the preceding considerations belong?; (ii) do they have any effect on epistemology?; (iii) what are the differences between the differential and the integral growth equations?; (iv) what is more advantageous: to start from the former or from the latter?

8.2.8. Work out and discuss the following generalization of the equation for the growth of knowledge presented in the previous Problem. Let the reference set be a set of N intercommunicating units (persons, professions, etc.) capable of learning and producing knowledge. Denote with

K_i the amount of knowledge (or of learning) of the i-th unit. Assume that the speed dK_i/dt at which the i-th unit learns (or contributes to knowledge) is a linear function of what is known by the same unit (i.e., $a_{ii}K_i$), of what the other members of the set know $\left(\text{i.e., } \sum_{j\neq i} a_{ij} K_j\right)$, and of the speed at which these other members learn or create $\left(\text{i.e., } \sum_{j\neq i} b_{ij}\, dK_j/dt\right)$:

$$dK_i/dt = \sum_{j=1}^{N} a_{ij} K_j + \sum_{j\neq i}^{N} b_{ij}\, dK_j/dt.$$

This is a system of N linear partial differential equations, which may apply to both learning and the growth of knowledge in any situation characterized by both competition and cooperation. The first term in the second-hand member of each equation represents knowledge transmission and is responsible for the cumulative aspect of the growth of knowledge. The second term in the right-hand member represents knowledge stimulation (e.g., by emulation) or inhibition (e.g., by discrimination and secrecy); it will represent processes of involution, stagnation or revolution, according to the signs of the coefficients b_{ij}. The $2N^2$ coefficients a_{ij} and b_{ij} are phenomenological summaries of certain properties of the units involved; these properties should occur explicitly at a deeper level of analysis. At any rate they summarize innate abilities, the properties of the communication channels and the methods of learning to stimulate and inhibit the growth of learning or of knowledge.

8.2.9. What kind of mathematical tool should be tried in an attempt to mathematize the theory of evolution in its present state? Justify your choice.

8.2.10. Illustrate and analyze the following statements by J. L. Synge, in *Geometrical Optics: A Introduction to* Hamilton's *Method* (London: Cambridge University Press, 1937), p. 1: "A 'perfect' scientific theory may be described as one which proceeds logically from a few simple hypotheses to conclusions which are in complete agreement with observation, to within the limits of accuracy of observation. [...] As accuracy of observation increases, a theory ceases to be 'perfect': modifications are introduced, making the theory more complicated and less 'useful' [...] in truth man has always created 'ideal' theories. Nature is much too complicated to be considered otherwise than in a simplified or idealized form, and it is inevitable that this idealization should lead to discrepancies between theoretical prediction and observation". *Alternate Problem:* Do the simplifications and pretences required by the building of mathematical "models" and/or the consequences of their initial assumptions justify the conventionalist (neo-Kantian) doctrine that all our theories are *fictions* in the sense that they are deliberate and self-contradictory falsifications of reality tested by their usefulness alone? For an exhaustive exposition of fictionalism, see H. VAIHINGER, *Die Philosophie des Als Ob*, 4th ed. (Leipzig: Meiner, 1920). Hint: Courage!

8.3. *Reconstruction (Formalization)

A theory is said to be *logically reconstructed*, or *fully formalized*, if its basis is stated exactly and exhaustively. The complete formalization of a theory consists in the explicit and complete symbolic (nonverbal) formulation of the theory's axioms and in the fullest possible statement, or else mention, of the theory's presuppositions and rules; the latter are stated in the theory's metalanguage, which usually is not a formalized language. Symbolically:

$$\textit{Formalization} = \textit{Symbolization} + \textit{Axiomatization} + \textit{Rules} + \\ + \textit{Presuppositions}.$$

Axiomatization comprises the enumeration (but not the explicit elucidation) of the primitive symbols, and the explicit definition of the derivative concepts. The rules to be laid down or mentioned are the norms of formation and transformation of complex formulas (the syntactical rules) and the rules of interpretation (the semantical rules). As to a theory's presuppositions, they are rarely mentioned outside formal science, yet it is desirable to do it if one wishes to avoid the puzzlement of many philosophers at the simplicity of the basis of some bulky theories.

Complete formalization is rarely attempted outside the formal sciences, which have provided the paradigm of formalization ever since EUCLID (wrongly) though he had entirely formalized geometry, and more particularly since HILBERT (also wrongly) thought that formalization can exhaust a theory and can therefore be final, i.e. incorrigible. The mere enumeration of the presuppositions of a theory can be quite a problem in the case of factual theories, which not only presuppose formal (logical and/or mathematical) tools but also a number of ideas belonging to other factual theories; moreover, presuppositions are often seen in the light of ulterior developments. But the fact that grapes are sour most of the time does not in the least diminish their value. The axiomatization of a theory's core is desirable if only because it favors the recognition of concept independence and of consistency and yields an exact criterion of proof. And formalization adds to these advantages the explicit enumeration of presuppositions and rules that might otherwise be accepted uncritically: by stating them explicitly we can keep them under control. The farther formalization is carried the more cogent the argument becomes and the better the nonempirical tests of theory adequacy can be given. In short, reconstruction or formalization amounts to our putting the cards on the table and inviting critical examination and improvement.

Let us exemplify the above ideas. Our first illustration will be a tiny mathematical theory, namely the theory of *quasiordering* or *congruence*. We shall expound only the *foundation* of the theory, i.e. we shall leave aside its definitions and theorems.

0. *Presuppositions*.

Elementary logic (propositional calculus and first order functional calculus with identity), and elementary set theory. This body of presuppositions provides, among others, the logical signs of the theory: the connectives & and $\rightarrow$; the universal quantifier $(x)_U$, read 'for every x in U'; and the auxiliary signs '(' and ')'.

1. *Primitives* (Alphabet).

A set constant $U=\{x, y, z, \ldots\}$.
A binary relation $\sim$.

2. *Formation Rules* (Syntax of formulas).

*FR*1. '$x \sim y$' is a well formed formula (wff).

*FR*2. Every formula composed of wff's and one or more logical connectives is a wff.

*FR*3. The universal generalization of a wff is a wff.

3. *Transformation Rules* (Syntax of derivation).

TR. If, in a wff, an individual variable is replaced by any other variable of the same type, a new wff (and precisely a trivial theorem) is obtained.

(The enumeration of the formation and transformation rules is redundant if the presuppositions of the theory are mentioned. The above rules have been stated here to conform to usage and to recall that whenever a system is taken for granted or presupposed all its rules of formation and transformation are got into the bargain.)

4. *Axioms* (Initial assumptions).

*Q*1. For every x in U, $x \sim x$ (congruence is reflexive); i.e.,

$$(x)_U (x \sim x). \qquad [8.5]$$

*Q*2. For every x, y and z in U, if $x \sim z$ and $y \sim z$, then $x \sim y$; i.e.

$$(x)_U (y)_U (z)_U (x \sim z \;\&\; y \sim z \rightarrow x \sim y). \qquad [8.6]$$

5. *Interpretations.*

Model 1: Congruence of plane triangles.

*IR*1. $I(U)=$ the set of plane triangles

*IR*2. $I(\sim)=$ is similar to.

Model 2: Congruence of rigid bodies.

*IR*3. $I(U)=$ the set of rigid solid rods

*IR*4. $I(\sim)=$ can be exactly juxtaposed to.

Model 3: Descent.

*IR*5. $I(U)=$ all people.

*IR*6. $I(\sim)=$ has an ancestor in common with.

The foundation of a formalized formal (logical or mathematical) theory consists of *six layers:* the presuppositions, the primitives, the syntactical rules of formation and transformation, the axioms, and the interpretation rules if any; all else in the theory is either definition or theorem. The foundation of a formalized factual theory can be given a simpler structure: the formation and transformation rules can be skipped because they are built into the underlying logical and mathematical theories. A good way of showing how a formalized factual theory is built is to exhibit an instance.

We shall take a fragment of the learning theory known as *reinforcement theory*: we shall first expound a sketch of the natural or naive theory and shall then attempt to formalize it. The central hypothesis of reinforcement theory is: "Responses are reinforced [i.e. made to be preferred by the organism] by reward and weakened by punishment". In order to build a theory out of this assumption we shall use the general framework of utility theory (see Sec. 8.1). In terms of this theory the above hypothesis may be reformulated thus: "The responses with highest utility [subjective value] are preferred". So far, this is just a strong version of the first axiom of the general utility theory expounded in Sec. 8.1. In order to introduce it into psychology we must refine and specify the concepts of preference and utility and render them objectively measurable. To this end, the utility $u(o, r)$ of the outcome o associated with the response r can be assigned a number, at least in the ordinal sense. Similarly, the utility $u(-o, r)$ of the same response associated with any other possible outcome (not-o) will be assigned another number. Preference will be quantified with the help of the probability concept: we shall say that the outcome o is preferred to any other possible outcome $-o$, if the (objective) probability of the response r chosen to achieve o increases from trial to trial. Calling $P(r, t)$ the probability of the response r being chosen by the organism on the t-th trial, the elucidation of "preference" becomes: r is *preferred* to $-r$ if and only if $P(r, t+1) > P(r, t)$. By introducing the probability concept we automatically secure the help of the probability theory, which ensures conceptual definiteness and deductive power; at the same time the empirical testability of the theory is enhanced since now preference is tantamount to probability increase, and for test purposes probability will be interpreted as long-term relative frequency, which is measurable.

The central hypothesis of reinforcement theory — "The alternatives with highest utility are preferred" — can then be written

$$u(o, r) > u(-o, r) \leftrightarrow P(r, t+1) > P(r, t). \qquad [8.7]$$

But this is still a semiquantitative (precisely, an ordinal) hypothesis: it does not state what exactly the change of the value of P from trial to trial is, and accordingly not much can be inferred from it. If we wish to formulate a richer theory we shall have to make the former more definite. This can be achieved by assuming some specific function relating the probabilities of sucessive trials. The simplest possible assumption is that the ratio of the probabilities of a given response on any two successive trials be constant, i.e. a number independent of the trial number. Let us then try this simple (therefore suspect)

Axiom.

$$u(o, r) > u(-o, r) \leftrightarrow P(r, t+1) = k P(r, t), \qquad 0 < k \leqq 1. \qquad [8.8]$$

This recursion formula enables us to compute the change in probability over any sequence of trials. In fact, for the 2nd trial we have $P(r, 2) = k P(r, 1)$ and for the third $P(r, 3) = k P(r, 2)$. Introducing the first equation into the second we get $P(r, 3) = k^2 P(r, 1)$. By using the technique of mathematical induction we prove the general

Theorem 1.

$$P(r, t) = k^{t-1} P(r, 1). \qquad [8.9]$$

The relative and cumulative change due to successive reinforcements is easily obtained from this formula by subtracting $P(r, 1)$ and dividing by the same number (in accordance with the definition of "relative change"):

Theorem 2.

$$\frac{P(r, t) - P(r, 1)}{P(r, 1)} = k^{t-1} - 1. \qquad [8.10]$$

With the help of the probability theory many other theorems can be established: for example, for the probability of alternative responses $r \vee r'$ and joint responses $r \& r'$. This is another example of how an entire factual theory can be generated by laying a single specific (extra-mathematical) assumption in the nest of a mathematical calculus.

The fact that, in the above form, the reinforcement theory is too simple to be true need not concern us here. A more realistic theory will be obtained by complicating the central hypothesis — e.g., by choosing more complex relations — or by replacing it by an altogether different assumption. (For one thing, $P(r, t+1)$ depends not only on $P(r, t)$ but also on the result of t.) And a deeper theory will be obtained if the probabilities are somehow related to biological concepts about, say, neural mechanisms. The above sketch of a theory was only meant as raw material for an exercise in formalization. A possible formalization is as follows.

0. *Presuppositions.*

Generic: elementary logic and elementary probability theory (which in turn presupposes elementary set theory and arithmetic).

Specific: Bernoullian utility theory or at least its specific primitives, chiefly among them the generic concept of utility, to be specified below.

1. *Specific Primitives.*

A set R of *responses* r of an organism.
A set O of *outcomes* o associated with a given response r.
A sequence T of *trials* t.
The *utility* $u(o, r)$ of an outcome o associated with a response r.
The (objective) *probability* $P(r, t)$ of a response r on the t-th trial.

2. *Axiom.*

$$(o)_O (r)_R (t)_T [u(o, r) > u(-o, r) \leftrightarrow P(r, t+1) = k P(r, t)], \quad 0 < k \leqq 1. \qquad [8.11]$$

3. *Interpretation in Empirical Terms.*

I. The numerical value of the probability $P(r, t)$ is approximated by the long-term relative frequency with which a large population of similar organisms (e.g., freshmen) choose the response r on the t-th trial. [In other words, the probabilities concerned are objective.]

Actually we should add interpretation assumptions for the four remaining primitive symbols of the theory; but this is not necessary as we have smuggled them into when listing the primitives, as is customary.

The fragment of theory basis we have just formalized contains no dimensional variables, such as "length" or "time", save perhaps "utility". Whenever dimensional variables do enter a theory, it is often claimed that an additional set of statements must be included in the foundation of the theory: namely, all the conventions regarding scales, units and standards of measurement (which will concern us in Ch. 13). We may call these the *pragmatic conventions* involved in the empirical test of a theory. We shall not count them among the theory's foundations. The ground for this decision is threefold. First, pragmatic conventions are irrelevant to the theory's objective meaning, i.e. to its reference: they are relevant only to the empirical test of the theory — and then they may change according to the kind of test while the theory stays unchanged. Second, the meaning and the truth value of a theory are invariant under changes of scale. Third, units are as conventional as standards: so much so that law formulas are required to be invariant with respect to changes of units. Thus, e.g., NEWTON's laws of motion were not reformulated when the French revolution introduced the decimal system in place of the various medieval systems in use in NEWTON's time. Particular units and standards do not enter a theory until data are introduced, i.e. until contacts are established with experience. For example, if we refer GALILEI's general law "$v=gt$" to some place at sea level on the Equator, we specify 'g' writing either 9.80 m/sec² or 31.83 ft/sec². For these reasons we postpone the discussion of pragmatic conventions to Ch. 13.

Generally speaking, then, we may regard the *foundation of a formalized factual theory* as made up of the following layers: the presuppositions, the list of primitives, the axioms, and the interpretation assumptions in factual and/or empirical terms. To unearth and critically examine such foundations as thoroughly as possible is the aim of *foundational research*, which is an intersection of science and philosophy — hence neglected by both scientists and philosophers.

A factual theory is (ideally) completed with the addition of definitions, theorems, and pragmatic conventions. In summary, the composition of a completely formalized theory is this:

<table>
<tr><td rowspan="4">Foundation</td><td>Presuppositions (generic and specific)</td><td></td></tr>
<tr><td>Primitives (specific undefined symbols)</td><td rowspan="2">Primitive Frame</td></tr>
<tr><td>Axioms (unproved initial assumptions)</td></tr>
<tr><td>Interpretation assumptions</td><td></td></tr>
<tr><td rowspan="3">Superstructure</td><td>Definitions</td><td></td></tr>
<tr><td>Theorems</td><td></td></tr>
<tr><td>Pragmatic conventions (e.g., on units)</td><td></td></tr>
</table>

The formal or nonempirical components of a factual theory are to be found among the presuppositions (logical and mathematical theories) and in the definitions. The remaining components of a factual theory have a factual content or reference even if indirect and hypothetical. What the referent of the theory is will be partially indicated by the interpretation hypotheses that assign a meaning to its primitives and eventually to some of its defined terms as well. The reference class or universe of discourse will be prominent among the primitives: it shows what the theory is about.

If the referents of the theory are not mentioned when the primitives are interpreted, a grave danger of misinterpretation (or even of multiple conflicting interpretation) arises. Thus, if an atomic theory is about the complex system constituted by an atom and a measurement instrument, then some of the latter's parameters (classical variables) should occur explicitly amongst the theory's primitives, and consequently in the basic equations (axioms) of the theory. It will not do to interpret some theorems in terms of entities and properties not mentioned in the list of primitives or in the definitions. In fact such an interpretation, not warranted by the original interpretation rules and bearing on symbols that as a matter of fact do not occur in the initial assumptions, will be arbitrary. In particular, if the observer is not introduced in an atomic theory *ab initio*, i.e. if the parameters and the interaction of the observer with the supposedly observed system do not appear in the basic equations, then the theory cannot be said to refer to an atomic object coupled with the observer — let alone conjured up by the observer. Similarly, the theorems of Newtonian mechanics cannot be interpreted in terms of the Aristotelian concepts of potency and natural place, because these concepts occur neither as primitives nor as defined concepts in Newtonian mechanics. Again, the vitalist has no right to interpret any theorem of scientific biology in terms of a life breath or of an immaterial organizing force: if he wants to use these concepts he has to build a fresh theory containing them — and showing how they are related to scrutable properties. Let us, in short, abide by the *Rule:* Do not smuggle referents that do not occur in the basic assumptions. This rule is a specification of the requirement of semantic closure laid down in Sec. 7.2. It seems obvious, yet its violation is the source of a number of philosophical misinterpretations of science, among them the subjectivist interpretations of the uncertainty relations of quantum mechanics (see Sec. 7.4).

The primitives of a factual theory must have a *factual* meaning but need not have an *operational* meaning in the sense that they represent observable or measurable properties. Thus, e.g., the action in mechanics, the potentials in field theory, and the ψ-function in atomic theory, have no operational meaning: they cannot be measured even indirectly, as is on the other hand the case of mass, electric resistance, or refractive index; only certain derivative symbols have observable or rather measurable correlates. A conventionalist would say that such high-brow constructs are only convenient source-functions out of which observation predicates can be derived. A realist would retort that, in addition to being compact sources of lower-level predicates, such constructs do have a factual import even though they lack an empirical referent. He might base his claim on the place such constructs play in the theory: all of them are handled as properties *of* physical systems, not as symbols hanging in the air. Thus, given the action, the kind of physical system referred to by the theory remains determined; to prescribe (hypothesize) the field potentials amounts to give the kind and state of the field; and when the ψ-function of a system is written down the state of a specific system is characterized. All these constructs, then, are coordinated to real referents from the start and are treated as source properties of these referents. The fact that they are remote from experience does not prove that they are fictions but rather that they are inadequate, as they stand, for the description of experience.

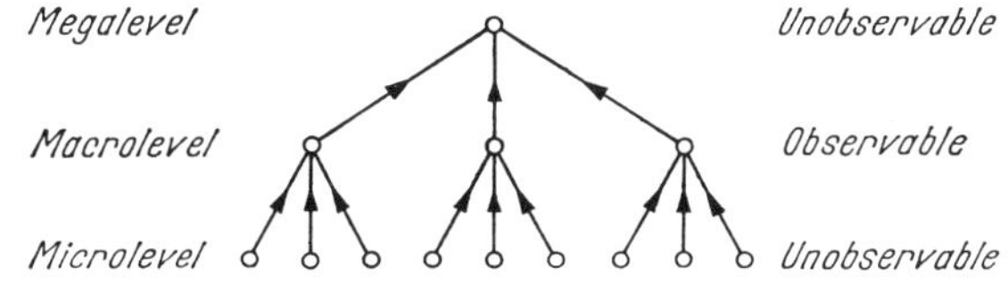

Fig. 8.9. Human experience: midway between electron and cosmos

The essential (unreplaceable) occurrence of constructs (nonobservational concepts) in the advanced theories of factual science is due to the structure of the world and man's place in it: to the fact that human experience has access to some middle-sized facts only — neither atomic nor cosmic. In fact, every phenomenon (experientiable fact) is either (i) the outcome of myriads of microscopic, unobservable facts, or (ii) one of a myriad of events that make up a fact in a giant unobservable system, such as society or the universe (see Fig. 8.9). In either case the objects themselves (the theories' referents), with all their properties, are beyond personal experience and must therefore be hypothesized. But in either case we control our constructs by observing the observable.

If observables do not commend themselves as primitives of objective, nonanthropocentric theories, how shall we select the unobservables in the bewildering wilderness of possible constructs? One criterion could be the choice of the best understood concepts. But understanding or intuitability is largely a personal and circumstantial matter; thus, to some the concept of set is crystal-clear whereas to others it is obscure. Moreover, if adopted as a selection criterion, no radically new concept would ever be introduced in science: in fact, the new is unfamiliar, and radical novelty can only be introduced at the basis, i.e. in the form of primitive concepts or primitive assumptions. Shall we then counsel the greatest freedom in the choice of primitives, as advocated by conventionalism? To some extent this is possible and desirable, particularly in formal science: any "excess" is preferable to the Philistine sticking to the familiar experiences. But complete freedom is not even found in the case of abstract games. Even here the choice of basic concepts is determined by the state of knowledge and by the prevailing outlook. Thus the concept of natural number, regarded as the primitive *par excellence* in 19th century mathematics, was later on debunked by the set concept.

In the factual sciences, at least, the choice of primitives seems to be guided by the following, largely tacit desiderata: (i) *maximum richness*, i.e. maximum number of possible relations with other concepts, so that rich systems of formulas can be built and a large number of terms can be defined in function of the primitives; (ii) *maximum degree of abstraction* (in the epistemological sense), i.e. greatest possible distance from immediate experience, in order that high level formulas with a high degree of generality can be constructed; (iii) *greatest depth*, in the sense of reference to fundamental or source properties that may account for the largest possible number of derivative properties. The latter may be related to, but is not identical with the logical requirement of concept richness: the depth of a concept is determined by the place of its referent in the objective level structure of reality. But we know very little about this. For example we do not know whether it is a feature of the world, rather than a historical accident, that leads to the consideration of time, position, and mass, as mutually independent primitives. We still do not understand why this should be so. Anyway, although we have no clear concept of concept depth, maximally deep concepts seem to be preferred as primitives. We are better at building theories than at finding out why we build them the way we do.

What can we expect from foundational research? In other words: what is the aim and outcome of theory reconstruction or formalization? The rational reconstruction of a theory rarely yields new important scientific results although it may give some side results and often does show basic mistakes in the natural theory concerned. Formalization has two chief aims, the one theoretical, the other metatheoretical. The main *theoretical aim* of formalization is to bring or enhance order and clarity, to remove redundancies, to point out gaps and, in general, to improve the logic of the theory. The *metatheoretical aim* of formalization is to facilitate the investigation of the theory itself, by exhibiting its logical structure, its pressuppositions, and its factual and empirical content. This will definitely facilitate the evaluation and the criticism as well as the eventual correction of the theory. Such is the practical value of formalization, quite aside from the pleasure the metascientist may derive from it.

Modern logic and mathematics have benefited enormously from formalization, but so far only a few isolated attempts have been made to formalize fragments of factual theories. Not even Newtonian mechanics has so far been axiomatized — let alone formalized — to every physicist's satisfaction. HILBERT'S sixth problem — the axiomatization of physical theories — remains unsolved. The great achievements of modern factual science have been won without formalization. From this we may infer that formalization is not necessary for scientific progress; that it is not sufficient either is shown by the fact that formalization is no substitute for either invention or test. But this does not prove that formalization is dispensable and that the whole of foundational research is a harmless pastime: child birth, too, can occur without the help of asepsis and conditioning, which does not prove the latter's worthlesness. Foundational research should speed up theoretical progress if only because once a theory has been formalized to some extent, but not before, can we guess what assumptions must be restricted, relaxed, altered, or abandoned altogether. Formalization does not replace conception and does not yield perfect theories, but it does yield the formulations most suitable for a critical examination, which is a prerequisite of scientific progress. In addition, it throws light on the nature of theory and can therefore be expected to expedite theory construction. Last, but not least, formalization is a constructive task philosophers can perform in cooperation with scientists: it should therefore contribute to instil a philosophical attitude in scientists and a scientific attitude in philosophers, and consequently to the cessation of the warfare between science and philosophy.

Problems

8.3.1. Formalize a simple logical or mathematical theory, such as the theory of identity. Thereupon check the interpretation (of, e.g., identity as empirical or phenomenal equality); that is, find out whether it fits the facts.

8.3.2. Geometry was traditionally defined as the science of figures, and almost every treatise on geometry has figures in it. Moreover, most proofs in elementary geometry employ figures. KANT seems to have inferred from this that figures are indispensable for geometrical reasoning. Not until the very end of the 19th century was it discovered that EUCLID had employed some propositions that fail to occur among his axioms and do not follow from them but are on the other hand built into certain figures employed by him. Does this establish that figures are essential to geometry and consequently that no part of geometry can be formalized? Or does it rather prove (i) that elementary geometry has concrete models and (ii) that the teaching of geometry is facilitated with visualizable models? *Alternate Problem:* Discuss the claim that formalization has an intimidating effect, that it paralyzes criticism, and that it would be better if scientific papers were written as pieces of autobiography.

See J. W. N. WATKINS, "Confession is Good for Ideas", in D. EDGE Ed., *Experiment* (London: BBC, 1964).

8.3.3. Report on one of the rare instances in which the presup positions of a factual theory are explicitly mentioned: namely, G. P MURDOCK's theory of kinship terminology, in his *Social Structure* (New York: MacMillan, 1949), pp. 130—138. *Alternate Problem:* At least three kinds of formulas have not been assigned a place in our treatment of formalized theories although they are relevant to theories: data, problems, and philosophical hypotheses. Why?

8.3.4. Physical theories do not specify the value of universal constants such as the velocity of light, the electron charge, or the action unit; yet whenever the theories are either applied or tested the precise numerical values of those constants are entered. Why? *Alternate Problem:* Speculate on the kind of primitive concepts you would choose if you were the size of an atom or the size of a galaxy (on the wild assumption that you would be entitled to speculate in either case).

8.3.5. Examine the following arguments pro and con the formal presentation of a scientific subject in high school and college teaching. (i) "The task of demonstrating a theorem is considerably facilitated if the axioms, definitions, and rules are explicitly enumerated and resorted to: the student is then given a starting point and a firm grasp". (ii) "A comprehension of the subject is gained only through establishing links with experience: the intuitive approach is therefore more efficient than the formal approach". Do these arguments bear on the same side of the question? Are they really incompatible? *Alternate Problem:* Would it be possible to explain either the boiling of water or the unrest in Latin America with theories containing no transobservational concepts?

8.3.6. Discuss the following evaluation of the opportunity of formalization. "Formalization is desirable only after a period of growth, test, and criticism of the given natural theory: in short, once we have made sure the theory is a good approximation. Premature formalization may be dangerous in suggesting that the theory is in fact finished and moreover backed by mathematics. This was, in fact, the effect of CARATHÉODORY's axiomatization of thermodynamics". *Alternate Problem:* Discuss the following theses. (i) Every system can be formalized entirely, either progressively or with one stroke. (ii) No system can be completely formalized.

8.3.7. Ever since EUCLID it has been thought that the best reconstruction of a scientific theory is its axiomatization. The last vigorous statement of this point of view was D. HILBERT's classical paper "Axiomatisches Denken", *Mathematische Annalen*, **78**, 405 (1918), recommended for its clarity. K. GÖDEL's incompleteness theorems have cast doubts on that view: we now know that set theory and arithmetic are not exhausted by axiom systems. From this some have concluded that axiomatization is not the best reconstruction and that nonaxiomatizable systems may be more faithful reconstructions of scientific theories. See G. KREISEL, "Models, Translations and Interpretations", in TH. SKOLEM, Ed., *Mathematical Interpretation of Formal Systems* (Amsterdam: North-Holland, 1955), p. 26. Discuss this view and relate it to the proposal made in the text, to axiomatize the core of factual theories (see Sec. 7.3). For GÖDEL's work see K. GÖDEL, *On Formally Undecidable Propositions of Principia Mathematica and Related Systems*, Introd. by R. B. BRAITHWAITE (Edinburgh and London: Oliver & Boyd, 1962).

8.3.8. Formalize a comparatively simple factual theory, such as ray optics, statics, classical thermodynamics, or the theory of direct current circuits. Hint: ask for a grant. *Alternate Problem:* Study the role of metanomological statements (see Sec. 6.7) in the construction and formalization of theories.

8.3.9. Energetism — the phenomenological approach popular in physics between 1890 and 1910, and favored by G. KIRCHHOFF, W. OSTWALD, P. DUHEM, E. MACH and others — held that "energy" ought to be taken as the basic (primitive) concept in every physical theory, displacing thereby position and time (favored by mechanists). Was there anything logically wrong with this idea? If so what was it, and what might the effect of a formalization of classical physical theories have been? *Alternate Problem:* Work out and illustrate the concept of arbitrary interpretation sketched in the text.

8.3.10. Discuss the proposal that the primitives of a factual theory be only observational concepts (hence operationally "definable"), which in turn secures the condition of observation statements for the axioms. See H. A. SIMON, "Definable Terms and Primitives in Axiom Systems", in L. HENKIN, P. SUPPES and A. TARSKI, Eds., *The Axiomatic Method* (Amsterdam: North-Holland, 1959). *Alternate Problem:* Is the choice of primitives influenced by philosophical ideas?

8.4. Reference and Evidence

A formal theory is a self-contained system in the sense that neither its meaning nor its truth, if any, depend on anything external to the system — save perhaps on its presuppositions, which are again formal.

A factual theory, on the other hand, is at least hoped to account, to a first approximation, for some aspect of reality. The extent to which the theory succeeds or fails to represent its object is probed in various ways, among which agreement with observation and experiment excels. In other words, a factual theory refers to a piece of reality and the adequacy of this reference (i.e. the theory's degree of truth) is tested by experience in conjunction with certain nonempirical criteria such as external consistency. A factual theory must then be considered both referentially and evidentially. Entire philosophies have been grounded on the failure to distinguish these two aspects.

Referentially (semantically) considered a factual theory points, in an immediate way, to a conceptual image or theoretical model which is in

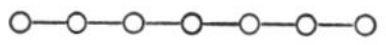

Fig. 8.10. One-dimensional model of a crystal. The lines represent elastic forces

turn supposed to refer to a real system. Thus, an ideal model of a crystal is a set of equally spaced particles attached to a massless elastic spring (see Fig. 8.10); and a rough model of the ecology of algae-eating rockfish is a two-component ecosystem (prey-predator) neglecting every other factor. *Evidentially* (methodologically) considered, the same theory points in an indirect way to a set of observed facts (the available evidence) and in a mediate way to a more comprehensive potential class of observable facts (see Fig. 8.11). The mediate or real referents of a theory need not be directly observable; they never are if the theory is deep enough.

For example, the *immediate referent* of particle dynamics is a set of mass-points; the mass-point is a theoretical concept that may in turn be regarded as a conceptual image of, say, a distant planet, which is in turn one of an unlimited number of *mediate referents* of particle dynamics. Reports on planet motion constitute part of the evidence of particle dynamics and any such future report is a further possible evidence. In this case the mediate referent of the theory happens to be observable; but a molecule, another possible referent of particle dynamics, is not observable. Again, the immediate referents of atomic theory are theoretical atoms, i.e. conceptual images (whether picturable or not) of real atoms, which are in turn the mediate referents of the theory. The evidential support of the atomic theory does not consist of reports on atomic behavior — there hardly are — but of reports on molar facts that, according to the theory, are the outcome of microscopic events. Similarly palaeontology refers, in an immediate way, to hypothetical reconstructions of extinct animals, and these models refer in turn to the real, but not directly known, extinct animals. And the evidential support of palaeontology does not consist of reports on extinct animals

— there are none — but of reports on their remains (mainly fossil bones) and tracks: these data are what palaeontological theories attempt to interpret (explain).

A theory's evidences are, in general, different from descriptions of the theory's referents. For one thing, the mediate referents of a theory are supposed to exist independently of the theory (which is not the case of the immediate referent or theoretical model); of course, this assumption

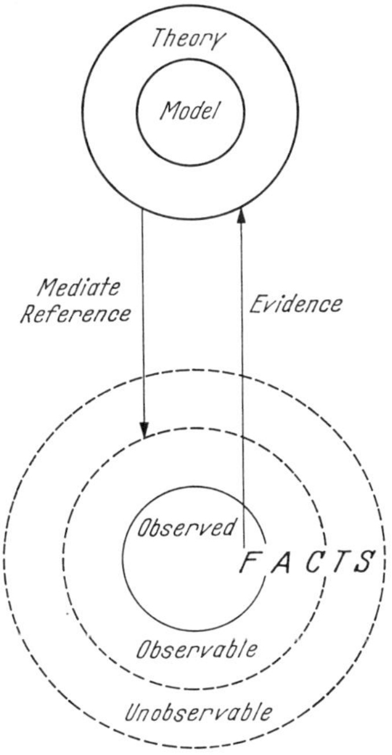

Fig. 8.11. Relation of theory to reality and experience. Experiential facts (observed and observable) are included in the wider set of facts. While the theory refers mediately to this wider set, only the subset of experiential facts gives it evidential support. The immediate reference of the theory to its own ideal model is not indicated in the diagram

may be false, as is so often the case with theories of "elementary" particles, but the assumption of independent existence is made if only for the sake of argument. On the other hand there is no evidence without theory: an observation report may or may not be relevant to a given theory, i.e. it may or may not be an evidence in favor or against a certain theory. The theory itself will determine this — usually with the help of other theories — because observation reports (data) must be interpreted in theoretical terms in order to become evidences. Evidences are not born but made: only a theory can transform a datum into an evidence — e.g., a report on a fossil bone into an evidence relevant to a theory concerning human phylogeny.

A way of summarizing the foregoing is as follows. Theoretical formulas are read in terms of interpretation assumptions of two kinds: referential and evidential. A factual theory is assigned a (core) *meaning* by a set of *referential interpretation assumptions (RIA)* that establish a correspondence between some of the theory's signs and its referents. Now, since a theory's immediate referent is an idealized model of a class of concrete referents (the theory's mediate referent, supposed to exist out there), there will be two kinds of referential interpretation assumptions: (i) type I hypotheses and conventions will establish correspondences between nonformal concepts and traits of the theory's model (which traits are further concepts rather than real things or properties); (ii) type II hypotheses will lay down correspondences between traits of the theoretical model and traits of the latter's supposedly real referent. For

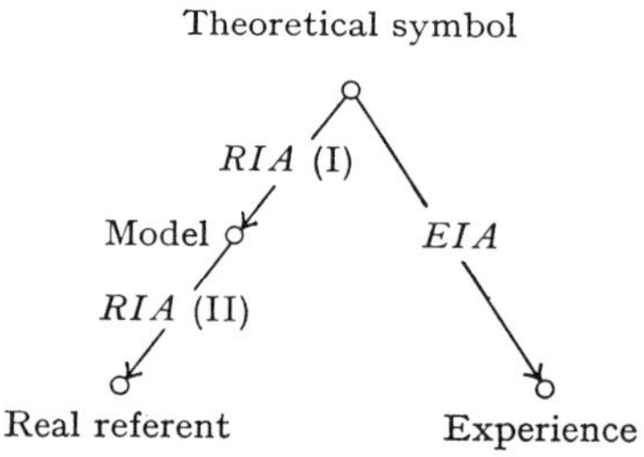

Fig. 8.12. Interpretation assumptions. *RIA* (I): type I referential interpretation rule; example: "'x' designates the position of a mass-point". *RIA* (II): type II referential interpretation assumption; example: "The mass-point idealizes a distant planet". *EIA* evidential interpretation assumption; example: "A non-twinkling light spot on the sky is a planet"

example, mechanics contains the concept of reference frame, to which a rigid body with certain kinematical properties is made to correspond. Since actually there are no rigid bodies we must content ourselves with semirigid systems as materializations (physical analogs) of the concept of frame of reference.

The referential interpretation assumptions, whether of the first or of the second kind, are necessary to ensure the theory's meaning but are not sufficient to secure its testability. In fact, the theory may well refer to alleged events that cannot be observed even indirectly. The testability of a theory is made possible (though not warranted) by a set of *evidential interpretation assumptions (EIA)* which link low level terms to observable entities and traits (such as visible semirigid bodies), as well as to observation and measurement devices and operations. Such evidential interpretation assumptions are not usually counted among the theory's formulas because they depend not only on the theory but also on (i) alternative theories and (ii) the available means of test. The differences between the various kinds of interpretation rules are represented in Fig. 8.12.

Reference and evidence must be distinguished and appraised if certain typical misunderstandings are to be avoided. In fact, neglect of reference and concentration on evidence may lead to subjectivism by way of a narrow empiricism; and forgetting the extremely indirect nature of most evidence as well as the idealizations involved in theoretical models leads to naive mirrorism, the view that theories are mirror images of reality. In either case the chief point of theorizing, namely the conceptual and testable representation of reality, is missed.

Many philosophers imagine that, by focusing on the test of theories, they can elude their objective reference; but testing presupposes objective reality, which the theory itself may not presuppose. In fact, save in the case of low-brow, i.e. underdeveloped theories, we know beforehand that our theories deal with idealizations (models), such as randomly communicating persons or freely moving bodies. Only the test and application of a theory presupposes the real existence of the latter's referent. If we say 'Every member of the set G satisfies approximately the van der Waals equation', we establish a relation between an entity which is real for the experimenter (namely, a member of the set G) and an object which is ideal both from the point of view of the experimenter and the theorist. (It is an ideal object but not altogether *fictitious:* it is not a free and aimless creation of the human mind but the outcome of a deliberate and imaginative effort to represent a real object.) If empirical tests did not presuppose objective reality they would not be empirical: computation or even contemplation might suffice.

Similarly, a theory of electrons does not presuppose the reality of the theoretical model it involves but rather the existence of certain largely unknown things, called electrons, which pose the problem of building better and better electron theories. There would be no point in building and improving electron theories if we did not believe that there are in nature certain entities which are at least roughly accounted for by such theories. Of course, such an existence hypothesis may be false; but this possibility does not eliminate the assumption and, moreover, the only way in which we could show that there are no electrons is by refuting every conceivable electron theory.

The statement that electron theories have a referent which is supposed to be real does not entail that electrons were first discovered and electron theories were then developed to sum up and systematize the evidence: electrons are not observed but inferred objects. This, in turn, does not entail that electrons were invented rather than discovered: it does entail that the electron discovery was an achievement of both experiment and hypothesis. A set of observable facts was first discovered; a hypothesis was then invented to explain the known facts; subsequently the further fact was discovered that the electron hypothesis met some of the tests; and this, in turn, prompted the expansion of the electron hypothesis into a theory capable of unifying the hitherto disconnected pieces of knowledge. A number of electron theories have since been proposed and all of them have the same mediate referent (the electron), of which they give different, increasingly refined and complex images (see Fig. 8.13).

The first electron theory was built with the help of classical mechanics and classical electromagnetic theory; although it involved a rather crude model of the electron (the small charged sphere), it performed marvels. It was eventually found unsatisfactory both as regards experimental evidence and in the light of a wider frame of physical theory: for instance, it did not account correctly for the splitting of spectral lines of the light emitted by atomic electrons embedded in a magnetic field

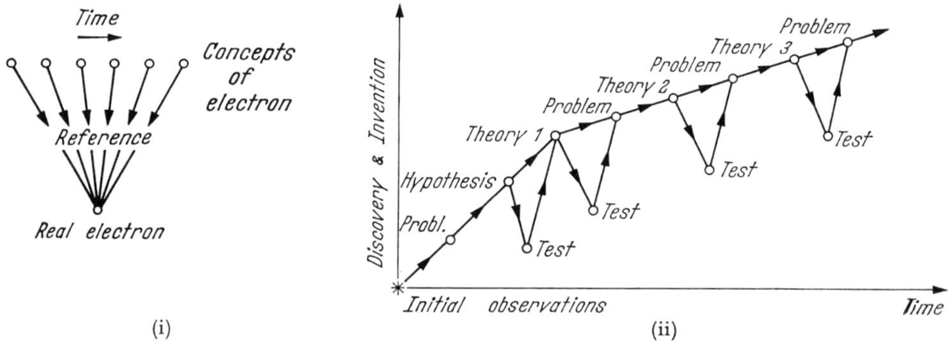

Fig. 8.13. (i) Electron theories build an immediate referent (theoretical model) each, but all these models have a single real referent. (ii) Building and mending successive electron theories

(empirical inadequacy), and it did not explain what could counteract the electrostatic repulsion of the electron's parts (inconsistency). At first the classical theory was mended to overcome these empirical and theoretical shortcomings, but with the (theoretically anticipated) discovery of the wave properties of "particles" new electron theories were eventually put forward, which encompassed some traits of the old theory. According to most of these new theories the electron has not only a mass and a charge and behaves somewhat like a wave packet, but has also a nonclassical intrinsic rotation (spin). Then, new theoretical developments (quantum electrodynamics) and new observations prompted the correction of the most accurate and complex of these theories (Dirac's). But there are indications that the last reforms will not last and even that there are no definitive reforms.

Three traits of this process should command our attention. First, observation by itself poses the problem of building theories and it tests them, but theories are not "abstracted" or "inferred" from observation reports — if only because these, the bits of evidence, may not refer to the same things as the theories do, just as a glance may suggest love without being love. Second, there is no reason to suppose that the process of adaptation of theory to fact will ever come to an end — unless, of course, mankind stops thinking: every theory about a given concrete object both omits some of its characteristics and adds ideas having no objective counterpart. In any case, there is never a one-to-one correspondence between a theory's components and the parts and aspects of the theory's mediate referent: the theory-fact correspondence is *global* rather than pointwise: the whole theory corresponds, in a more or less imperfect way, to the whole object and, moreover, the set of facts referred to by the theory need have no common part with the set of phenomena that test the theory. (The whole set of phenomena is just a proper subset of the class of all facts, but the set of experientiable facts relevant to a given theory need not be a subset of the class of facts referred to by the theory.) Third, usually the new theory does not abolish the former theories entirely but retains some of its components — concepts or even whole hypotheses — in a more or less modified form.

Theories are not "inferred" from data but data test theories through certain inferences that very often involve additional theories, as many as needed to bridge the gap between the theory's reference and its evidence. The problem of the empirical test of a factual theory is this: Given a set of theoretical predictions and a set of empirical data, infer ("conclude") whether the two sets match or not. Let us glance at this problem now; its detailed consideration will concern us in Ch. 15.

Theories — with the possible exception of psychological theories on observation — have no *observational content:* they deal with an ideal model which is often an allegoric or symbolic, seldom a literal or iconic, representation of the theory's mediate referent. Consequently theories cannot be directly compared with observational data. Before being able to compare a set of theoretical predictions with a set of empirical reports we must *render them comparable* by couching them in the same language. (This was done in Sec. 8.1 in relation with the theory demolition technique.) At first sight there are two ways of achieving this aim: either we translate the theoretical predictions into an observational language or, conversely, we translate the empirical findings into the language of the theory. Actually the two transformations are carried out at once: the rough empirical data are interpreted right away in theoretical terms and the theoretical predictions are interpreted in semiempirical terms so that both, data and predictions, meet on a level midway between the rarified atmosphere of constructs and the oppressing atmosphere of empirical concepts.

For example, when a pendulum is used (together with rational mechanics) to infer the strength of gravity or to record seismic waves, the measured length of the pendulum is regarded as an approximate value of the exact distance between the bob's center-of-mass (a theoretical concept) and the suspension "point" (another ideal object). We do not first state an ordinary language phrase couched in phenomenal terms and then translate it into a phrase containing theoretical terms but we start with sentences containing both theoretical and empirical terms, such as, e.g., 'The measured value of the distance between the suspension point and the bob's center of mass equals 100 cm to within 1 per cent'. In general, the empirical procedures of science, such as measurement, are never purely empirical but are mingled with fragments of theories, both factual and formal, and this is why they can be relevant to the theory that is being tested. On the other hand a report on what the experimenter is feeling while he is executing the experiment, i.e. a strictly phenomenalist report, will be irrelevant to the theory, hence useless for the purpose of testing it.

A nice illustration of the intertwining of empirical and theoretical elements in the actual practice of science is offered by seismology, the study of the elastic disturbances of Terra. The central substantive hypothesis of seismology is that earthquakes and minor disturbances generate sound waves that are propagated in the interior and over the surface of the planet. And the central strategic (methodological) hypothesis is that the analysis of these waves can yield information regarding the location and intensity of the disturbances' origin. On these hypotheses and with the help of other branches of physics seismographs are designed, built and operated in order to detect, amplify and record seismic waves. These records (seismograms) constitute the main observational material of seismology (see Fig. 8.14 (ii)). Now the chief task of seismology is not to pile up seismograms but to build conceptual pictures of the earth's unobserved interior, its various layers and faultings, as well as of the waves traveling in it. Seismograms are sources of information and tests of geophysical hypotheses — as long as they can be interpreted.

The interpretation of earthquake records (the seismological evidence) is made on the basis of (i) the theory of elasticity and, particularly, of the wave motion in elastic media; (ii) the theory of the seismograph (the detector, magnifier and recorder of waves), and (iii) a model of the earth's structure (core-mantle-surface layers) which, together with (i), provides a picture of the possible paths of waves, and which is partly established on seismological data. On this basis the seismograms are "read": they get transformed from data into evidence. Thus, the earliest of all waves, called P waves, are interpreted as longitudinal waves because the theory suggests (and laboratory measurements confirm) that longitudinal waves travel fastest. The next group of waves are interpreted as simply and doubly reflected P waves. After a while comes a group of S waves, which are interpreted as transverse waves because theory "allows" them to arrive only after the faster P waves. And so on (see Fig. 8.14 (ii) and (iii)). Of course, it is possible to check independently, in the laboratory rather than in the field, the theoretical results that in an elastic medium a perturbation will travel either as a longitudinal or as a transversal wave and that the former is faster than the latter. Also, the hypothesis that the S-P interval is an index of the epicenter-station distance can be

(i)

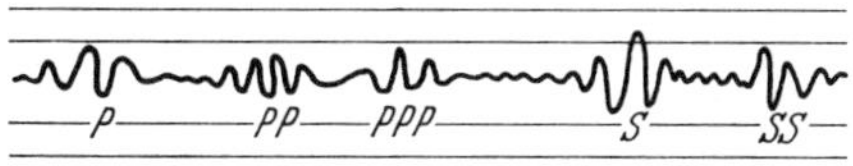

(ii)

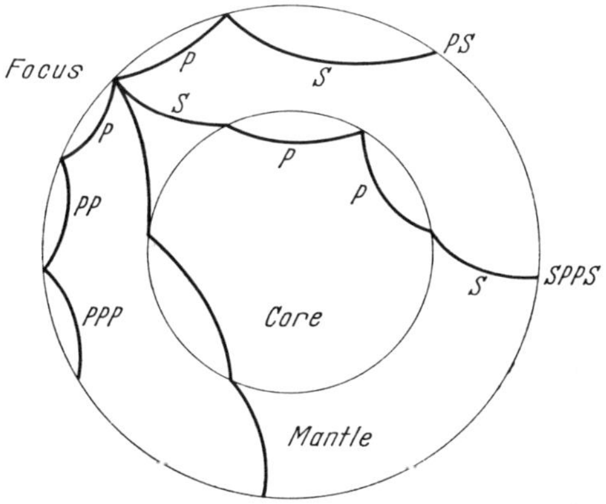

(iii)

Fig. 8.14. (i) The heart of a mechanical seismograph: a pendulum, indicator of seismic waves. (ii) The evidence: a seismogram. (iii) A model of the earth structure and of the sound rays in its interior; notice reflections at the earth's surface and refractions at the core boundary. All grossly oversimplified

tested as well. In conclusion, in order to "read" a seismogram so that it may become a set of data regarding an event (e.g., an earthquake) or an evidence relevant to a theory (e.g., about the inner structure of our planet), the seismologist employs elasticity theory and all the theories that may enter the design and interpretation of the seismograph. The nuclear physicist, the chemist and the physiologist proceed in a similar way.

Let us jump to general conclusions: (i) the data for testing theories are not *gathered* but rather *produced* with the help of the same and/or other theories; (ii) the theories involved in interpreting observed facts as evidences reconstruct (hypothetically) the *whole chain* between the reference-fact and the evidence-fact — between the cold and the sneeze, the feeling and the handshake. Compare this with the current account of theory testing as the comparison of raw (uninterpreted) data with theoretical predictions. This account falsely presupposes that data are searched for and gathered by ordinary experience, just like mushrooms: that experience is directly read (conceptualized) without the help of

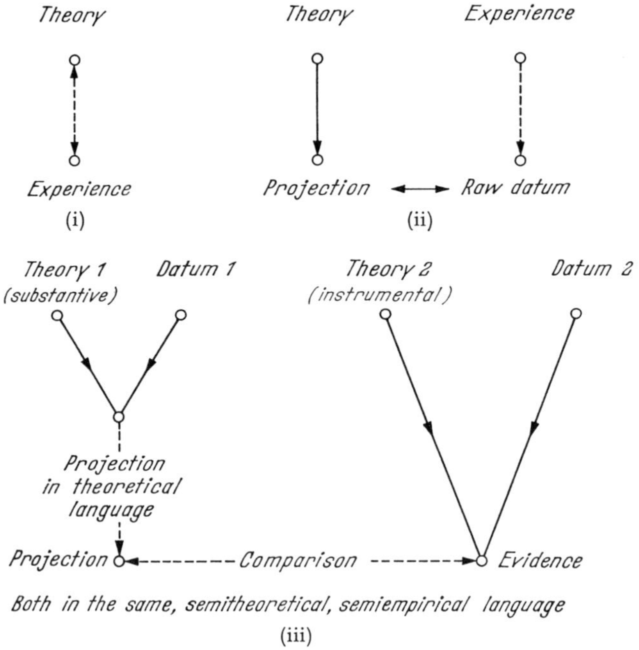

Fig. 8.15. Empirical vs. theoretical propositions. (i) *Naive view:* theory is inferred from data and yields data; (ii) *official view:* experience yields raw data which are confronted with theoretical projections; (iii) *actual procedure:* the evidence is not collected impartially but is produced with the help of theory, and theoretical projections are translated into a semiempirical language

theory. But there are no neutral experiences in science: every evidence is produced in the light of some theory (usually a set of fragments of theories) and is relevant to some theory or other.

In the best of cases a datum can be neutral with respect to the theory T1 that is being tested, but then it will somehow be related to a second theory T2 used in planning the observation, designing the instruments, or interpreting the results of observation. T2, a substantive theory in other contexts, plays now an *instrumental* role (see Fig. 8.15). For example, the visible track of an ionizing particle in a photographic plate will be an evidence relevant to the theory T1 concerning the behavior (e.g., the mode of decay) of ionizing particles of a given class on condition that the track be explained by a different theory T2 concerning the passage of ionizing particles through matter. And a document concerning slave labor in Antiquity becomes an evidence relevant to a sociological theory T1 concerning social relations in ancient society provided the document can be interpreted with the help of some economic theory T2 concerning the function of slave work in ancient society. (Incidentally, the enormous weight of interpretation in the mere selection and description of historical material has misled some historians into thinking that history cannot be objective, whereas the correct inference is that history, unlike chronicling, cannot be written without ideas concerning the mechanism of human action.) The selection and interpretation of empirical reports by means of theories is so "natural" in some cases — particularly in the sciences of man — that we are apt to miss it and take if for a direct "reading of facts". The intervention of theory is conspicuous when the facts under study are inaccessible to our experience — as in the cases of atomic and historical events — yet is never absent. If a datum is alien to theory then it cannot be relevant to theory.

In any case, suppose our factual theory has been contrasted to a set of empirical data. Whether it is now found to agree with them or not, unfavorable evidence is bound to be produced sooner of later. What then: shall we reject the theory altogether? Not if there is no truer theory in sight, one which predicted the new evidence. In the absence of such a better theory the first move is to try to patch up the theory: to save it by introducing suitable *ad hoc* hypotheses. We saw in Sec. 5.8 that there is nothing wrong with *ad hoc* hypotheses as long as they are testable in principle; moreover, on occasion an *ad hoc* hypothesis, far from being a mere temporary expedient to placate reality, turns out to be the germ of a new, richer theory. A dramatic illustration of this kind of event is provided by PLANCK's hypothesis of the quantum, originally aimed at removing the inconsistency between classical physics and certain empirical data concerning the spectrum of radiation in a hollow cavity. PLANCK assumed that the oscillators responsible for the emission and absorption of such a radiation had their energies restricted to an integral number of a basic amount of energy. This was sufficient to derive an empirically correct formula (PLANCK's blackbody formula). Many physicists protested against this *tour de force:* the quantization of energy was inconsistent with classical physics. The oddity of PLANCK's tactics was that he changed none of the postulates of physical theories, but patched up one of its theorems. Had he changed some of the postulates he would have produced a new unified theory; as it was he produced an internally inconsistent theory — but one which was empirically confirmed. Subsequent physicists, dissatisfied with this logical inconsistency, invented new theories which incorporated PLANCK'S hypothesis of quantization either in the axiom basis (BOHR'S old quantum theory) or in some of the theorems derived from new axioms (the new quantum theories). In any case, PLANCK'S *ad hoc* hypothesis turned out to be true and fruitful beyond his own expectations; and, by being absorbed by the new theories, it ceased to be *ad hoc*.

In general, when a theory fails, the first attempt is to make small modifications in it, either readjusting some of its initial assumptions or adding some *ad hoc* hypotheses, whether consistent with the former or not. It is only when such alterations fail to fit all the available evidence, or when they outrage the scientist's love of consistency, unity and beauty, that radically new theories are tried — until some of the newcomers hits the mark. If, in addition to solving the problems the previous theory allowed to state but did not solve correctly, the new theory reaches deeper than the older and has repercussions in adjacent theories, we say its introduction amounts to a *scientific revolution*. But the depth of theories deserves a separate section.

Problems

8.4.1. Theories are often described as *data systematizations*. Why then do not catalogues, tables, diagrams and systematic classifications qualify as theories? *Alternate Problem:* Discuss the following fragment from E. MACH'S *History and Root of the Principle of Conservation of Energy* (1872; Chicago: Open Court, 1911), p. 57: "In the investigation of nature, we always and alone have to do with the finding of the best and simplest rules for the derivation of phenomena from one another [...]. Perhaps one might think that rules for phenomena, which cannot be perceived in the phenomena themselves, can be discovered by means of the molecular theory. Only that is not so. In a complete theory, to all details of the phenomenon details of the hypothesis must correspond, and all rules for these hypothetical things must also be directly transferable to the phenomena. But then molecules are merely a valueless image". For the opposite view, according to which the aim of theorizing is the construction of a conceptual image of the external world, see L. BOLTZMANN, *Populäre Schriften* (Leipzig: Barth, 1905), Chs. 5 and 12.

8.4.2. Comment on the following statement by E. NAGEL, *Principles of the Theory of Probability*, in *International Encyclopedia of Unified Science* (Chicago: University of Chicago Press, 1939) I/6, p. 416: Theories "function primarily as means for effecting transitions from one set of statements to other sets, with the intent of controlling natural changes and of supplying predictions capable of being checked through manipulating directly experienceable subject matter. Accordingly, in their actual use in science, theories serve as *instruments* in specific contexts, and in this capacity they are to be characterized as good or bad, effective or ineffective, rather than as true or false or probable". If theories were just convenient tools or accounting systems, would it be possible to explain why some of them are useful and others not? And would they contain interpretation rules with no operational import? For instance, would mechanics make reference to point masses or rather to operations with levers and springs?

8.4.3. Decision theory allows us to compute the risk of *accepting a false hypothesis* (or of rejecting a true hypothesis). Does decision theory fit in the instrumentalist philosophy of science, which regards hypotheses and theories as acceptable (or unacceptable) rather than as true (or false)? *Alternate Problem:* Work out the distinction between the substantive and the instrumental role played by theories.

8.4.4. H. REICHENBACH and other philosophers have maintained that a scientist trying to prove something about anything (e.g., kangaroos) need not assume its objective existence. Discuss this thesis. *Alternate Problem:* Examine the kind of evidence relevant to any atomic theory.

8.4.5. Take any theory and disclose its referential and its evidential interpretation rules. *Alternate Problem:* Do the problems of the reference and the test of a theory arise if the theory is regarded as a language?

8.4.6. Comment on the claim that the heliostatic system of COPERNICUS and PTOLEMY'S various geostatic theories were equivalent, or just two "different modes of speech". See, e.g., J. PETZOLD, *Das Weltproblem von positivistischen Standpunkte aus* (Leipzig: Teubner, 1906), p. 37, and H. REICHENBACH, *The Rise of Scientific Philosophy* (Berkeley and Los Angeles: University of California Press, 1951), p. 107.

8.4.7. Examine the following views concerning the ontological import of scientific theories. (i) *Phenomenalism:* theories are summaries, or at most idealizations of experience. (ii) *Instrumentalism* (conventionalism and pragmatism): theories are predictive devices, hence tools for action. (iii) *Naive realism:* theories are direct representations of reality. (iv) *Critical realism:* theories are symbolic and partial representations of systems of traits of pieces of reality. See E. NAGEL, *The Structure of Science* (New York: Harcourt, Brace & World, 1961), Ch. 6, and K. R. POPPER, *Conjectures and Refutations* (New York: Basic Books, 1963), Ch. 3.

8.4.8. The astrophysicist and philosopher A. S. EDDINGTON claimed that he could derive empirical data with the sole help of general principles. Is this logically possible, i.e. is it possible to deduce singulars from generals alone? If not, what becomes of *apriorism* in respect of factual science? And what is the moral concerning the relevance of logic to the philosophy of science?

8.4.9. Analyze the argument of H. POINCARÉ, in *Science and Hypothesis*, in favor of *conventionalism*. Here it goes in a nutshell: (i) Physical theories are mathematically statable. (ii) The axioms of any mathematical theory are laid down by stipulation; any choice among them is a matter of handiness. (iii) Therefore the axioms of any physical theory can be convenient, but not more true to fact than those of any other mathematical theory. Recall Sec. 7.4. *Alternate Problem:* Are interpretation rules sufficient to determine the meaning of a factual theory? See M. BUNGE, *Intuition and Science* (Englewood Cliffs, N. J.: Prentice-Hall, 1962), pp. 72ff. and *Foundations of Physics* (Berlin-Heidelberg-New York: Springer, 1967), pp. 66ff.

8.4.10. Fossil remains of new species — e.g., transition groups — appear suddenly. Such gaps in the fossil record have been regarded as an evidence against Darwinism and in favor of alternative theories, such as that of macroevolution. Examine how Darwinists explain the scarcity or even the total lack of *evidence* for certain facts the theory *refers* to. Is such an explanation consistent with the view that theories are data packages? *Alternate Problem:* Study the following *paradox of empirical evidence*. What gives a theory some empirical justification is a set of data. But data are evidences in favor of a theory just in case they can somehow be inserted in it, i.e. if they are to some extent acceptable to the theory. Thus, e.g., the acceleration of a particle in a void is evidence in favor of field theories but is irrelevant to alternative theories. Empirical confirmation would therefore seem to be circular. Is this so? In case it is, can we jump outside the circle?

8.5. Depth

The wave theories of light are deeper than ray optics and the reflex theories of behavior are deeper than the stimulus-response theories. In effect, the ray theory does not investigate the nature of the light ray and it does not inquire into the mechanism of light reflection and refraction; the wave theories, on the other hand, explain among other things how rays are built up (namely, by wave interference). This they achieve by hypothesizing a "mechanism" of propagation and interference, which they describe with the help of high-level constructs such as "amplitude", "wavelength" and "phase". Similarly the reflex theories of behavior attempt to explain precisely those associations which the purely behaviorist account is contented with establishing. This they achieve by introducing nonobservational concepts (hypothetical constructs) such as "reflex arc", "excitation", and "inhibition".

The scientific theories in any field can be ordered in respect of depth. For example, we may build the following serial orders by interpreting '>' as the relation "deeper than":

Quantum optics > Electromagnetic optics > Mechanical wave optics > Ray optics.

Relativistic quantum mechanics > Nonrelativistic quantum mechanics > Classical dynamics > Classical kinematics.

Nonequilibrium statistical mechanics > Equilibrium statistical mechanics > Classical thermodynamics > Elementary heat theory.

Theory of reaction mechanisms > Classical chemical kinetics.

Synthetic evolution theory > DARWIN's theory of evolution > Pre Darwinian evolution theories.

The depth of a theory depends on the depth of the *problems* it sets out to solve: the deeper the problems it succeeds in solving the deeper the theory. Take, for example, the problem system of bird migration. One can ask what are the actual migration routes of the various species or one may take a further step and ask how do birds choose the right direction to reach their goal and how do they keep that direction during flight. The first question calls for an empirical description but the latter requires a theory and indeed a deep theory, one dealing with the possible "mechanism" of choice and maintenance of flight direction. A number of "mechanisms" have been proposed, such as guiding by the earth's magnetic field, by the Coriolis force associated with the earth's spinning, and finally by sun light. All such theories of bird navigation, no matter whether they are true or false, are deep theories going beyond appearances and analyzing the action of certain physical stimuli on the bird's sense organs: they do not merely state certain behavior patterns (e.g., the formation of the migration habit) but attempt to disclose its "mechanism". Even deeper theories are needed in this field, theories taking into account the innate component of bird orientation and explaining it at the level of cellular biology (genetic information coding).

The deeper theories are the more *specific*, hence the more informative. As a consequence of this greater definiteness or commitment they are also better testable; and, being better testable, they are more apt both to acquire and to lose a good empirical foundation: that is, the set of data relevant to them is more varied and therefore the chance that unfavorable evidence emerges is enhanced. Think of the classical electromagnetic theory of nearby action (FARADAY-MAXWELL's theory); among other things it predicted the existence of electromagnetic waves — a feat that was beyond the capability of the theories of action at a distance, which were simply unconcerned with relations among neighboring points in a continuum, which alone can give rise to a traveling ripple. Or think of DARWIN's evolution theory, which did not just assert that species evolve — a hypothesis entertained by many before DARWIN — but proposed a definite mechanism (spontaneous variation plus natural selection), the existence of which could be tested, e.g. by artificial selection. In turn the contemporary or synthetic theory of evolution, based on genetics, is deeper than DARWIN's because it explains variation — as an outcome of genic and chromosomic mutation — whereas DARWIN took variations for granted. LAMARCK's environmentalist theory, too, contained mechanisms supposedly responsible for evolution, such as the atrophy of disused organs, the inheritance of acquired characters, and the inherent tendency to perfectioning. But these hypotheses fail to explain how changes in the individual organism during its lifetime are imprinted in its reproductive cells so that these can transmit the information to the offspring. Hence quite apart from the truth or even the testability of LAMARCK's theory, DARWIN's was deeper.

Theory depth can be characterized by the possession of three attributes: the occurrence of *high-level constructs*, the presence of a "*mechanism*", and a *high explanatory power*. The three properties are intimately linked: it is only by introducing high-brow (transempirical) concepts that unobservable "mechanisms" can be hypothesized, and only what is hypothesized to occur in the depths can explain what is observed at the surface. We formulate this idea in the following *Definition:* A theory $T1$ is *deeper* than a theory $T2$ if and only if (i) $T1$ includes higher-level constructs (unobservables) than $T2$ does (epistemological aspect); (ii) these constructs occur in hypothetical "mechanisms" underlying the facts referred to by $T2$ (ontological aspect); and (iii) $T1$ logically explains $T2$, i.e. $T1 \vdash T2$ (logical aspect).

Since the less deep theories are the closer to phenomena, they are often called *phenomenological*. Those theories which, on the other hand, hypothesize definite "mechanisms", whether observable or not, may be called *representational*. If such mechanisms are mechanical strictly speaking, then the theory will be *mechanistic*. But these are only extremes: in between there lie a number of *semi*phenomenological — or, alternatively, semirepresentational — theories.

The adjective 'phenomenological' is misleading in this context because it conveys the idea that the less deep theories are strictly descriptive of phenomena and have no explanatory power. But no theory proper is purely descriptive: if it lacks propositions from which descriptive propositions can be deduced it is not a theory; and the strong propositions from which the descriptions derive have not, as a rule, the same referents as the latter: whereas the former may refer to unobserved facts the latter refer to observable facts, whence the derivation may require additional theories (see Sec. 8.4). Phenomenological theories do not satisfy then the demand of phenomenalist philosophy, according to which transempirical concepts are to be avoided because they are meaningless. And phenomenological theories are even farther from satisfying the requisite of employing sense data alone: even psychological theories, if scientific, are built with objective elements rather than with what appears to the subject (the "phenomena" of philosophers); they may refer to subjective elements — if they are psychological theories — but only as long as these are objectifiable for the researcher, who will try to account for them in a nonphenomenalist way. For these and other reasons the name *black box theory* seems preferable to 'phenomenological theory'.

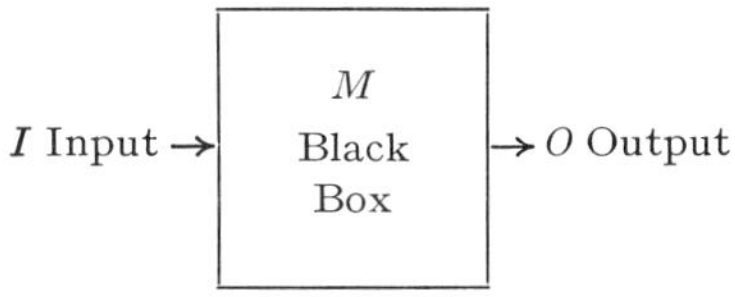

Fig. 8.16. A black box theory regards its referent as a block devoid of structure. Its over-all behavior is accounted for by peripheral variables I and O eventually linked by auxiliary (intervening) variables M

A black box theory treats its object or subject matter as if it were a system devoid of internal structure: it focuses on the system's behavior and handles the system as a single unit (see Sec. 5.4). A black box theory will accordingly account for over-all behavior in terms of relations among global variables such as net causes (inputs) and net effects (outputs); these will be mediated by referent-less intervening variables (see Fig. 8.16). The black box resembles the annual report that the company manager submits to the shareholders: both the theory and the report speak about incomes and outcomes, net gains and net losses, and even of over-all trends; but they do not explain the *processes* at work (in the company or in the box).

A black box theory can symbolically be summed up in the following equation relating the input I and the output O:

$$O = MI, \qquad [8.12]$$

where M (a variable, an operator, or a function) *mediates* or *intervenes* between the external variables I and O; if it happens to be a function, M will map the set of input values into the set of output values, i.e. $M: I \to O$. The mediating symbol 'M' epitomizes the properties of the box but, in the black box theory, M is not derived on the basis of such properties: M is a purely syntactical link between a column of values for the input I and a column of values for the output O (or between several columns if I and O are sets of variables). It is only in the context of representational theories that M becomes a complex symbol or a derivative variable concerning the constitution and structure of the box — in short, it represents the "*mechanism*" mediating between the input and the output. In other words, M is an uninterpreted variable in the black box theory, and it becomes an interpreted variable (a hypothetical construct) in the corresponding translucid (representational)

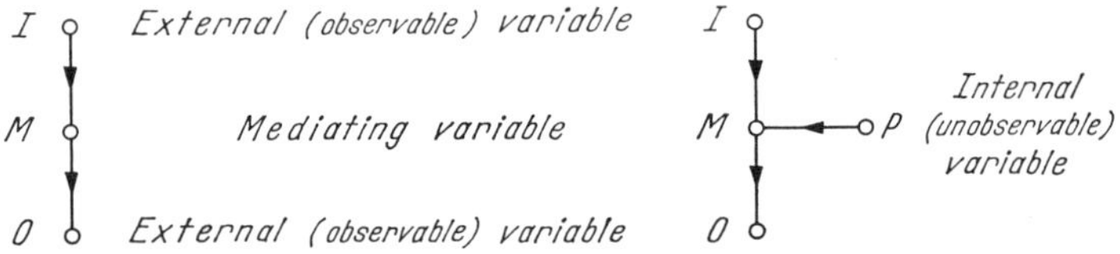

Fig. 8.17. The roles of mediating symbols M in (i) black box theories and (ii) translucid box theories. In (i) M is a syntactical link between observables; in (ii) M derives from internal variables P

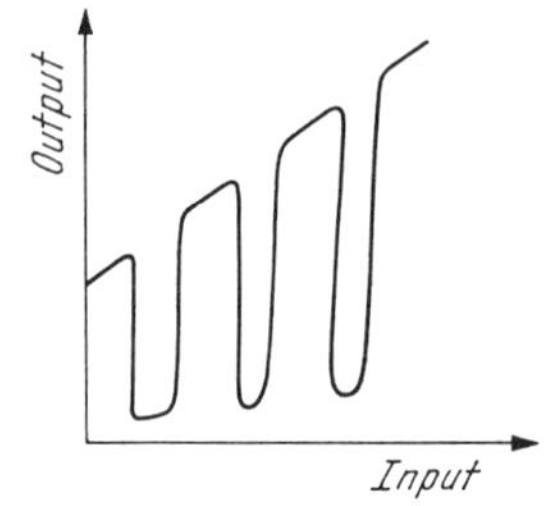

Fig. 8.18. An input-output curve crying for an explanation of the absorption minima

theory that eventually absorbs the former. Needless to say, M need not represent a *mechanical* mechanism or even a *picturable* one. And, although M is not empirically or operationally meaningful (it represents no property that can be manipulated), it is assigned a referent in the translucid box theory: namely, a set of internal variables that characterize the box' inners. In short, whereas in black box theories the mediating symbol M is factually meaningless, it acquires a factual meaning (not an operational one, however) in translucid box theories (see Fig. 8.17). The status of a given variable as intervening variable or as hypothetical construct is therefore relative to the theory in which it occurs.

An input-output relation commits us to no specific mechanism but poses the problem of hypothesizing the possible mechanisms responsible for it. (If anyone refuses to work on such a problem he is entitled to; what he should not do is to discourage others in the name of some philosophical tenet.) Take the generic case depicted in Fig. 8.18, in which the gross output increases with the input except in certain intervals in which part of the input is absorbed by the system (atom, person, economy, or whatever it is). We may decide to stop at this point or we may take it as the starter of a new research cycle, by asking what the absorption mechanism is: our decision will strongly depend on our willingness to work and on our philosophical outlook. If we are not behaviorists we shall proceed to analyze the system (conceptually or empirically) into components and ask what their mutual relations are: in this way possible mechanisms will begin to emerge below the surface. A celebrated case of this kind was the explanation of the FRANCK-HERTZ experiment concerning the shelling of a gas by a beam of subatomic particles: the absorption minima recorded by the experimenters were explained by BOHR's atomic theory as energy transfers from the bombarding particles to the gas' atoms. A more recent illustration of this kind of analysis of global data is the input-output analysis of a country's economy (W. LEONTIEFF), in which the transactions among the different sectors of the economy are inquired into. In either case an analysis of a black box into a system of smaller black boxes in performed and the analysis is accepted as partially true not just if it is consistent with the gross input-output relation but if it is compatible with some of the known relevant laws and if it can help predict otherwise unpredictable effects.

Black box theories are dominant in the data-fitting stage of theory construction, i.e. in the period in which data are *systematized* rather than *interpreted*. The completion of the picture will require the interpretation of the mediating symbol 'M' as a "mechanism", which will in turn force the introduction of concepts and hypotheses regarding what is not observed, i.e. the workings of the box' inners. Take for example osmosis, that conspicuous process in living matter. The net effect of osmosis is the passage of liquid (e.g., water) through a membrane to a region of higher concentration. This net effect can be accounted for by a black box theory involving only the concepts of concentration, pressure, and temperature. A deeper theory will involve, in addition, a model for the diffusion (unobservable process) of molecules (unobservable entities) through the (unobservable) holes of the membrane; and also, eventually, a model for the (unobservable) chemical reactions that in some cases occur at the interface. Such a deeper theory is needed to explain the selective osmosis occurring through living membranes — a process we must understand if we wish to improve health.

The greater depth of representational theories explains why, with few exceptions, the general trend in the history of science has been the supplementation of black boxes with translucid boxes. Such a supplementation of superficial by deep theories in gradual: what we call a translucid box is only a bunch of smaller black boxes which will have to be analyzed in turn in terms of still smaller black boxes. Whether there is a limit to this process, a limit set by the nature of things (unanalyzable material atoms) or not, is opinionable matter. At the present time both hypotheses, the one postulating the existence of ultimate (not further analyzable) atoms and the one of unlimited complexity, seem equally tenable. The differences between the two hypotheses are, for the time being, purely pragmatic: the hypothesis of the unlimited complexity is both less intuitive and more fruitful than the hypothesis of ultimate atomicity: the former suggests looking for complexity where none is yet apparent — it generates fruitful problems.

Despite the greater richness of representational theories there have been and still are stanch partisans of phenomenological theories. This is due not only to a definite philosophical bias, namely the sticking to empirical data and the distrust of theory. Black boxes have virtues of their own that explain why they are not *replaced* by translucid box theories but rather *supplemented* by them. In fact, black boxes have, among others, the following properties: they are

1. *highly general,* in the sense that they are consistent with an unlimited number of specific mechanisms and are applicable to systems of different kinds (recall information theory);

2. *global* or wholistic: they pay no attention to either details or inners but focus on over-all traits and trends (recall thermodynamics);

3. *epistemologically simple,* i.e. economical as regards the use of transcendent or nonobservational concepts (recall *S-R* psychological theories);

4. *accurate,* since by adjusting and readjusting the values of the parameters (which are not all of them assigned definite properties as referents) they may cover more data than representational theories, in which the parameters cannot be arbitrarily adjusted because they are supposed to represent objective properties (recall the kinetic theories of nuclear physics);

5. *safe:* by keeping silent about mechanisms they take fewer risks (recall electric circuit theory).

These advantages explain why we still employ, whenever possible, ray optics alongside wave optics, classical thermodynamics alongside statistical mechanics, and so on — particularly in applied science. But from another point of view the above are disadvantages. In fact, generality in the sense of lack of specificity is just a sign that the kind of system has been neglected; the global or nonlocal character is paid for by ignorance concerning the inner structure; simplicity in the sense of lack of hypothetical constructs is a mark of conceptual and ontological shallowness; and the safety resulting from the failure to make detailed commitments is not more valuable than the wisdom of the taciturn. The above properties explain why black box theories are favored by technologists: for the sake of action a versatile, global, simple-minded, adjustable, and safe theory is preferable to a deeper but, by the same token, more complex and less dependable theory. The preference of pragmatists for that kind of theory has the same explanation since for the pragmatist there is no pure science, every piece of knowledge being nothing but a tool for action.

In addition to the above mentioned double-edged properties, black box theories have the following definite disadvantages over translucid box theories:

1. *low content:* they are less complete and definite than the corresponding representational theories and consequently they have both a

2. *low testability* — since they make few and prudent wagers — and a

3. *low heuristic power:* by not perforating deep enough the layer of appearances they cannot guide research in the exploration of deeper layers.

Methodologically, black box theories are valuable as *testers* of the corresponding representational theories (which may, however, disclose slight or even gross inaccuracies in them). Pragmatically they are valuable because they are usually handy, whereas representational theories, for being more complex, are more difficult to handle. But the progress of knowledge consists, to a large extent, in burrowing inside black boxes, i.e. in transcending the phenomenological approach, which is usually (not always) best suited for both a preliminary stage in theorizing and for technology. The preference for translucid boxes over black boxes in the field of pure science does not entail the rejection of black box theories. What should be criticized, because it is an obscurantist attitude, is *black-boxism*, i.e. the philosophy that commands the abandonment of translucid boxes (representational theories) and would accordingly have us destroy the most advanced sections of contemporary science. This philosophy is favored by practically oriented people, who conceive of scientific theories as just tools for establishing economical relations among data. This kind of pragmatism, which is interested in scientific theories as means rather than as ends, is short-sighted and self-defeating, because the deeper a theory the more useful it can be in the long run. Representational theories are ultimately the more rewarding, even practically — though in an indirect way — because they help to explain *how things work* — and this knowledge may be needed to improve our mastering over things.

The above discussion suggests that theory construction is always dominated by some *approach* or other. An approach may be characterized as a broad, nonspecific way of considering a subject. It involves both a point of view and an aim. ("Approaching", is then, a fourth degree relation: w approaches x with the aim y and with the viewpoint z.) The study of one and the same set of facts may be approached in a number of ways. Thus, e.g., a spectacular fall in the stock market may be approached by the economist with the aim of reading the state of the country's economic ailments on the view that financial events reflect economic states; the social psychologist may approach the same event in order to gather data relevant to the view that market fluctuations are the outcome of subjective estimates, rumor diffusion, and panic contagion; and the historian may be interested in studying the event as a consequence of, say, a long string of unsound investments and as the origin of governmental control devices. Each of the three specialists will focus on a particular aspect of the same fact because every one of them has a different target and a different viewpoint. The different approaches may be mutually consistent and, moreover, mutually supplementing. If all of them are taken up in studying a given set of facts an *interdisciplinary approach* is said to be adopted. (But there is no interscience — e.g., no space science.)

As with the study of facts by means of existing theories so with the construction of theories to explain facts: this task is always done with a more or less definite approach, and this approach includes philosophical tenets: they hide both in the selection of the target and in the viewpoint. Thus, e.g., the black box approach consists in regarding systems "from the outside", without burrowing in their insides. In this case the aim may be to obtain an inexpensive (or even a cheap) systematization with the properties mentioned a while ago (e.g., wide coverage, simplicity, and safety). And the underlying philosophy, if operant at all, may be either conventionalism or a narrow empiricism according to which theories are built with bits of experience. To state explicitly and, whenever possible, to justify both the aim and the underlying philosophy, is good strategy and honest practice. When the kind of approach is explicitly stated the ensuing theory may be improved on by — among other possible moves — shifting the approach.

The variety of approaches goes to explain the frequent variety of theories concerning one and the same set of data. This variety contrasts with the popular view that science — unlike the incoherent chorus of philosophies — is monolithic. Differences of opinion among scientists are much more common than is usually conceded — only, they are fruitful. Discrepancies in the scientific community are a source of progress: this is the whole difference between scientific and political disputes. They are a source of progress because they consist of different *ideas*, very often mutually incompatible; because they are not normally tied to vested interests and last, but not least, because there are acknowledged means for settling disputes at least in the long term. Pseudosciences can afford to be monolithic — though not in the sense that they are formally coherent and semantically unitary; by contrast, the scientific spirit thrives on discrepancies, both among ideas and among ideas and data. The uniformization of scientific opinion over every issue would be as lethal to research as the fragmentation into unreconcilable schools. Both free dissent and spontaneous agreement are necessary for the successful pursuit of truth. Dissent, because it is a symptom that the problem at hand may not have been solved, even provisionally; agreement, because it is a green light to the search for new problems. The new problems that a (temporarily accepted) theory can give rise to are problems of theory application and of theory test.

This concludes our treatment of scientific systematization. The applications and tests of scientific systems will be examined in Volume II: *The Search for Truth.*

Problems

8.5.1. Do phenomenological (black box) theories contain exclusively observational concepts? And could you mention one theory without theoretical concepts? *Alternate Problem:* Towards the end of the last century the elastic theory of light, with its odd mechanical aether, was discredited and its downfall dragged first the whole class of mechanical models, then all claims to a true depiction of reality. This trend, successfully exploited by conventionalism, was accentuated by the constitution of thermodynamics as a phenomenological (mechanism-independent) theory and its successful applications to chemistry, as well as by MAXWELL's gradual abandonment of FARADAY's elastic tubes of electric forces. Thereupon the kinetic theory of matter, the atomic models of molecules (e.g., KEKULÉ's benzene ring) and other theories and hypotheses came to be regarded as just *mechanical analogies* with no factual reference: they were to be interpreted as metaphors rather than as literal depictions of reality: the most that could be said is that things happened *as if* there were atoms — but, of course, it was bad taste to take such fictions for realities. A few years later the reality of atoms was shown almost conclusively. More exactly, exact measurements were made that had been predicted by atomistic theories alone (such as EINSTEIN's and SMOLUCHOWSKI's theories of Brownian movement) and which were unthinkable on alternative systems. Whereupon more and more detailed theories concerning the structure of atoms, then of nuclei, and finally of the component particles themselves, were put forward. Draw some conclusion and try to explain why black-boxism, far from having been defeated by the above developments, is so much alive in contemporary microphysics.

8.5.2. COPERNICUS' revolutionary treatise *De revolutionibus orbium coelestium* (1543) contained an anonymous preface — since attributed to

A. Osiander — which intended to make the work palatable to the prevalent tenets. Ptolemy's methodological views — which were conventionalist rather than realist — were expounded in this preface, which stated that the aim of astronomy is to account for appearances by means of hypotheses the sole function of which should be to enable the astronomer the exact computation of the apparent motions rather than the representation of reality. For a story of this phenomenological approach to astronomy see P. Duhem, *ΣΩΖΕΙΝ ΤΑ ΦΑΙΝΟΜΕΝΑ, Essai sur la notion de théorie physique* (Paris: Hermann, 1908), especially pp. 77ff., and R. M. Blake, "Theory of Hypothesis among Renaissance Astronomers", in E. H. Madden, Ed., *Theories of Scientific Method* (Seattle: University of Washington Press, 1960). Was the conventionalist view consistent with the heliostatic theory propounded by Copernicus? If so, what was the point of remodeling astronomy if at that time the two views were equally confirmed by observation?

8.5.3. Report on any of the following works: (i) L. Boltzmann, *Populäre Schriften* (Leipzig: Barth, 1905), Chs. 9 to 11; (ii) R. Dugas, *La théorie physique au sens de Boltzmann* (Neuchatel: Ed. du Griffon 1959), Chs. iv—xii. (iii) A D'Abro, *The Decline of Mechanism* (New York: Van Nostrand, 1939), Ch. XI, on phenomenological theories. (iv) K. R. Popper, *The Open Society and its Enemies*, 4th ed. (London: Routledge & Kegan Paul, 1962), I, Ch. 3, Sec. vi on methodological nominalism vs. methodological essentialism. (v) J. M. Blatt and V. F. Weisskopf, *Theoretical Nuclear Physics* (New York: Wiley, 1952), pp. 313 and 517ff., on the channel picture of nuclear reactions. (vi) K.W. Spence, "Types of Constructs in Psychology", in M. H. Marx, Ed., *Psychological Theory* (New York: MacMillan, 1951). (vii) E. C. Tolman, "The Intervening Variable", *ibid.* (viii) K. McCorquodale and P. E. Meehl, "Operational Validity of Intervening Constructs", *ibid.* (ix) M. H. Marx, "Hypothesis and Construct", *ibid.* (x) E. R. Hilgard, "Intervening Variables, Hypothetical Constructs, Parameters, and Constants", *American Journal of Psychology*, **71**, 238 (1958). (xi) W. Weaver, "The Imperfections of Science", *Proceedings of the American Philosophical Society*, **104**, 419 (1960). (xii) M. Bunge, "A General Black Box Theory", *Philosophy of Science*, **30**, 346 (1963). (xiii) M. Bunge, "Phenomenological Theories", in *The Critical Approach*, Ed. by M. Bunge in honor of Karl Popper (New York: Free Press, 1964).

8.5.4. Set up a parallel between the black box approach and the "formal" or "symbolic" handling of mathematical entities and operations (as practiced by Euler and Heaviside). *Alternate Problem:*

Experimental and theoretical research on problem solving is usually approached either by focusing on (i) the relations between environmental variables (those which determine the task) and performance, or (ii) the inner processes that the subject is supposed to undergo — hence on internal variables such as intelligence and memory. Are these approaches mutually exclusive? Which is more likely to bear immediate fruit? And which is the deeper and therefore the more difficult but at the same time the more promising in the long run?

8.5.5. Do phenomenological scientific theories, phenomenalism (the epistemological doctrine) and phenomenology (E. HUSSERL's doctrine) have anything in common? *Alternate Problem:* Discuss the claim that the more general a theory is the less deep it is bound to be.

8.5.6. Is the difference between intervening or mediating variables and hypothetical constructs intrinsic or contextual? Examine, e.g., the role of the acceleration of gravity in mechanics (constant acceleration) and in gravitation theory (field strength).

8.5.7. Radical infallibilism (fallibilism) implies that science advances in the sense of decreasing (increasing) falsifiability. Link this with the phenomenological/representational tension and establish whether, as a matter of fact, there is a consistent tendency in either direction.

8.5.8. Examine the "paradox of theorizing" and the "theoretician's dilemma" discussed by C. G. HEMPEL in H. FEIGL, M. SCRIVEN and G. MAXWELL, Eds., *Minnesota Studies in the Philosophy of Science*, II (Minneapolis: University of Minnesota Press, 1958), pp. 49—50. Would the "paradox" and the "dilemma" arise were it not on the presupposition that the ultimate aim of scientific theory is to systematize the phenomena that can be directly observed? Do the "paradox" and the "dilemma" arise on the assumption that the ultimate aim of theorizing in factual science is to depict and understand a reality that happens to be mostly hidden to the senses and cannot consequently be accounted for in terms of observation concepts alone?

8.5.9. The mathematician V. VOLTERRA proposed (1931) the following axiom system to account for the dynamics of competing populations (in particular predators and preys):

$$dN_i/dt = a_i N_i + \sum_{j \neq i} b_{ij} N_i N_j; \qquad i, j = 1, 2, \ldots n; \qquad b_{ij} = -b_{ji},$$

where 'N_i' designates the size of the i-th population and 'b_{ij}' measures the strength of the interaction between the species concerned. Is this a phenomenological or a representational theory? Or does it prehaps overflow the dichotomy?

8.5.10. Compare C. L. HULL's approach to learning theory with that of W. K. ESTES and C. J. BURKE. The former employed constructs such as excitatory potential, habit strength, primary drive, incentive motivation, and inhibition (all of them treated as intervening variables rather than as hypothetical constructs), which are all eschewed by the latter authors in favor of response and stimulus. See, e.g., D. LEWIS, *Quantitative Methods in Psychology* (New York: McGraw-Hill, 1960), pp. 500ff. What, if anything, is basically wrong with HULL's theories: that they contain unobservable variables or rather that they treat some of them as intervening variables with no relation to physiological variables? And what, if anything, is wrong with the other approach: that it uses probability (e.g., the probability that a certain fraction of the stimuli will elicit a given set of overt responses) or that it refuses to analyze the probabilities into deeper level variables?

Bibliography

ABRO, A. D': The decline of mechanism, part I. New York: D. van Nostrand Co. 1939.

BOLTZMANN, L.: Populäre Schriften, chs. 5 and 9 to 12. Leipzig: Johann Ambrosius Barth 1905.

BRAITHWAITE, R. B.: Scientific explanation, chs. III and IV. Cambridge: Cambridge University Press 1953.

BUNGE, M.: Physics and reality, Dialectica **19**, 195 (1965).

— The maturation of science. In: I. LAKATOS and A. MUSGRAVE, (eds.), Proceedings of the Internat. Colloquium in the Philosophy of Science, London, 1965, vol. III. Amsterdam: North-Holland 1967.

— Foundations of Physics. Berlin-Heidelberg-New York 1967.

CAMPBELL, N. R.: Foundations of science [originally Physics: The elements 1920], ch. VI. New York: Dover 1957.

CARNAP, R.: The methodological character of theoretical concepts. In: H. FEIGL and M. SCRIVEN (eds.), Minnesota studies in the philosophy of science, I. Minneapolis: University of Minnesota Press 1956.

CHURCHMAN, C. W., R. L. ACKOFF, and E. L. ARNOFF: Introduction to operations research, ch. 7. New York and London: John Wiley & Sons 1957.

COLEMAN, J. S.: Introduction to mathematical sociology. New York: Free Press 1964.

DUHEM, P.: The aim and structure of physical theory, part I. 1914; New York: Atheneum 1962.

EINSTEIN, A.: Physics and reality (1936). In: Out of my later years. New York: Philosophical Library 1950.

FEIGL, H.: Principles and problems of theory construction in psychology. In: W. DENNIS (ed.), Current trends of psychological theory. Pittsburgh: University of Pittsburgh Press 1951.

Gross, L. (Ed.): Symposium on sociological theory, chs. 8 to 14. Evanston (Ill.): Row, Peterson & Co., 1959.

Hull, C. L.: Hypothetico-deductive method of theory construction. In L. Stolurow (ed.), Readings in learning. New York: Prentice-Hall 1953.

Hutten, E.: The language of modern physics. London: Allen & Unwin 1956.

Marx, M. H., (Ed.): Psychological theory: Contemporary readings. New York: MacMillan 1951.

Maxwell, G.: The ontological status of theoretical entities. In: H. Feigl and G. Maxwell (eds.), Minnesota studies in the philosophy of science, III. Minneapolis University of Minnesota Press 1962.

Merton, R. K.: Social theory and social structure, 2nd ed., part I. Glencoe (Ill.): Free Press 1957.

Nagel, E.: The structure of science, chs. 5 and 6. New York: Harcourt, Brace & World 1961.

Popper, K. R.: Three views concerning human knowledge. In: Conjectures and refutations. New York: Basic Books 1963.

Simon, H. A., and A. Newell: Models: Their use and limitations. In D. White (ed.), The state of the social sciences. Chicago: Chicago University Press 1956.

Smart, J. J. C.: Theory construction. In: A. Flew (ed.), Logic and language, 2nd Ser. Oxford: Blackwell 1955.

— Philosophy and scientific realism. London: Routledge & Kegan Paul; New York: Humanities Press 1963.

Tisza, L.: The conceptual structure of physics. Rev. Modern Phys. **35**, 151 (1963).

Törnebohm, H.: A logical analysis of the theory of relativity. Stockholm: Almqvist & Wiksell 1952.

Truesdell, C.: Experience, theory, and experiment. Proceedings of the Sixth Hydraulics Conference, Bulletin 36, University of Iowa Studies in Engineering 1956.

Watson, W. H.: Understanding physics today. Cambridge: Cambridge University Press 1963.

Author Index

Subject Index

Druck der Universitätsdruckerei H. Stürtz AG.. Würzburg